AF477861

Mechanisms of Circadian Systems in Animals and Their Clinical Relevance

Raúl Aguilar-Roblero • Mauricio Díaz-Muñoz
María Luisa Fanjul-Moles
Editors

Mechanisms of Circadian Systems in Animals and Their Clinical Relevance

 Springer

Editors
Raúl Aguilar-Roblero
Instituto de Fisiología Celular
Universidad Nacional Autonoma de México
México, Distrito Federal, México

Mauricio Díaz-Muñoz
Instituto de Neurobiología
Universidad Nacional Autónoma de México
Juriquilla, Querétaro, México

María Luisa Fanjul-Moles
Facultad de Ciencias
Universidad Nacional Autónoma de México
México, Distrito Federal, México

ISBN 978-3-319-08944-7   ISBN 978-3-319-08945-4 (eBook)
DOI 10.1007/978-3-319-08945-4
Springer Cham Heidelberg New York Dordrecht London

Library of Congress Control Number: 2014951521

Foreword

The solar cycle of light and dark is the most pervasive recurring stimulus experienced by living organisms. Mechanisms of this adaptation have diverged throughout evolution, but the fundamental mechanisms are quite similar throughout the plant and animal kingdoms. The foundation of each is the presence of specific sets of genes, called "clock genes" whose products regulate the timing of cellular, physiological, and behavioral events into 24-h cycles, circadian rhythms. These are the critical components of this adaptation to the solar cycle. Circadian rhythms are characterized by two properties. The first is generation by internal clocks with a period approximating 24 h. The second is entrainment of the cycles by external stimuli, predominantly light. Hence, the period of entrained rhythms is precisely 24 h. In mammals, including humans, cellular circadian rhythms are ubiquitous throughout the body. Coordinated cellular function within tissues and organs is provided by a specific network of brain regions designated the circadian timing system (CTS). Prior to the 1970s the location of these regions was unknown, but a series of studies then identified the suprachiasmatic nucleus (SCN) of the hypothalamus as a likely circadian pacemaker and a retinohypothalamic tract (RHT), originating from retinal ganglion cells to terminate in the SCN, as the primary entrainment pathway. The conclusions at the time were that the SCN drives rhythms in the brain and other organs through neural and hormonal signals. Further studies over the last 20 years have modified this view to incorporate observations that cellular rhythms are expressed ubiquitously. These necessitate a new formulation that coordinated circadian rhythms in physiological and behavioral functions are generated throughout the body but entrained by a brain circadian timing system that provides a generalized entrainment to the solar cycle.

This volume, edited by Aguilar-Roblero, Fanjul, and Díaz-Muñoz of the Division de Neurociencias, Universidad Autónoma Nacional de México, contains two valuable contributions to our understanding of circadian rhythms and their disorders. The first is a much more complete description of the neurobiology of circadian function than that outlined above. And, the second is a detailed account of disorders of the circadian timing system and their pathophysiology. There are, of course, several related questions. Why individual chapters? The major reason is that the authors represent a "who's who" of this field ensuring that the quality of the chapters is high, and expertise trumps the unanimity of authorship by a single set of writers. Why a book in a field so vibrant and rapidly moving? Will it not be outdated soon after its publication? There are two answers to this question. It is valuable to the field to have a timely assessment of our status as background for future work. The second is why a book? Why not chapters in electronic form? For those of us whose tenure in the field precedes the electronic era, a book has a special meaning. It is satisfying to hold in one's hands, and turning pages. And, it looks impressive in one's library. I can recommend this book without hesitation, and I look forward to its publication. I visualize myself in an easy chair, turning the pages and, perhaps, slowly consuming a bottle of stout in the process, and enjoying it all.

Pittsburgh, PA, USA   Robert Y. Moore

Preface

Biological rhythms have been known to man since the ancient Greeks, but they became a topic of systematic scientific interest only since the last century. As a matter of fact, it has been only until the dawn of the present century when its links to Physiology and Medicine have called the attention of a broader academic audience and the general population. The reason is mainly due to the recent advances on the molecular genetics of circadian rhythms and the discovery of its cross-links—mainly at the gene regulation level—with other systems involved in control of metabolic networks, cell cycle, as well as behavioral and mental processes.

This book provides a wide perspective on the organization of circadian systems along the animal kingdom from invertebrates to mammals. At the same time, it addresses some aspects of the physiological and clinical relevance of circadian regulation for human beings. It is organized in three parts preceded by a brief introduction that presents the basic concepts of circadian rhythms. In the first part *Circadian Systems* the authors outline a broad perspective of the timing system organization in a variety of animal groups including Crustacean, Insects, Fish, Birds and Primates. The next part *Mechanism of Circadian Oscillation* focuses on one of the most studied mammalian models, the Rodents. In this part the role of the suprachiasmatic nuclei as a circadian clock is reviewed from the cellular to the system level. The final part *Clinical Relevance of Circadian Rhythmicity* provides a perspective to integrate our knowledge on circadian rhythmicity with current human physiology and it reviews recent advances of molecular regulation of circadian rhythms in relation to metabolism, nutrition, cell cycle regulation, cancer, and neurological and mental disorders.

The book is aimed to graduate students in Neuroscience, Biology, Psychology, and Medicine interested in an integrative perspective of circadian rhythms. It also provides some contemporary advances on circadian system organization for experts in circadian biology. It is important to stress the integrative approach of each of the contributions, which distinguishes this book from other related texts. Finally, it is appropriate to mention that the contributors are active and distinguished scientists widely recognized in the field of Chronobiology from Argentina, Brazil, México, Netherlands, Japan, Sweden, and the USA.

We hope to call the attention of young science students to a fascinating area of Biology in order to contribute to the advance of this yet emerging field of science, and to correct or complete some of the hypothesis here presented. This is an invitation to join us in the adventure of understanding time in biology.

México, D.F. México   Raúl Aguilar-Roblero
Juriquilla, Querétaro, México   Mauricio Díaz-Muñoz
México, D.F. México   María Luisa Fanjul-Moles

Contents

Contributors

Raúl Aguilar-Roblero División de Neurociencias, Instituto de Fisiología Celular, Universidad Nacional Autónoma de México, México, DF, México

Charles N. Allen Oregon Institute of Occupational Health Sciences, Oregon Health & Science University, Portland, OR, USA

John Fontenele Araujo Departamento de Fisiologia, Centro de Biociências, Universidade Federal do Rio Grande do Norte, Natal, RN, Brazil

Adrian Báez-Ruíz Departamento de Neurobiología Celular y Molecular, Instituto de Neurobiología, Universidad Nacional Autónoma de México, Querétaro, QRO, México

Ivana L. Bussi Chronobiology Lab, National University of Quilmes (UNQ)/ CONICET, Bernal, Quilmes, Buenos Aires, Argentina

Mario Caba Centro de Investigaciones Biomédicas, Universidad Veracruzana, Xalapa, VER, México

Carolina Carrijo Departamento de Fisiologia, Centro de Biociências, Universidade Federal do Rio Grande do Norte, Natal, RN, Brazil

Leandro P. Casiraghi Chronobiology Lab, National University of Quilmes (UNQ)/CONICET, Bernal, Quilmes, Buenos Aires, Argentina

Vincent M. Cassone Department of Biology, University of Kentucky, Lexington, KY, USA

Juan J. Chiesa Chronobiology Lab, National University of Quilmes (UNQ)/ CONICET, Bernal, Quilmes, Buenos Aires, Argentina

Christopher S. Colwell Laboratory of Circadian and Sleep Medicine, Department of Psychiatry and Biobehavioral Sciences, University of California Los Angeles, Los Angeles, CA, USA

Mauricio Díaz-Muñoz Departamento de Neurobiología Celular y Molecular, Instituto de Neurobiología, Universidad Nacional Autónoma de México, Querétaro, QRO, México

Gabriela Domínguez-Monzón División de Neurociencias, Instituto de Fisiología Celular, Universidad Nacional Autónoma de México, México, DF, México

José M. Duhart Chronobiology Lab, National University of Quilmes (UNQ)/ CONICET, Bernal, Quilmes, Buenos Aires, Argentina

María Luisa Fanjul-Moles Lab. Neurofisiología Comparada de Invertebrados, Departamento de Ecología y Recursos Naturales, Facultad de Ciencias, Universidad Nacional Autónoma de México, México, DF, México

Diego A. Golombek Chronobiology Lab, National University of Quilmes (UNQ)/ CONICET, Bernal, Quilmes, Buenos Aires, Argentina

Horacio O. de la Iglesia Department of Biology, Program in Neurobiology and Behavior, University of Washington, Seattle, WA, USA

Robert P. Irwin Oregon Institute of Occupational Health Sciences, Oregon Health & Science University, Portland, OR, USA

Claudia Juárez-Portilla Centro de Investigaciones Biomédicas, Universidad Veracruzana, Xalapa, VER, México

Nathan J. Klett Oregon Institute of Occupational Health Sciences, Oregon Health & Science University, Portland, OR, USA

Takashi Kudo Laboratory of Circadian and Sleep Medicine, Department of Psychiatry and Biobehavioral Sciences, University of California Los Angeles, Los Angeles, CA, USA

Dika Kuljis Laboratory of Circadian and Sleep Medicine, Department of Neurobiology, University of California Los Angeles, Los Angeles, CA, USA

Shailesh Kumar Department of Neuroscience, Howard Hughes Medical Institute, University of Pennsylvania Perelman School of Medicine, Smilow Center for Translational Research, Philadelphia, PA, USA

Michael L. Lee Department of Biology, Program in Neurobiology and Behavior, University of Washington, Seattle, WA, USA

Dawn H. Loh Laboratory of Circadian and Sleep Medicine, Department of Psychiatry and Biobehavioral Sciences, University of California Los Angeles, Los Angeles, CA, USA

Gabriella Lundkvist Department of Neuroscience, Karolinska Institutet, Stockholm, Sweden

Luiz Menna-Barreto Escola de Artes, Ciências e Humanidades, Universidade de São Paulo, São Paulo, SO, Brazil

Enrique Meza Centro de Investigaciones Biomédicas, Universidad Veracruzana, Xalapa, VER, México

Stephan Michel Department of Molecular Cell Biology, Leiden University Medical Center, Leiden, The Netherlands

Manuel Miranda-Anaya Unidad Multidisciplinaria de Investigación, Facultad de Ciencias, Universidad Nacional Autónoma de México, Querétaro, QRO, México

Mykhaylo G. Moldavan Oregon Institute of Occupational Health Sciences, Oregon Health & Science University, Portland, OR, USA

Christian Molina-Aguilar Departamento de Neurobiología Celular y Molecular, Instituto de Neurobiología, Universidad Nacional Autónoma de México, Querétaro, QRO, México

Elvira Morgado Facultad de Biología, Universidad Veracruzana, Xalapa, VER, México

Natalia Paladino Chronobiology Lab, National University of Quilmes (UNQ)/ CONICET, Bernal, Quilmes, Buenos Aires, Argentina

Julio Prieto-Sagredo Lab. Neurofisiología Comparada de Invertebrados, Departamento de Ecología y Recursos Naturales, Facultad de Ciencias, Universidad Nacional Autónoma de México, México, DF, México

Daniel Quinto-Muñoz División de Neurociencias, Instituto de Fisiología Celular, Universidad Nacional Autónoma de México, México, DF, México

Kathiane Santana Departamento de Fisiologia, Centro de Biociências, Universidade Federal do Rio Grande do Norte, Natal, RN, Brazil

Analyne M. Schroeder Laboratory of Circadian and Sleep Medicine, Department of Psychiatry and Biobehavioral Sciences, University of California Los Angeles, Los Angeles, CA, USA

José Segovia Departamento de Fisiología, Biofísica y Neurociencias, Centro de Investigación y de Estudios Avanzados, del Instituto Politécnico Nacional (CINVESTAV), México, DF, México

Amita Sehgal Department of Neuroscience, Howard Hughes Medical Institute, University of Pennsylvania Perelman School of Medicine, Smilow Center for Translational Research, Philadelphia, PA, USA

Shigenobu Shibata Laboratory of Physiology and Pharmacology, School of Advanced Science and Engineering, Waseda University, Shinjuku-ku, Tokyo, Japan

Christiane Andressa da Silva Departamento de Fisiologia, Centro de Biociências, Universidade Federal do Rio Grande do Norte, Natal, RN, Brazil

Rae Silver Department of Psychology, Barnard College, Columbia University, New York, NY, USA

Ann-Judith Silverman Department of Pathology and Cell Biology, Columbia University Medical Center, New York, NY, USA

Yu Tahara Laboratory of Physiology and Pharmacology, School of Advanced Science and Engineering, Waseda University, Shinjuku-ku, Tokyo, Japan

Eleonora Trajano Instituto de Biociências, Universidade de São Paulo, São Paulo, SP, Brazil

Olivia Vázquez-Martínez Departamento de Neurobiología Celular y Molecular, Instituto de Neurobiología, Universidad Nacional Autónoma de México, Querétaro, QRO, México

Stefan Waliszewski Centro de Investigaciones Biomédicas, Universidad Veracruzana, Xalapa, VER, México

Abbreviations

4-AP	4-Aminopyridine
5-HT	5-Hydroxytryptamine, serotonin
AA	Androgen aromatase
AAADC	Aromatic amino acid decarboxylase
AANAT	Arylalkylamine-N-actyltransferase
AAV	Adeno-associated virus
ACC	Acetyl-CoA carboxylase
ACL	ATP citrate lyase
AD	Alzheimer's disease
Ae1	Na^+ transporter
AR	Androgen receptors
ARC	Arcuate nucleus
ARNTL	Aryl hydrocarbon receptor nuclear translocator-like, transcription factor of basic helix-loop-helix PAS domain
ATP	Adenosine triphosphate
Atpa1a	Na^+ pump
AVP	Arginine-vasopresin peptide
BACHD	Bacterial artificial chromosome which contains the full-length huntingtin protein, transgenic mouse model of Huntington Disease
BAPTA-AM	1,2-bis-(o-Aminophenoxy)-ethane-*N,N,N′,N′*-tetraacetic acid, tetra-acetoxymethyl ester, intracellular Ca^{2+} chelator
Bax	bcl-2-like protein 4, pro-apoptotic factor
BBR2	Gastrin-releasing peptide receptor
Bcl-2	B-cell lymphoma 2, anti-apoptotic regulator
Bcl-xL	B-cell lymphoma-extra large, anti-apoptotic regulator
Bcrp	Breast cancer resistance protein
BD	Bipolar disorder
bHLH-PAS	Basic helix-loop-helix, protein structural motif of "clock genes" *per, arnt, sim*
Bim	Bcl-2-like protein 11, pro-apoptotic factor

BK	Large-conductance Ca^{2+} activated K^+
Bmal-1	Gene of BMAL-1 transcription current, binds on the basic helix-loop-helix promoter domain
BMAL-1	Brain muscle ARNT-like 1, transcription factor of "clock genes"
BMI	Body mass index
BNST	Bed nucleus of the stria terminalis
BPAD	Bipolar affective disorder
Ca^{2+}	Calcium ion
CAG	Codon that codes for the amino acid glutamine, trinucleotide repeat region in the gene encoding huntingtin
CalR	Calretinin, calcium-binding protein
cAMP	Cyclic-adenosine monophosphate
CBT	Core body temperature
CCG	Clock-controlled gene
CDC2	Cell division cycle protein 2 also known as Cyclin-dependent kinase 1
CDKs	Cyclin-dependent kinases
C-FOS	Transcription factor
c-fos	Early expression gene, proto-oncogene, gene of the factor of the DNA AP1 (activator Protein-1) binding site
cGMP	Cyclic-guanine monophosphate
CJL	Chronic jet-lag
CK1	Casein kinase 1
CK1δ	Casein kinase 1 delta
CK1ε	Casein kinase 1 epsilon
CK2	Casein kinase 2
Clk	Gene clock
CLOCK-BMAL1	Transcription factor dimmer of mammalian "clock genes", binds on the basic helix-loop-helix promoter domain
cM	Caudal mesopallium
c-myc	Gene of a regulatory cell proliferation protein, oncogene
Co	Cochlear nuclei
CRE	cAMP-responding element
CREB	CRE-binding protein
CRH	Corticotrophin-releasing hormone
CRY	Cryptochrome protein
Cry	Cryptochrome gene
CRY1	Cryptochrome 1 protein
*cry*1	Cryptochrome 1 gene
CRY2	Cryptochrome 2 protein
cry2	Cryptochrome 2 gene
CSNK1ε	Human casein kinase 1 epsilon
CT	Circadian time

CYC	Cycle, protein, *Drosophila* "clock gene" transcription factor homolog to mammalian BMAL-1
DA	Dopamine
DBP	D site of albumin promoter (albumin D-box) binding protein, transcription factor of the PAR bZIP family (Proline and Acidic amino acid-Rich basic leucine ZIPper)
Dbp	D site of albumin promoter (albumin D-box) binding protein gene
DBT	Doubletime (*Drosophila* homolog of mammalian casein kinase 1ε)
dCRY	*Drosophila* cryptochrome protein
DD	Constant darkness
DEC	Negative regulator of transcription factor CLOCK/BMAL1
Dec1	Gene of negative regulator of transcription factor CLOCK/BMAL1
DEC2	Negative regulator of transcription factor CLOCK/BMAL1
Dec2	Gene of negative regulator of transcription factor CLOCK/BMAL1
DEN	Diethyl-nitrosamine (carcinogenic agent)
DHEA	Dehydroepiandrosterone
dim LL	Constant dim light
Dio2	Type 2 iodothyronine deiodinase
Dio3	Type 3 iodothyronine deiodinase
DLM	Dorsolateral thalamus
DMH	Dorsomedial hypothalamus
DNQX	6,7-Dinitroquinoxaline-2,3-dione (AMPA and Kainate receptor antagonist)
Dra	Na^+ transporter
dTIM	*Drosophila* timeless protein, negative regulator of transcription factor CLK/CYC
ECG	Electrocardiogram
E_{Cl}	Chloride equilibrium potential
ECM	Extracellular matrix
ECoG	Electrocorticogram
EMG	Electromyogram
EOG	Electro-oculogram
Ep	Ectopallium
EPSC	Excitatory postsynpatic currents
ER	Estrogen receptor
ERG	Electroretinogram
ERR	Estrogen-related receptor
ERα	Estrogen receptor alpha
ERβ	Estrogen receptor beta
EX	Eye enucleation
FAA	Food anticipatory activity
FAD+	Oxidized flavin adenine dinucleotide
FASN	Fatty acid synthase
FASPS	Familial advance sleep-phase syndrome
FD	Forced desynchrony

fDR, FDR	Fast delayed rectifier potassium current
FEO	Food-entrainable oscillator
FEPO	Food-entrainable peripheral oscillator
FH	Fumarate hydratase
FOS	Transcription factor of the DNA AP1 (activator Protein-1) binding site, proto-oncogene
Frq	Gene frequency, produce protein FRQ negative regulator of transcription factor
Fura-2	Aminopolycarboxylic acid used as Ca^{2+} indicator dye
fura2 AM	Aminopolycarboxylic acid acetoxymethyl (AM) ester
G1	Growth 1/Gap 1 phase of the cell cycle
G1-S	Checkpoint between G1 phase and the S phase in which the cell is cleared for progression into the synthesis phase (S) of the cell cycle in which DNA is replicated
G2-M	Checkpoint between pre-mitotic phase (G2) and the mitotic (M) phase of the cell cycle
GABA	Gamma-aminobutyric acid, neurotransmitter
$GABA_A$	Ionotropic GABA receptor
$GABA_B$	Metabotropic GABA receptor
GAD	Glutamate decarboxylase
GAD65	65 kDa glutamate decarboxylase
GAD67	67 kDa glutamate decarboxylase
GAT	High affinity GABA transporter
GFAP	Glial fibrilliary acidic protein
GHT	Geniculo-hypothalamic tract
GLUT1	Glucose transporter 1, facilitated glucose transporter member 1
GLUT2	Glucose transporter 2, facilitated glucose transporter member 2
GLUT3	Glucose transporter 3, facilitated glucose transporter member 3
Glut5	Fructose transporter
GM-CSF	Granulocyte-macrophage colony-stimulating factor, cytokine that functions as a white blood cell growth factor
GnRH	Gonadotrophin-releasing hormone
GRP	Gastrin-releasing peptide
GSK3β	Glycogen synthase kinase 3β
GT	Gastrointestinal tract
H3	Histone H3
H3 Ser-10	Ser10 in histone 3 tail
HB	Hofbauer-Buchner
HCN	Hyperpolarization-activated conductance gated by cAMP
HD	Huntington's disease
HDAC	Histone deacetylase complex
HDAC3	Histone deacetylase 3
HIOMT	Hydroxyindole-O-methyltransferase
HPA	Hypothalamic–pituitary–adrenal
HR	Heart rate

HRV	Heart rate variability
HTT	Huntingtin protein
HVC	Hyperstriatum ventrale (pars caudalis), avian high vocal center
I_A	A type K^+ current
IDH1	Isocitrate dehydrogenase 1
IDH2	Isocitrate dehydrogenase 2
IFN-α	Interferon 1 alpha
IGF	Insulin-like growth factor
IGL	Intergeniculate leaflet
I_h	Hyperpolarization-activated conductance
IHDA	Incertohypothalamic dopaminergic neurons
IL	Interleukin
IMEL	$2[_{125}I]$-iodomelatonin
IP_3R	Inositol (1,4,5)-triphosphate receptor
IPSC	Inhibitory postsynaptic currents
Jarid1a	Lysine-specific demethylase 5A
$K_{(Ba)}$	Barium-sensitive K^+ current
KCC	Potassium-chloride cotransporter
Kcnma1	Gene of BK, calcium-activated potassium channel subunit alpha-1
LARK	*Drosophila* RNA-binding protein of the RNA recognition motif (RRM)
LD	Light–dark cycle
LEO	Light-entrainable oscillator
LHA	Lateral hypothalamic area
LKB	Serine/threonine kinase STK11
LL	Constant light conditions, constant bright light conditions
LMAN	Lateral magnocellular nucleus of the anterior nidopallium
LPS	Lipopolysaccharide
LS	Lateral septum
MAPK	Mitogen-activated protein kinase
MBH	Mediobasal hypothalamus
Mct1	Monocarboxylate transporter 1
MD	Major depression
MDD	Major depressive disorder
Mdr1	Multidrug resistance 1
Mel_{1A}	Melatonin receptor 1_A gene nomenclature
Mel_{1b}	Melatonin receptor 1_B gene nomenclature
Mel_{1C}	Melatonin receptor 1_C gene nomenclature
mENK	met-enkephalin
mGluR	Metabotropic glutamate receptors
miR-132	MicroRNA-132 modulates cholinergic signaling and inflammation in human inflammatory bowel disease. De-repression of FOXO3a death axis by microRNA-132 and -212 causes neuronal apoptosis in Alzheimer's disease. In humans it is localized in 17p13.3
miRNA-134	MicroRNA found to be increasingly expressed in schizophrenia

miR-142-3p	MicroRNA expressed at high levels in mature hematopoietic cells and has a crucial role during T-lymphocyte development. In humans it is localized in 14q32.31
miR-219	MicroRNA regulated by the circadian genes CLOCK and BMAL1. It has also been linked with NMDA receptor signaling in humans, and it has been suggested that deregulation of this miRNA can lead to the expression of mental disorders such as schizophrenia. In humans it is localized in 6p21.32
miR-24	miR-24 has been shown to suppress expression of two crucial cell cycle control genes, E2F2 and Myc, in hematopoietic differentiation and also to promote keratinocyte differentiation by repressing actin-cytoskeleton regulators PAK4, Tsk5, and ArhGAP19. In humans it is localized in 9q22.32 and 19p13.13
miR-29a	MicroRNA-29a regulates tumor necrosis factor receptor associated factor-4: an adapter protein overexpressed in metastatic prostate cancer. In humans it is localized in 19p13.13
miR-30a	RNA gene affiliated with the miRNA class. miR30a target fibroblast growth factor (FGF20), Diseases associated with miR30a include chronic myeloid leukemia, and esophageal cancer and parkinson disease. In humans it is localized in 6q13
miRNA	MicroRNA
MK-801	(+)-5-Methyl-10,11-dihydro-5H-dibenzo[a,d]cyclohepten-5,10-imine, dizocilpine, NMDA receptor antagonist
MLd	Lateral dorsal mesencephalic nuclei
MLL1	Mixed-lineage leukemia-1 histone-lysine N-methyltransferase
Mll3	Mixed-lineage leukemia-3 histone-lysine N-methyltransferase
mPer2	Mouse period gene 2
Mrp2	Multidrug resistance-associated protein 2
mSCN	Medial suprachiasmatic nuclei, avian
MT	Meal timing
MT1	Melatonin receptor 1, pharmacological nomenclature equivalent to Mel_{1A}
MT2	Melatonin receptor 2, pharmacological nomenclature equivalent to Mel_{1b}
NAD+	Oxidized nicotinamide adenine dinucleotide, coenzyme involved in redox reactions
NADH	Reduced nicotinamide adenine dinucleotide, coenzyme involved in redox reactions
NADP+	Oxidized nicotinamide adenine dinucleotide phosphate, coenzyme involved in redox reactions
NADPH	Reduced nicotinamide adenine dinucleotide phosphate, coenzyme involved in redox reactions
nBOR	Nucleus of the basal optic root, avian
NE	Norepinephrine

NF-YA	Nuclear factor Y-A, alpha subunit of the trimeric NFY transcription factor that binds to CCAAT (cytosine-cytosine-adenosine-adenosine-thymidine) box motifs in several genes
NGVB	National Gene Vector Biorepository
Nhe3	Na$^+$ transporter
Nherf1	Na($^+$)/H($^+$) exchanger regulatory factor
NK	Natural killer (cell), cytotoxic lymphocyte from the innate immune system
NKCC1	Na$^+$–K$^+$–Cl$^-$ cotransporter 1
NKT	Natural killer T (cell), T lymphocyte involved in cell-mediated immunity by linking the adaptive immune system with the innate immune system
NMDA	*N*-Methyl-D-aspartate
nNOS	Neuronal nitric oxide synthase
NO	Nitric oxide
NPAS2	Neuronal PAS domain-containing protein 2, transcription factor for "clock genes," binds on the basic helix-loop-helix promoter domain
NPY	Neuropeptide Y
NREM	Non-rapid eye movement (sleep)
NT	Neurotensin
NTS1	Neurotensin receptor type 1
NTS2	Neurotensin receptor type 2
OPN1	Iodopsin, avian rodopsin
OPN2	Rhodopsin
OPN4	Melanopsin
OPN5	Neuropsin, UV-sensitive photoreceptor protein
ORX	Orexin
OS	Oxidative stress
OT	Oxytocin
Ov	Thalamic nucleus ovoidalis, avian
OX7	Mouse monoclonal antibody directed against the Thy-1.1
p21WAF1/CIP1	Cyclin-dependent kinase inhibitor 1A
PACAP	Pituitary adenylate cyclase activating polypeptide
PAI-1	Plasminogen activator inhibitor-1
PAR-bZIP	PAR domain basic leucine zipper
PAS	Protein structural motif of "clock genes" from *per, arnt, sim*
pDNA	Plasmid DNA
PDP1	*PAR* domain protein 1
Pept1	H($^+$)/peptide cotransporter 1
per	Period gene
PER	Period protein
per1	Period 1 gene
PER1	Period 1 protein
Per1-luc	Transgenic mouse containing *per1* promoter and luciferase gene

per2	Period 2 gene
PER2	Period 2 protein
PER2-LUC	Transgenic mouse containing the *per2* gene with a knock-in of luciferase gene
per3	Period 3 gene
PER3	Protein period 3
PGC-1	Peroxisome proliferator-activated receptor-γ coactivator 1
PGC-1α	Proliferator-activated receptor gamma coactivator 1-alpha
PHDA	Periventricular hypophysial dopaminergic population
PINX	Surgically remotion of the pineal gland
PK2	Prokineticin 2
PKG	cGMP-dependent protein-kinase
PKM2	Pyruvate kinase M2
POA	Preoptic area
PP1	Protein phosphatase 1
PPAR	Peroxisome proliferator-activated receptor
PPARα	Peroxisome proliferator-activated receptor alpha
PPAR-γ	Peroxisome proliferator-activated receptor gamma
PRC	Phase response curve
PRL	Prolactin
Puma	p53 Upregulated modulator of apoptosis, pro-apoptotic factor
PVN	Paraventricular hypothalamic nuclei
PVT	Thalamic paraventricular nucleus
Q_{10}	Thermic coefficient, refers to the factor by which any metabolic process changes by an increase of $10°$ C
R6/2 CAG150+	R6/2 CAG150+ mouse model of HD with more than150 CAG repeats
RA	Robust nucleus of the archipallium, avian
RAR	Retionic acid response (element)
RB	Retinoblastoma-associated protein
RCA	Retrochiasmatic area
RCLs	Replication competent lentiviruses
REM	Rapid eye movement
rev-erb	Gene of an orphan member of the nuclear hormone receptor family
REV-ERB	Orphan member of the nuclear hormone receptor protein, transcriptional repressor of the basic helix-loop-helix promoter domain
rev-erb α	*nr1d1* gene, an orphan member of the nuclear hormone receptor family
rev-erb β	*nr1d2* gene, an orphan member of the nuclear hormone receptor family
RF	Restricted feeding
RHT	Retinohypothalamic tract

rigi	Mouse ortholog of *per* found by screening of mouse c-DNA libraries
ror	The RAR-related orphan receptor gene
ROR	Protein of the RAR-related orphan receptor
RORE	Retinoic acid-related orphan receptor response element
RORα	Receptor-related orphan receptor alpha
rorγ	RAR-related orphan receptor gamma gene
ROS	Reactive oxygen species
Rt	Nucleus rotundus, avian
RyR	Intracellular calcium channel ryanodine receptor
RyR2	Intracellular calcium channel ryanodine receptor type 2
SAD	Seasonal affective disorder
SCN	Suprachiamatic nuclei
SD	Sleep deprivation
SDH	Succinate dehydrogenase
SERCA	Sarco/endoplasmic reticulum calcium ATPase
SF	Scheduled food
SGG	Glycogen synthase kinase 3, protein kinase shaggy
shaggy	Gene of the protein kinase shaggy, also known as gsk3 or zw3
SIM	*Drosophila* single-minded protein
siRNA	Small interference RNA
Sirt1	NAD+-dependent histone deacetylase sirtuin-1
SIRT3	NAD+-dependent histone deacetylase sirtuin-3
SIRT9	NAD+-dependent histone deacetylase sirtuin-9
SK	Small conductance $K^+_{(Ca)}$ current
SLIMB	Protein supernumerary limbs, an E3 ligase and targeted for ubiquitination-mediated proteosomal degradation
SNP	Single nucleotide polymorphisms
SON	Supraoptic nucleus
sPVZ	Subparaventricular zone
SWS	Slow-wave sleep
T3	Triiodothyronine
T4	Thyroxine
TEA	Tetraethylammonium
TeO	Optic tegmentum
TGF-α	Transforming growth factor-alpha
TH	Tyrosine hydroxylase
TIDA	Tuberoinfundibular dopaminergic area
Tim	Timeless gene
TIM	Timeless protein
TNF	Tumor necrosis factor
TP53	Tumor protein p53
TrH	Tryptophan hydroxylase
TRH	Thyrotrophin-releasing hormone

TTL	Transcriptional–translational feedback loop
TTX	Tetrodotoxin
Tyf	*Drosophila* gene twenty-four
TYF	Twenty-four protein
V1aR	Vasopressin receptor 1a
VEP	Visually evoked potentials
VGAT	Vesicular GABA transporter
VIP	Vasoactive intestinal peptide
vLGN	Ventral lateral geniculate nucleus
VNTR	Variable number of tandem-repeat
VPAC2R	VIP receptor 2
VRI	Vrille protein
VRI/PDP1	Vrille and PAR domain protein 1 dimmer
vSCN	Ventral suprachiasmatic nuclei
wee-1	Gene of a nuclear protein kinase
WEE-1	Nuclear protein kinase, phosphorylates the amino acids Tyr15 and Thr14 of CDC2
XOSG	X-organ sinus gland complex
ZT	Zeitgeber time
α	Alpha, active interval of the rest–activity cycle
α-MSH	Alpha-melanocyte-stimulating hormone
$\gamma ENac$	Epithelial Na$^+$ channel
ρ	Rho, rest interval of the rest-activity cycle
τ	Tau, period of a free-running rhythm
ϕ	Phi, phase of a circadian cycle
ϕ_{pi}	Photoinducible phase
ϕ_e	External phase
ϕ_ι	Internal phase
ψ	Psi, phase relation
$\psi_{dawndusk}$	Phase relationship between the dawn oscillator to the dusk oscillator
$\psi_{\iota e}$	Entrained phase relationship

Chapter 1
Introduction to Circadian Rhythms, Clocks, and Its Genes

Raúl Aguilar-Roblero

Abstract From conception to death, life encompasses innumerable processes in continuous change and in many instances repetition; after all, life on earth has to cope from its origin with a highly cyclic environment. Thus, it is not surprising that cyclic phenomena are found in all living organisms and at all levels of organization. When this processes repeat in a regular manner, we refer to them as rhythmic. Biological rhythms occur in a wide range of frequencies, from cycles per seconds to cycles per year. Of particular interest are those rhythmic functions which repeat daily. These are the circadian rhythms whose organization and relevance are the interest of the present book. In this brief introduction, a general perspective of circadian rhythmicity and its mechanisms will be presented; we will start by a brief summary of the basic concepts and its biological relevance; then we will provide an account of the main ideas which lead to the characterization of circadian clocks among different species and levels of organization; and finally, we will provide a brief account on the characterization of the molecular mechanism underlying circadian oscillators, the so-called clock genes, and its distribution in mammalian tissues.

1.1 Basic Concepts of Circadian Rhythms

Circadian rhythms are periodic variations shown by most organisms, which repeat every 24 h and present three main characteristics: (1) they are produced by processes endogenous to the organisms, generally referred to as biological clocks; (2) under natural conditions, they align to external cycles from the environment by means of processes known as entrainment or synchronization; and (3) the speed of the clock (period) is temperature compensated, so the time to complete each cycle is not significantly affected by changes in environmental temperature. In the following sections, we will refer to each of these characteristics.

R. Aguilar-Roblero (✉)
División de Neurociencias, Instituto de Fisiología Celular,
Universidad Nacional Autónoma de México, Circuito exterior S/N,
Ciudad Universitaria, Coyoacán, 04510 Mexico, D.F., Mexico
e-mail: raguilar@ifc.unam.mx

© Springer International Publishing Switzerland 2015
R. Aguilar-Roblero et al. (eds.), *Mechanisms of Circadian Systems in Animals and Their Clinical Relevance*, DOI 10.1007/978-3-319-08945-4_1

1.1.1   Free Running

The persistence of rhythmicity in the absence of external periodic time signals reveals the endogenous nature of these variations, and it is the main characteristic of circadian rhythmicity; otherwise the rhythm should be referred to as diurnal (Halberg 1960). The relevance of circadian rhythmicity in biology as an adaptive process to a perpetual cyclic environment has been recognized since the first half of the last century (Pittendrigh 1993; Bünning 1969). From this perspective, circadian rhythmicity reflects the ability of organisms to generate an estimate or measure of time—*biological time*—independently of geophysical cycles.

The endogenous nature of circadian rhythmicity can only be observed in laboratory constant conditions, in the absence of external periodic signal such as illumination, temperature, and humidity, and the rhythms thus observed are referred to as *free running* (Aschoff 1965). Free-running rhythmicity has a period (which is the time taken for each cycle to repeat) slightly different from 24 h (hence the term *circadian*); this is the endogenous period of the rhythm under observation (Fig. 1.1).

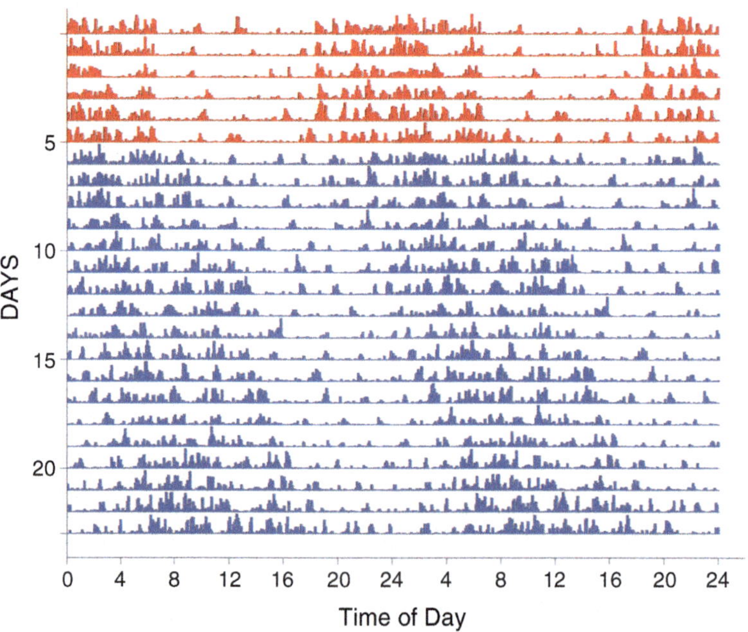

Fig. 1.1 Double plot actogram of locomotor activity in a rat kept under controlled environmental conditions. From day 1 to 5, white light (250 lux) was on at 07:00 h and off at 19:00 h. Dim red light (4 lux) was continuously on (LD). On day 6, white light was kept off until the end of the recording (DD). Each *line* is a histogram of 5 min bins of activity of two consecutive days; the 2nd day of each row is plotted again in the next row. During the first 5 days, the locomotion rhythm is entrained to the LD cycle (*red*); activity increases when white light goes off and decreases shortly before lights are on. On day 6, the rhythm starts to free run (*blue*) with a period of approximately 24:30 h. Room temperature was kept constant at 22 °C (±1°)

Free-running rhythms manifest the integration of processes able to measure time within the organism itself. Such internal processes have been called circadian clock, pacemaker, or oscillator, and although each term has very specific connotations on its meaning, they are often used as synonymous. Finally, it is important to realize that behavioral or other manifestations of circadian rhythms are considered as the overt expression (or handles) of a circadian clock.

1.1.2 Entrainment

In natural conditions, circadian rhythms are aligned with the light–dark cycle, other external cycle's secondary to the earth's rotation or periodic events such as food availability and biological or social cues. The adaptive nature of the circadian rhythms depends on entrainment, because it allows the organism to align the biological and geophysical times. Entrainment, also known as synchronization, refers to the processes to match the period of the circadian rhythm to the period of the external cycle (Fig. 1.1) and to establish a constant time relation on a particular event of the circadian rhythm (such as awakening) and a specific event of the environmental cycle (such as dawn or dusk). Thus, when entrained, a particular phase of the circadian rhythm is locked to a particular phase of the external cycle, and both rhythms have equal periods (Pittendrigh 1993). When entrainment is studied under light–dark cycles, light has shown to speed up the clock(s), thus shortening the period of rhythmicity; in turn, during darkness, the clock(s) slows down, thus lengthening the period, so the complete cycle has an average of 24 h (Daan and Pittendrigh 1976b).

1.1.2.1 The Phase Response Curve

Entrainment has also been studied by applying discrete light pulses (minutes to hours) at different times of the circadian cycle to animals in constant conditions and measuring the response of the free-running rhythm, which is known as a phase response curve (PRC). This approach has shown that the same stimulus applied at different times of a circadian cycle has different effects on the phase (see Box 1.1 for definitions) of the rhythm due to the dynamical response of the underlying clock(s) (Pittendrigh 1984; Daan and Pittendrigh 1976a). The PRC has three characteristic zones (Fig. 1.2) in different species: (1) the dead zone occurs during the subjective day and is characterized by the lack of response to light, therefore the phase of activity observed after the light pulse occurs at the time it was expected from projecting the phase before the light pulse; (2) the delay zone occurs early during the subjective night and is characterized by phase delays, that is, when activity after the light pulse occurs later than expected; and (3) the advance zone occurs late during the subjective night and is characterized by phase advances, when activity after the light pulse occurs earlier than expected from projecting the phase before the light pulse. Phase shifts reach a steady state after some transitory cycles and

Box 1.1 Parameters Related to Rhythmic Phenomena

The period of a cyclic phenomenon refers to time lapse for a complete cycle to occur; in natural conditions, the period of the circadian rhythmicity is 24 h (but see also Sect. 1.1.1). The amplitude refers to the change in the intensity of the oscillatory variable from its highest value (peak) to its lowest (trough); in the context of the cosinor analysis (which is used to fit cyclic phenomena to a cosine function), it also may refer to the change in the intensity of the variable from the mean value of the variable throughout the cycle (mesor) to the highest value of the best fitting cycle (acrophase). Finally, the phase of the rhythm refers to the time at which any particular value of the cyclic variable occurs, for example, the time of waking or sleep initiates correspond each to a particular phase of the sleep–wake cycle. Also the time at which the peak or acrophase occurs is also a clear-cut phase reference for cyclic or rhythmic phenomena. Finally, since each rhythmic variable has specific phase descriptors, the time relation between two variables is best referred to as phase relation; for example, the relation between waking and breakfast in a particular subject could be described as -1 h relation of breakfast with respect to awakening (a delay of 1 h after waking to have breakfast).

maintain the new phase until perturbed by another phase-shifting stimulus (Pittendrigh 1984). Phase shifts are measured (in minutes or hours) by subtracting the phase of the rhythm before the stimulus to the phase in steady state after the stimulus. It is worth noticing that although day and night are absent in constant free-running conditions, we can still discriminate a subjective day and a subjective night, which depends on the circadian organization of the species under study, either active during the day or during the night. Moreover, the length of a free-running rhythm differs from the period of the entrained rhythm (24 h in natural conditions); therefore, each hour of the free-running day will also differ from 60 min. To estimate the circadian time (CT), the free-running period is divided in 24 segments or circadian hours. The PRC depicts in the abscissa the CT at which the light pulse is applied, while the ordinate depicts the magnitude of the induced phase shift, usually delays as negative values and advances as positive ones. The analysis of the PRC provides information about the dynamics of the circadian clock(s) which generates overt rhythmicity in each of the species studied, and nowadays is a valuable tool to unravel the cellular and molecular mechanisms involved in the functioning of the clock(s).

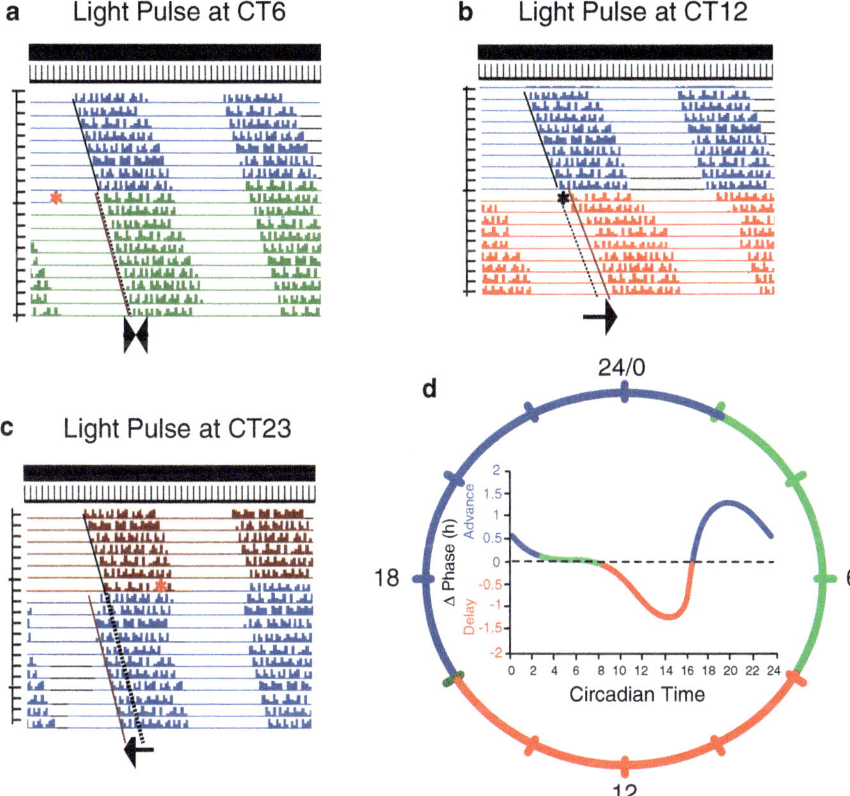

Fig. 1.2 Diagram of the phase response curve (PRC) to light pulses for a nocturnal rodent. The schematic actograms illustrate the phase responses to light pulses applied at different circadian times as indicated. The basal free-run recording is shown in *blue* (**a**, **b**) *dark red* (**c**); the effect on rhythmicity of a light pulse (*red asterisk*) at different circadian times (CT) is shown in *green* (at CT6), *red* (at CT12), or *blue* (at CT 23). The *arrows* below each actogram indicate the change on the activity onset after the light pulse (*dotted lines*) with respect to the recordings before the light pulse was applied (*continuous line*). The PRC is shown inside the 24 h dial; each color on the curve and the dial corresponds to the characteristic regions of the PRC: dead zone (*green*), the delay zone (*red*), and the advance zone (*blue*)

1.1.2.2 Masking

Complete cycles of light–dark or high–low temperature may also affect the behavioral expression of rhythmicity beyond the clock itself, therefore masking its phase and period; such effects of external factors on the expression of rhythmic behaviors are referred to as masking effects (Aschoff and von Goetz 1988; Rietveld et al. 1993).

1.1.3 Temperature Compensation

Temperature compensation was inferred by Pittendrigh as an essential property of circadian rhythms necessary to prevent the speed of the clock to be influenced by the temperature in the environment. For a detailed description on how this hypothesis was conceived and proved, we refer the reader to Pittendrigh (1993). The thermic coefficient or Q_{10} refers to the factor by which any metabolic process changes by an increase of 10 °C in the environment; thus, a Q_{10} of 1 indicates the process is not altered, while a Q_{10} of 2 means the speed of the process duplicates. The period of free-running rhythms has a Q_{10} close to 1 which indicates that it remains almost constant in a wide range of temperatures. This property was demonstrated in practically all species studied, and it is considered as a fundamental characteristic of circadian rhythmicity. The mechanisms involved in such property are still under investigation.

1.2 The Search for Circadian Clocks

During the 1970s circadian clocks became concrete biological entities in different species. In invertebrates the possible locus of circadian clocks was located in the brain of the silk moth (Truman 1972) and the eye of the *Aplysia* (Jacklet and Geronimo 1971) and *Bulla* (Block and Wallace 1982), while in vertebrates the clocks were identified in the pineal gland in birds (Gaston and Menaker 1968) and in the suprachiasmatic nuclei (SCN) in the rat hypothalamus (Moore and Eichler 1972; Stephan and Zucker 1972) (Fig. 1.3). All these studies stimulate the search for studying the mechanisms of circadian clocks at systemic, cellular, and molecular levels. When these and other studies were published, there was an intense debate whether they indicate or not the presence of a circadian oscillator or they were merely part of the clockwork or gear mechanism. It was thus necessary to outline the minimal criteria to be met in order to positively identify a circadian clock; such criteria included the following: circadian rhythmicity shall be completely disrupted when the putative cells are ablated; rhythmicity will be restored when the putative cells are transplanted from an intact donor to a lesion host; brief stimulation of the putative cells will induce phase shifts in overt rhythmicity; and when isolated from the organism, the putative cells shall continue to exhibit oscillations with a circadian period (Menaker et al. 1978). Although the criteria are enunciated here as referring to a cell or group of cells, the criteria are also valid to identify molecular components of the circadian oscillators such as genes, enzymes, and cell signaling processes.

Besides the characteristics summarized previously, a conceptual black-box model of circadian clocks was developed. The model involved an oscillator (the actual clock mechanism) which allowed the measurement of biological time; sensorial receptors which input to the oscillator and allows entrainment to environmental cycles; and output pathways which transmit the time signal from the

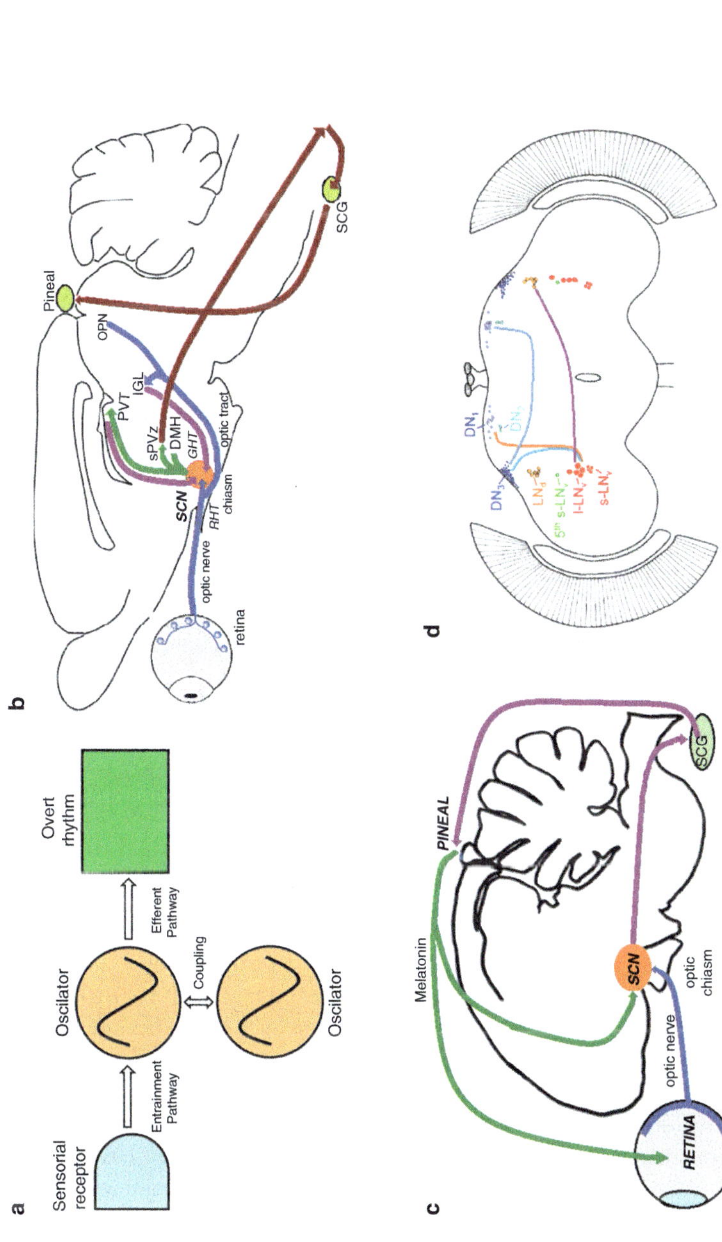

Fig. 1.3 Schematic of circadian systems in different species. (**a**) Conceptual model of the circadian system according to Eskin (Skinogram). (**b**) Rodent. (**c**) Avian. (**d**) *Drosophila*. Abbreviations in **b** and **c**: *SCN* suprachiasmatic nuclei (circadian clock), *sPVZ* subparaventricular zone, *DMH* dorsomedial hypothalamus, *PVT* thalamic paraventricular nucleus, *IGL* intergeniculate leaflet, *OPN* olivary pretectal nucleus, *SCG* superior cervical ganglion, *RHT* retino-hypothalamic tract, *GHT* geniculo-hypothalamic tract. Abbreviations in **d**: *DN* neuron clusters (1–3) in the dorsal brain, *LNd* dorsal–lateral brain, *l-LNv* large ventral lateral neurons, *s-LNv* small ventral–lateral neurons, *5th* fifth small ventral–lateral neuron

oscillator to the effector systems of the organism which will generate the overt rhythms (Eskin 1979). As circadian clocks were identified in different animals, the concept of the circadian clock was modified to that of circadian system, which differentiates among the time-measuring entity (the oscillator or clock itself) and the inputs and output elements of the clock. Furthermore, although the original circadian system referred to a single oscillator, evidence suggested that there might be several oscillators in an organism; in such case, in order to maintain its internal synchronization, it would be necessary to postulate signaling processes to couple the different oscillators among themselves and sustain the temporal organization of the individual (Fig. 1.3).

1.2.1 Unraveling the Molecular Circadian Clock

The general structure of a biological oscillator, circadian or otherwise, was proposed in the early 1980s from the analysis of a number of well-characterized biological and chemical oscillations. Thus, it was demonstrated that a system with a delayed negative feedback loop behaves as an oscillator. The delayed element of the system is necessary to set the system to oscillate, and its kinetics determines the periodicity of the oscillation (Friesen and Block 1984). It is worth noting that in the absence of a significant delay, the feedback loop will keep a constant output of the system (Wiener 1948). On the other hand, experiments involving pharmacological blockade of gene transcription or its translation into proteins in *Bulla gouldiana* were able to stop the circadian clock in a reversible manner (Khalsa et al. 1992); these evidences lead to propose that the circadian clock comprised a transcription–translation loop (Block et al. 1995). These two hypotheses had a great influence in our notion of how circadian oscillators were organized at the cellular and molecular level.

The first step in the search for the molecular substrate of circadian rhythms occurred in 1971 when Konopka and Banzer described an arrhythmic mutation in *Drosophila melanogaster*; they identified the affected locus and call it *per* from period (of rhythmicity). A decade later, the gene was cloned (Reddy et al. 1984), and about 1990, Hardin et al. demonstrated that the Per protein feeds back to the *per* gene to regulate its own mRNA level. In 1973 mutants of *Neurospora crassa* with altered circadian periods were identified and named frequency mutants, since the name period was already taken (Feldman and Hoyle 1973). Some years later, the gene *frq* was identified and cloned; a negative feedback loop was described, where the Frq protein regulates its own mRNA transcription (Aronson et al. 1994). Some years later, the *clock* mutation was identified in the mouse and the gene was cloned (King et al. 1997) and was named after the acronym of circadian locomotor output cycles kaput. Interestingly, mouse *clock* mRNA levels did not show circadian oscillations as in the *Drosophila* ortholog (Shearman et al. 1999). On the other hand, homology screens of mouse c-DNA libraries to *Drosophila per* gene led to the finding of a mouse ortholog which was named RIGI, while independently a second ortholog for *Drosophila per* was characterized in mice (Shearman et al. 1997; Sun

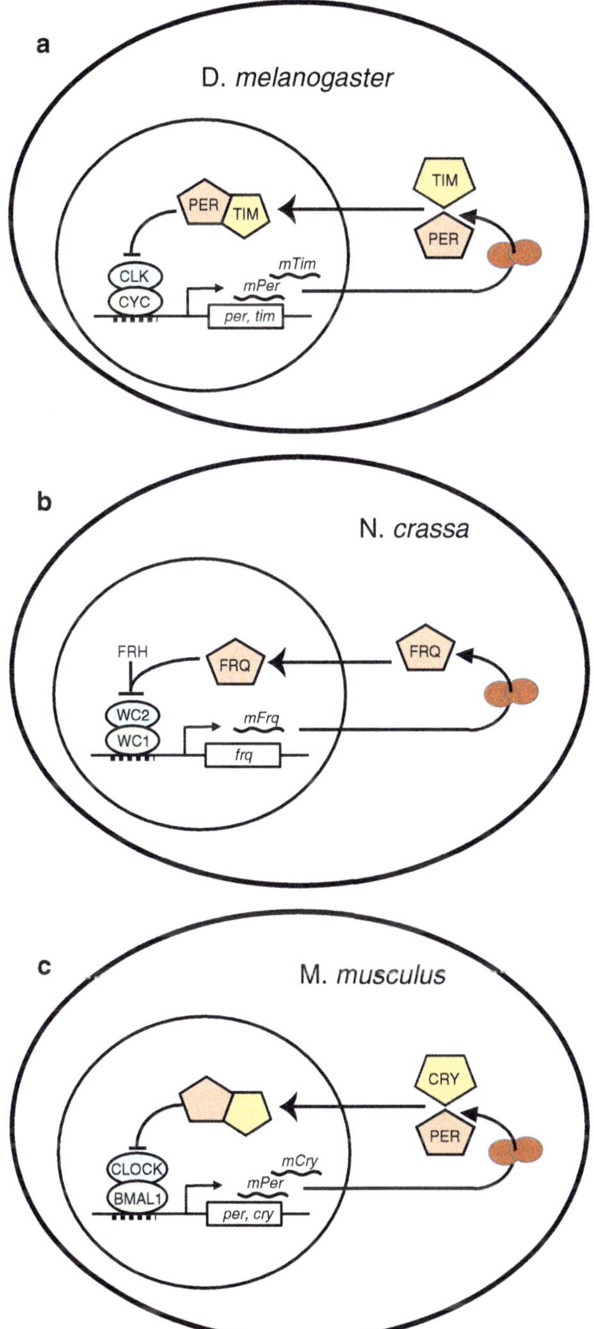

Fig. 1.4 The core molecular circadian clock in different species. Symbols: ↦, transcription; ~, messenger RNA; →, translation; →, translocation to the nucleus; ⊥, repression. Abbreviations: *CLK* clock, *CYC* cycle, *PER* period, *TIM* timeless, *WC1* white collar 1, *WC2* white collar 2, *FRQ* frequency, *FRH* RNA helicase, *CRY* cryptochrome, *BMAL1* brain and muscle ARNT-like 1 protein. *Uppercase letters* indicate proteins, *lowercase* indicates DNA, and *upper- and lowercase* indicates RNA; *ellipses* indicate transcription factors; *pentagons* indicate proteins inhibiting its own transcription; *dotted line* indicates the promoter for the gene indicated in the box; *orange circles*, ribosomes

et al. 1997). These findings together with the contributions of other groups lay down the basis for the current knowledge on circadian clock genes (Fig. 1.4). Nowadays it has been shown that most circadian oscillators consist of delayed feedback transcription–translation loops of genes that regulate its own expression at least in eukaryotic cells (Lowrey and Takahashi 2004; Dunlap 1999; Hardin et al. 1990; Zheng and Sehgal 2012).

During the last 10 years, we have begun to understand the molecular basis of the circadian oscillators beyond the transcription–translation feedback loop. At the posttranslational level, the phosphorylation of "clock proteins" was the first process involved as a crucial step to set the speed of the clock in *Drosophila* (Chiu et al. 2011) and mammals (Lee et al. 2011), and recently, it has been suggested that the posttranslational feedback loop based on phosphorylation–dephosphorylation processes could also be a relevant part of the core molecular clock (Brown et al. 2012). Phosphorylation promotes dimerization of "clock proteins" necessary to translocate to the nuclei and inhibit its own transcription but also tag such proteins to ubiquitination and hydrolysis. Epigenetic regulation has also been found to participate in the transcription of "clock genes" and "clock-regulated genes" (Sassone-Corsi 2010; Feng and Lazar 2012; Sahar and Sassone-Corsi 2013). Epigenetic regulation involves histone phosphorylation and acetylation which allows transcription by remodeling chromatin and exposing DNA segments to transcription machinery, histone acetylation also induces chromatin remodeling and may stimulate or repress gene expression, and DNA methylation induces chromatin compaction and suppresses gene expression (Jenuwein and Allis 2001). These processes allow reversible responses to changes in the environment which can affect many physiological processes, including development, aging, and metabolism (Christensen and Marsit 2011). The activation of clock-controlled genes by CLOCK-BMAL1 has been shown to be coupled to histone modifications; thus, deacetylases SIRT1 and HDAC3 or methyltransferase MLL1 has been shown to be recruited in a circadian manner to these promoters (Etchegaray et al. 2003; Naruse et al. 2004). Clock also possesses histone acetyltransferase activity directed to histone H3 and also has acetylation activity on its partner BMAL1 (Hirayama et al. 2007) and the glucocorticoid receptor (Nader et al. 2009). Finally, H3 Ser-10 phosphorylation is involved in the transcriptional response to light in the SCN (Crosio et al. 2000).

1.3 Closing Remarks

In this book, we will provide a wide perspective on the organization and physiological relevance of circadian systems from invertebrates to mammals. It is organized in three sections: in the first section "Circadian Systems," the authors will outline the circadian system organization, including molecular, cellular, and systemic aspects in "Insects, Crustacean, Fish, Birds, and Primates"; the second section "Mechanism of Circadian Oscillation" will focus on mammalian models, and the role of the suprachiasmatic nuclei as a circadian clock will be reviewed from the cellular to the system level in rodents. The food-entrainable oscillator will also be reviewed in the lactating rabbit model; finally, the last section "Clinical Relevance of Circadian Rhythmicity" will cover different aspects of circadian rhythms in relation with health and disease.

Acknowledgments We thank José Luis Chávez, Ana María Escalante, and Francisco Pérez for assistance. This work was partially supported by grants from CONACyT 128528 and PAPIIT IN204811.

References

Aronson BD, Johnson KA, Loros J, Dunlap JC et al (1994) Negative feedback defining a circadian clock: autoregulation of the clock gene frequency. Science 263:1578–1584

Aschoff J (1965) Circadian rhythms in man. Science 148:1427–1432

Aschoff J, von Goetz C (1988) Masking of circadian activity rhythms in hamsters by darkness. J Comp Physiol A 162:559–562

Block GD, Wallace SF (1982) Localization of a circadian pacemaker in the eye of a mollusc, Bulla. Science 217:155–157

Block G, Geusz M, Khalsa S et al (1995) Cellular analysis of a molluscan retinal biological clock. Ciba Found Symp 183:51–66

Brown SA, Kowalska E, Dallmann R (2012) (Re)inventing the Circadian Feedback Loop. Dev Cell 22:477–487

Bünning E (1969) Common features of photoperiodism in plants and animals. Photochem Photobiol 9:219–928

Chiu JC, Ko HW, Edery I (2011) NEMO/NLK phosphorylates PERIOD to initiate a time-delay phosphorylation circuit that sets circadian clock speed. Cell 145:357–370

Christensen BC, Marsit CJ (2011) Epigenomics in environmental health. Front Genet 2:84. doi:10.3389/fgene.2011.00084

Crosio C, Cermakian N, Allis CD et al (2000) Light induces chromatin modification in cells of the mammalian circadian clock. Nat Neurosci 3:1241–1247

Daan S, Pittendrigh CS (1976a) A functional analysis of circadian pacemakers in nocturnal rodents: II. The variability of phase response curves. J Comp Physiol 106:253–266

Daan S, Pittendrigh CS (1976b) A functional analysis of circadian pacemakers in nocturnal rodents: III. Heavy water and constant light: homeostasis of frequency? J Comp Physiol 106:267–290

Dunlap JC (1999) Molecular bases for circadian clocks. Cell 96:271–290

Eskin A (1979) Identification and physiology of circadian pacemakers. Fed Proc 38:2570–2572

Etchegaray JP, Lee C, Wade PA et al (2003) Rhythmic histone acetylation underlies transcription in the mammalian circadian clock. Nature 421:177–182

Feldman JF, Hoyle MN (1973) Isolation of circadian clock mutants of *Neurospora crassa*. Genetics 75:605–613

Feng D, Lazar MA (2012) Clocks, metabolism, and the epigenome. Mol Cell 47:158–167

Friesen WO, Block GD (1984) What is a biological oscillator? Am J Physiol 246:R847–R853

Gaston S, Menaker M (1968) Pineal function: the biological clock in the sparrow? Science 160:1125–1127

Halberg F (1960) Temporal coordination of physiologic function. Cold Spring Harb Symp Quant Biol 25:289–310

Hardin PE, Hall JC, Rosbash M (1990) Feedback of the *Drosophila* period gene product on circadian cycling of its messenger RNA levels. Nature 343:536–540

Hirayama J, Sahar S, Grimaldi B (2007) CLOCK-mediated acetylation of BMAL1 controls circadian function. Nature 450:1086–1090

Jacklet JW, Geronimo J (1971) Circadian rhythm: population of interacting neurons. Science 174:299–302

Jenuwein T, Allis CD (2001) Translating the histone code. Science 293:1074–1080

Khalsa S, Whitmore D, Block GD (1992) Stopping the circadian pacemaker with inhibitors of protein synthesis. Proc Natl Acad Sci U S A 89:10862–10866

King DP, Zhao Y, Sangoram AM et al (1997) Positional cloning of the mouse circadian clock gene. Cell 16:89641–89653

Konopka RJ, Benzer S (1971) Clock mutants of *Drosophila melanogaster*. Proc Natl Acad Sci U S A 68:2112–2116

Lee HM, Chen R, Kim H et al (2011) The period of the circadian oscillator is primarily determined by the balance between casein kinase 1 and protein phosphatase 1. Proc Natl Acad Sci U S A 108:16451–16456

Lowrey PL, Takahashi JS (2004) Mammalian circadian biology: elucidating genome-wide levels of temporal organization. Annu Rev Genomics Hum Genet 5:407–441

Menaker M, Takahashi JS, Eskin A (1978) The physiology of circadian pacemakers. Annu Rev Physiol 40:501–526

Moore RY, Eichler VB (1972) Loss of a circadian adrenal corticosterone rhythm following suprachiasmatic lesions in the rat. Brain Res 42:201–206

Nader N, Chrousos GP, Kino T (2009) Circadian rhythm transcription factor CLOCK regulates the transcriptional activity of the glucocorticoid receptor by acetylating its hinge region lysine cluster: potential physiological implications. FASEB J 23:1572–1583

Naruse Y, Oh-hashi K, Iijima N et al (2004) Circadian and light-induced transcription of clock gene Per1 depends on histone acetylation and deacetylation. Mol Cell Biol 24:6278–6287

Pittendrigh CS (1984) Circadian systems: entrainment. In: Aschoff J (ed) Handbook of behavioral neurobiology, vol 4. Plenum, New York, pp 95–124

Pittendrigh CS (1993) Temporal organization: reflections of a Darwinian clock-watcher. Annu Rev Physiol 55:16–54

Reddy P, Zehring WA, Wheeler DA et al (1984) Molecular analysis of the period locus in *Drosophila melanogaster* and identification of a transcript involved in biological rhythms. Cell 38:701–710

Rietveld WJ, Minors DS, Waterhouse JM (1993) Circadian rhythms and masking: an overview. Chronobiol Int 10:306–312

Sahar S, Sassone-Corsi P (2013) The epigenetic language of circadian clocks. In: Kramer A, Merrow M (eds) Circadian clocks, vol 217, Handbook of experimental pharmacology. Springer, Berlin, pp 29–44

Sassone-Corsi P (2010) The year in circadian rhythms. Mol Endocrinol 24:2081–2087

Shearman LP, Zylka MJ, Weaver DR et al (1997) Two period homologs: circadian expression and photic regulation in the suprachiasmatic nuclei. Neuron 19:1261–1269

Shearman LP, Zylka MJ, Reppert SM et al (1999) Expression of basic helix-loop-helix/PAS genes in the mouse suprachiasmatic nucleus. Neuroscience 89:387–397

Stephan FK, Zucker I (1972) Circadian rhythms in drinking behavior and locomotor activity of rats are eliminated by hypothalamic lesions. Proc Natl Acad Sci U S A 69:1583–1586

Sun ZS, Albrecht U, Zhuchenko O et al (1997) A putative mammalian ortholog of the *Drosophila* period gene. Cell 90:1003–1011

Truman JW (1972) Physiology of insect rhythms. II: The silkmoth brain as the location of the biological clock controlling eclosion. J Comp Physiol 81:99–114

Wiener N (1948) Cybernetics: of control and communication in the animal and the machine. MIT Press, Cambridge, MA, pp 212

Zheng X, Sehgal A (2012) Speed control: cogs and gears that drive the circadian clock. Trends Neurosci 35:574–585

Part I
Circadian Systems

Chapter 2
Oxidative Stress and Its Role in the Synchronization of Circadian Rhythms in Crustaceans: An Ecological Perspective

María Luisa Fanjul-Moles and Julio Prieto-Sagredo

Abstract This work reviews concepts regarding the endogenous circadian clock and the relationship between oxidative stress (OS), light, and entrainment in different organisms, particularly in crayfish. In the first section, the molecular control of circadian rhythms in invertebrates, particularly in *Drosophila*, is reviewed, and this model is contrasted with recent reports on the circadian genes and proteins in crayfish. Second, the redox mechanisms and signaling pathways that participate in the entrainment of the circadian clock in different organisms such as zebrafish, its relationship with the redox state, and synchronization by cryptochromes is discussed. Finally, the relationship between metabolism, ROS signals, and transcription factors, such as HIF-1-alpha in crayfish, as well as the possibility that HIF-1-alpha participates in the regulation of circadian control genes in crustaceans, particularly in crayfish, is discussed.

2.1 Redox State as an Entrainment Pathway

Different environmental cyclic cues, or *zeitgeber*, are able to synchronize the circadian rhythms, with light being one of the most important. Circadian rhythms are entrained by light to adapt to the daily solar cycles, and the 24-h light–dark cycle (LD) is considered the most important zeitgeber for synchronization. In nature, the intensity and spectrum of light change seasonally throughout the daily cycle, specifically near dawn and dusk. The photo-entrainment of the circadian clock depends on these two light parameters, producing physiological and behavioral rhythms that match the particular features of the environmental cycles, such as the time of sunrise and sunset, organizing the internal temporal order of the animal according to external time (Bradshaw and Holzapfel 2010).

M.L. Fanjul-Moles (✉) • J. Prieto-Sagredo
Lab. Neurofisiología Comparada de Invertebrados, Departamento de Ecología y Recursos Naturales, Facultad de Ciencias, Universidad Nacional Autónoma de México, Av. Universidad 3000, 04510 México, D.F., México
e-mail: mfajul@gmail.com

© Springer International Publishing Switzerland 2015
R. Aguilar-Roblero et al. (eds.), *Mechanisms of Circadian Systems in Animals and Their Clinical Relevance*, DOI 10.1007/978-3-319-08945-4_2

In addition to light, other extrinsic factors, including temperature, activity, and food availability, can advance or delay (phase shift) circadian rhythms, thereby synchronizing the clock with the external environment (Pittendrigh 1981). This coupling of environmental cues to the clock is termed entrainment. Another class of stimuli based on arousal or activity can also reset an animal's circadian clock in a manner distinct from light. The mechanism underlying these nonphotic phase shifts is unknown, although suppression of canonical clock genes and immediate early genes has been implicated (Antle et al. 2008).

Thus, functioning as an endogenous clock, the circadian system entrains the overt rhythms to the seasonal photoperiodic changes. Daily exogenous fluctuations in the external environment such as light and temperature also produce oxidative stress (OS) in a predictable manner. Meanwhile, daily and endogenous fluctuations produce reactive oxygen species (ROS) such as superoxide anions and hydroxide peroxide as a consequence of metabolism and behavior. These ROS may function as intracellular second messengers (Finkel 1998). In animals, circadian and exogenous daily variations in metabolism, locomotor, and brain activities result in corresponding oscillations of the redox state. External and internal induced oxidative stresses may perturb overt rhythms and the proper internal synchronization, that is, the internal temporal order of physiological processes, and thus alter their physical fitness. In addition, animals are able to defend themselves against the periodic rise in ROS by means of compensatory antioxidative rhythms (Hardeland et al. 2003).

2.2 Circadian Photopigments and Their Response to Light

Notably, in some animals peripheral clocks are directly light responsive (Whitmore et al. 2002). In a variety of organisms, photopigments undergo forward light-induced reactions involving electron transfer to the excited state, flavin, to generate radical intermediates, which correlate with the biological activity of the intermediates (Sancar 2003).

The light-dependent signaling state in cryptochrome (CRY), flavin, and folate-containing blue-light photoreceptors, proposed as circadian extra-retinal photoreceptors in invertebrates such as *Drosophila* (Stanewsky 2003), is also present in crustaceans. In the crayfish *Procambarus clarkii* and *Cherax destructor* (Fanjul-Moles et al. 2004; Sullivan et al. 2009), CRY is involved in the entrainment of the clock (Fig. 2.1). Cryptochromes are found throughout the biological kingdom, ranging from protists to plants, animals, and humans (Stanewsky 2003; Lin and Todo 2005). They are evolutionarily derived from photolyases, or DNA-repair enzymes. Both cryptochromes and photolyases exhibit a high degree of structural homology and bind the same flavin adenine dinucleotide (FAD) and folate light-absorbing cofactors. Although belonging to the DNA photolyase/cryptochrome protein family and being highly similar in amino acid sequence (Todo 1999; Sancar 2000), they differ in their biological function. Cryptochromes have lost the ability to repair DNA, but they play a key role in controlling the circadian clock in plants and animals.

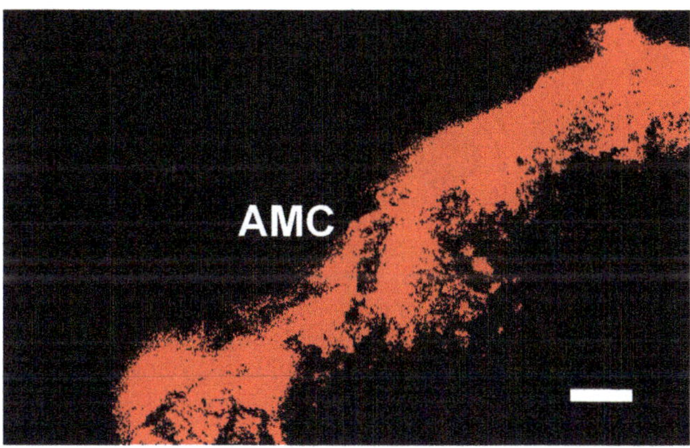

Fig. 2.1 Confocal image of cryptochrome (CRY) immunoreactivity in the *Procambarus clarkii* brain at 19:00. The image corresponds to optical sections of a whole-mount preparation. Neurons of the protocebral anterior medial cluster (AMC) express strong CRY immunoreactivity. *Bars* 100 μm. (Modified from Fanjul-Moles et al. 2004)

The photochemistry and the signaling pathways of cryptochromes are just beginning to be understood. Recent attention has focused primarily on the pathway of flavin photoreduction as a possible mechanism of photoreceptor activation (Müller and Ahmad 2011). Studies with isolated proteins have shown that these can be photoreduced from flavin in the oxidized state to the radical state by light in the presence of a reducing agent. The photoreduction of flavin correlates with its biological activity, as shown by an action spectra with maximal activity at a 450-nm peak with no activity above 500 nm (Hoang et al. 2008; VanVickle-Chavez and Van Gelder 2007; Kay et al. 2003; Yu et al. 2010). In the presence of flavin-containing oxidases, light drives the production of intracellular ROS, such as H_2O_2 (Uchida et al. 2010). The excess production of ROS has deleterious effects because ROS can react with various cellular targets to cause OS (Dalton et al. 1999). However, as already mentioned, light-induced ROS can also take on a signaling role by stimulating intracellular signal pathways that lead to transcriptional activation, including the transactivation of genes.

Cryptochromes were originally identified in plants and have orthologues and paralogues among insects and vertebrates (Yu et al. 2010). The role of cryptochromes in the circadian clock differs among species. In mammals, CRY1 and CRY2 represent the core of the circadian oscillator with a light-independent function (Lin and Todo 2005); however, in *Drosophila* and other insects, CRYs were originally thought to act as circadian photoreceptors. Recently, in non-drosophilid insects, a second cry (cry2) gene has been identified in some species that encodes for a light-insensitive protein; this is more similar in sequence to the mammalian CRYs that act as transcriptional repressors (Zhu et al. 2005). Other insects, such as bees and beetles, possess only a mammalian-like CRY (Yuan et al. 2007).

Drosophila CRY also appears to have a tissue-specific pleiotropic role, because in clock neurons that generate rhythmic locomotor behavior it acts as a photoreceptor

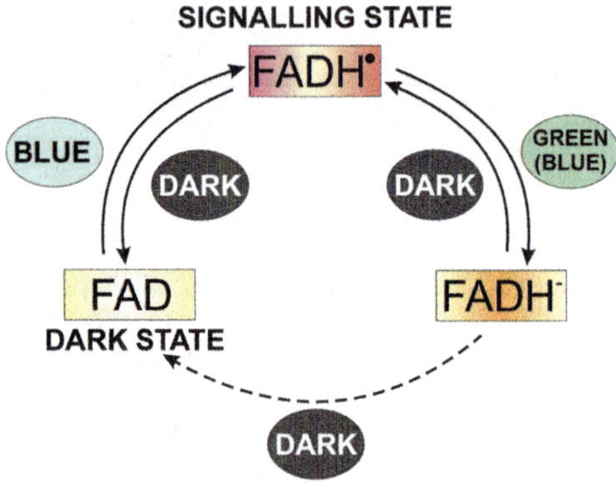

Fig. 2.2 The photocycle of plant cryptochromes. In the dark, the flavin chromophore is in its oxidized state. *Blue light* induces conversion to a metastable semi-quinone redox state that is the activated signaling state. *Green light* causes further reduction to the fully reduced redox state of flavin, which is inactive in signaling. In the *dark*, fully reduced flavin reoxidizes to the fully oxidized form and can be reactivated by *blue light*. The photocycle of plant cryptochromes is different from DNA photolyases, in which only the fully reduced redox state is catalytically active. (Reproduced with permission from Hoang et al. 2008)

and mediates light synchronization of the circadian clock by promoting the light-dependent degradation of dTIM (Emery et al. 1998; Peschel et al. 2009). After absorbing a photon, dCRY undergoes a conformational change involving its C-terminal domain and binds to dTIM, which is then tagged for ubiquitination and proteasomal degradation. The mechanism by which dCRY initiates the cascade of events that leads to dTIM degradation remains unclear (Dubruille et al. 2009). However, the homology between *Drosophila* CRY and photolyases, or light-dependent enzymes utilizing a flavin cofactor to repair DNA, suggests that flavin-dependent reduction/oxidation redox reactions may be involved in dCRY functions (Froy et al. 2002). Cryptochrome-bound flavin is found in an oxidized redox state in vivo, and light activation results in flavin photoreduction to a radical intermediate that represents the likely signaling state. The derived photocycle of animal cryptochromes is therefore similar to the reaction mechanism of plant cryptochromes (Fig. 2.2). Both photocycles involve the reduction of flavin, leading to a cycling between the reduced (active) and oxidized (inactive) redox forms (Berndt et al. 2007; Hoang et al. 2008).

In *Drosophila* it has been proposed that CRY activation involves intramolecular electron transfer and presumably subsequent conformational changes; thus, the cellular redox status also regulates the transfer of photic information and CRY stability (Lin et al. 2001). Using a microarray analysis, Sathyanarayanan et al. (2008) identified three thioredoxin domain-containing redox molecules (GstE7, Txl, and CG11790) and a cytochrome p450, CYP49a1, that modulate light-dependent CRY and TIM degradation. These results suggest that cellular redox status and electron transfer modulate the light-dependent activation of CRY, which in turn affects the subsequent transmission of the light signal to TIM and the degradation of CRY itself, with the subsequent clock resetting.

In zebrafish cells, the light-induced redox changes stimulating intracellular mitogen-activated protein kinase (MAPK) signaling that transduces photic signals to zCry1a gene transactivation (Hirayama et al. 2009). Importantly, light also drives the production of intracellular ROS, such as H_2O_2, that leads to an altered redox status and increases intracellular catalase activity by stimulating catalase transcription, an event which occurs after the maximum expression of the zCry1a gene has been reached (Uchida et al. 2010). This increased catalase activity diminishes light-induced cellular ROS levels, resulting in decreased zCry1a transcription and creating a negative feedback loop. Thus, this altered redox state triggers the transduction of photic signals that regulate and synchronize the circadian clock (Uchida et al. 2010) (Fig. 2.3).

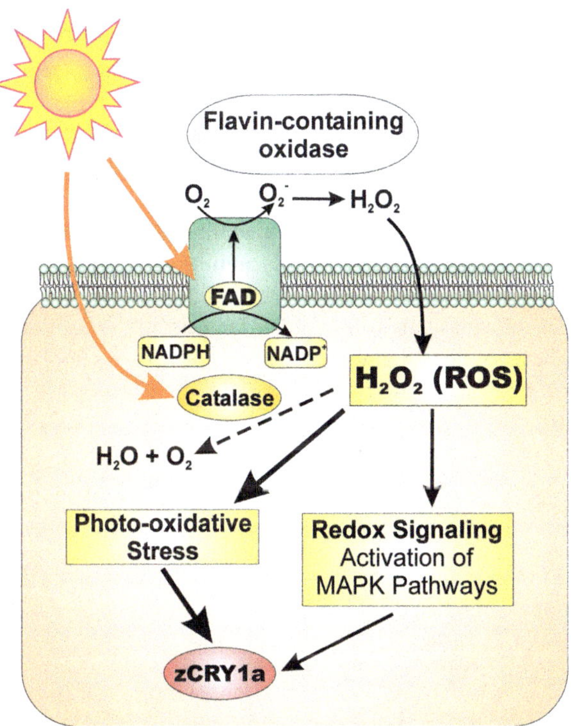

Fig. 2.3 Potential molecular mechanism underlying light-dependent redox signaling in zebrafish. In the presence of flavin-containing oxidases, light drives the production of intracellular reactive oxygen species (ROS) such as H_2O_2. Excess ROS production has deleterious effects because ROS can react with various cellular targets to cause photo-oxidative stress. However, light-induced ROS can also take on a signaling role by stimulating MAPK pathways that lead to transcriptional activation, including transactivation of the zCry1a gene. Light also increases catalase transcription and thus intracellular catalase activity, resulting in H_2O_2 degradation and decreased photo-oxidative stress. This reduction in ROS also leads to decreased zCry1a expression, thus creating a negative feedback loop that directly impinges on the circadian clock. (Reproduced with permission from Hirayama et al. 2009)

2.3   Cryptochromes, Circadian Rhythms, and Redox Mechanisms in Crustaceans

In crustaceans, some behavioral studies have reported rhythmic activity–rest changes as well as synchronization or entrainment of circadian rhythms by blue light (Fanjul-Moles et al. 1992; Bernal-Moreno et al. 1996; Miranda-Anaya and Fanjul-Moles 1997; Aguzzi et al. 2009, 2011). Recently, it has been reported that positive phototaxic circadian rhythmicity is stronger under blue light in the parasite *Argulus japonicus* (Branchiura) (Yoshizawa and Nogami 2008), suggesting the presence of cryptochromes in the circadian photoreceptor system of different groups of crustaceans. The circadian system of genus *Procambarus* is considered to be distributed throughout a multioscillatory system in a hierarchical nature (Fanjul-Moles and Prieto-Sagredo 2003). It is a complex model, which includes three pairs of coupled oscillators such as the retina, the eyestalk X-organ–sinus gland complex (XOSG), putative brain pacemakers, and probably a pair of extra-retinal photoreceptors in the caudal abdominal ganglion. A pair of extra-retinal photoreceptors in the brain (BPR) are involved in light-dependent entrainment (Strauss and Dircksen 2010) (Fig. 2.4), which in crayfish, similar to insects, appears to be mediated by CRYs (Miranda-Anaya and Fanjul-Moles 1997) (Fig. 2.5).

Immunochemical, biochemical, and behavioral studies using a *Drosophila* anti-CRY antibody (Fanjul-Moles et al. 2004; Escamilla-Chimal and Fanjul-Moles 2008; Sullivan et al. 2009) have shown the presence and circadian rhythm of CRY in the brain of crayfish, as well as its role in this organism in the activity of rhythm synchronization. The molecular characteristic of this protein has not been characterized.

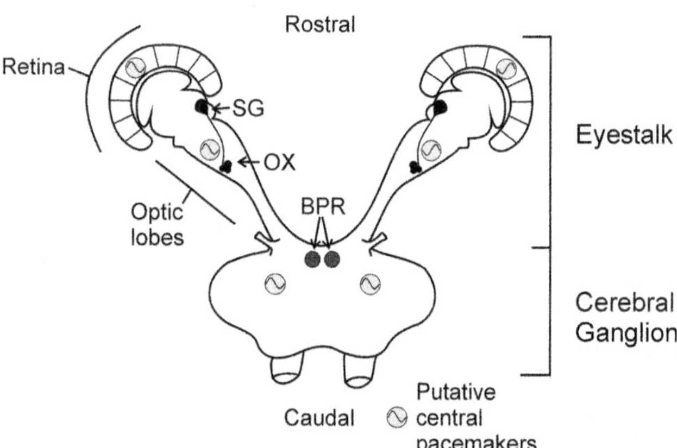

Fig. 2.4 Schematic summary of known circadian oscillators and their functional correlations in crayfish. Established physiological oscillators or pacemakers (*grey circles*) exist in the retina, the X-organ–sinus gland complex, and likely the brain, whereas the brain photoreceptor (BPR) is not an oscillator but is responsible for light-driven entrainment of locomotory rhythms. *BPR* brain photoreceptor, *SG* sinus gland, *XO* X organ

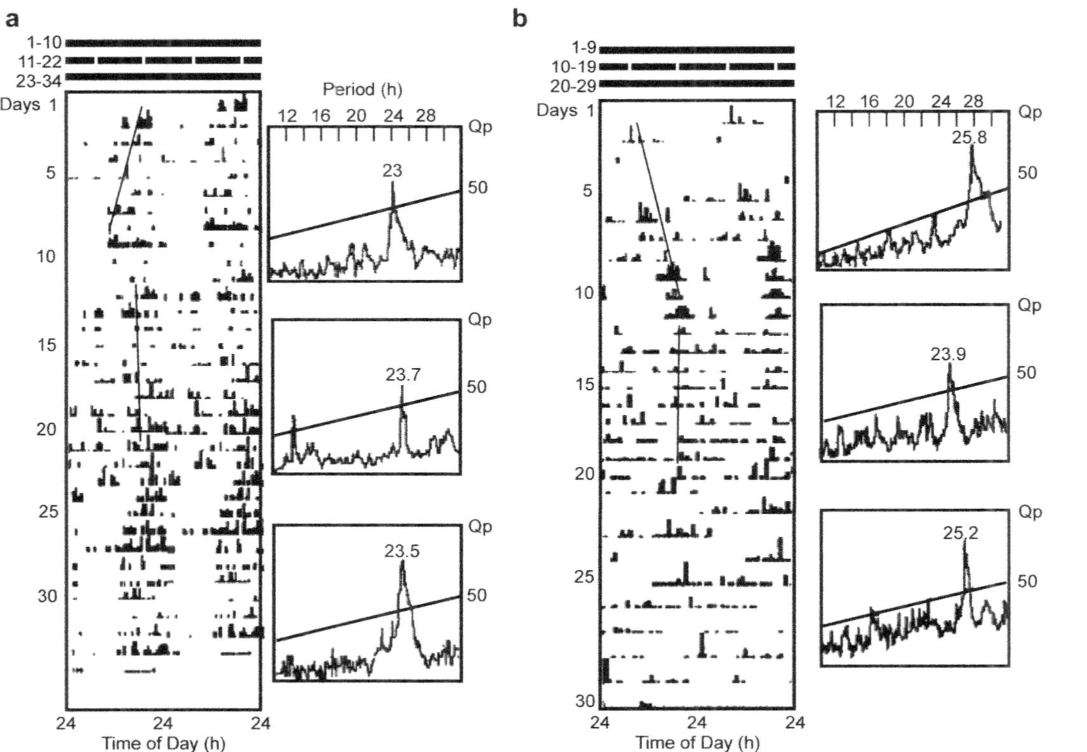

Fig. 2.5 Locomotor activity record and its associate periodograms of two juvenile crayfish, intact (**a**) and retina ablated (**b**), that were exposed to red Skeleton photoperiod (SP) (two 30-min pulses of 640 nm and 25 W m^2). Note the "lights-on" peak during the entrainment condition in the intact crayfish is not present in retina-deprived crayfish. Note the rhythm splitting as well as the change of the rhythm period value during the DD post entrainment. (**a**) A 19:00 pulse light occurring at CT 14 during the 11th day evokes a phase delay. (**b**) A 19:00 light pulse occurring at CT 13 produces an advance. *Regression line* periods are 22.8 and 24.1 h (**a**) and 25.3 and 23.88 h (**b**), respectively. (Modified from Miranda-Anaya and Fanjul-Moles 1997)

Based on both the homologies and the molecular mass detected by Western blot analysis (60 kDa), and considering that the peptide sequence within the C-terminus of *Drosophila* sp. used to generate the antibody was specific to the *Drosophila melanogaster* and *Drosophila pseudoobscura* CRY 1, this protein seems to function as a photopigment in the brain of crayfish (Escamilla-Chimal and Fanjul-moles 2008; Sullivan et al. 2009). Recently, Mazzotta et al. (2010) isolated and cloned a new CRY gene from the Antarctic krill (*Euphausia superba*). The EsCRY gene appears to be an orthologue of mammalian-like CRYs and clusters with the insect CRY2 subfamily. Importantly, these results suggest the presence of two CRYs in crustaceans, similar to mammals and some insects. Both CRYs could act either as a photopigment or as a transcription factor participating in the transcription–translation loop (TTL) of the core of the clock.

Both the mechanisms of the photo-activation of CRY and the different pathways involved in the resetting of the clock are unknown in crustaceans; however, the results of some experiments provide hints toward ROS and redox mechanisms underlying these processes. Fanjul-Moles et al. (2009) demonstrated bi- and uni-modal daily and circadian rhythms in all glutathione (GSH) parameters, substrates, and enzymes in the putative pacemakers of crayfish, including the optic lobe, brain, and retina, as well as an apparent direct effect of light on these rhythm conditions, especially in the retina.

In a series of experiments we used two different species, *Procambarus clarkii* and *Procambarus digueti*, adapted with a different geographical distribution and different latitudes, 27° and 17° N, with different luminosity and daylength throughout the year. During the summer *P. clarkii* is exposed to photoperiods of as long as 16 h whereas in southern latitudes the maximal daylight for *P. digueti* is about 13 h. Using animals of both species collected in the field, we performed different eco-physiological studies in our laboratory (Prieto-Sagredo et al. 2000; Fanjul-Moles and Prieto-Sagredo 2003).The results of different experiments showed daily and circadian changes in OS produced by different light conditions are counterbalanced by antioxidant circadian rhythms of glutathione and related enzymes detected in liver and hemolymph. The robust antioxidant circadian rhythms of *P. clarkii* are able to entrain to all conditions, resetting to lights on or off. However, the weak circadian glutathione system of *P. digueti* did not entrain to LD cycles, showing a random distribution of phases (Fig. 2.6). This observation suggests that the photoperiodic history of each species determines the adaptive abilities of entrainment to extreme light conditions, probably by means of the OS control of circadian antioxidative mechanisms. This finding, of course, indicates interesting ecological implications in the process of light synchronization. Moreover, we found circadian differences between liver and hemolymph glutathione (GSH) oscillatory systems that indicate differences in the coupling strength of both compartments to the crayfish circadian system. These results led us to propose the hemolymph as a passive system responding to an oscillatory driving force depending on various self-sustaining oscillators, one of which could be the liver, proposed elsewhere as a peripheral clock in mammals (Stokkan et al. 2001).

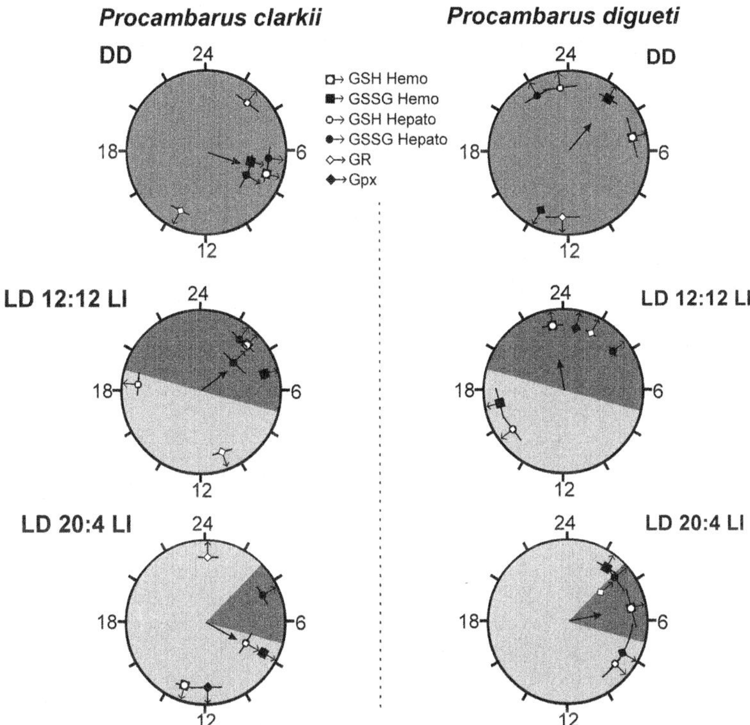

Fig. 2.6 Estimated acrophases of the glutathione daily rhythms of *Procambarus clarkii* and *P. digueti* entrained to the control and experimental conditions. The mean vector indicates the clustering direction of the rhythms acrophase. *Horizontal bars* indicate the mean ± 6 SE. V test indicated, for *P. clarkia*, $P < 0.05$ in LD 12:12 LI and LD 20:4 LI. (Modified from Fanjul-Moles and Prieto-Sagredo 2003)

Herein we investigated the retina and optic lobe–brain (OL-B) circadian GSH system and its ability to handle reactive oxygen species (ROS) produced as a consequence of metabolic rhythms and light variations. We characterized daily and antioxidant circadian variations of the different parameters of the glutathione system, including GSH, oxidized glutathione (GSSG), glutathione reductase (GR), and glutathione peroxidase (GPx), as well as metabolic and lipoperoxidative circadian oscillations in the retina and optic lobe–brain complex (OL-B) proposed elsewhere as central pacemakers of crayfish, determining internal and external GSH system synchrony. The results of our experiments demonstrated bi- and unimodal daily and circadian rhythms in all GSH cycle parameters, substrates, and enzymes in OL-B and retina, as well as an apparent direct effect of light on these rhythms, especially in the retina. The luminous condition appears to stimulate the GSH system to antagonize ROS and lipid peroxidation (LPO) daily and circadian rhythms occurring in both structures, oscillating with higher LPO under dark conditions.

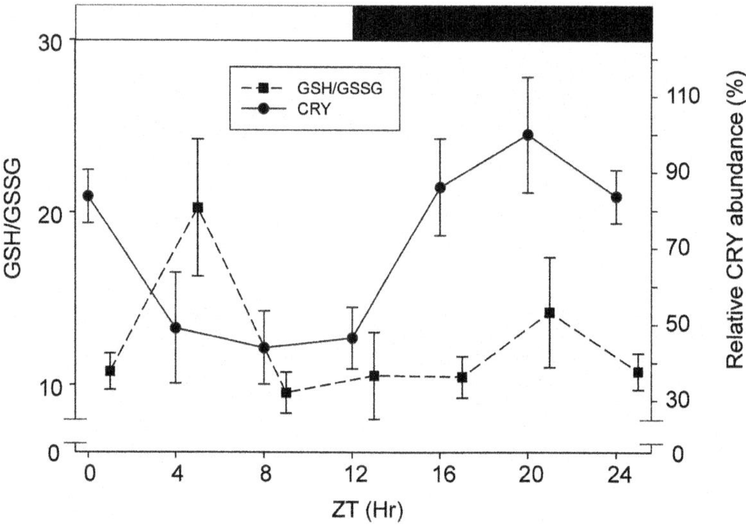

Fig. 2.7 Relationship between GSH/GSSG ratio and CRY abundance in the brain of *P. clarkii* submitted to a 24-h LD cycle. Light drives photo-oxidative stress, as is shown by the low GSH/GSSG ratio, and the decrease in CRY abundance suggests CRY degradation. However, importantly light seems to upregulate glutathione reductase (not shown), which increases GSH and the GSH/GSSG ratio. This reduction in reactive oxygen species (ROS) leads to greater CRY degradation. The abundance increments of CRY occur in the dark phase (from ZT 12 to ZT 20) when the brain showed maximal oxidation (not shown). *ZT* zeitgeber time. (Modified from Escamilla-Chimal and Fanjul-Moles 2008 and Fanjul-Moles et al. 2009)

These results confirmed that light producing photo-oxidation is a factor that determines ROS in crayfish, as was previously reported (Duran-Lizarraga et al. 2001). Putative pacemakers in the retina and eyestalk–brain complex showed higher GSH/GSSG ratio mean values in LD than in DD. The increment in this ratio in the retina and eyestalk coincides with the photophase, indicating that OS produced by light is antagonized by the rapid transformation of GSSG into GSH. This result suggests that the antioxidant defense system (ADS) is upregulated at the mid-photophase level, between ZT4 and ZT6 at the same time of the CRY higher decrement (Fanjul-Moles et al. 2004, 2009; Escamilla-Chimal and Fanjul-Moles 2008) (Fig. 2.7). This response suggests a photo-oxidative redox stress signal similar to those of the model of light-induced signaling cascades potentially involved in the control of the circadian clock of zebrafish (Uchida et al. 2010). In crayfish, the redox signal might be the result of the direct effect of ROS produced by the light onset of the LD cycle on CRY to reset the clock. This ROS increment seems activate the glutathione system, especially glutathione reductase (GR), directly or indirectly, through an unknown signaling pathway (Fanjul-Moles et al. 2009). It has been proposed that light-induced ROS, especially in response to UVB, is involved in the activation of mitogen-activated protein kinase (MAPK) downstream antioxidant-response effectors (Assefa et al. 2005), and certain MAPK signal regulation by GSH has been described recently in the mouse (Limón-Pacheco et al. 2007). Many reports have indicated that the GSH/GSSG ratio is an important "sensor" for ADS regulation (Dalton et al. 1999; Dalle-Donne et al. 2009; Maciel et al. 2010), occurring in parallel with ROS increment. These changes promote the oxidation of protein cysteinyl thiols, which activate or deactivate specific enzymes in the signaling cascades (for a review, see Maciel et al. 2010).

As already mentioned, the photic input to the clock directly activates MAPK signaling cascades in zebrafish cells. The light-induced activation of these pathways controls the expression of two evolutionarily related genes, z64Phr and zCry1a, revealing that light-dependent DNA repair and entrainment of the circadian clock share common regulatory pathways (Hirayama et al. 2009). We are exploring similar possibilities in crayfish, where we have already found the presence and cycling of MAPK proteins, such as extracellular-signal-regulated kinases (ERK), in some of the putative pacemakers of this animal, such as brain and eyestalk (unpublished data), as well as the presence of PKA and its possible participation in the circadian regulation of cAMP response element-mediated circadian gene expression (unpublished data).

2.4 Synchronization and Metabolism

Another stimulus able to reset the phase of the circadian system depends on the modulation of the redox state through metabolism (Bianchi 2008). Experimental evidence supports the link between metabolism and circadian rhythms in mammals. The timing of metabolism can be influenced by the circadian system via systemic cues from the suprachiasmatic nuclei (SCN) or peripheral oscillators. A relationship between both processes by means of the AMP/ATP ratio, sensed mainly by adenosine monophosphate-dependent protein kinase (AMPK) or glucose and fatty acid sensors, has been shown (Albrecht 2012). Furthermore, it is well documented that the circadian clock controls the level of many cellular and circulating metabolites, and it is accepted that there is a cyclic relationship between the circadian clock and metabolism wherein the rhythm impacts metabolism and metabolism feeds back to impinge upon the rhythm (Roennnenberg and Merrow 1999). Interestingly, AMPK impacts the mammalian circadian clock mechanisms in various ways. It can directly phosphorylate CRY1, leading to destabilization and degradation of this core clock protein, and consequently affecting the negative limb of the circadian clock mechanism, or it may modulate PER2 protein stability via an indirect mechanism involving casein kinase ε (CK1ε). (Lamia et al. 2009) Thus, AMPK appears to be another potential regulator of the coupling and interaction between metabolism and the circadian clock, as it is in the case of light.

Thus, it is plausible to hypothesize that essentially any parameter reliably changing as a function of metabolic activity, such as ROS, is a candidate for the direct entrainment of the molecular core clock. ROS and antioxidants influence the expression of a number of genes in eukaryotes. Studies from different organisms have demonstrated that the direct effect of the redox state depends on NAD cofactors for the transcriptional and translational control of the molecular clock. Experiments in mammalian cell culture revealed that the reduced forms NADH and NADPH stimulated the binding of CLOCK/BMAL1 and NPAS2/BMAL1 to their cognate E-box sequences, although the oxidized forms NAD+ and NADP+ strongly inhibited its binding (Rutter et al. 2001). The reduced cofactors NADH and NADPH have daily rhythms in plants (Roennnenberg and Merrow 1999) and regulate the activity of clock-like transcription factors in animals, strongly enhancing their DNA binding (Rutter et al. 2001). In crayfish the presence and oscillation of glutathione, and of the related enzymes in the circadian system of crayfish as already mentioned, indicate the presence and cycling of NADPH as an electron donor to reduce the system and perhaps to regulate the clockwork in this crustacean.

One mechanism by which the clock's output pathways are predicted to be rhythmically controlled is through transcription factors or signaling molecules that are themselves components of the oscillator. These direct outputs may in turn regulate downstream clock-controlled genes (CCGs) in a complex web of events. For example, in invertebrates, such as the fly, the positive oscillator components dCLK and CYC bind to E-box elements in gene promoters and mediate rhythmic transcription of negative components of the oscillator PER and TIM, as well as some clock outputs, such as the distal pigment hormone (PDH) gene. A similar situation could exist in crustaceans, where a CCG, such as PDH (Farca-Luna et al. 2010; Strauss et al. 2011) or the crustacean hyperglycemic hormone (CHH), have been proposed as putative clock outputs (Fanjul-Moles and Prieto-Sagredo 2003). In crayfish, metabolic changes, such as in the production of CHH and its relationship to glucose and lactate rhythmic concentrations, show evidence that circadian rhythms correlate with the expression of the hypoxia-inducible factor-alpha (HIF-1-alpha) at dawn and dusk (Velázquez-Amado et al. 2012).

The transcription factor HIF-1-alpha is a heterodimer composed of a regulated protein, HIF-1-alpha, and its constitutive partner, HIF-1-alpha-b. This protein is the most prominent and best described transcription factor that activates the hypoxia-induced expression of target genes involved in cellular and physiological responses such as oxygen transport, iron metabolism, glycolysis, glucose uptake, and growth factor signaling (Bozek et al. 2010). The gene coding for HIF-1-alpha is found to be clock controlled in the mouse (Panda et al. 2002); furthermore, HIF-1-alpha participates in metabolism (Bozek et al. 2007; Bozek et al. 2009). Thus, it has been proposed that this transcription factor is connected to the circadian clock. In mammals, HIF-1-alpha has been described as a factor with an overrepresented number of binding sites in the promoters of CCG (Reinke et al. 2008). Novel target genes activated by HIF-1-alpha are being constantly identified, and targets include genes with protein products involved in many functions including energy metabolism and hormonal control (Dröge 2002). However, in addition to regulating the response of different animals to hypoxia, HIF-1-alpha is considered part of the mechanisms that regulate the transcriptional output of the circadian clock in several organisms. In mammals, HIF-1-alpha has been described as a factor with a number of overrepresented binding sites in the promoters of CCGs, and its mRNA and promoter activity in human cells has recently been reported (Fujita et al. 2010).

HIF-1-alpha regulates the metabolic adjustment to hypoxia in many crustacean species, such as crabs (Head 2010), crayfish (Velázquez-Amado et al. 2012), shrimp (Kodama et al. 2012), and *Daphnia* (Gorr et al. 2004). Hypoxia and hyperoxia present during physiological states of the life cycle in crustaceans, as well as extreme changes in the environment that challenge these organisms, generating an excess of ROS; these physiological states as dormancy are sometimes rhythmic. Thus, the ability to reduce the metabolic rate during exposure to environmental stress, termed metabolic rate suppression, is thought to be an important component to enhance survival in many crustaceans and related to both HIF-1-alpha and the output of the clock (Gorr et al. 2010). Recently, it has been suggested that the oscillation of HIF-1 is involved in the circadian pacemaker of *P. clarkii* (Velázquez-Amado et al. 2012). Light/dark 12:12 cycles induced greater HIF-1 expression at ZT 13 than at ZT 0100. Thus, these cyclic values persist in darkness and are inversely related with the minimal and maximal ROS concentration at both hours, oxidative stress resulting from light–dark cycles with a long photoperiod (20:4 LD) that activated HIF-1a in all the putative pacemakers of crayfish not withstanding. HIF-1 has a rhythmic expression in the retina and eyestalk and seems to be related to metabolism and antioxidant rhythms (Fanjul-Moles et al. 2009; Duran-Lizarraga et al. 2001), suggesting that HIF-1-alpha is a possible mediator between hypoxic and circadian pathways that may regulate target genes in crustaceans in a manner that is similar to its effects in mammals.

2.5 Conclusions and Perspectives

Evidence indicates that the circadian rhythms of crayfish are controlled by a distributed circadian system that includes four pairs of coupled oscillators (the retina, the eyestalk, XO-SG, brain pacemakers, and the six abdominal ganglia) and two extra-retinal circadian photoreceptors sensitive to blue light, which are involved in photic entrainment. In the crayfish *P. clarkii*, histochemical and biochemical studies have demonstrated that these oscillators express clock proteins similar to those found in the *Drosophila* TTL. In addition, the brain and the sixth abdominal ganglion extra-retinal receptors express CRY. It has been proposed that the blue light-induced photo-entrainment of some rhythms in crayfish, particularly ERG amplitude and activity rhythms, are mediated by CRY. The changes produced by light intensity and photoperiod length in ROS and the antioxidant rhythms of crayfish inhabiting different latitudes suggest that a photo-oxidative redox signal may be contributing to this animal's entrainment to latitudinal photoperiodic changes using similar signaling pathways as aquatic vertebrates to activate CCG transcription. A necessary step is to design further experiments to characterize both clock genes and transcription factors involved in the expression and synchronization of circadian rhythms in crustaceans, particularly in crayfish, exploring the relationship between redox status and light entrainment to determine the signaling pathways involved in the mechanisms of the metabolic and luminous synchronization in this interesting animal group. Undoubtedly, new and ecological comparative experiments will help to clarify this phenomenon.

Acknowledgments This work was partially supported by PAPIIT IN 218811 and CONACYT 178526.

References

Aguzzi J, Sanchez-Pardo J, García JA et al (2009) Day-night and depth differences in haemolymph melatonin of the Norway lobster, *Nephrops norvegicus*. Deep Sea Res Part I Oceanogr Res Pap 56:1894–1905

Aguzzi J, Company JB et al (2011) Activity rhythms in the deep-sea: a chronobiological approach. Front Biosci 16:131–150

Albrecht U (2012) Timing to perfection: the biology of central and peripheral circadian clocks. Neuron 74:246–260

Antle MC, Tse F, Koke SJ et al (2008) Non-photic phase shifting of the circadian clock: role of the extracellular signal-responsive kinases I/II/mitogen activated protein kinase pathway. Eur J Neurosci 28:2511–2518

Assefa Z, Van Laethem A, Garmyn M, Agostinis P (2005) Ultraviolet radiation-induced apoptosis in keratinocytes on the role of cytosolic factors. Biochim Biophys Acta 1755:90–106

Bernal-Moreno JA, Miranda-Anaya M, Fanjul-Moles ML (1996) Phase shifting the ERG amplitude circadian rhythm of juvenile crayfish by caudal monochromatic illumination. Biol Rhythm Res 27:299–301

Berndt A, Kottke T, Breitkreuz H et al (2007) A novel photoreaction mechanism for the circadian blue light photoreceptor Drosophila cryptochrome. J Biol Chem 282(17):13011–13021

Bianchi MM (2008) Collective behavior in gene regulation: metabolic clocks and cross-talking. FEBS J 275:2356–2363

Bozek K, Kiełbasa SM, Kramer A et al (2007) Promoter analysis of mammalian clock controlled genes. Genome Inform 18:65–74

Bozek K, Relogio A, Kielbasa M et al (2009) Regulation of clock-controlled genes in mammals. PLoS One 4(3):e4882

Bozek K, Rosahl AL, Gaub S et al (2010) Circadian transcription in liver. Biosystems 102:61–69

Bradshaw WE, Holzapfel CM (2010) What season is it anyway? Circadian tracking vs photoperiodic anticipation in insects. J Biol Rhythms 25(3):155–165

Dalle-Donne I, Rossi R, Colombo G et al (2009) Protein glutathionylation: a regulatory device from bacteria to humans. Trends Biochem Sci 34:85–96

Dalton P, Shertzer HG, Puga A (1999) Regulation of gene expression by reactive oxygen. Annu Rev Pharmacol Toxicol 39:67–101

Dröge W (2002) Free radicals in the physiological control of cell function. Physiol Rev 82:47–95

Dubruille R, Murad A, Rosbash M et al (2009) A constant light-genetic screen identifies KISMET as a regulator of circadian photoresponses. PLoS Genet 5(12):e1000787. doi:10.1371/journal.pgen.1000787

Duran-Lizarraga ME, Prieto-Sagredo J, Gonsebatt M et al (2001) Crayfish *Procambarus clarkii* shows circadian variations in different parameters of the GSH cycle. Photochem Photobiol 74:350–355

Emery P, So WV, Kaneko M et al (1998) CRY, a Drosophila clock and light-regulated cryptochrome, is a major contributor to circadian rhythm resetting and photosensitivity. Cell 95:669–679

Escamilla-Chimal EG, Fanjul-Moles ML (2008) Daily and circadian expression of cryptochrome during the ontogeny of crayfish. Comp Biochem Physiol A Mol Integr Physiol 151:461–470

Fanjul-Moles ML, Prieto-Sagredo J (2003) The circadian system of crayfish: a developmental approach. J Microsc Res Tech 60:291–301

Fanjul-Moles ML, Miranda-Anaya M, Fuentes-Pardo B (1992) Effect of monochromatic light upon the ERG circadian rhythm during ontogeny in crayfish *Procambarus clarkii*. Comp Biochem Physiol A 102:99–106

Fanjul-Moles ML, Escamilla-Chimal EG, Gloria-Soria A et al (2004) The crayfish *Procambarus clarkii* CRY shows daily and circadian variation. J Exp Biol 207:1453–1460

Fanjul-Moles ML, Prieto-Sagredo J, López DS et al (2009) Crayfish *Procambarus clarkii* retina and nervous system exhibit antioxidant circadian rhythms coupled with metabolic and luminous daily cycles. Photochem Photobiol 85:78–87

Farca-Luna AJ, Heinrich R, Reischig T (2010) The circadian biology of the marbled crayfish. Front Biosci 2:1414–1431

Finkel T (1998) Oxygen radicals and signaling. Curr Opin Cell Biol 10:248–253

Froy O, Chang DC, Reppert SM (2002) Redox potential. Differential roles in dCRY and mCRY1 functions. Curr Biol 12(2):147–152

Fujita S, Koyama Y, Higashimoto M et al (2010) Regulation of circadian rhythm of human vascular endothelial growth factor by circadian rhythm of hypoxia inducible factor-1: implication for clinical use as anti-angiogenic therapy. Ann Cancer Res Ther 18:28–36

Gorr TA, Cahn JD, Yamagata H et al (2004) Hypoxia-induced synthesis of hemoglobin in the crustacean Daphnia magna is hypoxia inducible factor-dependent. J Biol Chem 34: 36038–36047

Gorr TA, Wichmann D, Hu J et al (2010) Hypoxia tolerance in animals: biology and application. Physiol Biochem Zool 83:733–752

Hardeland R, Coto-Montes A, Poeggeler B (2003) Circadian rhythms, oxidative stress, and antioxidative defense mechanisms. Chronobiol Int 20:921–962

Head JM (2010) The effects of hypoxia on hemocyanin regulation in *Cancer magister*: possible role of hypoxia-inducible factor. J Exp Mar Biol Ecol 386:77–85

Hirayama J, Miyamura N, Uchida Y et al (2009) Common light signaling pathways controlling DNA repair and circadian clock entrainment in zebrafish. Cell 8:2794–2801

Hoang N, Schleicher E, Kacprzak S et al (2008) Human and Drosophila cryptochromes are light activated by flavin photoreduction in living cells. PLoS Biol 6:e160

Kay CW, Schleicher E, Kuppig A et al (2003) Blue light perception in plants. Detection and characterization of a light-induced neutral flavin radical in a C450A mutant of phototropin. J Biol Chem 278:10973–10982

Kodama K, Rahman MS, Horiguchi T et al (2012) Assessment of hypoxia inducible factor-1α mRNA expression in mantis shrimp as a biomarker of environmental hypoxia exposure. Biol Lett 8:278–281

Lamia KA, Sachdeva UM, DiTacchio L et al (2009) AMPK regulates the circadian clock by cryptochrome phosphorylation and degradation. Science 326:437–440

Limón-Pacheco JH, Hernández NA, Fanjul-Moles ML et al (2007) Glutathione depletion activates mitogen-activated protein kinase (MAPK) pathways that display organ-specific responses and brain protection in mice. Free Radic Biol Med 43:1335–1347

Lin C, Todo T (2005) The cryptochromes. Genome Biol 6:220. doi:10.1186/gb-2005-6-5-220

Lin FJ, Song W, Meyer-Bernstein E et al (2001) Photic signaling by cryptochrome in the Drosophila circadian system. Mol Cell Biol 21:7287–7294

Maciel FE, Geihs MA, Monserrat JM et al (2010) Antioxidant defense system rhythms in crustaceans and possible roles for melatonin. Front Biosci 12:1448–1459

Mazzotta GM, De Pittà C, Benna C et al (2010) A cry from the krill. Chronobiol Int 27:425–445

Miranda-Anaya M, Fanjul-Moles ML (1997) Nonparametric effects of monochromatic light on the activity rhythm of juvenile crayfish. Chronobiol Int 14:25–34

Müller P, Ahmad M (2011) Light-activated cryptochrome reacts with molecular oxygen to form a flavin-superoxide radical pair consistent with magnetoreception. J Biol Chem 286:21033–21040

Panda S, Antoch MP, Miller BH et al (2002) Coordinate transcription of key pathways in the mouse by the circadian clock. Cell 109:307–320

Peschel N, Chen KF, Szabo G et al (2009) Light dependent interactions between the Drosophila circadian clock factors cryptochrome, jetlag, and timeless. Curr Biol 19:241–247

Pittendrigh C (1981) Circadian rhythms and entrainment. In: Aschoff J (ed) Handbook of behavioral neurobiology, vol 4, Biological rhythms. Plenum, New York, pp 95–124

Prieto-Sagredo J, Ricalde-Recchia I, Durán-Lizarraga ME et al (2000) Changes in hemolymph glutathione status after variation in photoperiod and light-irradiance in crayfish *Procambarus clarkii* and *Procambarus digueti*. Photochem Photobiol 71(4):487–492

Reinke H, Saini C, Fleury-Olela F et al (2008) Differential display of DNA binding proteins reveals heat-shock factor 1 as a circadian transcription factor. Genes Dev 22:331–345

Roennnenberg T, Merrow M (1999) Circadian clocks and metabolism. J Biol Rhythms 14: 449–459

Rutter J, Reick M, Wu LC et al (2001) Regulation of clock and NPAS2 DNA binding by the redox state of NAD cofactors. Science 293:510–514

Sancar A (2000) Cryptochrome: the second photoactive pigment in the eye and its role in circadian photoreception. Annu Rev Biochem 69:31–37

Sancar A (2003) Structure and function of DNA photolyase and cryptochrome blue-light photoreceptors. Chem Rev 103:2203–2237

Sathyanarayanan S, Zheng X, Kumar S et al (2008) Identification of novel genes involved in light-dependent CRY degradation through a genome-wide RNA screen. Genes Dev 22:1522–1533

Stanewsky R (2003) Genetic analysis of the circadian system in *Drosophila melanogaster* and mammals. J Neurobiol 54:111–147

Stokkan KA, Yamazaki S, Tei H et al (2001) Entrainment of the circadian clock in the liver by feeding. Science 291(5503):490–493

Strauss J, Dircksen H (2010) Circadian clocks in crustaceans: identified neuronal and cellular systems. Front Biosci 15:1040–1074

Strauss J, Zhang Q, Verleyen P et al (2011) Pigment dispersing hormone in *Daphnia* interneurons, one type homologous to insect clock neurons displaying circadian rhythmicity. Cell Mol Life Sci 68:3403–3423

Sullivan JM, Genco MC, Marlow ED et al (2009) Brain photoreceptor pathways contributing to circadian rhythmicity in crayfish. Chronobiol Int 26:1136–1168

Todo T (1999) Functional diversity of the DNA photolyase/blue light receptor family. Mutat Res 434:89–97

Uchida Y, Hirayama J, Nishina H (2010) Common origin: signaling similarities in the regulation of the circadian clock and DNA damage responses. Biol Pharm Bull 33:535–544

VanVickle-Chavez SJ, Van Gelder RN (2007) Action spectrum of Drosophila cryptochrome. J Biol Chem 282:10561–10566

Velázquez-Amado RM, Escamilla-Chimal EG, Fanjul-Moles ML (2012) Daily light-dark cycles influence hypoxia-inducible factor-1 and heat shock protein levels in the pacemakers of crayfish. Photochem Photobiol 88:81–89

Whitmore D, Foulkes NS, Sassone-Corsi P (2002) Light acts directly on organs and cells in culture to set the vertebrate circadian clock. Nature 404:87–91

Yoshizawa K, Nogami S (2008) The first report of phototaxis of fish ectoparasite *Argulus japonicus*. Res Vet Sci 85:128–130

Yu X, Liu H, Klejnot J et al (2010) The cryptochrome blue light receptors. Arabidopsis Book 8:e0135

Yuan Q, Metterville D, Briscoe AD et al (2007) Insect cryptochromes: gene duplication and loss define diverse ways to construct insect circadian clocks. Mol Biol Evol 24:948–955

Zhu H, Yuan Q, Briscoe AD et al (2005) The two CRYs of the butterfly. Curr Biol 15:R953–R954

Chapter 3
Control of Rest–Activity Behavior by the Central Clock in *Drosophila*

Shailesh Kumar and Amita Sehgal

Abstract Our current understanding of circadian rhythms comes from advances in many different species. Remarkably, basic features of these rhythms and their underlying mechanisms are conserved across organisms, thereby allowing a unique cross-disciplinary approach. On the molecular and cellular level, the contributions of research done in *Drosophila* are unquestionable. The molecular clock mechanism and some of the clock genes first deciphered in *Drosophila* have now even been implicated in human circadian disorders. Here we review the basic elements of the circadian system that drives rest–activity behavior in flies. We describe the molecular basis of the clock, the mechanisms by which the clock entrains to environmental signals, and the brain clock circuit, in which different groups of clock neurons subserve distinct aspect of the behavioral rhythm. Studies in these different areas are providing a comprehensive understanding of how rhythmic behavior is generated.

3.1 Introduction

The geophysical cycles caused by the daily rotation of the earth create abiotic selection pressure in the form of light: dark and temperature cycles on almost all organisms (reviewed in Edery 2000; Takahashi et al. 2008a, b; Allada and Chung 2010; Bell-Pedersen et al. 2005). These abiotic pressures are imposed on biotic needs such as finding mates, shelter, and food and avoiding predators, and they influence organisms' behavior in such a way that distinct behavioral and physiological processes are temporally separated. Adaptation to the daily fluctuations in the environment is facilitated by an endogenous circadian system, which helps to align molecular, cellular, and physiological processes relative to one another (reviewed in Takahashi et al. 2008a, b; Sharma 2003). These endogenous timekeepers are ubiquitous in nature and found in diverse organisms, ranging from bacteria to mammals,

S. Kumar (✉) • A. Sehgal
Department of Neuroscience, Howard Hughes Medical Institute, University of Pennsylvania
Perelman School of Medicine, Smilow Center for Translational Research,
Philadelphia, PA 19104, USA
e-mail: kumars@mail.med.upenn.edu; amita@mail.med.upenn.edu

© Springer International Publishing Switzerland 2015
R. Aguilar-Roblero et al. (eds.), *Mechanisms of Circadian Systems in Animals and Their Clinical Relevance*, DOI 10.1007/978-3-319-08945-4_3

indicating their critical role in survival and fitness (reviewed in Sharma 2003; Kumar et al. 2005; Johnson and Golden 1999; Doherty and Kay 2010; Yerushalmi and Green 2009). Around 400 BC, in one of the very first documentations of circadian rhythms of physiology, Androsthenes (historian of Alexander the great) observed that the position of specific tree leaves changes during the night and day. Leaves display a horizontal position during the day and a more vertical position at night. However, an empirical demonstration of circadian clocks came much later, in 1729, when French astronomer de Mairan found that leaves of a heliotrope, *Mimosa pudica*, drooped during the nighttime and opened up in the morning. Interestingly he showed that the drooping of leaves continued even in constant darkness in a time-dependent manner, suggesting that it was driven by an endogenous circadian clock (de Mairan 1729). Nevertheless, the existence of internal clocks was not confirmed until the 1940s when it was found that the internal circadian period varies from species to species and does not match the period of any geomagnetic cycle created by the earth's rotation. Subsequently, it became clear that intrinsic genetic factors drive these rhythms. In 1971, Konopka and Benzer conducted a genetic screen to identify genes required for circadian behavior in the fruit fly, *Drosophila melanogaster*. This classic study, which identified the first behavioral gene *period* (*per*), initiated the era of neurogenetics of behavior. Fruit fly populations display a rhythmic pattern in their eclosion (act of emergence of adults from pupal cases) in the presence of light–dark cycles and under constant dark conditions. Konopka and Benzer's genetic screen revealed that mutations in the *per* gene could modify the rhythmic pattern of eclosion, indicating that *per* is essential for circadian rhythms. However, the underlying molecular mechanisms were unknown until about a decade later, when two groups independently isolated the *per* gene (Reddy et al. 1984; Zehring et al. 1984; Bargiello and Young 1984; Baylies et al. 1987; Bargiello et al. 1987). Analysis of the protein revealed that PER contains a region shared with the *Drosophila* single-minded (SIM) and the mammalian aryl hydrocarbon receptor nuclear translocator (ARNT) proteins, hence termed a PAS domain after the three proteins in which it was originally found. This domain was later shown to mediate protein–protein interactions (Huang et al. 1993). In addition PER contains a stretch of Thr-Gly repeats that are implicated in mediating adaptation to different latitudinal clines (Sawyer et al. 1997).

In a major breakthrough, it was found that *per* transcript and protein levels cycle in a circadian manner, and the PER protein is required for cyclic expression of its own mRNA (Hardin et al. 1990). PER functions in the circadian clock through a feedback loop where it represses its own transcription and thus produces a cyclic pattern in *per* mRNA expression (Hardin et al. 1992; Zeng et al. 1994). Subsequently, a second circadian clock gene, *timeless*, was discovered and shown to also negatively regulate itself and to interact with PER (Sehgal et al. 1994, 1995; Myers et al. 1995; Gekakis et al. 1995). Indeed, PER and TIM are largely co-regulated and they function together to generate an autoregulatory loop that lies at the heart of the clock mechanism in *Drosophila*. Similar transcriptional–translational feedback loops (TTLs) were found to drive circadian oscillations of core clock genes in other multicellular organisms such as fungi, animals, and plants (reviewed in Young and

Kay 2001; Zheng and Sehgal 2008). Besides controlling their own transcription, these key clock genes also regulate and disseminate oscillatory information to a second tier of genes, which are referred to as clock-controlled genes (ccgs) (reviewed in Dunlap 1999). These ccgs are not part of the core clock; rather they contribute towards rhythmic outputs in different cells, tissues, and body organs. In fact, recent genomic studies have shown that most of the transcriptome oscillates with about 24 h cycle in variety of tissues (McDonald and Rosbash 2001; Ueda et al. 2002; Claridge-Chang et al. 2001).

Across all taxa, the general and a simplistic model for a circadian clock is that multiple proteins regulate their expression by inhibiting their own transcription. In such a model, the clock proteins do not appear until a certain time of the day; subsequently they undergo progressive posttranslational modifications, thus introducing a critical delay in their accumulation and entry into the nucleus to become competent negative regulators. Timely degradation of the proteins ensures the end of each cycle and is critical for maintaining circadian cycling as well as a precise circadian period. Although the autoregulating clock components are clearly essential, several other genes/proteins also play a key role in maintenance of the loop. These proteins may be part of additional molecular steps that appear to impart delays in feedback loops, which will be discussed in detail in Sects. 3.3 and 3.7 (described below).

3.2 Interlocked Feedback Loops

In the core feedback loop, *per* and *tim* transcription is initiated by the binding of bHLH–PAS containing transcription factors CLK and CYC to E-boxes in the *per* and *tim* promoter regions (Allada et al. 1998; Darlington et al. 1998; Rutila et al. 1998). *per* and *tim* mRNAs build up from ZT4 to 18 where ZT refers to zeitgeber time (ZT0=lights-on and ZT12=lights-off under standard laboratory light/dark conditions) (Fig. 3.1). After lights-off the PER and TIM proteins accumulate in the cytoplasm where they form a complex with each other and also with other proteins that mainly regulate their stability in the cytoplasm (Fig. 3.1). There is a lag of approximately 6–8 h between the accumulation of the mRNAs and that of the proteins. This delay in the accumulation of PER and TIM is not completely understood, but could arise from several molecular steps (described in detail below under Sect. 3.3). Briefly, DOUBLETIME (DBT), a homolog of mammalian casein kinase 1ε, phosphorylates PER in the cytoplasm and causes rapid degradation of PER, and so PER is not stabilized until TIM binds to the PER–DBT complex in the cytoplasm (Fig. 3.1) (Kloss et al. 1998; Price et al. 1998; Gekakis et al. 1995; Zeng et al. 1996). Recent studies suggest translational control of PER accumulation, but this control does not appear to be rhythmic, as one would expect for a mechanism that introduces a delay at a specific time of day (described in detail below; Chiu et al. 2011; Chiu et al. 2008; Lim et al. 2011; Lim and Allada 2013). PER and TIM proteins are also modified by other kinases such as CK2 and Shaggy (SGG) (Fig. 3.1). *Drosophila* ortholog of glycogen

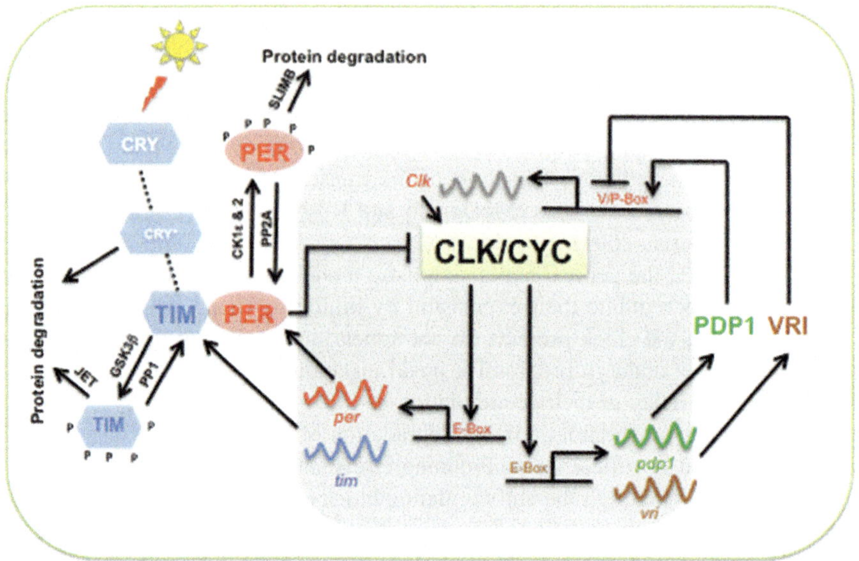

Fig. 3.1 The molecular clock in *Drosophila*: The schematic depicts the daily cycle that drives cyclic expression of *per* and *tim* mRNAs and PER and TIM proteins. Protein expression is tightly controlled by the indicated kinases and phosphatases. CRY is shown to reset the clock through its action on TIM. The interlocked loop that drives rhythmic expression of *Clk* mRNA is also shown. Note that while *Clk* mRNA cycles, the protein does not, so this loop may serve more to maintain stable levels of CLK rather than oscillations of clock components

synthase kinase 3β (GSK3β) affects largely the subcellular localization of these proteins (Akten et al. 2003; Martinek et al. 2001; Lin et al. 2002). Furthermore, there are distinct phosphatases that influence the subcellular location and stability of these proteins: protein phosphatase 2A (PP2A) and 1 (PP1) dephosphorylate PER and TIM (Sathyanarayanan et al. 2004; Fang et al. 2007) (Fig. 3.1).

After their translocation into the nucleus, around ZT18–22, the PER–TIM complex along with DBT binds CLK directly and represses transcription by promoting phosphorylation of CLK and releasing the CLK/CYC heterodimer from E-Boxes (Lee et al. 1998, 1999; Yu et al. 2006; Menet et al. 2010). The following morning, after lights-on, TIM is degraded (Hunter-Ensor et al. 1996; Lee et al. 1996; Myers et al. 1996). At this time, PER is again phosphorylated by DBT, but this time in the nucleus. Phosphorylated PER is recognized by SLIMB (an E3 ligase and targeted for ubiquitination-mediated proteosomal degradation) (Grima et al. 2002; Naidoo et al. 1999; Kloss et al. 2001). Subsequently, the newly formed hypophosphorylated CLK builds up and heterodimerizes with CYC to bind again to the E-box of PER to begin a new molecular cycle of the *per* and *tim* transcription.

In addition to this major feedback loop, most clocks contain a second transcription–translation loop in which one of the positive factors is regulated through negative feedback. In case of *Drosophila*, the positive factor regulated in this manner is CLK heterodimers bind to E-boxes present in the *PAR domain protein* (*Pdp1e*) and *Vrille*

(*Vri*) and drive rhythmic expression of both these genes (Blau and Young 1999; Cyran et al. 2003) (Fig. 3.1). *Vri* and *Pdp1e* mRNA levels reach peak levels ~ ZT14, but the PDP1 protein product lags by about 6 h. *vri* mRNA and protein accumulate in phase with one another and VRI binds to the VRI/PDP1 box present in the *Clk* promoter region at ~ ZT14–ZT18 and represses *Clk* mRNA levels (Cyran et al. 2003). PDP1e levels rise several hours later and replace VRI on the V/P box to activate *Clk* transcription (Cyran et al. 2003; Zheng et al. 2009). This is facilitated by the action of PER–TIM. Around ZT18-4, the PER–TIM complex feeds back to inhibit CLK/CYC-dependent transcriptional activity, which results in reduced *vri* mRNA and protein expression and thus allows PDP1e to activate *Clk* transcription (Cyran et al. 2003). Around, ZT0–4 when the PER–TIM complex is degraded, a new cycle of *vri* and *pdp1* transcription is initiated. VRI and PDP thus generate a loop to regulate a positive component of the major loop and thereby maintain ~24 h molecular oscillations (Fig. 3.1).

Mutants that abolish the transcriptional activity of CLK or CYC (*Clk^{Jrk}* or *Cyc^0*) display peak levels of *Clk* mRNA, even though loss of CLK decreases expression of the *Clk* activator, PDP1, suggesting that there are additional activators of the *Clk* gene (Glossop et al. 1999). Genetic analyses of the *Clk* promoter region also support the idea that <u>*Clk*</u> can be transcriptionally regulated by transcription factors other than PDP1 and VRI (Gummadova et al. 2009).

3.3 Posttranscriptional Regulation

For most cycling components, rhythmic expression is brought about through transcriptional regulation. As noted above, a significant proportion of the genome is expressed cyclically, although the specific genes regulated in this manner may vary from tissue to tissue. However, in recent years it has become clear that circadian regulation can be exerted at several other levels also. These include intermediate steps from transcript to protein and can help cells adjust protein levels without altering the transcription process. Following mRNA synthesis, splicing and occasionally formation of multiple isoforms can occur. mRNAs are translocated into the cytoplasm where they undergo additional modification or degradation. Some of these steps include capping of the 5′ end, deadenylation of the 3′ end of the mRNA, splicing, and RNA editing. These molecular steps together can regulate stability of the mRNA and/or its translation.

3.4 mRNA Splicing and Stability

One of the most common posttranscriptional processes that regulates gene expression is alternative splicing, and, not surprisingly, this is involved in circadian regulation. For instance, alternative splicing in the eighth intron of the *per* gene allows for seasonal adaptation (Collins et al. 2004; Majercak et al. 2004). Shorter photoperiods as well as cold temperature induce alternative splicing events that affect PER expression such that *per* mRNA and protein accumulate earlier and to higher levels. On the behavioral level, the evening activity peak occurs earlier under cold temperature or short photoperiod conditions and the afternoon siesta is reduced (Majercak et al. 1999, 2004). Similar mechanisms have also been reported in *Neurospora* where *per's* functional homolog, *frequency* (*frq*), also undergoes alternate splicing to adapt to seasonal variations, suggesting a conserved role of such molecular events (Colot et al. 2005; Diernfellner et al. 2005). The underlying mechanism is unknown, but it is speculated that *trans*-factors that regulate splicing are also controlled by seasonal variations. Thus, it appears that PER is the main determinant for integrating ambient day length and temperature information to the circadian clock.

A few earlier studies suggested additional posttranscriptional regulation of clock genes. For instance, the half-life of *per* mRNA in *Drosophila* varies throughout the circadian cycle (So and Rosbash 1997). Subsequent studies from mammalian labs showed that the mRNA stability of *per2* and *cry1* changes over time (Woo et al. 2009, 2010) and that regulated polyadenylation plays a critical role in circadian regulation. Using a genome-wide screen to identify mRNAs whose polyA tail exhibits circadian oscillations, Kojima et al. (2010) showed that the rhythmicity in polyA tails of several mRNAs positively correlates with rhythmic protein expression.

3.5 Role of miRNAs

Another potential route for posttranscriptional regulation is through miRNA-mediated RNA processing. miRNAs are about 22 nucleotides long/short noncoding RNAs that regulate expression of target genes by directly binding to the target transcripts and affecting RNA stability and/or translational efficiency (Bushati and Cohen 2007). miRNAs are generally characterized by their complementarity to 3′ untranslated regions (UTRs) of mRNA targets (Jackson and Standart 2007). A screen for cyclically expressed miRNAs in adult *Drosophila* heads from wild type and *cyc⁰¹* flies identified two miRNAs that are under circadian control—miR 263a and miR 263b (Yang et al. 2008). Predictions based upon specific algorithms indicated several core clock genes as potential targets, such as *per*, *Clk*, *dbt*, *clockwork orange* (*cwo*), and *twins* (*tws*) (Yang et al. 2008).

Using a different biochemical and genetic approach, Kadener et al. (2009) inhibited the miRNA biogenesis pathway in clock cells and followed up with tiling array analyses to identify novel miRNAs involved in the molecular clockwork. Three clock genes, *Clk*, *vri*, and *cwo*, were found to be associated with AGO1 (a catalytic unit of the RISC complex associated with miRNA biogenesis). Furthermore, reporter assays performed on the 3′UTR of the *Clk* gene identified *bantam* as an important miRNA, which is bound to this region as a part of the RISC complex (Kadener et al. 2009). *Bantam* was previously reported to have an essential role in stem cell maintenance in optic lobe development (Li and Padgett 2012).

In vivo analyses of *bantam* also showed that it is required for proper maintenance of rest–activity rhythms.

Another miRNA implicated in circadian rhythms is miR-279, which was identified through a forward genetic screen (Luo and Sehgal 2012). Overexpression or knockdown of this miRNA impairs rest–activity rhythms without affecting the core clock; thus miR-279 functions as an output molecule. Specifically miR-279 regulates activity–rest rhythms through its direct effects on *Unpaired* (*Upd*), a ligand of the JAK/STAT pathway (Luo and Sehgal 2012). Together, these studies identify miRNAs mediated posttranscriptional regulation that has a key impact on the core clock or on the output in flies. Thus, miRNAs serve as major molecular players in circadian timekeeping, and future experiments warrant the identification and characterization of additional miRNAs in circadian regulation.

3.6 Translational Control

A recent mammalian study has shown that ~20 % of soluble proteins in the liver are regulated in a circadian manner, but their corresponding transcripts do not cycle (Reddy et al. 2006). This suggests an important role of posttranslational regulation in circadian rhythms. Such regulation has not been characterized in any detail in flies, but a few studies have indicated the possible relevance of translational control in the molecular oscillations of clock components. Translational control typically involves RNA-binding proteins, and one such protein, LARK, is implicated in the regulation of *Drosophila* circadian rhythms (McNeil et al. 1998; Newby and Jackson 1995; Huang et al. 2007). The *lark* mRNA cycles robustly and its expression is under control of the circadian clock (McNeil et al. 1998); however, the role of this gene is not fully understood. Through experiments to identify potential targets of LARK, Huang et al. (2007) identified *Ecdysone-induced protein 74EF* (*E74*). E74 mutants were shown to exhibit arrhythmic eclosion behavior, which corroborated earlier findings implicating *lark* in eclosion output. Mammalian LARK shuttles between the nucleus and the cytoplasm to control translation of target RNAs (Kojima et al. 2007; Lai et al. 2003; Lin and Fu 2007), and it is likely that *Drosophila* LARK does so as well. Interestingly many LARK targets contain miRNA binding sites (Ying et al. 2006), so it is possible that LARK acts through miRNAs to regulate the translation of specific target RNAs. A similar role has been shown for another RNA-binding protein, fragile-X mental retardation protein (FMRP), which has an important circadian function (Sofola et al. 2008; Dockendorff et al. 2002; Inoue et al. 2002).

Recent studies have shed light on the translational control of clock proteins. One particular study in flies showed that TWENTY-FOUR (TYF) is required for robust PER translation in the central clock neurons (Lim et al. 2011). TYF unlike LARK is not an RNA-binding protein, but is associated directly with polyA-binding protein (PABP) and indirectly with the 5′-cap-associating translation initiation factor eIF4G. Based on co-immunoprecipitation assays, *per* mRNA and, to somewhat lesser extent, *tim* mRNA was found with TYF. Interestingly in vivo, the loss of *tyf* affected only PER proteins but not the *per* mRNA. This study also showed that TYF was required mainly in the central clock neurons in the brain (Lim et al. 2011). Based on cell culture and in vivo data, it appears that following transcription, *per* mRNA is bound by a translational inhibitor, and TYF, together with PABP, eIF4G, and a few other factors, recognizes this inhibitor and displaces it. Taken together TYF ensures efficient translation of PER, although, as noted above, it is not yet linked to the delayed accumulation of PER relative to its mRNA.

3.7 Posttranslational Regulation

A transcriptional–translational loop would normally be complete in much less time than 24 h. For instance, subsequent to their synthesis, the PER and TIM proteins are translocated into the nucleus to repress their own transcription by inhibiting CLK/CYC transcriptional activity. These biochemical steps could be completed quite rapidly, and yet, they account for a significant proportion of a circadian cycle. Clearly, additional molecular steps introduce delays in the transcriptional–translational loop. The *per* and *tim* genes and their products not only cycle in their total abundance but also their phosphorylation state oscillates over the course of the day. This posttranslational modification of PER and TIM is very crucial for driving protein cycling and overt circadian rhythms, as the constitutive expression of *per* and *tim* mRNA can restore protein oscillations and behavioral rhythms in a significant proportion of *per;tim* double mutants (Yang and Sehgal 2001). In addition to phosphorylation, de-phosphorylation, formation of protein–protein complexes, and ubiquitination steps may also help to provide critical delays in feedback loops.

As mentioned above, CLK and CYC transcriptional factors heterodimerize and activate *per* and *tim* transcription during the day and into the early night. Any amount of TIM produced before sunset is degraded by light as TIM is sensitive to light; thus, TIM does not accumulate until dusk. Given that PER is unstable in the absence of TIM, due to its constant phosphorylation and degradation by DOUBLETIME (DBT) (Price et al. 1998), PER accumulation follows that of TIM. While DBT phosphorylates PER at several sites, phosphorylation at a specific serine residue (S47) is critical for targeting PER for proteosomal degradation (Chiu et al. 2008). S47-phosphorylated PER is recognized by the SLIMB protein, which is a component of the Skp1/Cullin/F-box protein (SCF) ubiquitin ligase complex, and targeted to the proteasome (Ko et al. 2002). Recent detailed studies have shown that PER phosphorylation involves several complex regulatory steps that affect either its stability or activity. For example, phosphorylation of the S589 residue is critical for PER stability and activity since it is involved in subsequent phosphorylation events, including phosphorylation at S47 (Kivimae et al. 2008). S589 phosphorylation inhibits DBT-mediated phosphorylation at S47 (Ko et al. 2002), thus producing more stable hypophosphorylated PER. However, this hypophosphorylated PER is less effective as a transcriptional repressor, and it is only upon de-phosphorylation of S589 that DBT-mediated phosphorylation generates a less stable but more transcriptionally activity form of PER (Kivimae et al. 2008). Furthermore, a recent study also shows that S589 does not act by itself in inhibiting phosphorylation at S47; rather it is part of a phospho-cluster which is important for modification of PER in a hierarchical manner. DBT action on this cluster is regulated by NEMO/NLK, a proline-directed kinase that phosphorylates PER at S596 to initiate a cascade of phosphorylation (Chiu et al. 2011).

Recent identification of two additional phospho-sites on PER, at 610 and 613, and their interaction with the *per^S* domain further strengthens the idea that the *per^S* domain acts as a major regulatory center (Garbe et al. 2013). Overall, these studies strongly indicate how the interplay of different kinases at phospho-clusters can generate a biochemical timer in *Drosophila* circadian clocks, since mutations at key phosphorylation sites lead to faster or slower clocks.

Posttranslational modification of TIM involves phosphorylation by GSK3β, which is called SGG in flies. Overexpression of SGG leads to period shortening accompanied by elevated nuclear TIM and PER levels, and reducing SGG levels results in period lengthening (Martinek et al. 2001). TIM can be directly phosphorylated by SGG under in vitro conditions and hypomorphic *sgg* mutants display reduced TIM phosphorylation (Martinek et al. 2001). Another important kinase, whose function in the molecular clock is somewhat unclear, is casein kinase 2 (CK2). CK2 is composed of two α and two β subunits, and independent mutations in these subunits result in circadian phenotypes in locomotor activity behavior (Akten et al. 2003; Smith et al. 2008). *tik* and *andante* are hypomorphic mutations of CK2 in the α and β subunits respectively that affect both PER and TIM stability and nuclear entry (Smith et al. 2008; Akten et al. 2003). These mutants exhibit long periodicity, reduced PER and TIM levels, delayed nuclear translocation, and longer circadian period under DD conditions (Smith et al. 2008; Akten et al. 2003). In vitro phosphorylation assays have shown CK2 can directly phosphorylate PER. On the other hand, reductions of CK2 activity in flies lead to elevated TIM levels even in the absence of PER, suggesting CK2 mainly targets TIM (Meissner et al. 2008). It is possible that CK2 acts as a priming kinase by phosphorylating PER and TIM and rendering them as imminent targets for DBT and SGG (Blau 2003).

The positive elements of the circadian clock are also posttranslationally regulated. Overall CLK levels do not cycle, but there are oscillations in CLK phosphorylation (Lee et al. 1998). CLK phosphorylation is coincident with the nuclear entry of the PER repression complex (PER–TIM–DBT) and so is implicated in the negative feedback mechanism (Yu et al. 2006, 2009; Kim and Edery 2006; Menet et al. 2010). It is thought that DBT (catalytically active or inactive) is required to recruit an unknown kinase that phosphorylates CLK (Yu et al. 2009). A recent study has shown that NEMO functions in clock cells to reduce CLK levels and to lengthen circadian period, but whether it directly phosphorylates CLK or not is not known (Yu et al. 2011). Another recent study reported that CK2 also targets CLK for phosphorylation leading to more stable but less transcriptionally active CLK (Szabo et al. 2013). Collectively these posttranslational steps are also interlocked across the different loops and provide key timing mechanisms in the molecular clock.

3.8   Anatomical Organization of *Drosophila* Circadian Clock

The molecular oscillations that give rise to circadian behavioral rhythms take place in distinct subsets of a total 150 neurons that are a small fraction of the ~100,000 neurons in the *Drosophila* brain (Fig. 3.2). These so-called clock neurons express canonical clock genes and are entrained by light. Some of the earliest studies by Handler and Konopka (1979) indicated that a small number of clustered neurons in the brain regulate adult locomotor activity rhythms (Fig. 3.2). The locations of these neurons were later determined by immunostaining of the *per* or *tim* gene products as well as their promoter-driven reporter activities (Zerr et al. 1990; Ewer et al. 1992; Kaneko et al. 1997; Kaneko and Hall 2000). However, *per* and *tim* mRNA and protein are also expressed outside the CNS, mainly in the compound eyes and ocelli, antennae, bristles, fat bodies, ovaries, prothoracic glands, and wings, where they generate peripheral clocks (reviewed in Helfrich-Förster 2002; Myers et al. 2003). *per* and *tim* genes are also expressed in glial cells, which lie in close proximity to clock neuron arborizations and are known to contribute to rhythmic behavior in a calcium-dependent manner (Helfrich-Förster 1995; Ng et al. 2011). However, they are not sufficient enough to maintain robust circadian rhythmicity in the absence of distinct clock neurons (Ewer et al. 1992).

In adults, these neurons are bilaterally situated between the optic lobes and the central brain and broadly categorized into two subsets—lateral and dorsal neurons (Fig. 3.2; reviewed in Nitabach and Taghert 2008). The lateral neurons can be further divided into two subgroups on each side of the brain hemisphere. One group consisting of 6–8 is located dorsolaterally and is referred as LN_d neurons. The other group consists of five small (s-LN_vs) and four large (l-LN_vs) neurons that are located ventrolaterally (Fig. 3.2) (Kaneko and Hall 2000). The small and large LNvs are defined by their small and large cell bodies. The dorsal neurons, as the name suggests, are located dorsally and they are composed of ~15 DN1, 2 DN2, and ~40

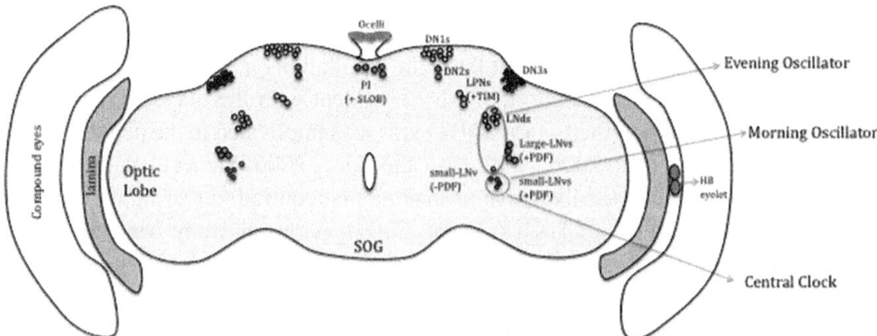

Fig. 3.2 The clock circuit in the brain. The different groups of PER or TIM expressing neurons in the brain are shown. Where functions have been well established, these are indicated. The neurons designated "central clock" are the small LNvs. These are necessary and sufficient to drive a rhythm in constant darkness

DN3 cells (Fig. 3.2). The DN1 and DN2s are medium-sized neurons and are located on the posterior side of the dorsal brain, while the DN3s are small sized that lie laterally in the dorsal brain region (Helfrich-Förster 2003). Besides these LN and DN clusters of neurons, there are additional neuronal clusters, called LPNs (lateroposterior neurons), that are PER and TIM positive (Fig. 3.2) (Kaneko and Hall 2000; Helfrich-Förster 2003). Several lines of evidence have shown that LN$_v$s are the key pacemaker clock neurons, which are necessary and sufficient for maintaining circadian rhythms of rest–activity, in constant darkness (Frisch et al. 1994; Vosshall and Young 1995; Stoleru et al. 2004; Grima et al. 2004). On the other hand, the DNs have been shown to modulate and fine-tune the circadian rhythms under different environmental conditions (Murad et al. 2007; Stoleru et al. 2007).

The exact location and arborizations of the LN$_v$s neurons (five small and four large in each brain hemisphere) could be defined through their expression of a distinct neuropeptide, pigment dispersing factor (PDF; Renn et al. 1999), with one exception that the fifth s-LN$_v$ does not express this peptide. Indeed, expression of the PDF peptide and the use of *per-* and *tim-*driven reporters established the composition of the clock network specifically the small and l-LNvs, LNds, DN1s, DN2s, and DN3s (Kaneko and Hall 2000). Except for the l-LNvs, the other clock neurons project towards the dorsal protocerebrum (Helfrich-Förster 2003). The l-LNvs connect to the accessory medulla via the posterior optic tract in the central brain (Fig. 3.3). Dorsal projections from the s-LNvs and projections to the optic lobes from the l-LNvs indicate that a common circadian output signal is relayed to different parts of the brain (Fig. 3.3) (Helfrich-Förster 2003). The dorsal area has more neuronal connections than other parts of the brain, suggesting that the circadian output signal may be transferred to different clusters of neurons in this region for timing neurophysiological functions (Fig. 3.3).

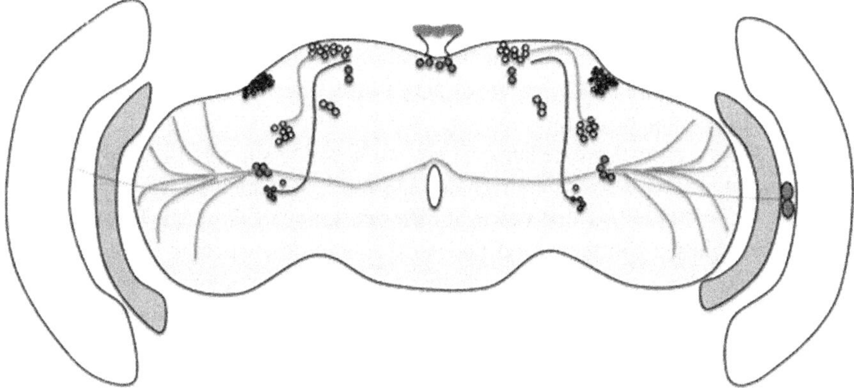

Fig. 3.3 Wiring of the clock circuit. The diagram depicts known projections of specific clock neurons. As might be evident, these have not been very well studied for the non-LNv clock cells. Within the LNv cluster, the l-LNvs project to the optic lobes and to the contralateral side and the s-LNvs project dorsally. In addition, the LNvs receive photic input from the Hofbauer–Buchner eyelet shown in the medulla

These 150 clock neurons expressing canonical clock genes are intricately networked and give rise to robust behavioral rhythms (Fig. 3.3). Earlier studies had indicated that LNvs might be the key pacemaker neurons for driving endogenous rhythms. *Disconnected* (*Disco*) mutants lacking LNs only but with intact DNs exhibit arrhythmic rest–activity as well as eclosion behavior (Dushay et al. 1989). In addition, genetic studies where PER function was rescued only in the LN clusters (all three subsets), but not in DNs, restored rhythmic behavior in *per* null flies (Frisch et al. 1994). Conversely, PER expression only in DNs did not result in robust rhythmic behavior (Ewer et al. 1992). Additionally, recent genetic analyses have definitively identified the s-LNvs as the main pacemaker neurons that drive circadian rhythms under constant dark (DD) conditions (Fig. 3.3), while l-LNvs remain largely dispensable under these conditions (Grima et al. 2004). For instance, rescuing PER function only in small LNvs but not in l-LNvs is sufficient to restore rhythmicity in flies lacking PER under DD conditions (Grima et al. 2004).

In LD conditions, the s-LNvs serve a special role in maintaining the morning peak of locomotor activity (Grima et al. 2004; Stoleru et al. 2004). The l-LNvs as described above have not been assigned any circadian function; however, recent reports strongly suggest that they are involved in detecting light input from visual and perhaps circadian photoreceptors (Sheeba et al. 2008; Shang et al. 2008). In brief these studies indicate that the l-LNvs are important for diurnal behavior, light-dependent arousal, as well as circadian phase advances (Shang et al. 2008). These subsets of neurons are ideally located in such a way that they receive light input signals from optic lobes and HB eyelets (extra photoreceptor important for entrainment) (Helfrich-Förster et al. 2002). Presumably, these neurons integrate rhodopsin-based light information along with information from the circadian photoreceptor, CRY (which is expressed in these neurons), to the circadian clock. Additionally the l-LNvs along with the DN2s apparently form a distinct circuit of unknown function (Stoleru et al. 2005) (Fig. 3.3).

The fifth s-LNv is similar in size to other s-LNvs but lacks PDF expression. This particular LNv and the LNds regulate the evening component of activity rest behavior under LD conditions (Grima et al. 2004; Stoleru et al. 2004). Thus, the PDF positive s-LNvs constitute the morning (M) oscillator and the fifth LNds + LNds make up the evening (E) oscillator. These studies also support the two-oscillator (M-E) model proposed by Pittendrigh and Daan (1976) which allows adaptation to the different photoperiods that occur at different times of the year. Briefly, the M oscillator exhibits a shorter period under LL conditions, whereas the E oscillator shows period lengthening and thus helps to adapt to the longer photoperiod by broadening the phase of activity. Then what role do DNs play? Although the s-LNvs are sufficient for maintenance of rhythms under DD conditions, the flies lack robust rest–activity behavior and show long period (Murad et al. 2007). The DN1s are presumably important for driving rhythms under constant light conditions (LL) (Murad et al. 2007; Stoleru et al. 2007). The DN2s lie in close proximity to the dorsal projections from the s-LNvs and are somewhat unusual in their molecular expression. The PER expression profile is out of phase when compared to the rest of the clock neurons, but whether this has any impact on circadian behavior is not yet known (Kaneko et al. 1997; Stoleru et al. 2005). The role of DN3s has also been studied. These neurons are not important for driving rhythms in DD, but under LD conditions, they are critical for the maintenance of the evening peak of rest–activity behavior (Veleri et al. 2003). Taken together, the overall role of the DN cluster is to provide robustness to the core pacemaker cells under different environmental conditions (Murad et al. 2007; Stoleru et al. 2007).

The lateroposterior neurons (LPNs) are another group of circadian neurons that are TIM positive but lack PER and do not quite belong to any other subset of clock neurons (Yoshii et al. 2005; Shafer et al. 2006). LPNs may function together with DN2s in temperature entrainment of activity–rest pattern (Miyasako et al. 2007). We can say that the regulation of circadian behavior emerges from a very complex network of clock neurons in which distinct group of cells have specialized roles in determining the rest–activity pattern under different environmental conditions. Furthermore, flexibility in the networking of clock neurons appears to facilitate adaptation to different environmental signals. In recent years, the power and advent of newer *Drosophila* genetic tools has added much valuable information to our understanding of the circadian regulation of rest–activity behavior.

3.9 Entrainment of Circadian Rhythms

How do circadian clocks respond to the outside world? To cope with the fluctuating external environment, circadian clocks have evolved sophisticated molecular pathways that help them to adapt to these conditions. Of the various environmental signals—light, temperature, humidity, social cues, and magnetic field—light is the most important zeitgeber ("time giver" in German) for entrainment of circadian rhythms. We will review a few key studies that elucidated the mechanisms by which circadian neurons synchronize to external signals.

Entrainment is achieved through classical ocular organs such as compound eyes, ocelli, or nonclassical pathways like the Hofbauer–Buchner (HB) eyelet and the blue-light photoreceptor CRYPTOCHROME (CRY) (Helfrich-Förster 2005). The compound eyes, ocelli, and HB eyelets use different sets of rhodopsins, whereas the canonical clock neurons express CRY. All of these photoreceptors contribute towards entrainment of the circadian clock. Flies missing one or two of these photoreceptors still retain the capability to synchronize to LD cycles, although they display impairments in specific circadian responses to light (Emery et al. 2000; Helfrich-Förster et al. 2001, 2002; Stanewsky et al. 1998; Klarsfeld et al. 2004). However, eliminating all photoreceptors including CRY renders the clock completely blind to the external environment, as it starts free-running even under entraining conditions (Helfrich-Förster et al. 2001; Klarsfeld et al. 2004).

The anatomical pathways that carry signals from the compound eyes and the ocelli to central clock neurons are not completely understood. The HB eyelets, which are located within the optic lobes, send axonal projections to the accessory medulla and have also been shown to make connections with LN$_v$s (Malpel et al. 2004;

Helfrich-Förster et al. 2002). The precursor of HB eyelets, Bolwig's organs in larvae, is connected to the LNvs, and it gates the circadian rhythm of light avoidance, which requires the larval pacemaker neurons (Malpel et al. 2002; Helfrich-Förster 2002; Mazzoni et al. 2005).

CRY belongs to a family of flavoproteins and is expressed in most circadian neurons. Emery et al. (1998) found that cry^b hypomorph mutants (a point mutation in the flavin-binding domain) render flies rhythmic in constant light (typically flies lose rhythms under these conditions) and significantly less sensitive to the light pulses (Emery et al. 1998; Stanewsky et al. 1998). Conversely, flies overexpressing CRY proteins showed increased behavioral responses to light pulses (Stanewsky et al. 1998; Emery et al. 1998, 2000; Egan et al. 1999). Expression of CRY solely in the LN_vs in a cry^b genetic background also substantially rescues the light responses, but completely restoration of the photic response is achieved through expression in a broader set of clock neurons (Emery et al. 2000). Expressing CRY just in the compound eyes does not rescue the circadian light response in cry^b flies, suggesting that CRY acts in clock neurons. At the molecular level, CRY upon activation by light undergoes a conformational change and binds to TIM. This step, which leads to the degradation of TIM and subsequently CRY, is crucial in the daily resetting of the molecular clock at dawn (Suri et al. 1998). As a result PER becomes available for phosphorylation by DBT and is also degraded and a new cycle is initiated. We now know that degradation of TIM involves a specific E3 ligase, an F-box protein JETLAG (JET) (Lin et al. 2001; Koh et al. 2006), which promotes ubiquitination and subsequent proteosomal degradation of TIM (Naidoo et al. 1999; Koh et al. 2006; Peschel et al. 2006). SGG is another important protein whose overexpression mimics the behavioral phenotype of cry^b in constant light (LL) conditions (Stoleru et al. 2007). Apparently, SGG binds with and stabilizes CRY and makes it unavailable for binding with TIM. This step results in loss of TIM degradation and persistent rhythmicity under LL conditions (Stoleru et al. 2007). As noted above, other photo-transduction pathways can also entrain circadian rhythms. Double mutants cry^b and $norpA^{P41}$, which lack CRY as well as a phospholipase C required for vision in ocular photoreceptors, can still entrain to LD cycles, indicating yet another mode of photic input into the clock (Helfrich-Förster et al. 2001). However, loss of function of the $glass$ (gl^{60j}) gene together with cry^b renders flies completely blind (Helfrich-Förster et al. 2001). gl^{60j} flies lack all known photoreceptors including those in compound eyes, ocelli, as well as HB eyelets (Helfrich-Förster et al. 2001; Veleri et al. 2007). Furthermore, blocking synaptic transmission between the HB eyelet and its potential target cells, the LN_vs, results in defects in behavioral resetting particularly under extreme photoperiods (Veleri et al. 2007). This study established the HB eyelet's role as an additional photoreceptor in the circadian clock.

Flies have also evolved molecular mechanisms to fine-tune their circadian response to light. For example, a natural polymorphism in the tim gene results in an ls-tim (long) or s-tim (short) isoform of TIM, which exhibit differential sensitivity to light (Peschel et al. 2006). ls-TIM binds poorly with CRY because of an extra 23 amino acids in its N-terminus and hence is less sensitive to light, whereas s-TIM strongly binds with CRY and exhibits more light sensitivity (Peschel et al. 2006; Tauber et al. 2007; Sandrelli et al. 2007).

Temperature is another important external cue that can synchronize the daily behavior of flies. NorpA has been implicated in thermal entrainment but the cells mediating this are largely unknown (Glaser and Stanewsky 2005). Furthermore, a recent paper also showed a role of chordotonal mechanosensory organs in temperature entrainment (Sehadova et al. 2009). These organs are found in peripheral tissues that send signals to the central clock neurons to synchronize daily activity in the presence of temperature cycles. This function is dependent upon expression of the *nocte* gene in the chordotonal organs (Sehadova et al. 2009). While the focus of this chapter is on mechanisms underlying the control of behavior by central clock neurons, we note that many peripheral clocks can independently entrain to environmental signals. For instance, isolated explants of different tissues entrain to temperature cycles (Glaser and Stanewsky 2005). Similarly, CRY is expressed in peripheral tissues that harbor a *bona fide* circadian clock such as the Malpighian tubules and eyes, and it may serve to entrain these clocks as well constitute part of the clock mechanism (Ivanchenko et al. 2001; Collins et al. 2006).

3.10 Coupling Between Clock Cell and Their Outputs

There is ample evidence for the existence of several different neuronal subsets and oscillators in the fly brain, but the question remains, how do these groups talk to each other? Several studies have shown the existence of a hierarchical structure in the clock network, based upon alteration of membrane excitability or peptide signaling (reviewed in Nitabach and Taghert 2008). The LNvs are the master pacemakers that send signals, primarily peptidergic (see below), to other clock neurons, but it appears that some of the other neurons also affect the LNvs. The DN1 and DN3 neurons are thought to be glutamatergic and they send projections to the s-LNvs (Hamasaka and Nassel 2006). Manipulating the expression of the glutamate receptor in the LNvs results in alter circadian behavior in DD and substantial effects on the rest–activity pattern in LD (Collins et al. 2012; Hamasaka et al. 2007). The LNvs are also positive for the 5-HT receptor, which affect larval neurophysiology by regulating calcium levels (Hamasaka and Nassel 2006). Altering the 5-HT receptor in the LNvs reduces light entrainment of rest–activity behavior in adult flies (Yuan et al. 2005). Another neurotransmitter gamma-aminobutyric acid (GABA) also alters the output of LNvs by affecting intracellular calcium levels (Hamasaka et al. 2005). In addition, recent studies using genetic, pharmacological, and/or anatomical approaches have also indicated a role for other neurotransmitters such as histamine and acetylcholine (Hamasaka and Nassel 2006; Wegener et al. 2004).

Central clock neurons utilize primarily a peptidergic signal, pigment dispersing factor (PDF), to regulate locomotor activity. In one of the earliest studies, Helfrich-Förster (1995) showed that crustacean pigment dispersing hormone (PDH) specifically labels LNvs, four large- and four small-sized LNvs in each brain hemisphere in adults and only small LNvs in larvae. It is generally believed that PDF release is gated in a circadian manner at the tip of the dorsal projection from the s-LNvs (Park et al. 2000). Based upon the analyses of flies lacking PDF (*Pdf^{01}*), it has been suggested that it is required for controlling the morning peak in LD and free-running rhythms in DD conditions (Renn et al. 1999; Blanchardon et al. 2001). Overexpression of this peptide leads to complex rhythms including arrhythmia in DD conditions (Helfrich-Förster et al. 2000). Loss of the G protein-coupled receptor (GPCR) for PDF largely recapitulates the behavioral phenotype of Pdf^{01} mutants as well as flies whose LNvs are genetically ablated, which suggests that PDF signaling is solely dependent upon this receptor (Renn et al. 1999; Hyun et al. 2005; Lear et al. 2005b; Mertens et al. 2005). The anatomic location of this receptor is not clearly known, but data available through immunostaining and mRNA in situ hybridization suggest that it is expressed in the majority of clock neurons as well as some other cells including pars intercerebralis (PI) neurons (Hyun et al. 2005; Lear et al. 2005b; Mertens et al. 2005). The PI neurons are functionally similar to the vertebrate hypothalamus that has diverse physiological functions such as metabolism, sleep, growth, and reproduction (Foltenyi et al. 2007; Terhzaz et al. 2007; Rulifson et al. 2002). Other peptides such as neuropeptide F (NPF) and neuropeptide-like precursor 1 (NPLP1) have also been reported to affect rest–activity behavior in LD and DD conditions (Lee et al. 2006; Hermann et al. 2012; Baggerman et al. 2002). NPF is expressed in the LN_ds and fifth LNv (Pdf negative), and loss of this receptor ($npfR1$) affects the evening peak of activity under LD conditions and free-running behavior under DD conditions (Hermann et al. 2012; He et al. 2013).

3.11 Future Directions

Considerable work has been done to understand the circadian clockwork of *Drosophila* at the genetic, anatomical, and molecular levels; however, little has been done to elucidate how core clock neurons transmit timing signals to other parts of the CNS. There are indications of an involvement of cAMP phosphodiesterases and PKA in the circadian regulation of rest–activity behavior (Levine et al. 1994; Majercak et al. 1997; Park et al. 2000). And specific other molecules have been indicated in circadian output, e.g., NF1, NA (Williams et al. 2001; Lear et al. 2005a). Further studies are warranted to understand how pervasive the circadian system is in controlling other higher brain functions. Interactions between the PDF circuit and sleep/arousal circuits have been established, suggesting that clock cells do not just keep time but also modulate specific behavioral states. Circadian regulation is also implicated in learning and memory and could also contribute to mood and motivation. We should expect to see more detailed studies on these fronts as more sophisticated genetic tools are developed and utilized in the circadian biology field.

References

Akten B, Jauch E, Genova GK et al (2003) A role for CK2 in the *Drosophila* circadian oscillator. Nat Neurosci 6:251–257

Allada R, Chung BY (2010) Circadian organization of behavior and physiology in *Drosophila*. Annu Rev Physiol 72:605–624

Allada R, White NE, So WV et al (1998) Mutant *Drosophila* homolog of mammalian Clock disrupts circadian rhythms and transcription of period and timeless. Cell 93:791–804

Baggerman G, Cerstiaens A, De Loof A et al (2002) Peptidomics of the larval *Drosophila melanogaster* central nervous system. J Biol Chem 277:40368–4037465

Bargiello TA, Young MW (1984) Molecular genetics of a biological clock in *Drosophila*. Proc Natl Acad Sci U S A 81:2142–2146

Bargiello TA, Saez L, Baylies MK et al (1987) The *Drosophila* clock gene per affects intercellular junctional communication. Nature 328:686–691

Baylies MK, Bargiello TA, Jackson FR (1987) Changes in abundance or structure of the per gene product can alter periodicity of the *Drosophila* clock. Nature 326:390–392

Bell-Pedersen D, Cassone VM, Earnest DJ et al (2005) Circadian rhythms from multiple oscillators: lessons from diverse organisms. Nat Rev Genet 6:544–556

Blanchardon E, Grima B, Klarsfeld A et al (2001) Defining the role of *Drosophila* lateral neurons in the control of circadian rhythms in motor activity and eclosion by targeted genetic ablation and PERIOD protein overexpression. Eur J Neurosci 13:871–888

Blau J (2003) A new role for an old kinase: CK2 and the circadian clock. Nat Neurosci 6:208–210

Blau J, Young MW (1999) Cycling vrille expression is required for a functional *Drosophila* clock. Cell 99:661–671

Bushati N, Cohen SM (2007) MicroRNA functions. Annu Rev Cell Dev Biol 23:175–205

Chiu JC, Vanselow JT, Kramer A (2008) The phospho-occupancy of an atypical SLIMB-binding site on PERIOD that is phosphorylated by DOUBLETIME controls the pace of the clock. Genes Dev 22:1758–1772

Chiu JC, Ko HW, Edery I (2011) NEMO/NLK phosphorylates PERIOD to initiate a time-delay phosphorylation circuit that sets circadian clock speed. Cell 145:357–370

Claridge-Chang A, Wijnen H, Naef F et al (2001) Circadian regulation of gene expression systems in the *Drosophila* head. Neuron 32:657–671

Collins BH, Rosato E, Kyriacou CP (2004) Seasonal behavior in *Drosophila melanogaster* requires the photoreceptors, the circadian clock, and phospholipase C. Proc Natl Acad Sci U S A 101:1945–1950

Collins B, Mazzon EO, Stanewsky R et al (2006) *Drosophila* CRYPTOCHROME is a circadian transcriptional repressor. Curr Biol 16:441–449

Collins B, Kane EA, Reeves DC et al (2012) Balance of activity between LN(v)s and glutamatergic dorsal clock neurons promotes robust circadian rhythms in *Drosophila*. Neuron 74:706–718

Colot HV, Loros JJ, Dunlap JC (2005) Temperature-modulated alternative splicing and promoter use in the Circadian clock gene frequency. Mol Biol Cell 16:5563–5571

Cyran SA, Buchsbaum AM, Reddy KL et al (2003) vrille, Pdp1, and dClock form a second feedback loop in the *Drosophila* circadian clock. Cell 112:329–341

Darlington TK, Wager-Smith K, Ceriani MF et al (1998) Closing the circadian loop: CLOCK-induced transcription of its own inhibitors per and tim. Science 280:1599–1603

de Mairan J (1729) Observation botanique. Histoire de l'Academie Royale des Sciences 35–36

Diernfellner AC, Schafmeier T, Merrow MW et al (2005) Molecular mechanism of temperature sensing by the circadian clock of Neurospora crassa. Genes Dev 19:1968–1973

Dockendorff TC, Su HS, McBride SM et al (2002) *Drosophila* lacking dfmr1 activity show defects in circadian output and fail to maintain courtship interest. Neuron 34:973–984

Doherty CJ, Kay SA (2010) Circadian control of global gene expression patterns. Annu Rev Genet 44:419–444

Dunlap JC (1999) Molecular bases for circadian clocks. Cell 96:271–290

Dushay MS, Rosbash M, Hall JC (1989) The disconnected visual system mutations in *Drosophila melanogaster* drastically disrupt circadian rhythms. J Biol Rhythms 4:1–27

Edery I (2000) Circadian rhythms in a nutshell. Physiol Genomics 3:59–74

Egan ES, Franklin TM, Hilderbrand-Chae MJ et al (1999) An extraretinally expressed insect cryptochrome with similarity to the blue light photoreceptors of mammals and plants. J Neurosci 19:3665–3673

Emery P, So WV, Kaneko M et al (1998) M. CRY, a *Drosophila* clock and light-regulated cryptochrome, is a major contributor to circadian rhythm resetting and photosensitivity. Cell 95:669–679

Emery P, Stanewsky R, Helfrich-Förster C et al (2000) *Drosophila* CRY is a deep brain circadian photoreceptor. Neuron 26:493–504

Ewer J, Frisch B, Hamblen-Coyle MJ et al (1992) Expression of the period clock gene within different cell types in the brain of *Drosophila* adults and mosaic analysis of these cells' influence on circadian behavioral rhythms. J Neurosci 12:3321–3349

Fang Y, Sathyanarayanan S, Sehgal A (2007) Post-translational regulation of the *Drosophila* circadian clock requires protein phosphatase 1 (PP1). Genes Dev 21:1506–1518

Foltenyi K, Greenspan RJ, Newport JW (2007) Activation of EGFR and ERK by rhomboid signaling regulates the consolidation and maintenance of sleep in *Drosophila*. Nat Neurosci 10:1160–1167

Frisch B, Hardin PE, Hamblen-Coyle MJ et al (1994) A promoterless period gene mediates behavioral rhythmicity and cyclical per expression in a restricted subset of the *Drosophila* nervous system. Neuron 12:555–570

Garbe DS, Fang Y, Zheng X et al (2013) Cooperative interaction between phosphorylation sites on PERIOD maintains circadian period in *Drosophila*. PLoS Genet 9:e1003749

Gekakis N, Saez L, Delahaye-Brown AM et al (1995) Isolation of timeless by PER protein interaction: defective interaction between timeless protein and long-period mutant PERL. Science 270:811–815

Glaser FT, Stanewsky R (2005) Temperature synchronization of the *Drosophila* circadian clock. Curr Biol 15:1352–1363

Glossop NR, Lyons LC, Hardin PE (1999) Interlocked feedback loops within the *Drosophila* circadian oscillator. Science 286:766–768

Grima B, Lamouroux A, Chélot E et al (2002) The F-box protein slimb controls the levels of clock proteins period and timeless. Nature 420:178–182

Grima B, Chelot E, Xia R et al (2004) Morning and evening peaks of activity rely on different clock neurons of the *Drosophila* brain. Nature 431:869–873

Gummadova JO, Coutts GA, Glossop NR (2009) Analysis of the *Drosophila* Clock promoter reveals heterogeneity in expression between subgroups of central oscillator cells and identifies a novel enhancer region. J Biol Rhythms 24:353–367

Hamasaka Y, Nassel DR (2006) Mapping of serotonin, dopamine, and histamine in relation to different clock neurons in the brain of *Drosophila*. J Comp Neurol 494:314–330

Hamasaka Y, Wegener C, Nassel DR (2005) GABA modulates *Drosophila* circadian clock neurons via GABAB receptors and decreases in calcium. J Neurobiol 65:225–240. doi:10.1002/neu.20184

Hamasaka Y, Rieger D, Parmentier ML et al (2007) Glutamate and its metabotropic receptor in *Drosophila* clock neuron circuits. J Comp Neurol 505:32–45

Handler AM, Konopka RJ (1979) Transplantation of a circadian pacemaker in *Drosophila*. Nature 279:236–238

Hardin PE, Hall JC, Rosbash M (1990) Feedback of the *Drosophila* period gene product on circadian cycling of its messenger RNA levels. Nature 343:536–540

Hardin PE, Hall JC, Rosbash M (1992) Circadian oscillations in period gene mRNA levels are transcriptionally regulated. Proc Natl Acad Sci U S A 89:11711–11715

He C, Cong X, Zhang R et al (2013) Regulation of circadian locomotor rhythm by neuropeptide Y-like system in *Drosophila melanogaster*. Insect Mol Biol 22:376–388

Helfrich-Förster C (1995) The period clock gene is expressed in central nervous system neurons which also produce a neuropeptide that reveals the projections of circadian pacemaker cells within the brain of Drosophila melanogaster. Proc Natl Acad Sci U S A 92(2):612–616

Helfrich-Förster C (2002) The circadian system of *Drosophila melanogaster* and its light input pathways. Zoology (Jena) 105:297–312

Helfrich-Förster C (2003) The neuroarchitecture of the circadian clock in the brain of *Drosophila melanogaster*. Microsc Res Tech 62:94–102

Helfrich-Förster C (2005) Neurobiology of the fruit fly's circadian clock. Genes Brain Behav 4:65–76

Helfrich-Förster C, Täuber M, Park JH et al (2000) Ectopic expression of the neuropeptide pigment-dispersing factor alters behavioral rhythms in *Drosophila melanogaster*. J Neurosci 20:3339–3353

Helfrich-Förster C, Winter C, Hofbauer A et al (2001) The circadian clock of fruit flies is blind after elimination of all known photoreceptors. Neuron 30:249–261

Helfrich-Förster C, Edwards T, Yasuyama K et al (2002) The extraretinal eyelet of *Drosophila*: development, ultrastructure, and putative circadian function. J Neurosci 22:9255–9266

Hermann C, Yoshii T, Dusik V et al (2012) Neuropeptide F immunoreactive clock neurons modify evening locomotor activity and free-running period in *Drosophila melanogaster*. J Comp Neurol 520:970–987

Huang ZJ, Edery I, Rosbash M (1993) PAS is a dimerization domain common to *Drosophila* period and several transcription factors. Nature 364:259–262

Huang Y, Genova G, Roberts M et al (2007) The LARK RNA-binding protein selectively regulates the circadian eclosion rhythm by controlling E74 protein expression. PLoS One 2:e1107

Hunter-Ensor M, Ousley A, Sehgal A (1996) Regulation of the *Drosophila* protein timeless suggests a mechanism for resetting the circadian clock by light. Cell 84:677–685

Hyun S, Lee Y, Hong ST et al (2005) *Drosophila* GPCR Han is a receptor for the circadian clock neuropeptide PDF. Neuron 48:267–278

Inoue S, Shimoda M, Nishinokubi I et al (2002) A role for the *Drosophila* fragile X-related gene in circadian output. Curr Biol 12:1331–1335

Ivanchenko M, Stanewsky R, Giebultowicz JM (2001) Circadian photoreception in *Drosophila*: functions of cryptochrome in peripheral and central clocks. J Biol Rhythms 16:205–215

Jackson RJ, Standart N (2007) How do microRNAs regulate gene expression? Sci STKE 2007:re1

Johnson CH, Golden SS (1999) Circadian programs in cyanobacteria: adaptiveness and mechanism. Annu Rev Microbiol 53:389–409

Kadener S, Menet JS, Sugino K et al (2009) A role for microRNAs in the *Drosophila* circadian clock. Genes Dev 23:2179–2191

Kaneko M, Hall JC (2000) Neuroanatomy of cells expressing clock genes in *Drosophila*: transgenic manipulation of the period and timeless genes to mark the perikarya of circadian pacemaker neurons and their projections. J Comp Neurol 422:66–94

Kaneko M, Helfrich-Förster C, Hall JC (1997) Spatial and temporal expression of the period and timeless genes in the developing nervous system of *Drosophila*: newly identified pacemaker candidates and novel features of clock gene product cycling. J Neurosci 17:6745–6760

Kim EY, Edery I (2006) Balance between DBT/CKIepsilon kinase and protein phosphatase activities regulate phosphorylation and stability of *Drosophila* CLOCK protein. Proc Natl Acad Sci U S A 103:6178–6183

Kivimae S, Saez L, Young MW (2008) Activating PER repressor through a DBT-directed phosphorylation switch. PLoS Biol 6:e183

Klarsfeld A, Malpel S, Michard-Vanhée C et al (2004) Novel features of cryptochrome-mediated photoreception in the brain circadian clock of *Drosophila*. J Neurosci 24:1468–1477

Kloss B, Price JL, Saez L et al (1998) The *Drosophila* clock gene double-time encodes a protein closely related to human casein kinase I epsilon. Cell 94:97–107

Kloss B, Rothenfluh A, Young MW et al (2001) Phosphorylation of period is influenced by cycling physical associations of double-time, period, and timeless in the *Drosophila* clock. Neuron 30:699–706

Ko HW, Jiang J, Edery I (2002) Role for Slimb in the degradation of *Drosophila* Period protein phosphorylated by Doubletime. Nature 420:673–678

Koh K, Zheng X, Sehgal A (2006) JETLAG resets the *Drosophila* circadian clock by promoting light-induced degradation of TIMELESS. Science 312:1809–1812

Kojima S, Matsumoto K, Hirose M et al (2007) LARK activates posttranscriptional expression of an essential mammalian clock protein, PERIOD1. Proc Natl Acad Sci U S A 104:1859–1864

Kojima S, Gatfield D, Esau CC et al (2010) MicroRNA-122 modulates the rhythmic expression profile of the circadian deadenylase Nocturnin in mouse liver. PLoS One 5:e11264

Konopka RJ, Benzer S (1971) Clock mutants of *Drosophila melanogaster*. Proc Natl Acad Sci U S A 68:2112–2116

Kumar S, Mohan A, Sharma VK (2005) Circadian dysfunction reduces lifespan in *Drosophila melanogaster*. Chronobiol Int 22:641–653

Lai MC, Kuo HW, Chang WC et al (2003) A novel splicing regulator shares a nuclear import pathway with SR proteins. EMBO J 22:1359–1369

Lear BC, Lin JM, Keath JR et al (2005a) The ion channel narrow abdomen is critical for neural output of the *Drosophila* circadian pacemaker. Neuron 48:965–976

Lear BC, Merrill CE, Lin JM et al (2005b) A G protein-coupled receptor, groom-of-PDF, is required for PDF neuron action in circadian behavior. Neuron 48:221–227

Lee C, Parikh V, Itsukaichi T et al (1996) Resetting the *Drosophila* clock by photic regulation of PER and a PER-TIM complex. Science 271:1740–1744

Lee C, Bae K, Edery I (1998) The *Drosophila* CLOCK protein undergoes daily rhythms in abundance, phosphorylation, and interactions with the PER-TIM complex. Neuron 21:857–867

Lee C, Bae K, Edery I (1999) PER and TIM inhibit the DNA binding activity of a *Drosophila* CLOCK-CYC/dBMAL1 heterodimer without disrupting formation of the heterodimer: a basis for circadian transcription. Mol Cell Biol 19:5316–5325

Lee G, Bahn JH, Park JH (2006) Sex- and clock-controlled expression of the neuropeptide F gene in. Proc Natl Acad Sci U S A 103:12580–12585

Levine JD, Casey CI, Kalderon DD et al (1994) Altered circadian pacemaker functions and cyclic AMP rhythms in the *Drosophila* learning mutant dunce. Neuron 13:967–974

Li Y, Padgett RW (2012) bantam is required for optic lobe development and glial cell proliferation. PLoS One 7:e32910

Lim C, Allada R (2013) ATAXIN-2 activates PERIOD translation to sustain circadian rhythms in *Drosophila*. Science 340:875–879

Lim C, Lee J, Choi C et al (2011) The novel gene twenty-four defines a critical translational step in the *Drosophila* clock. Nature 470:399–403

Lin S, Fu XD (2007) SR proteins and related factors in alternative splicing. Adv Exp Med Biol 623:107–122

Lin FJ, Song W, Meyer-Bernstein E et al (2001) Photic signaling by cryptochrome in the *Drosophila* circadian system. Mol Cell Biol 21:7287–7294

Lin JM, Kilman VL, Keegan K et al (2002) A role for casein kinase 2alpha in the *Drosophila* circadian clock. Nature 420:816–820

Luo W, Sehgal A (2012) Regulation of circadian behavioral output via a MicroRNA-JAK/STAT circuit. Cell 148:765–779

Majercak J, Kalderon D, Edery I (1997) *Drosophila melanogaster* deficient in protein kinase A manifests behavior-specific arrhythmia but normal clock function. Mol Cell Biol 17:5915–5922

Majercak J, Sidote D, Hardin PE et al (1999) How a circadian clock adapts to seasonal decreases in temperature and day length. Neuron 24:219–230

Majercak J, Chen WF, Edery I (2004) Splicing of the period gene 3′-terminal intron is regulated by light, circadian clock factors, and phospholipase C. Mol Cell Biol 24:3359–3372

Malpel S, Klarsfeld A, Rouyer F (2002) Larval optic nerve and adult extra-retinal photoreceptors sequentially associate with clock neurons during *Drosophila* brain development. Development 129:1443–1453

Malpel S, Klarsfeld A, Rouyer F (2004) Circadian synchronization and rhythmicity in larval photoperception-defective mutants of *Drosophila*. J Biol Rhythms 19:10–21

Martinek S, Inonog S, Manoukian AS et al (2001) A role for the segment polarity gene shaggy/GSK-3 in the *Drosophila* circadian clock. Cell 105:769–779

Mazzoni EO, Desplan C, Blau J (2005) Circadian pacemaker neurons transmit and modulate visual information to control a rapid behavioral response. Neuron 45:293–300

McDonald MJ, Rosbash M (2001) Microarray analysis and organization of circadian gene expression in *Drosophila*. Cell 107:567–578

McNeil GP, Zhang X, Genova G et al (1998) A molecular rhythm mediating circadian clock output in *Drosophila*. Neuron 20:297–303

Meissner RA, Kilman VL, Lin JM et al (2008) TIMELESS is an important mediator of CK2 effects on circadian clock function in vivo. J Neurosci 28:9732–9740

Menet JS, Abruzzi KC, Desrochers J et al (2010) Dynamic PER repression mechanisms in the *Drosophila* circadian clock: from on-DNA to off-DNA. Genes Dev 24:358–367

Mertens I, Vandingenen A, Johnson EC et al (2005) PDF receptor signaling in *Drosophila* contributes to both circadian and geotactic behaviors. Neuron 48:213–219

Miyasako Y, Umezaki Y, Tomioka K (2007) Separate sets of cerebral clock neurons are responsible for light and temperature entrainment of *Drosophila* circadian locomotor rhythms. J Biol Rhythms 22:115–126

Murad A, Emery-Le M, Emery P (2007) A subset of dorsal neurons modulates circadian behavior and light responses in *Drosophila*. Neuron 53:689–701

Myers MP, Wager-Smith K, Wesley CS et al (1995) Positional cloning and sequence analysis of the *Drosophila* clock gene, timeless. Science 270:805–808

Myers MP, Wager-Smith K, Rothenfluh-Hilfiker A et al (1996) Light-induced degradation of TIMELESS and entrainment of the *Drosophila* circadian clock. Science 271:1736–1740

Myers EM, Yu J, Sehgal A (2003) Circadian control of eclosion: interaction between a central and peripheral clock in *Drosophila melanogaster*. Curr Biol 13:526–553

Naidoo N, Song W, Hunter-Ensor M et al (1999) A role for the proteasome in the light response of the timeless clock protein. Science 285:1737–1741

Newby LM, Jackson FR (1995) Developmental and genetic mosaic analysis of *Drosophila* m-dy mutants: tissue foci for behavioral and morphogenetic defects. Dev Genet 16:85–93

Ng FS, Tangredi MM, Jackson FR (2011) Glial cells physiologically modulate clock neurons and circadian behavior in a calcium-dependent manner. Curr Biol 21:625–634

Nitabach MN, Taghert PH (2008) Organization of the *Drosophila* circadian control circuit. Curr Biol 18:R84–R93109

Park JH, Helfrich-Förster C, Lee G et al (2000) Differential regulation of circadian pacemaker output by separate clock genes in *Drosophila*. Proc Natl Acad Sci U S A 97:3608–3613

Peschel N, Veleri S, Stanewsky R (2006) Veela defines a molecular link between Cryptochrome and Timeless in the light-input pathway to *Drosophila's* circadian clock. Proc Natl Acad Sci U S A 103:17313–17318

Pittendrigh CS, Daan S (1976) A functional analysis of circadian pacemakers in nocturnal rodents. V. Pacemaker structure: a clock for all seasons. J Comp Physiol 106(3):333–355

Price JL, Blau J, Rothenfluh A et al (1998) double-time is a novel *Drosophila* clock gene that regulates PERIOD protein accumulation. Cell 94:83–95

Reddy P, Zehring WA, Wheeler DA et al (1984) Molecular analysis of the period locus in *Drosophila melanogaster* and identification of a transcript involved in biological rhythms. Cell 38:701–710

Reddy AB, Karp NA, Maywood ES et al (2006) Circadian orchestration of the hepatic proteome. Curr Biol 16:1107–1115

Renn SC, Park JH, Rosbash M et al (1999) A pdf neuropeptide gene mutation and ablation of PDF neurons each cause severe abnormalities of behavioral circadian rhythms in *Drosophila*. Cell 99:791–802

Rulifson EJ, Kim SK, Nusse R (2002) Ablation of insulin-producing neurons in flies: growth and diabetic phenotypes. Science 296:1118–1120

Rutila JE, Suri V, Le M et al (1998) CYCLE is a second bHLH-PAS clock protein essential for circadian rhythmicity and transcription of *Drosophila* period and timeless. Cell 93:805–814

Sandrelli F, Tauber E et al (2007) A molecular basis for natural selection at the timeless locus in *Drosophila melanogaster*. Science 316:1898–1900

Sathyanarayanan S, Zheng X, Xiao R et al (2004) Posttranslational regulation of *Drosophila* PERIOD protein by protein phosphatase 2A. Cell 116:603–615

Sawyer LA, Hennessy JM, Peixoto AA et al (1997) Natural variation in a *Drosophila* clock gene and temperature compensation. Science 278:2117–2120

Sehadova H, Glaser FT, Gentile C et al (2009) Temperature entrainment of *Drosophila's* circadian clock involves the gene nocte and signaling from peripheral sensory tissues to the brain. Neuron 64:251–266

Sehgal A, Price JL, Man B et al (1994) Loss of circadian behavioral rhythms and per RNA oscillations in the *Drosophila* mutant timeless. Science 263:1603–1606

Sehgal A, Rothenfluh-Hilfiker A, Hunter-Ensor M et al (1995) Rhythmic expression of timeless: a basis for promoting circadian cycles in period gene autoregulation. Science 270:808–810

Shafer OT, Helfrich-Förster C, Renn S et al (2006) Reevaluation of *Drosophila melanogaster's* neuronal circadian pacemakers reveals new neuronal classes. J Comp Neurol 498:180–193

Shang Y, Griffith LC, Rosbash M (2008) Light-arousal and circadian photoreception circuits intersect at the large PDF cells of the *Drosophila* brain. Proc Natl Acad Sci U S A 105:19587–19594

Sharma VK (2003) Adaptive significance of circadian clocks. Chronobiol Int 20:901–919

Sheeba V, Fogle KJ, Kaneko M et al (2008) Large ventral lateral neurons modulate arousal and sleep in *Drosophila*. Curr Biol 18:1537–1545

Smith EM, Lin JM, Meissner RA et al (2008) Dominant-negative CK2alpha induces potent effects on circadian rhythmicity. PLoS Genet 4:e12

So WV, Rosbash M (1997) Post-transcriptional regulation contributes to *Drosophila* clock gene mRNA cycling. EMBO J 16:7146–7155

Sofola O, Sundram V, Ng F et al (2008) The *Drosophila* FMRP and LARK RNA-binding proteins function together to regulate eye development and circadian behavior. J Neurosci 28:10200–102052

Stanewsky R, Kaneko M, Emery P et al (1998) The cryb mutation identifies cryptochrome as a circadian photoreceptor in *Drosophila*. Cell 95:681–692

Stoleru D, Peng Y, Agosto J et al (2004) Coupled oscillators control morning and evening locomotor behaviour of *Drosophila*. Nature 431:862–868

Stoleru D, Peng Y, Nawathean P et al (2005) A resetting signal between *Drosophila* pacemakers synchronizes morning and evening activity. Nature 438:238–242

Stoleru D, Nawathean P, Fernández MP et al (2007) The *Drosophila* circadian network is a seasonal timer. Cell 129:207–221

Suri V, Qian Z, Hall JC et al (1998) Evidence that the TIM light response is relevant to light-induced phase shifts in *Drosophila melanogaster*. Neuron 21:225–234

Szabo A, Papin C, Zorn D et al (2013) The CK2 kinase stabilizes CLOCK and represses its activity in the *Drosophila* circadian oscillator. PLoS Biol 11:e1001645

Takahashi JS, Hong HK, Ko CH et al (2008a) The genetics of mammalian circadian order and disorder: implications for physiology and disease. Nat Rev Genet 9:764–775

Takahashi JS, Shimomura K, Kumar V (2008b) Searching for genes underlying behavior: lessons from circadian rhythms. Science 322:909–912

Tauber E, Zordan M, Sandrelli F et al (2007) Natural selection favors a newly derived timeless allele in *Drosophila melanogaster*. Science 316:1895–1898

Terhzaz S, Rosay P, Goodwin SF et al (2007) The neuropeptide SIFamide modulates sexual behavior in *Drosophila*. Biochem Biophys Res Commun 352:305–310

Ueda HR, Matsumoto A, Kawamura M et al (2002) Genome-wide transcriptional orchestration of circadian rhythms in *Drosophila*. J Biol Chem 277:14048–14052

Veleri S, Brandes C, Helfrich-Förster C et al (2003) A self-sustaining, light-entrainable circadian oscillator in the *Drosophila* brain. Curr Biol 13(20):1758–1767, 3006997 [pii]

Veleri S, Riege D, Helfrich-Förster C et al (2007) Hofbauer-Buchner eyelet affects circadian photosensitivity and coordinates TIM and PER expression in *Drosophila* clock neurons. J Biol Rhythms 22:29–42

Vosshall LB, Young MW (1995) Circadian rhythms in *Drosophila* can be driven by period expression in a restricted group of central brain cells. Neuron 15:345–360

Wegener C, Hamasaka Y, Nassel DR (2004) Acetylcholine increases intracellular Ca2+ via nicotinic receptors in cultured PDF-containing clock neurons of *Drosophila*. J Neurophysiol 91:912–923

Williams JA, Su HS, Bernards A et al (2001) A circadian output in *Drosophila* mediated by neurofibromatosis-1 and Ras/MAPK. Science 293:2251–2256

Woo KC, Kim TD, Lee KH et al (2009) Mouse period 2 mRNA circadian oscillation is modulated by PTB-mediated rhythmic mRNA degradation. Nucleic Acids Res 37:26–37

Woo KC, Ha DC, Lee KH et al (2010) Circadian amplitude of cryptochrome 1 is modulated by mRNA stability regulation via cytoplasmic hnRNP D oscillation. Mol Cell Biol 30:197–205

Yang Z, Sehgal A (2001) Role of molecular oscillations in generating behavioral rhythms in *Drosophila*. Neuron 29:453–467

Yang M, Lee JE, Padgett R et al (2008) Circadian regulation of a limited set of conserved microRNAs in *Drosophila*. BMC Genomics 9:83

Yerushalmi S, Green RM (2009) Evidence for the adaptive significance of circadian rhythms. Ecol Lett 12:970–981

Ying S, Chang DC, Miller JD et al (2006) The microRNA: overview of the RNA gene that modulates gene functions. Methods Mol Biol 342:1–18

Yoshii T, Heshiki Y, Ibuki-Ishibashi T et al (2005) Temperature cycles drive *Drosophila* circadian oscillation in constant light that otherwise induces behavioural arrhythmicity. Eur J Neurosci 22:1176–1184

Young MW, Kay SA (2001) Time zones: a comparative genetics of circadian clocks. Nat Rev Genet 2:702–715

Yu W, Zheng H, Houl JH et al (2006) PER-dependent rhythms in CLK phosphorylation and E-box binding regulate circadian transcription. Genes Dev 20:723–733

Yu W, Zheng H, Price JL et al (2009) DOUBLETIME plays a noncatalytic role to mediate CLOCK phosphorylation and repress CLOCK-dependent transcription within the *Drosophila* circadian clock. Mol Cell Biol 29:1452–1458

Yu W, Houl JH, Hardin PE (2011) NEMO kinase contributes to core period determination by slowing the pace of the *Drosophila* circadian oscillator. Curr Biol 21:756–7618

Yuan Q, Lin F, Zheng X, Sehgal A (2005) Serotonin modulates circadian entrainment in *Drosophila*. Neuron 47:115–127

Zehring WA, Wheeler DA, Reddy P et al (1984) P-element transformation with period locus DNA restores rhythmicity to mutant, arrhythmic *Drosophila melanogaster*. Cell 39:369–376

Zeng H, Hardin PE, Rosbash M (1994) Constitutive overexpression of the *Drosophila* period protein inhibits period mRNA cycling. EMBO J 13:3590–3598

Zeng H, Qian Z, Myers M et al (1996) A light-entrainment mechanism for the *Drosophila* circadian clock. Nature 380:129–135

Zerr DM, Hall JC, Rosbash M et al (1990) Circadian fluctuations of period protein immunoreactivity in the CNS and the visual system of *Drosophila*. J Neurosci 10:2749–2762

Zheng X, Sehgal A (2008) Probing the relative importance of molecular oscillations in the circadian clock. Genetics 178:1147–1155

Zheng X, Koh K, Sowcik M et al (2009) An isoform-specific mutant reveals a role of PDP1 epsilon in the circadian oscillator. J Neurosci 29:10920–10927

Chapter 4
Biological Rhythmicity in Subterranean Animals: A Function Risking Extinction?

Luiz Menna-Barreto and Eleonora Trajano

Abstract In this chapter, we discuss evidence of regression of circadian locomotor activity in exclusively subterranean species (troglobites), having fishes as models, by comparing such findings with observations on related epigean (surface) species, that may also form self-sustained subterranean (troglophilic) populations. These results favor the hypothesis of regression of a function which may have lost its adaptive value for species permanently isolated in hypogean habitats for many generations—regression similar to the reduction of eyes and dark pigmentation, typical of troglobites in general. Recent data on feeding behavior of blind catfish compared to epigean congeners suggest a process of partial regression, affecting locomotion but not feeding, due perhaps to the persistence of regular food availability in the otherwise continuously dark cave environment. Among non-troglobitic subterranean animals, trogloxenes present regular, cyclical movements between hypogean and epigean habitats, whereas troglophiles may move between these habitats, promoting genetic connectivity between surface and subterranean populations, but without following well-defined rhythmic patterns.

4.1 Introduction: The Subterranean Habitat as a Natural Laboratory

Chronobiology has evolved deeply rooted in demonstrations of the persistence of circadian rhythms in constant conditions, mostly in laboratory settings. Humans and other species kept in controlled conditions (typically light, temperature) show regular rhythms, both under cycles close to 12:12 h or in constant conditions. Although not always clearly stated, subjects of these studies were brought from their habitats where environmental cycles were present; in some cases, animals were raised in the

L. Menna-Barreto (✉)
Escola de Artes, Ciências e Humanidades, Universidade de São Paulo, São Paulo, SP, Brazil
e-mail: menna@usp.br

E. Trajano
Instituto de Biociências, Universidade de São Paulo, São Paulo, SP, Brazil

© Springer International Publishing Switzerland 2015
R. Aguilar-Roblero et al. (eds.), *Mechanisms of Circadian Systems in Animals and Their Clinical Relevance*, DOI 10.1007/978-3-319-08945-4_4

lab from birth also under light cycles. Although we find several human studies with blind people, the subjects had all been under other environmental cycles, such as social interaction, feeding, etc. Animal studies involving constant conditions from birth, such as that published by Cambras and Díez-Noguera (1988) with rats, are scarce and in our view deserve closer attention. Observations of circadian rhythmicity in species living in the arctic region do not qualify in this case, because there are changes in light in the circadian range both in summer and winter. Exclusively subterranean species (troglobites) may be of interest because they have evolved in isolation under constant darkness for many generations, never leaving the subterranean realm.

In general, troglobites originate from the so-called troglophilic populations, i.e., subterranean populations of epigean (surface) animals that are able to successfully colonize and establish self-sustained populations in subterranean habitats, such as caves (defined as subterranean spaces large enough for human access). Once genetically isolated by allopatric (vicariant events such as extinction of the epigean population or geographic barriers) or parapatric processes, differentiation leads to a progressive accumulation of autapomorphic (exclusive) character states; some of them, at some point, hamper the life in the surface, even if the barrier no longer exists. Among these autapomorphies, researchers focus their attention on the troglomorphisms, character states that may be directly linked to the selective regime typical of subterranean habitats: permanent absence of light, thus, of photoperiods and photoautotrophy, and the tendency to environmental stability. Many troglomorphisms are regressive characters related to the absence of light: structures and behaviors that become functionless under this condition (e.g., visual organs, melanic pigments) may be progressively lost due to the accumulation of neutral mutations (Wilkens 2010); these include the most conspicuous and acknowledged characteristics of troglobites that are the reduction of eyes and dark pigmentation.

In any subterranean community, we find, side by side, troglobites, troglophiles, and also trogloxenes, which are organisms habitually found in caves but which must return periodically to the surface in order to complete their life cycle, bats being the most familiar example (Barr 1968; Trajano 2012). Troglobites should not leave their subterranean dwelling; otherwise they would not survive to complete their life cycle; troglophiles may leave but without a predictable cyclical pattern; trogloxenes-trogloxenes must leave, following a periodical pattern that depends on the species biology. This natural situation composes an attractive scenery, some kind of laboratory where the demonstration of presence/absence of circadian rhythmicity may shed some light in the evolutionary history of such ecosystem.

Questions such as whether species living in constant darkness for generations show regression of circadian rhythms may begin to be answered with comparative studies of subterranean organisms and their epigean relatives. Regression of a phenotype may be linked to the absence of adaptive consequences; thus, organisms living in caves lack pigmentation or eyes such as an eyeless and depigmented cave fish, but have they also lost their circadian timing system?

4.2 Circadian Rhythmicity in Subterranean Fish Under Free-Running and Light/Dark Conditions: Locomotor Activity

Fish are conspicuous subterranean animals, relatively easy to observe in the natural habitat and in the laboratory, being excellent subjects for chronobiological studies. Moreover, certain troglobitic species, such as the Mexican blind tetra characin *Astyanax jordani* (nomenclature following Proudlove 2010 and Strecker et al. 2012), and the Atlantic molly, *Poecilia mexicana* cave form, also from Mexico, are quite common in their habitat and successfully hybridize with their epigean (surface) congeners, constituting privileged laboratory models for genetic studies. The first one has been intensively investigated at the molecular level, unfortunately not so for the individual and populational level.

Detailed chronobiological studies with troglobitic fishes have focused on relatively few taxa: not surprisingly, the blind Mexican tetras, the Indian *Indoreonectes evezardi* (=*Nemacheilus evezardi*) (Cypriniformes: Nemacheilidae; Proudlove 2010), and several Brazilian species. Preliminary data have also been gathered for Thai cave fish.

Brazil is known by its rich and diversified subterranean ichthyofauna, with more than 29 species of troglobitic species so far recognized (Trajano and Bichuette 2010; Cordeiro and Bichuette unpubl. data), distributed among three orders and seven families, occurring in habitats as diverse as fast-running base-level streams, upper tributaries, and slow-moving waters in the phreatic zone forming lakes and pools inside caves. These fishes showed a high diversity in morphological specialization, from species with slightly but significantly reduced eyes and pigmentation in comparison to epigean congeners, to those homogeneously depigmented and eyeless, with all degrees of variation between these extremes, without a taxonomic correlation. Such diversity would be a consequence of differences in the time of isolation in subterranean habitats, among other factors (see Trajano 2007, for a discussion on age of troglobites), offering unique opportunities for comparative studies.

Except for the highly specialized Brazilian characin, *Stygichthys typhlops* (Fig. 4.1), known from phreatic waters accessible through artificial wells in eastern Brazil, *A. jordani*, widely distributed in northeastern Mexico and so far known from 25 karst localities ranging as far north as the Rio Grande system between Mexico and the United States, is the only troglobitic representative of the large Afro-Neotropical order Characiformes, currently with more than 2,000 recognized species. Such unusual success as cavernicoles may be explained by the particular features of its putative epigean sister species, *Astyanax mexicanus* (=*A. fasciatus*), which, unlike most *Astyanax* species, would present preadaptations to the subterranean way of life, such as crepuscular activity, ability to feed in darkness, and a chemically stimulated spawning behavior (Wilkens 1988). It is noteworthy that the nominal species *Astyanax jordani* would encompass at least three distinct molecular lineages (Strecker et al. (2012), four or more based on osteological studies.

Fig. 4.1 White "piaba," *Stygichthys typhlops* (Teleostei: Characiformes), troglobitic fish from phreatic waters in Minas Gerais State, eastern Brazil, showing a high degree of specialization to the subterranean life, including fully regressed eyes and melanic pigmentation. Photo: Dante Fenolio

In spite of the abundant studies on the troglobitic tetra characins, mostly concerned with genetic aspects, not much has been done from a chronobiological perspective. In the few published reports of studies performed under constant (free-running) conditions, epigean *Astyanax mexicanus* fish are compared with their cave relatives *A. antrobius* (from El Pachon cave, a junior synonym of *A. jordani*, Proudlove 2010, referred to by later authors as *A. mexicanus* cave form) and *A. jordani* sensu stricto (from La Cueva Chica, the type locality, also referred to by later authors as *A. mexicanus* cave form) fish; the first is completely devoid of eyes, the second has intermediate characteristics of eyes and pigmentation (Erckens and Martin 1982a). As expected for an epigean species, in *Astyanax mexicanus* free-running activity rhythms were detected under constant darkness. The authors also applied several LD cycles (12:12 h, 6:6 h, 4:4 h, 16:8 h, etc.), all shown to entrain locomotor activity. On the other hand, in the phylogenetically old *A. antrobius*, although activity was entrained by all applied LDs, the circadian rhythm of total activity disappeared immediately after the transition from LD to DD (no residual oscillations). The activity responses to changing environmental conditions were not as uniformly quick as in *A. mexicanus*, but the system hardly needed a swing-in time to be synchronized by the imposed LD. The authors concluded that the internal clock of *A. antrobius* was simplified in relation to its epigean ancestor: the passive system has developed into an extremely passive one, incapable of synchronizing; thus, the circadian oscillator was subject to regression, but it was not completely lost (Erckens and Martin 1982b). *Astyanax jordani* from La Chica seems to be intermediate also in this aspect, because one or two residual oscillations were observed after a transition from LD (12:12) to DD (Erckens and Weber 1976).

We have studied the temporal pattern of locomotor activity of 11 Brazilian species, the highly specialized characid (Characiformes), *Stygichthys typhlops*

Fig. 4.2 The blind catfish from Iporanga, *Pimelodella kronei* (Teleostei: Siluriformes: Heptapteridae), first troglobite discovered and described in Brazil, showing a moderate degree of troglomorphism: partially pigmented individual without externally visible eyes. Photo: Dante Fenolio

Fig. 4.3 *Rhamdia enfurnada* (Teleostei: Siluriformes: Heptapteridae), troglobitic catfish endemic to a single cave in south Bahia State, northeast Brazil, with a great intrapopulation variability in the degree of reduction of eyes and melanic pigmentation: individual with very reduced but still externally visible eyes and intermediate pigmentation. Photo: Dante Fenolio

(Fig. 4.1), of uncertain affinities within Tetragonopterinae, and ten siluriforms (catfishes and armored catfishes): the heptapterids *Pimelodella kronei* (Fig. 4.2), the first discovered and described Brazilian troglobite and to now the most intensively studied; *P. spelaea* and *Rhamdia enfurnada* (Fig 4.3) (all Rhamdiini); *Rhamdiopsis*

Fig. 4.4 *Rhamdiopsis krugi* (dorsal view) (Teleostei: Siluriformes: Heptapteridae), highly troglomorphic catfish from phreatic waters, found in several caves from central Bahia State, northeast
Brazil. Due to the total absence of melanic pigmentation, internal organs, such as the brain, may
be seen by transparence. Photo: Dante Fenolio

Fig. 4.5 Another highly specialized *Rhamdiopsis* catfish (Teleostei: Siluriformes: Heptapteridae;
undescribed species), from a phreatic waterbody in a cave from north Bahia State, northeast Brazil.
Photo: Dante Fenolio

krugi (Fig. 4.4) (cited as *Imparfinis* sp. in Trajano & Menna-Barreto 1995 and cave
fish from Chapada Diamantina in Trajano et al. 2005), *Rhamdiopsis* undescribed sp.
from Toca do Gonçalo cave, Bahia State (Fig. 4.5) (cited as *Taunayia* sp. in Trajano
and Menna-Barreto 2000) and *Rhamdiopsis* undescribed sp. from Salitre cave,
Minas Gerais State; the trichomycterids *Trichomycterus itacarambiensis*, *T. dali*, and

T. rubbioli; and the Callichthyidae, *Aspidoras* cave form (Trajano and Menna-Barreto 1995, 1996, 2000; Trajano et al. 2005, 2009, 2012). Among these, *P. spelaea* has the least reduced eyes when compared with the epigean congeners—the eye structure is apparently normal, but the size is significantly smaller, followed by *A. albater* cave form and *Rhamdiopsis* sp. from Salitre cave. On the other end of the range, *S. typhlops* (Fig. 4.1), *R. krugi* (Fig. 4.4), and *Rhamdiopsis* sp. from Toca do Gonçalo (Fig. 4.5) are the most troglomorphic species, homogeneously devoid of any traces of pigment or visual structures. *Rhamdia enfurnada* provides a good example of an intermediate evolutionary stage, with a mosaic distribution of character states, illustrating the independence of regression of eyes and melanic pigmentation (Trajano and Bichuette 2010). In Fig. 4.3, we show an individual with intermediate condition of both eyes and melanic pigmentation. *P. kronei* is so far known from six localities, the largest population being that of its type locality, the Areias de Cima cave. It is a relatively specialized troglobite, with ca. 90 % of the individuals from the Areias de Cima population without externally visible eyes (the remaining ones with vestigial eyes); the melanic pigmentation phenotypes show a normal distribution within this population, consistently with the notion of an independent evolution of these characters.

For those fish species also differ in their habitat. Although, as for the degree of troglomorphism, no taxonomic correlation was observed, the more specialized species are found in the upper phreatic zone connected to the surface through caves (*R. krugi* and *Rhamdiopsis* sp. from Toca do Gonçalo) or artificial wells (*S. typhlops*), typically a very stable habitat. The population of *T. dali* inhabits a submerged cave, apparently buffered against environmental fluctuations, but its morphology indicates that it was not isolated in the subterranean habitat for as long as the former three. On the other hand, species such as *P. kronei*, *Aspidoras albater* cave form, *Rhamdiopsis* sp. from Salitre cave, *Trichomycterus itacarambiensis*, and *R. enfurnada* inhabit at base-level streams, subject to periodic changes in water level, which may be very drastic in very seasonal areas, as is the case for the latter two.

For comparative studies, the ideal situation is to compare the troglobites to their epigean sister species and then to at least one close outgroup, but this requires phylogenies for the taxa of interest, which are not available for most of the studied groups. Moreover, it is necessary to collect and maintain both epigean and troglobitic specimens. We were able to fulfill such conditions for the blind catfish from Iporanga, *P. kronei*, with its epigean putative sister species, *P. transitoria* (Trajano and Menna-Barreto 2005; Duarte and Menna-Barret unpubl. data on feeding rhythms—see below), which forms troglophilic populations in the Alto Ribeira karst area, São Paulo State, one of which is syntopic with the troglobitic species in the Areias de Cima cave (such syntopy would be due to secondary dispersion of *P. transitoria* into the cave). We also compared *Rhamdiopsis* sp. from Toca do Gonçalo to the epigean *Taunayia bifasciata* (Trajano and Menna-Barreto 2000). The epigean catfish *P. transitoria* (seven specimens monitored) and *Taunayia bifasciata* (two specimens) exhibited strong, significant free-running circadian components of locomotor activity. In contrast, a great deal of variation regarding the expression of the free-running components of activity was observed in the studied blind catfish (six specimens from Areias cave, three from Bombas resurgence), with two Areias individuals showing no significant circadian rhythms. *Rhamdiopsis* sp. from Toca do Gonçalo presented much weaker periodicities: only one out of five studied specimens exhibited a significant circadian rhythm in surface activity, but not in bottom and total activity, and two were completely arrhythmic, not even showing ultradian rhythms. It is noteworthy that differences in temporal patterns of surface and bottom activity were also observed on *Astyanax* fish. As expected for siluriforms, generally nocturnal and quimioriented, thus preadapted to the subterranean life, peaks of activity occurred during the dark phases (Trajano and Menna-Barreto 2000: Fig. 4.1).

Individual variability regarding the presence versus absence of significant circadian components of locomotor activity was observed in all studied species, with a loose correlation with the degree of troglomorphism. Among the three most specialized species, *Stygichthys typhlops* has proceeded further towards the regression of the mechanism of time control (Trajano et al. 2009).

This individual variability in the free-running circadian rhythmicity may be due to the following factors: (1) a relatively short time in isolation in the subterranean habitat, insufficiently long to genetically stabilize changes in the timing systems throughout the populations, and (2) time signals being present that act as *zeitgebers*, such as temperature cycles for *P. kronei* (Trajano and Menna-Barreto 1995), *R. enfurnada*, and *Rhamdiopsis* sp. from Salitre cave (Trajano et al. 2009) or LD cycles for *R. krugi* from Poço Encantado cave (Trajano et al. 2005).

Several species were also studied under LD cycles, 12:12 h (Trajano and Menna-Barreto 2000; Trajano et al. 2012). In the highly troglomorphic *Rhamdiopsis* sp. from Toca do Gonçalo and *R. krugi*, activity was entrained by these cycles, but no residual oscillations were observed, indicating a possible masking effect. Such entrainment was not observed for *S. typhlops*, which represents a further step towards the loss of temporal mechanisms.

The comparison between the temporal patterns of locomotor activity between troglobitic fish and their epigean close relatives (*Astyanax antrobius* × *A. jordani* × *A. mexicanus*; *Pimelodella kronei* × *P. transitoria*; *Rhamdiopsis* sp. from Toca do Gonçalo × *T. bifasciata*), studied under the same protocols, provides good evidence for the hypothesis of evolutionary regression of circadian timing system in troglobitic species, either affecting the oscillator(s) itself (themselves) or due to an uncoupling between the oscillators affecting the circadian component of locomotor activity. Regression of the retina and, at least for some of the studied species, possibly also of the photoreceptor of the pineal organ may be involved in the disorganization of the circadian system verified in several troglobitic species. These results favor the notion of external ecological factors as the main factors stabilizing the selection for circadian rhythms. The progressive loss of circadian locomotor rhythmicity in different troglobitic fishes in parallel (but not necessarily correlated) with other regressive characters indicates that similar processes may be involved in such regression.

Troglomorphic traits such as reduced eyes and melanic pigmentation and some behaviors as cryptobiotic habits and photophobic reactions frequently present a mosaic distribution in troglobitic populations, as observed in the presently studied fishes. Except for the three most specialized troglobites, putatively phylogenetically old troglobites living in very stable environments, we did not observe any clear correlation between the degree of reduction of eyes and pigmentation and that of circadian rhythms. Apparently, the environmental characteristics of the habitat are another important factor, besides time of isolation in the subterranean realm, influencing the pace of regression evolution of characters involved in time control mechanisms; thus, the classical troglomorphisms alone are not good predictors for the degree of reduction of rhythmicity in troglobites.

As in the case of the eyes, pigmentation and other light-related characters that become functionally neutral in the continuously dark subterranean environment, the regression of the circadian timing system of locomotor activity (and possibly other circadian functions as well) evidenced for several troglobitic species may be due to the loss of selection, allowing for the accumulation of random mutations affecting those mechanisms. In the absence of stabilizing selection eliminating such mutations in subterranean animals, as it does for epigean species, the ecologically selected circadian rhythms present in the epigean ancestors may have been lost by troglobites or are in the process of regression in the case of more recent troglobites.

Finally, an example of a behavioral trait in troglobitic fish subject to the influence of *zeitgebers* other than LD cycles (daily temperature cycles, in the case) is provided by the armored catfish, *Ancistrus cryptophthalmus* (Loricariidae), from the State of Goiás, Central Brazil. Part of the population found in Angélica cave lives in the aphotic zone, but not far from the cave sinkhole (input of an epigean river). We observed a daily variation in the number of catfishes exposed on the rocky substrate, which is higher during the morning and conspicuously decreases in the afternoon, possibly as a response to an increase of 1 °C in water temperature observed along a period of 4 h (from 10:00 to 14:00 h) (E. Bessa & E. Trajano, pers.obs.). This may be a consequence of a daily fluctuation in hiding habits, which would be synchronized by the 24-h cycle of environmental temperature.

In spite of the lack of a significant circadian component in locomotor activity, we have shown the presence of faster (ultradian) rhythms in several studied catfish (Trajano and Menna-Barreto 1995, 1996, 2000). The presence of ultradian components in locomotor activity is suggestive of an action of uncoupled oscillators (Granados-Fuentes and Herzog 2013), thus unable to coalesce into a circadian pattern.

4.3 What About a Possible Role of Food Availability as a *Zeitgeber*?

The lack of a circadian rhythm in locomotor activity fits the hypothesis that in a constantly dark environment, circadian expressions are no longer needed for adaptation/survival. This logic applies to rhythms synchronized by the light/dark cycle but may not be true for other time cues possibly present in the cave environment, for instance, food. Food is generally scarce in caves and, if offered at more or less regular intervals, certainly applies for a role as zeitgeber. According to Friedrich (2013), biological clock genes are likely to be generally conserved in cave species, regulating rhythmic behaviors in response to nonvisual cycling variables. And indeed food entrainment has been described by Cavallari et al. (2011), with periodicity outside the circadian range (nearly bi-circadian, 47 h) and not linked to light reception in a Somalian cavefish showing extreme troglomorphic phenotype. Rather interestingly, those authors found mutations in nonvisual opsin photoreceptors (Melanopsin (Opn4m2) and TMT (teleost multiple tissue)-opsin, providing evidence for a light-sensing function but not expressed within the circadian range. Spieler (1992) has shown that feeding may entrain rhythms in several fish species, specifically locomotor activity but also agonistic behavior and circulating cortisol. Food anticipatory activity is a conspicuous evidence of an endogenously triggered oscillation with a clear adaptive significance (Boujard and Leatherland 1992). Several researchers have been arguing for a food-entrainable oscillator (FEO) acting independently of the now classic light-entrainable oscillator (LEO), and indeed, results obtained in three species of fish by Purser and Chen (2001) and Spieler (1992) and Sánchez-Vázquez et al. (1996) point in that direction. Given continuous access to sufficient food, fish would not need to anticipate mealtime, but with restricted amount and availability, synchronization becomes vitally relevant, also maximizing food utilization by the digestive system. Of course differences between fish under different food availability schedules may also be interpreted in an ecological context, reminding us of the brilliant contribution of Rijnsdorp and Daan (1981), in their description of a temporal niche to explain both prey and predator survival in the same spatial niche. In the relatively stable cave environment, which signals could acquire zeitgeber properties with any adaptive/survival value? Volpato and Trajano (2006) suggest that cyclical guano deposition by animals daily but not continually present in caves, such as bats and some birds, might apply for that role. The temporal pattern of guano deposition would thus act in the circadian range and effectively entrain cavefish rhythms. In a loach species, Biswas and Ramteke (2008) reported that periodic feeding at 18:00 h synchronized swimming activity rhythm; in that paper, they suggest that periodic restricted feeding could act as a powerful zeitgeber of circadian rhythms in subterranean organisms.

4.3.1 Some Evidence of a Food-Entrainable Circadian Oscillator in Cave Fish

An alternative way to demonstrate a possible role for an endogenous character of food entrainment, besides regular (circadian) feeding windows as discussed above, is to record spontaneous feeding. Thus, in an as yet unpublished observation, Leandro Duarte, a former student at our laboratory, studied *Pimelodella kronei* and *P. transitoria* catfish in aquaria equipped with sensors that could be triggered by indwelling fish (one per aquarium) and deliver one pellet of food (0.23 ± 0.005 g) for each touch of the sensor, which the fish learned to do with their snout after 3 weeks. Each touch was recorded building a time series later submitted to statistical (spectral) analysis for the detection of periodic components. We found circadian feeding activity for *P. transitoria* both under 12:12 h light/dark cycle (LD) and constant darkness (DD), whereas *P. kronei* exhibited free-running circadian rhythmicity under DD in three of the four individuals tested. Under LD, two out of four fish showed rhythmicity in the circadian range. These results indicate that the regression of the circadian component of locomotor activity is not necessarily accompanied by the equivalent regression of circadian feeding rhythmicity in cave fish.

Living for generations in a constant environment may have rendered the circadian timing system relatively weaker. This weakness may be interpreted in the light of current models of the organic timing systems as a result of the interaction (coupling) of multiple oscillators (Tosini and Menaker 1998; Saper et al. 2005). Long-term isolation from the light/dark environmental signals would compose a scenery where light-entrainable oscillators are no longer able to act as transducers and thus synchronize activity rhythms. The persistence of circadian feeding activity under constant darkness in troglobitic fish, on the other hand, may be interpreted as an evidence that regular food availability is acting as a *zeitgeber* or, alternatively, that feeding rhythms are taking longer to show regression. Studies with more species of cave fish are certainly needed, but for the moment, we tend to adopt the first explanation.

4.4 Troglophiles and Trogloxenes: Movements From and To Caves

A basic difference between these two categories is that individual trogloxenes <u>must</u> leave caves periodically in order to complete their life cycles, whereas troglophiles <u>may</u>, if the opportunity presents itself (for instance, as a consequence of dispersion movements, bringing the individual near to contacts with the surface), leave the cave habitat. For the latter, once near the surface, random movements (especially during the night) and/or attraction to resources that are more abundant in the surface, particularly food, eventually would bring the individual to get in contact with the epigean population; conversely, epigean individuals, also due to dispersion movements and/or attraction to the sheltered subterranean environment, move in the opposite direction This may be simple from the conceptual point of view, but not so easy to apply in the field: when an animal is observed moving between subterranean and epigean habitats, how can we establish whether it is doing so because it must or because it may do so?

The best approach to this question is based on chronobiological protocols. In general, trogloxenes must leave caves because subterranean resources, mainly food, are insufficient for the biological needs of the species. Therefore, movements between hypogean and epigean habitats follow regular, cyclical patterns. Their periodicity would depend on individual physiological requirements and locomotor abilities; thus, it may vary not only among taxa but also with season, sex, and age. Bats, the "classical" trogloxenes, are known to emerge from caves daily due to their high energetic demands. As well, del Castillo et al. (2009) observed that brachycephalid frogs, *Eleutherodactylus longipes*, use the zones near the entrances of a Mexican cave as shelter during the day phase, emerging regularly at night, at least during the dry season (the authors refer to these frogs as troglophiles, without mention to any classification system, but they are clearly trogloxenes according to the modern concepts). This pattern is probably common among anurans from arid and semiarid regions and highly seasonal climates, with pronounced dry seasons.

Variation with age and season has been shown for camel crickets (Orthoptera: Rhaphidophoridae—a group that includes well-studied trogloxenes also in Europe), genera *Ceuthophilus* and *Hadenoecus*, in the United States (Lavoie et al. 2007), and for the opilionid *Acutisoma spelaeum* (cited as *Goniosoma spelaeum*), in São Paulo State, southeast Brazil (Gnaspini et al. 2003). Contrasting to troglophiles, many trogloxenic invertebrates present infradian patterns of movements from and to hypogean corresponding to seasonal use of caves.

On the other hand, studies focusing on movements of troglophiles between surface and subterranean environment, using chronobiological protocols in the field, seem to be extremely rare, but this may be an artifact. The absence of patterns may be interpreted by researchers as failure to detect them, as in Hoenen and Marques (1998), and not as evidence supporting the status of troglophile, thus considered unworthy publishing. Hoenen and Marques (op. cit.) studied the cricket, *Strinatia brevipennis*, inside a small cave from south São Paulo, southeast Brazil, and found out that individual crickets did not exhibit circadian rhythms in their movements to and from the entrance zone, as expected for a troglophile (the status of epigean species with troglophilic populations is unquestionable for this species).

4.5 Final Comments

Troglobitic species, especially those evolving in highly stable habitats such as phreatic waters, are excellent models for testing hypotheses concerning the function of rhythmicity. Moreover, intra-taxon variation in the degree of morphological specialization to the subterranean life, as observed in Brazilian troglobitic fishes, offers good opportunities to study the evolution of regression of time control mechanisms in organisms living in the absence of *zeitgebers*, such as photoperiods and temperature cycles.

On the other hand, chronobiological protocols used in field studies are fundamental for distinguishing trogloxenes from troglophiles, a central question in subterranean biology.

Acknowledgments We are deeply indebted to Dante Fenolio, author of the cave fish photographs, for the authorization to use these images. ET is partially supported by the Conselho Nacional de Desenvolvimento Científico e Tecnológico (CNPq) (fellowship 302956/2010-7).

References

Barr TC Jr (1968) Cave ecology and the evolution of troglobites. Evol Biol 2:35–97
Biswas J, Ramteke AK (2008) Timed feeding synchronizes circadian rhythm in vertical swimming activity in cave loach, *Nemacheilus evezardi*. Biol Rhythm Res 39:405–412

Boujard T, Leatherland JF (1992) Circadian rhythms and feeding time in fishes. Environ Biol Fish 35:109–131

Cambras T, Díez-Noguera A (1988) Changes in motor activity during the development of the circadian rhythm in the rat. J Interdisipl Cycle Res 19:65–74

Cavallari N, Frigato E, Vallone D et al (2011) A blind circadian clock in cavefish reveals that opsins mediate peripheral clock photoreception. PLoS Biol 9(9):e1001142. doi:10.1371/journal.pbio.1001142

del Castillo AE, Castaño-Meneses G, Dávila-Montes MJ et al (2009) Seasonal distribution and circadian activity in the troglophilic long-footed robber frog, *Eleutherodactylus longipes* (Anura: Brachycephalidae) at Los Riscos Cave, Queretáro, Mexico: field and laboratory studies. J Cave Karst Stud 71:24–31

Erckens W, Martin W (1982a) Exogenous and endogenous control of swimming activity in *Astyanax mexicanus* (Characidae, Pisces) by direct light response and by a circadian oscillator. I. Analysis of the time-control systems of an epigean river population. Z Naturforsch 37:1253–1265

Erckens W, Martin W (1982b) Exogenous and endogenous control of swimming activity in *Astyanax mexicanus* (Characidae, Pisces) by direct light response and by a circadian oscillator. II. Features of time-controlled behaviour of a cave population and their comparison to an epigean ancestral form. Z Naturforsch 37:1266–1273

Erckens W, Weber F (1976) Rudiments of an ability for time measurement in the cavernicolous fish *Anoptichthys jordani* Hubbs and Innes (Pisces, Characidae). Experientia 32:1297–1299

Friedrich M (2013) Biological clocks and visual systems in cave-adapted animals at the dawn of speleogenomics. Integr Comp Biol 53:50–67

Granados-Fuentes D, Herzog ED (2013) The clock shop: coupled circadian oscillators. Exp Neurol 243:21–27

Hoenen SM, Marques M (1998) Circadian patterns and migration of Strinatia brevipennis (Orthoptera: Phalangopsidae) inside a cave. Biol Rhythm Res 29:480–487

Lavoie KH, Helf KL, Poulson TL (2007) The biology and ecology of North American cave crickets. J Cave Karst Stud 69:114–134

Proudlove G (2010) Biodiversity and distribution of the subterranean fishes of the world. In: Trajano E et al (eds) Biology of subterranean fishes. Science Publ., Enfield

Purser GJ, Chen WM (2001) The effect of meal size and meal duration on food anticipatory activity in greenback flounder. J Fish Biol 58:188–200

Rijnsdorp A, Daan S (1981) Hunting in the kestrel, *Falco tinnunculus* and the adaptive significance of daily habits. Oecologia 50:391–406

Sánchez-Vázquez FJ, Madrid JA et al (1996) Demand feeding and locomotor circadian rhythms in the goldfish, *Carassius auratus*: dual and independent phasing. Physiol Behav 60:665–674

Saper CB, Jun Lu J et al (2005) The hypothalamic integrator for circadian rhythms. Trends Neurosci 28:152–157

Spieler RE (1992) Feeding-entrained circadian rhythms in fishes. In: Ali MA (ed) Rhythms in fishes. Plenum, New York, pp 137–147

Strecker U, Hausdorj B, Wilkens W (2012) Parallel speciation in Astyanax cave fish (Teleostei) in Northern Mexico. Mol Phylogenet Evol 62:62–70

Tosini G, Menaker M (1998) Multioscillatory circadian organization in a vertebrate, *Iguana iguana*. J Neurosci 18:1105–1114

Trajano E (2007) The challenge of estimating the age of subterranean lineages: examples from Brazil. Acta Carsologica 36:191–198

Trajano E (2012) Ecological classification of subterranean organisms. In: White WB, Culver DC (eds) Encyclopedia of caves. Academic Press, Waltham, pp 275–277

Trajano E, Bichuette ME (2010) Subterranean fishes of Brazil. In: Trajano E, Bichuette ME (eds) Biology of subterranean fishes. Science Publ., Enfield, pp 331–355, color plates

Trajano E, Menna-Barreto L (1995) Locomotor activity pattern of Brazilian cave catfishes under constant darkness (Siluriformes, Pimelodidae). Biol Rhythm Res 26:341–353

Trajano E, Menna-Barreto L (1996) Free-running locomotor activity rhythms in cave-dwelling catfishes, *Trichomycterus* sp., from Brazil (Teleostei, Siluriformes). Biol Rhythm Res 27:329–335

Trajano E, Menna-Barreto L (2000) Locomotor activity rhythms in cave catfishes, genus *Taunayia*, from eastern Brazil (Teleostei: Siluriformes: Heptapterinae). Biol Rhythm Res 31:469–480

Trajano E, Duarte L, Menna-Barreto L (2005) Locomotor activity rhythms in cave fishes from *Chapada Diamantina*, northeastern Brazil (Teleostei: Siluriformes). Biol Rhythm Res 36:229–236

Trajano E, Carvalho MR, Duarte L et al (2009) Comparative study on free-running locomotor circadian rhythms in Brazilian subterranean fishes with different degrees of specialization to the hypogean life (Teleostei: Siluriformes, Characiformes). Biol Rhythm Res 40:477–489

Trajano E, Ueno JCH, Menna-Barreto L (2012) Evolution of time control mechanisms in subterranean organisms: cave fish under light-dark cycles (Teleostei: Siluriformes, Characiformes). Biol Rhythm Res 43:191–203

Volpato GL, Trajano E (2006) Biological rhythms: the physiology of tropical fishes. In: Val AL, Val VM, Randall DJ (eds) Fish physiology, vol 21. Elsevier, San Diego, pp 101–153

Wilkens H (1988) Evolution and genetics of epigean and cave *Astyanax fasciatus* (Characidae, Pisces). Evol Biol 23:271–367

Wilkens H (2010) Genes, modules and the evolution of cave fish. Heredity 105:413–422

Chapter 5
Avian Circadian Organization

Vincent M. Cassone

Abstract In birds, biological clock function pervades all aspects of biology, controlling daily changes in sleep: wake, visual function, song, migratory patterns and orientation, as well as seasonal patterns of reproduction, song, and migration. The molecular bases for circadian clocks are highly conserved, and it is likely the avian molecular mechanisms are similar to those expressed in mammals, including humans. The central pacemakers in the avian pineal gland, retinae, and SCN dynamically interact to maintain stable phase relationships and then influence downstream rhythms through entrainment of peripheral oscillators in the brain controlling behavior and peripheral tissues. Birds represent an excellent model for the role played by biological clocks in human neurobiology; unlike most rodent models, they are diurnal, they exhibit cognitively complex social interactions, and their circadian clocks are more sensitive to the hormone melatonin than are those of nocturnal rodents.

5.1 Introduction

Each morning, and especially in the spring, we are greeted by a cacophony of small birds singing a dawn chorus. In eastern North America, spring mornings are sometimes defined by the merry roundelay of the American robin, *Turdus migratorius*, the varied staccato whistles of the Northern cardinal, *Cardinalis cardinalis*, the *hey-hey* of the white-breasted nuthatch, *Sitta carolinensis*, or even the cheery chirping of the introduced house sparrow, *Passer domesticus*. In the backdrop, we may hear the doleful *ooh-wah-hoo-hoo* of the aptly named mourning dove, *Zenaida macroura*, as the bass section above croons with the honking of migrating Canada geese, *Branta canadensis*. There is no particular order of who sings or who calls first, and the orchestration is peripatetic at best, seemingly random, although many of these garden songsters are reacting to each other's songs. And yet, there is a coordination of the rhythm and timbre of this dawn chorus. These birds all possess an internal

V.M. Cassone (✉)
Department of Biology, University of Kentucky, Lexington, KY 40506, USA
e-mail: Vincent.Cassone@uky.edu

© Springer International Publishing Switzerland 2015
R. Aguilar-Roblero et al. (eds.), *Mechanisms of Circadian Systems in Animals and Their Clinical Relevance*, DOI 10.1007/978-3-319-08945-4_5

biological clock that is coincidentally entrained to the identical environmental signal, the rising of the morning sun, and, in turn, these internal clocks are tuned to the expression of clocks by their intraspecific and extraspecific neighbors.

While these appear to be the melodious embrace of the warming sun, they are, in fact, a cold war, defining territory for breeding and foraging in anticipation of reproductive success (Marler 2004; Williams 2004). In no other group of animals are the seasonal changes in reproductive function so obvious to the casual observer. We hear them stake their claims. We see them build their nests, incubate the eggs, and raise and fledge their young.

At certain times of year, small songbirds fatten for their annual migrations and, at certain times of day, dusk usually, become increasingly agitated as they gather for their vernal and autumnal treks to breeding and wintering grounds (Gwinner 1989, 2003). These birds typically eschew their nightly drifts into slumber during this time, sleeping little or not at all, a phenomenon called *Zugunruhe*, as they migrate during the night, avoiding the gauntlet of diurnal predators as they cross vast areas of our continent.

Each of these processes and more are strictly timed to a time of day and to a time of year (Cassone and Westneat 2012). They are not restricted to eastern North America, either, as these processes are repeated time and time again throughout the world, albeit at different times of year, depending on the latitude and local environment (Kumar et al. 1996, 2002, 2010). The question that arises is, "Why do birds so strictly time so many of their behavioral and physiological functions, and how do they accomplish it?" In essence, the child-like question, "Why does the sparrow sing on spring mornings?," is also a scientific question that is beginning to be answered, and these likely entail an understanding of the biological clock or clocks that underlie all rhythmic processes. Specifically, understanding of the molecular, physiological, and behavioral mechanisms underlying the temporal coordination of these complex processes and behaviors in birds will tell us more about human chronobiology as well, because like humans and unlike the standard laboratory rodent models for biological clocks, birds exhibit a complex orchestration of circadian behavior that controls daily patterns of sleep: wake, visual sensitivity, cognition, and social behavior. Further, study of the mechanisms underlying annual cycles of reproduction, migration, and metabolism in birds will provide clues to anticipated ecological changes due to climatic disruption. In essence, birds are images in our own mirrors, and we should pay attention to them more than current biomedical science might prefer.

5.2 Biological Rhythms and the Clocks That Control Them

Biological rhythms and the endogenous clocks that control them are fundamental properties of nearly all living organisms, ranging from cyanobacteria to humans (Bell-Pedersen et al. 2005). As diverse as the organisms that express biological rhythms, the formal properties of these rhythms are remarkably conserved (Pittendrigh 1993). These biological rhythms are functionally tied to environmental cycles they estimate; of these, we will concentrate in this review on two-*circadian rhythms* and *circannual cycles*.

The daily 24-h cycle of day and night imposes a rhythmic cascade of positive and negative selective pressures on nearly all organisms on Earth (Pittendrigh 1993). Daily light cycles provide the energy for photosynthesis, to warm water to a consistently liquid state and to maintain ambient temperatures for life. It also provides daily cycles in deleterious teratogenic, carcinogenic, and desiccating wavelengths. It is therefore no surprise that most free-living organisms, if not all, have adapted to these cycles through the expression of endogenously generated circadian (*circa = approximately; dian = a day*) oscillations that entrain to local time through the process of *entrainment*. Rhythmic processes cannot be identified as *circadian* unless they are experimentally observed to persist for at least 1 or 2 cycles, preferably more, when the organism in question is experimentally placed in constant environmental conditions of either constant darkness (DD) or constant dim light (dimLL), Constant high light (LL), may have other effects, frequently abolishing circadian rhythms altogether and/or damaging photoreceptive elements in the system (Aschoff 1979). In fact, many circadian rhythms persist for weeks, months, or years in constant environmental conditions.

In this scenario, organisms will repeatedly express patterns of behavior, physiology or biochemical processes with a period, τ, of close to but rarely exactly 24 h (Fig. 5.1). These endogenously driven rhythms must then be *entrained* to the relevant environmental cycle, typically the light:dark cycle (LD) of day and night, such that the internal phase, ϕ_i, of the organism's clock corresponds appropriately to the external phase, ϕ_e, of the LD cycle. Thus, a diurnal bird's locomotor activity pattern entrains to the LD cycle so that activity onset ϕ_i corresponds approximately to dawn ϕ_e, maintaining a stable phase relationship, ψ_{ie}.

Similarly, annual environmental cycles correspond to *circannual rhythms* expressed by many organisms when placed experimentally in constant photoperiods of 12 h of light and 12 h of dark (LD12:12) or another constant photoperiod (Gwinner 1989, 2003). Under these conditions many organisms, including birds, will express cycles of approximately 365 days. These in turn are believed to be entrained to the annual cycle by changes in photoperiod (Gwinner 1989) and/or a physiological proxy, such as the duration of the hormone melatonin (see below). Circannual cycles are not as well understood as are circadian rhythms, but they are likely linked physiologically (Gwinner and Brandstätter 2001; Gwinner et al. 1997).

The role of the circadian clock in annual cycles has been known for some time (Bünning 1969; Follett et al. 1992; Konishi et al. 1987; Menaker and Eskin 1967; Rowan 1926). In many species of birds, exposure to photoperiods of longer than 11.5 h/day results in the rapid induction of the hypothalamo–hypophysial–gonad axis, causing development and growth of testes and ovarian follicles. Although there are differences among species of birds and between birds and other taxa, neither the absolute length of the photoperiod, length of the scotoperiod (the duration of the dark phase), or their ratio is the proximal causes of gonadal induction. Rather, it is the circadian ϕ at which light impinges on photoreceptive elements that causes reproductive changes. For example, male Japanese quail, *Coturnix japonica*, and

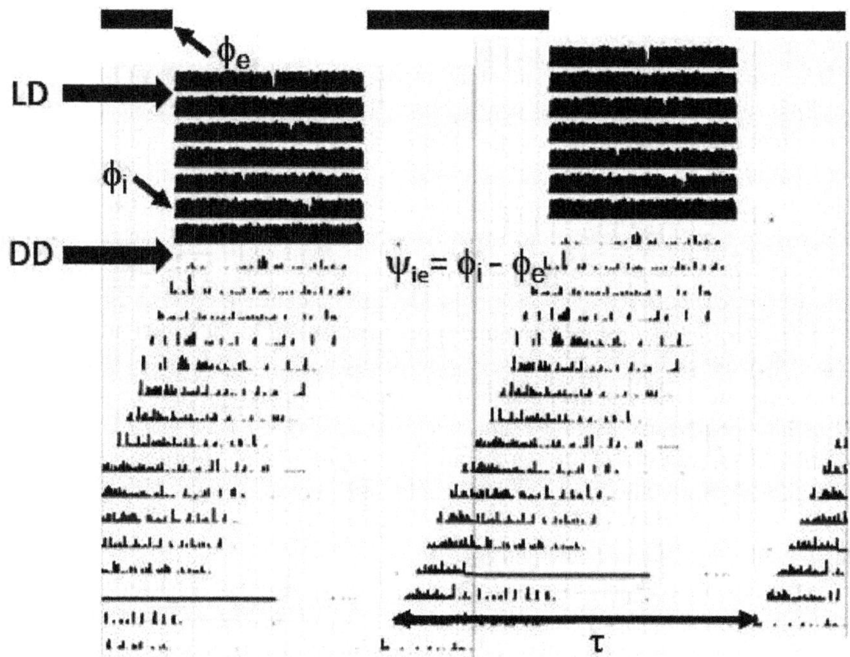

Fig. 5.1 Actogram of locomotor activity from a single zebra finch, *Taeniopygia guttata*. The *top bars* indicate the times during which lights are off (black) vs. on (white). These are plotted in a 48 h time span in order to "double-plot" the data. The time off lights on, ϕ_e, indicated by the *arrow*, is a phase reference determined by the investigator. The internal phase, ϕ_i, is determined to be the activity onset. The relationship between ϕ_i and ϕ_e is called ψ_{ie}. This relationship may change depending on the time of year and physiological condition of the bird. The internal period, τ, is indicated here as the average interval between activity onsets

white-crowned sparrows, *Zonotrichia leucophys*, which are maintained in LD 6:18 will exhibit regressed testes. However, if the last hour of the 6 h. photoperiod is extended each long night to a specific "photoinducible phase," ϕ_{pi}, usually 11–12 h following the onset of the short photoperiod, reproductive activity is commenced. Thus, birds exposed to LD 6:18 or L5: D1: L1: D17 (a single 1 h light pulse interrupting the night) will retain regressed gonads, but if the 1 h pulse occurs 5 h later (L5:D6:L1:D12), gonads will recrudesce (Follett et al. 1974; Menaker and Eskin 1967; Schleussner and Gwinner 1989; Sharp 2005; Yoshimura 2010). The same total amount of light (and dark) is present for each 24 h, but the effect is dramatically different. It is the *timing* of light coinciding with an internal process that induces or prevents reproductive activity, suggesting a circadian clock underlies photoperiodic time measurement. Nanda and Hamner (1958) had shown this to be the case in plants through an elegant series of experiments in which soy bean plants exposed to a 6 h photoperiod coupled with scotoperiods of varying lengths that were multiples of a 24 h cycle (e.g., LD6:18; LD6:42 or LD6:66) did not flower.

However, when plants were exposed to 6 h photoperiods followed by scotoperiods of lengths that did not resonate with 24 h (e.g., LD6:6; LD6:30 or LD6:54) flowering was observed. Similar studies in several species of birds showed that seasonal changes in both gonadal recrudescence and regression are regulated by a circadian clock (Follett and Pearce-Kelly 1991; Follett et al. 1992; Kumar et al. 1996).

These and other observations have led to two competing models for a role of circadian clocks in photoperiodic time measurement. In one, the "external coincidence model" proposed by Bünning (1969) suggests that light has two complementary roles. First, light entrains the circadian clock with a stable ψ_{ie} such that the photoinducible phase, ϕ_{pi}, is also maintained with a stable ψ approximately 11.5 h following the beginning of the photoperiod (lights on). When light coincides with the ϕ_{pi} either because the photoperiod is long under natural conditions or through exposure to an experimental light pulse, the reproductive axis is induced. The competing notion, the "internal coincidence model" proposed by Pittendrigh (cf. Pittendrigh 1993), stems from the observation that many circadian systems behave as if they are composed of at least two oscillators, one entrained to dawn and the other to dusk. In the internal coincidence model the phase relationship between the dawn oscillator to the dusk oscillator, $\psi_{dawndusk}$, induces seasonal changes in reproduction. As we will see below, each of these models has support from different experimental systems in birds. However, until specific structures and/or molecules become associated with these models, they are essentially untestable at the physiological level.

5.3 Molecular Genetics of Circadian Clock Function

Circadian rhythms are regulated by a highly conserved set of genes, collectively called "clock genes," whose products are believed to dynamically interact to elicit rhythmic patterns of transcription, translation, biochemical and physiological processes, and behavior (Fig. 5.2) (Bell-Pedersen et al. 2005; Eckel-Mahan and Sassone-Corsi 2013). In animals ranging from *Drosophila* to humans, the central core of this gene network can be broadly characterized as "positive elements" *clock* and *bmal1* and "negative elements" *Period 1 (Per1), period 2 (per2), period 3 (per3)*, and the cryptochromes *cryptochrome 1 (cry1)* and *cryptochrome 2 (cry2)*. In contrast to mammals, birds do not express a *per1* and have been shown to only express only *per2* and *per3* (Bailey et al. 2003, 2004; Yasuo et al. 2004; Yasuo and Yoshimura 2009; Yoshimura et al. 2000). *Clock* and *bmal1* are transcribed and then translated in the cytoplasm, where they dimerize and reenter the nucleus and activate transcription of the negative elements through the activation of E-box promoter elements. The *pers* and *crys* in turn are transcribed and translated in the cytoplasm, where the PER proteins are targeted for proteosomal proteolysis by a series of protein kinases, most notably casein kinase 1ε (CK11ε) and CK11δ. This process slows the accumulation of the cytoplasmic PER and thereby increases the period of the molecular cycle. In the cytoplasm, PER and CRY proteins form oligomers that reenter the nucleus and interfere with the CLOCK/BMAL1-mediated activation.

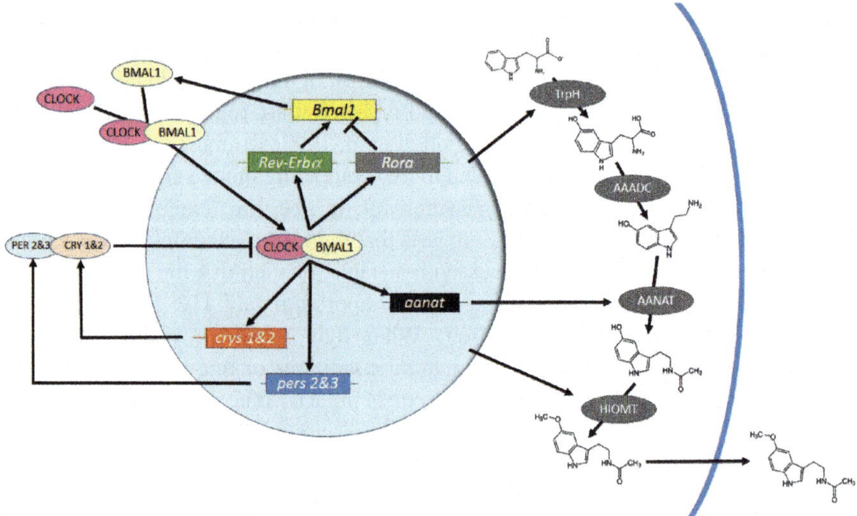

Fig. 5.2 Schematic of the molecular clockworks regulating circadian patterns of melatonin bio-synthesis in a pinealocyte or retinal photoreceptor. Positive elements CLOCK and BMAL1 enter the nucleus and activate expression of genes whose promoters contain an E-Box. Among these are the negative elements *period 2 & 3 (pers 2 & 3)* and *cryptochromes 1 & 2 (crys 1 & 2)*, *Rev-Erbα* and *Rora*, which form a secondary loop regulating *Bmal1* transcription, and output, clock-controlled genes such as arylalkylamine-*N*-actyltransferase (*aanat*). The *pers* and *crys* are trans-lated, form heterodimers with other components, such as the casein kinases, and reenter the nucleus to interfere with CLOCK/BMAL1 activation. Melatonin biosynthesis pathways are indi-cated on the *right*. Amino acid tryptophan is converted to 5-hydroxytryptophan by tryptophan hydroxylase (TrpH). Aromatic amino acid decarboxylase (AAADC) then converts 5-hydroxytryptophan to 5-hydroxytryptamine (5HT; serotonin), Then, during the night, AANAT converts 5HT to *N*-acetylserotonin, a substrate for hydroxyindole-*O*-methyltransferase (HIOMT), which produces melatonin itself. Presumably, melatonin diffuses out of the cell at this time, although a release mechanism may exist

A secondary cycle involving two genes containing E-box promoters, *Reverbα* and *rorα*, amplify the cycle by activating and inhibiting *bmal1* transcription respec-tively. Disruption and/or knock-out of these genes' action has profound effects on the expression of circadian rhythms in animals in which these technologies are possible (i.e., mice and *Drosophila*) ranging from changes in period to arrhythmicity.

Unfortunately, these technologies are not routinely available in birds just yet. However, there are several observations that tentatively link clock gene expression with avian rhythmic behavior. Noting population studies in human populations of single nucleotide polymorphisms (SNP) in the *clock* gene's 3′-UTR have revealed differences in times of sleep and sleep duration (Steinmeyer et al. 2012), have shown weak (5.9 %) to moderate (46.7 %) associations between awakening time of blue tits, *Cyanistes caeruleus*, with single nucleotide polymorphisms (SNPs) in *period 2* and *CK1ε*. However, these studies were conducted in free-living tits using radio frequency transponders under seminatural conditions, not under DD or dimLL, so that these effects cannot distinguish circadian clock effects from light-induced effects or effects on homeostatic regulation of sleep. Other studies have focused on allelic differences in C-terminal polyglutamine repeats in the CLOCK protein in breeding and migratory patterns of barn swallows, *Hirundo rustica*, suggesting negative selection on "deviant" genotypes (Caprioli et al. 2012). However, other studies in several different swallow species in the genus *Tachycineta* failed to show similar associations (Dor et al. 2012).

5.4 Extraocular Photoreception

In addition to photoreceptors in the lateral eyes shared by all vertebrate classes, it has been known for some time that non-mammalian vertebrates express functional photopigments within the brain that are critical for entrainment of both circadian rhythms and circannual cycles. In birds, these reside in the pineal gland, the preoptic area, the lateral septum, and the tuberal hypothalamus. Early studies by Benoit in the 1930s showed that domestic ducks, *Anas platyrhynchos*, that had been blinded (enucleated) continued to exhibit reproductive responses to changing photoperiod (Benoit and Assenmacher 1954), but work by Menaker and colleagues in the 1960s and 1970s in passerine birds clearly showed that the eyes are not necessary for circadian entrainment (Masuda et al. 1994; McMillan 1970) or seasonal control of reproduction (Menaker 1968; Menaker and Keatts 1968; Menaker and Underwood 1976; Underwood and Menaker 1970). In a classic series of experiments, Menaker's group demonstrated that enucleated house sparrows could entrain to a series of LD cycles of dimmer and dimmer illuminances (Menaker et al. 1970). Once birds were no longer capable of entrainment, they showed that the responsible photoreceptor resided inside the head by simply plucking feathers, and entrainment was reinstated. They then blocked entrainment by injecting India ink beneath the scalp (Menaker 1968).

Subsequent research has now identified at least four distinct structures within the brain that are functionally photoreceptive, containing several opsin-based photopigments and photoisomerases (Bailey and Cassone 2005; Foster and Hankins 2002; Li and Kuenzel 2008; Masuda et al. 1994). These include the pineal gland, which expresses a pineal-specific opsin, pinopsin (Bailey et al. 2003; Masuda et al. 1994; Max et al. 1995; Okano et al. 1994), as well as melanopsin (OPN4) (Bailey and Cassone 2005; Bailey et al. 2003; Chaurasia et al. 2005); and iodopsin (OPN1) (Bailey et al. 2003; (Masuda et al. 1994) and whose photoreceptive function will be discussed further below. In addition, neurons within the preoptic area express VA (vertebrate ancient) opsin (Davies et al. 2010, 2012; Soni and Foster 1997) and project to the tuberal hypothalamus, while the tuberal hypothalamus expresses a plethora of photoreceptive cells that appear to be divergent among avian species. For example, in domestic turkeys, *Meleagris gallopavo*, melanopsin-expressing cells in the premammillary nucleus (PMM) in the dorsal tuberal hypothalamus project directly to the median eminence (Kang et al. 2007, 2010; Kosonsiriluk et al. 2013).

In Japanese quail, *Coturnix coturnix*, CSF-contacting neurons in the mediobasal hypothalamus (MBH), which may be homologous to the PMM, express both OPN4 and neuropsin (OPN5) (Nakane and Yoshimura 2010; Nakane et al. 2010). In house sparrows, neurons within the arcuate nucleus express rhodopsin (OPN2) itself, in addition to OPN4 and OPN5 (Foster et al. 1985; Wang and Wingfield 2011). While it is not clear whether or which opsin-based photopigments are expressed in the lateral septal organ, illumination of this area of cerebrospinal fluid contacting neurons elicits a physiological response (Li et al. 2004; Li and Kuenzel 2008). Further, it is not clear whether each of these photoreceptive organs and/or their photopigments subserve mutually exclusive physiological processes or whether these overlap in their functions. In addition to opsin-based photopigments, all animals express flavin-based cryptochromes (Van Gelder 2001). While cryptochrome is the major photopigment responsible for photoentrainment in *Drosophila* (Emery et al. 1998), the multiple cryptochromes expressed by vertebrates have not been established as photoresponsive molecules. Even so, in birds, the role of cryptochromes in celestial orientation is light dependent (see below).

5.5 The Pineal Gland Is a Master Pacemaker for Avian Circadian Clocks

Searching for the location of the intracranial, extraretinal photoreceptors, Gaston and Menaker (1968) surgically removed the pineal gland (PINX) from house sparrows. While the birds retained their ability to entrain to LD, they became arrhythmic when placed in DD, demonstrating that the pineal gland is necessary for self-sustained circadian rhythmicity. However, the data also showed that the pineal gland is part of a system of circadian clock components, since PINX sparrows could anticipate the time of lights on in an LD cycle and because birds only gradually became arrhythmic over 5–15 days following transfer from LD to DD. Further, the effect of PINX is not universal among avian species. PINX of European starlings, *Sturnus vulgaris*, results in a range of behavioral changes ranging from arrhythmicity akin to those seen in house sparrows to slight disruption of behavioral locomotor rhythmicity (Gwinner and Brandstätter 2001). Circadian rhythms of locomotor behavior in columbiform and galliform birds are little or not affected at all by PINX (Ebihara et al. 1984; Underwood and Siopes 1984).

Even so, the pineal gland represents both the capacity for rhythmicity and time of day. In an elegant experiment, Zimmerman and Menaker (1979) transplanted pineal glands from two groups of house sparrows into the anterior chambers of the eye of PINX, arrhythmic sparrows maintained in DD. The first group of donor birds were entrained to an early LD cycle, with lights on at midnight, while the second set of donors were entrained to a late LD cycle, with lights on at 11 a.m. Transplantation restored circadian rhythms to both groups of recipients within 1 day. Moreover, birds that received pineal glands from early donors, exhibited an early ϕ_i while the recipients of late donor pineal glands exhibited a late ϕ_i. Thus, the pineal gland is not only necessary for circadian rhythmicity in these birds, but it contains a correlate that confers time of day to recipient birds.

Importantly, the data also suggested that the pineal gland must affect behavior through the secretion of a hormone, because 1 day is not thought to be sufficient for re-innervation of target tissues, wherever they are. That hormone was known even then to be the indoleamine melatonin from earlier work of Lerner and later of Axelrod, Klein, and their co-workers (cf. Klein et al. 1997), who explored the biochemical basis for melatonin biosynthesis in the pineal gland of the chick, *Gallus gallus domesticus*. Research from a large number of investigators have shown that pinealocytes, the photoreceptive, secretory cells of the avian pineal gland, take up the amino acid tryptophan, which is converted to 5-hydroxytryptophan by tryptophan hydroxylase (TrH; EC 1.14.16.4) (Chong et al. 2000) and then decarboxylated to produce serotonin (5HT) by aromatic L-amino acid decarboxylase (AAADC; EC 4.1.1.28). During the night in LD and subjective night in DD, 5HT is converted to *N*-acetylserotonin (NAS) by arylalkylamine (or serotonin)-*N*-acetyltransferase (AANAT; EC 2.3.1.87) (Bernard et al. 1997b). NAS is then converted to melatonin by hydroxyindole-*O*-methyltransferase (HIOMT; EC 2.1.1.4) (Voisin et al. 1992). The genes encoding each of these enzymes have been isolated, cloned, and sequenced in several avian species. In chick, at least, TrH, AANAT and HIOMT are regulated by both the molecular clockworks within the pinealocytes and directly by light at the transcriptional, translational, and post-translational levels, so that the enzymatic regulation of pineal melatonin is a dynamic, rhythmic process (Klein et al. 1997).

Avian pineal glands contain the circadian clockworks and photoreceptors to generate circadian patterns of melatonin biosynthesis in vitro as well as in vivo, which can be entrained to LD cycles directly (Binkley et al. 1977a). Pineal tissue and pinealocyte cultures express circadian patterns of AANAT activity (Binkley et al. 1977a; Kasal et al. 1979; Wainwright and Wainwright 1979) and melatonin efflux (Takahashi et al. 1980) such that levels are high during the night and low during the day in LD. These rhythms persist for 4–10 days in DD before damping to arrhythmicity. Exposure to light has three effects on cultured pineal rhythms: (1) Light inhibits melatonin biosynthesis. (2) Light increases amplitude and decreases damping, and (3) light phase-shifts the clock within pineal cells (Zatz and Mullen 1988a). In vivo, the avian pineal gland is innervated by post-ganglionic sympathetic nerves, and receives daily and circadian input through release of norepinephrine (NE) during the day and subjective day. Administration of NE to chick pineal glands in vivo and in vitro has two effects on pineal melatonin rhythms: (1) NE inhibits melatonin biosynthesis. (2) NE increases amplitude and decreases damping, but does *not* phase-shift the pineal circadian clock (Cassone and Menaker 1983; Cassone et al. 1986; Zatz and Mullen 1988b).

Using cDNA microarrays of chick pineal gland in vivo and in vitro and retinae in vivo, we have shown that many clock genes are expressed rhythmically in a fashion consistent with circadian patterns of clock gene expression in other model systems, such as *Drosophila* and mice (Bailey et al. 2003, 2004; Karaganis et al. 2008). In vivo, *bmal1*, *bmal2* and *clock* are expressed predominantly during late subjective day/early subjective night, while putative negative elements *per2* and *per3* (*per1* is not present avian genomes) are expressed during the subjective night to early subjective day. Interestingly, *cry1* and *cry2* are expressed during the early to mid-subjective day. In addition, transcripts associated with melatonin biosynthesis and photoreception are expressed rhythmically in patterns that are consistent with previous studies described above. Importantly, the 5′-flanking region of the chicken *aanat* gene contains an E-box (Chong et al. 2000), indicating the CLOCK-BMAL1 dimer may regulate *aanat* expression in the late subjective day/early subjective night. Transfection of COS cells with a reporter construct expressing luciferase and the chicken *aanat* promoter region in the presence of either chicken or human BMAL1 and CLOCK increases luciferase activity (Chong et al. 2000). Mutation of the E-box region dramatically decreases luciferase activity, suggesting that binding of the *aanat* E-box by CLOCK/BMAL1 heterodimers is a critical component of the regulation of melatonin biosynthesis (Chong et al. 2000; Kasal et al. 1979) Interestingly, in pineal glands, but not retinae, many transcripts associated with cytokine biosynthesis, immune function, and lymphopoiesis are both highly and rhythmically expressed in chick pineal gland. These data are similar to those obtained by in situ hybridization in Japanese quail (Yasuo and Yoshimura 2009; Yasuo et al. 2004).

Although in situ hybridization for clock genes suggest that these genes are enriched in the pineal gland, retinae, and other structures associated with clock function (Bailey et al. 2003, 2004; Wiltschko and Wiltschko 2013), quantitative real-time PCR for these transcripts indicate they are widely expressed in other parts of the brain and in the periphery. Daily and circadian patterns of *cry1*, *per3* and *bmal1* expression were observed in chick telencephalon, diencephalon, and optic tectum in the brain, as well as in heart and liver. PINX and EX of chicks decreases the amplitude of these clock gene rhythms but does not completely abolish them (Karaganis et al. 2009).

Interestingly, the photoreceptors in the retinae of the lateral eyes also synthesize and release melatonin in many vertebrate species. In fact, in Japanese quail and domestic pigeon, *Columba livia*, the retinae release almost as much melatonin into the systemic circulation as does the pineal gland and removal of this source by enucleation (EX) or retinectomy in addition to PINX results in arrhythmic circadian locomotor behavior, similar to the effects of PINX alone in passerine birds (Chabot and Menaker 1992; Ebihara et al. 1984; Underwood and Siopes 1984). Thus, the variability of the effects of PINX among birds may in part be due to this retinal component in some species and that it is not the pineal per se but rhythmic melatonin that is important for circadian locomotor behavior. To punctuate this view, rhythmic administration of melatonin to PINX house sparrows and European starlings or to EX/PINX pigeons (Chabot and Menaker 1992; Gwinner et al. 1997; Heigl and Gwinner 1995; Lu and Cassone 1993; Wang et al. 2012) restores a daily pattern of locomotor behavior. This synchronization of locomotor behavior by rhythmic melatonin administration represents entrainment of circadian clockworks in the PINX bird, since melatonin administration in a T-cycle different from 24 h results in systematic changes in the phase relationship (ψ) of melatonin to the onset of locomotor activity (Gwinner et al. 1997; Heigl and Gwinner 1995).

5.6   Sites of Melatonin Action in Birds

In the 1980s and 1990s, high affinity melatonin receptor binding using the radiola-beled agonist $2[^{125}I]$-iodomelatonin (IMEL) (Vakkuri et al. 1984) revealed high densities of IMEL binding in retinal, retinorecipient structures, and visual integra-tive structures in the avian brain as well as peripheral tissues (Dubocovich and Takahashi 1987; Rivkees et al. 1989) (Fig. 5.3). Binding affinity studies indicated kDs in the pM range with high specificity for melatonin itself. Brain structures that bind IMEL included retinorecipient structures in the visual suprachiasmatic nucleus (vSCN) of the circadian system, the ventrolateral and dorsal geniculate nuclei of the thalamofugal visual pathway, the optic tectum of the tectofugal path-way, and the nucleus of the basal optic root (nBOR) or the accessory optic path-way. In all species, integrative structures of the tectofugal pathway such as nucleus rotundus (Rt) and the ectopallium (Ep) also bind IMEL (Cassone et al. 1995; Rivkees et al. 1989). In some but not all species, hyperpallial structures, including the visual Wulst are sites of IMEL binding. In male passerine birds but not females,

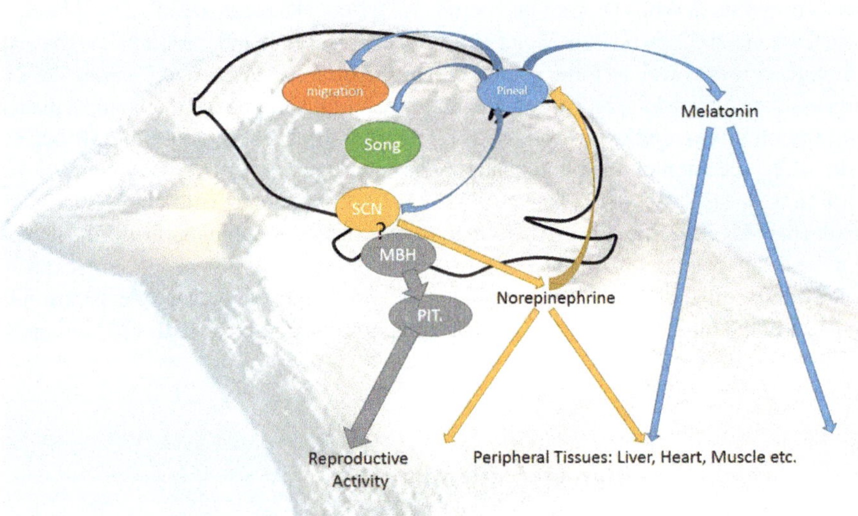

Fig. 5.3 A chorus of clocks. The coordination of circadian and circannual patterns of behavior and physiology is the composite of multiple circadian oscillators and their interactions. At the core are circadian pacemakers in the pineal gland and suprachiasmatic nuclei (SCN), whose interactions maintain each other's stability and self-sustainment through mutually inhibitory activity. Each of these pacemakers may influence downstream processes through entrainment of circadian oscilla-tors controlling and residing in brain (e.g., song, vision, and migration) and peripheral (e.g., liver, heart, and muscle) function. In birds, regulation of primary gonadal activity has been separated from this circadian system with circadian oscillators residing in the mediobasal hypothalamus itself. There is little evidence that these oscillators are affected by pineal melatonin, but it is an open question whether SCM oscillators influence MBH function in the circadian and seasonal control of reproduction

structures associated with bird song learning and control also revealed high affinity IMEL binding (Gahr and Kosar 1996; Whitfield-Rucker and Cassone 1996). These will be discussed in more detail below.

Reppert and colleagues were able to isolate and clone genes encoding two high affinity melatonin receptors; these were designated the Mel_{1A} and Mel_{1C} receptors (Reppert et al. 1995). Independent work isolated partial sequences encoding an ortholog of the Mel_{1B} receptor in the same year (Liu et al. 1995). Subsequent work has confirmed in birds that there are at least three melatonin receptors, the Mel_{1A}, Mel_{1B}, and the Mel_{1C} receptors (Reppert 1997). Pharmacologists working with mammals have named the Mel_{1A} and Mel_{1B} receptors MT1 and MT2 receptors, respectively (Dubocovich et al. 2010). However, mammals do not express the Mel_{1C} receptor subtype, and the Mel_{1A} and Mel_{1B} receptors have not been fully characterized pharmacologically in birds. In this review, we will employ the original nomenclature, the Mel_{1A}, Mel_{1B}, and the Mel_{1C} receptors. All three melatonin receptor subtypes represent seven-transmembrane domain, GTP-binding protein structures and all three are in the G_i GTP-binding protein category, although some cross-talk with G_q has been documented (Reppert 1997). The distributions of these three receptor subtypes are not uniform in chicks, zebra finches, and house sparrows. The Mel_{1A} receptor predominates in central nervous neurons and peripheral tissues (Karaganis et al. 2009; Natesan and Cassone 2002), while the Mel_{1B} receptor is expressed in inner retinal neurons and photoreceptors as well as other central nervous neurons (Natesan and Cassone 2002). In passerines, the Mel_{1B} receptor is the major receptor subtype in song control nuclei but the other two are expressed as well (Bentley et al. 2012; Jansen et al. 2005). The Mel_{1C} receptor, on the other hand, predominates in non-neuronal elements of the central nervous system (Reppert et al. 1995). Culture studies with chick astrocytes (Adachi et al. 2002) show 95–100 % diencephalic astrocytes express the Mel_{1C} receptor, while an overlapping 5–10 % expresses Mel_{1A}. Astrocytes do not appear to express Mel_{1B}. Intriguingly, neither IMEL binding nor strong melatonin receptor expression is present in the tuberal hypothalamus and/or hypophysis (Cassone et al. 1995; Reppert et al. 1995, Rivkees et al. 1989).

5.7 Avian Circadian Organization

Elimination of the pineal gland's photic and neural inputs results in damped circadian patterns of melatonin release, which also can be restored by rhythmic administration of light and/or NE (Cassone and Menaker 1983; Takahashi et al. 1980; Zatz and Mullen 1988a, b). Conversely, removal of the pineal pacemaker in passerine birds at least results in damped rhythmicity in behavioral and physiological output, suggesting whatever is left represents a damped circadian oscillator capable of entraining to light (Gaston and Menaker 1968). Whatever is left is likely at least in part the avian homolog for the mammalian hypothalamic nucleus (SCN) (Cantwell and Cassone 2006a, b), which is a master pacemaker for mammalian circadian organization (cf. Moore and Silver 1998).

In birds, two sets of structures have been associated with SCN function: the medial suprachiasmatic nuclei (mSCN) and the visual suprachiasmatic nuclei (vSCN) (Cantwell and Cassone 2006a, b; Cassone and Moore 1987; Yoshimura et al. 2001). These structures are connected via neuronal projections and are contiguous in terms of their cellular populations, especially in the distribution of astrocytes. The vSCN, but not the mSCN, expresses metabolic and electrical rhythmicity and receives retinohypothalamic (RHT) input. Further, the vSCN, but not the mSCN, contains melatonin receptor binding (Cassone et al. 1995; Lu and Cassone 1993; Reppert et al. 1995; Rivkees et al. 1989), and exogenous melatonin inhibits metabolic activity in the vSCN (Lu and Cassone 1993) but not in the mSCN. Finally, light activates *c*-fos expression in the vSCN, but not in the mSCN (King and Follett 1997). In quail, only the mSCN expresses clock gene rhythmicity (Yasuo and Yoshimura 2009; Yasuo et al. 2002), while in the house sparrow, both structures rhythmically express *per2* rhythms (Abraham et al. 2002). Lesions directed at the mSCN result in arrhythmicity similar to that observed following PINX (Takahashi and Menaker 1982). The most likely scenario is that the functions subsumed by the mammalian SCN are regulated through the integration of both mSCN and vSCN.

Each component, the pineal gland, retinae, and SCN, is integrated dynamically such that overt circadian organization is synchronized to environmental light cycles (LD) and such that internal processes are adaptively orchestrated (Cassone and Menaker 1984; Gwinner and Brandstätter 2001). Pineal (and retinal) melatonin, synchronized to LD cycles via endogenous photopigments, is secreted during the night and inhibits rhythmic metabolism and electrical activity of the vSCN. In turn, as the day approaches, oscillators within the pineal gland and retinae wane in their output, disinhibiting SCN activity. Oscillators within the mSCN and vSCN are active during the day, synchronized by LD via RHT input to the vSCN and possibly extraretinal input to the mSCN. One of the outputs of the vSCN at least is the rhythmic regulation of sympathetic activity, releasing norepinephrine (NE) within many peripheral targets. Among these is the pineal gland, where NE inhibits melatonin biosynthesis and release. It is not completely clear whether sympathetic NE synchronize circadian clockworks within the pineal gland, nor is it clear whether pineal melatonin affects clock gene expression in the SCN in vivo. However, in vitro, rhythmic melatonin administration synchronizes rhythms of both metabolic activity and the expression of both *per2* and *per3* (Paulose et al. 2009).

5.8 Circadian Regulation of Visual System Function

The presence of dense, high affinity melatonin receptors within the retina, retinorecipient structures of all four visual pathways, and visual integrative structures in the brains of multiple avian species (Cassone et al. 1995), strongly suggests visual sensitivity, accommodation, and more complex aspects of visual perception may be regulated on a circadian basis by melatonin. Circadian patterns in electroretinogram (ERG) and visually evoked potentials (VEP) recorded within the TeO show greater response amplitude during the day in LD and during subjective day in DD in both domestic chicks and pigeons (Lu et al. 1995; Peters and Cassone 2005; Wu et al. 2000). In chicks and pigeons, the implicit time of the A-wave, which reflects photoreceptor activity, and B-wave, which reflects inner retinal activation, are greater during the night and subjective night than during the subjective day, while light sensitivity is greatest during the subjective night, reflecting the activity of dark-sensitive rods (Peters and Cassone 2005; Wu et al. 2000). PINX of chicks reduces the amplitude of the circadian rhythm in ERG amplitude (McGoogan and Cassone 1999), and exogenous melatonin injections decreases ERG b-wave during the subjective day to levels similar to those described at night (McGoogan and Cassone 1999); (Peters and Cassone 2005). At this stage, it is not clear whether the effects of melatonin on avian visual system function directly influence physiological processes within visual system structures and/or regulate circadian clocks within the brain. The avian TeO expresses rhythms in clock gene expression in areas (Yasuo et al. 2004) superimposed with high affinity melatonin receptors (Cassone et al. 1995), but the physiological link has yet to be made. Further, it is not clear how higher order visual function is affected by the circadian clock and melatonin.

Interestingly, this pattern of circadian regulation of visual system function, at least at the ERG level is nearly identical in humans with high implicit times during the night and high b-wave amplitudes during the day (Hankins et al. 1998; Nozaki et al. 1983). Further, administration of exogenous melatonin to human volunteers decreases b-wave amplitude during the day (Gagné et al. 2009), similar to the situation in birds (McGoogan and Cassone 1999; Peters and Cassone 2005). The role of melatonin in retinal physiology and pathophysiology is an emerging area of research (Tosini et al. 2012), and birds provide an excellent model for human visual physiology in this regard.

5.9 Seasonal Cycles and Photoperiodism in Birds

Birds living in temperate zone latitudes generally restrict breeding to the spring and summer, maximizing the likelihood that young will be hatched during times at which food is plentiful (Gwinner 1989; Gwinner et al. 1997; Pittendrigh 1993). As such, many primary and secondary sexual characteristics in birds undergo dramatic changes in both form and function. In the short days of winter, gonadal activity and gonad size regress and become inactive, while gonads recrudesce, becoming more active, in response to increased photoperiod. If birds are maintained in long photoperiod, their reproductive systems become insensitive to the photostimulatory effects of long photoperiod and spontaneous regress. This process is called photorefractoriness, and birds remain photorefractory until they are placed in short days for some time to make them photosensitive again (Balthazart et al. 2010; Goldsmith et al. 1989; Kumar et al. 1996; Pittendrigh 1993).

In seasonally reproducing mammals such as hamsters and deer-mice, the nocturnal secretion of pineal melatonin is a critical signal that transduces time of year to the hypothalamo–hypophysial–gonadal axis to control reproductive function (Bartness et al. 1993; Goldman 2001). The duration of pineal melatonin biosynthesis faithfully reflects the duration of the scotoperiod; the duration is long during the long nights of winter and short during summer. These animals also exhibit a seasonal cycle of reproductive activity in which testes in males and estrus cyclicity in females increase as photoperiod increases in the spring. When photoperiod decreases, testes regress and females become anestrus. PINX in several species of hamsters prevents regression of gonads when the hamsters are transferred from long photoperiods to short photoperiods; their reproductive systems become blind to photoperiodic changes. When melatonin is infused into PINX Djungarian hamsters, *Phodopus sungorus*, long durations of melatonin, simulating winter, induce gonadal regression, while short durations, simulating summer, enable recrudescence. The sites for melatonin's activity in this process are a combination of melatonin receptors in the SCN and in the *pars tuberalis* of the hypophysis. Therefore, the *circadian* control of pineal melatonin regulates the *annual* control of reproduction. This brief summary of mammalian seasonality is necessarily short in a review about birds; extensive reviews of these mechanisms can be found elsewhere (cf. Goldman 2001; Hastings and Follett 2001).

Intriguingly, in spite of the fact that the rhythmic production of melatonin is critical for the expression of circadian locomotor rhythms in birds, melatonin does not affect seasonal changes in primary reproductive function in these species. As in mammals, pineal melatonin levels faithfully reflect the length of the scotoperiod both in vivo and in vitro (Binkley et al. 1977b; Brandstätter et al. 2000). However, PINX and/or EX of several species of birds has little effect on seasonal changes in gonad size or activity (Bentley 2001; Kumar et al. 2002; Schleussner and Gwinner 1989; Sharp 2005; Siopes 1983). Moreover, administration of exogenous melatonin of different durations has little effect on primary reproductive function (Cassone et al. 2008; Meddle and Follett 1997). This corresponds to the relative absence of melatonin receptor activity in the tuberal hypothalamus and hypophysis (Cassone et al. 1995), in stark contrast to the situation in seasonally reproducing mammals, where IMEL binding and melatonin receptor expression in *pars tuberalis* is a major site of melatonin action (Goldman 2001; Hastings and Follett 2001).

Yoshimura and colleagues have instead pointed to circadian clock function within the mediobasal hypothalamus (MBH) itself of Japanese quail controlling photoperiodic time measurement for reproductive function (Yoshimura 2010). Earlier studies had shown that lesion of the MBH blocked testicular recrudescence in response to lengthening photoperiods and illumination of this area had resulted in excitation of the tuberal hypothalamus and testicular growth (Foster et al. 1985; Kumar et al. 2010; Nakao et al. 2007). As stated above, the MBH of quail and the PMM of turkeys have been shown to express both OPN4 and OPN5 in cerebrospinal fluid (CSF) contacting neurons (Kang et al. 2010; Nakane et al. 2010; Nakao et al. 2007). Noting that PINX or EX or even SCN lesion had little effect on photoperiodic regulation of gonadal function, Yoshimura's group identified rhythmic expression of the clock genes in the MBH and hypothesized that this structure contained the circadian pacemaker associated with photoperiodic time measurement

(Nakao et al. 2008a; Ono et al. 2009). Using differential subtractive hybridization, they found type 2 iodothyronine deiodinase (*Dio2*) to be induced in the MBH by a light pulse associated with long-day induction. *Dio2* encodes an enzyme that catalyzes the conversion of inactive thyroxine (T4) to active triiodothyroine (T3) (Nakao et al. 2008a, b; Yasuo and Yoshimura 2009; Yasuo et al. 2006). Later, they showed that type 3 iodothyronine deiodinase (*Dio3*), which inactivates T3, was induced in the MBH by exposure to short days. The scenario they envision is that photoperiod is perceived by photopigments in the MBH that entrain a circadian oscillator within the MBH. In long days the circadian clock induces *Dio2*, while short days induce *Dio3*. Indeed, thyroidectomy induces gonadal recrudescence in starlings and Japanese quail (Dawson et al. 1986; Follett and Nicholls 1985) and injection of T3 into the MBH induces quail gonadal growth (Follett and Nicholls 1985; Yoshimura et al. 2003). These authors envision an external coincidence model in which circadian oscillators within the MBH are entrained by photoperiod colocalized in that structure. Gonadal recrudescence in starlings a ϕ_{pi}, *Dio2* is induced, enabling a metabolic cascade in response to T3 hormone, and gonadal induction occurs. It is not clear, at this stage, what molecular components link the circadian clock to *Dio2* or *Dio3*. The MBH of quail and the PMM of turkeys rhythmically express clock genes (Ikegami and Yoshimura 2012; Ikegami et al. 2009; Leclerc et al. 2010). In humans, *Dio2* is regulated by CCAAT/enhancer-binding proteins, and *per2* is known to be a target of these transcription factors as well (Gegear et al. 2010; Thoennissen et al. 2012). Perhaps, this link may provide clues to an analogous mechanism in birds.

5.10 Avian Song Production and Perception

In parallel to primary reproductive processes, both the probability of a male bird to sing in response to a given stimulus as well as the size and complexity of the song control nuclei within its brain vary depending upon the time of year (Dawson et al. 1986; Marler 2004; Whitfield-Rucker and Cassone 2000). The song control system of oscine passeriform birds is a specialized network of brain nuclei involved in singing and song learning. This system receives auditory input from ascending, primary auditory pathways beginning in the cochlear nuclei (Co), which project to the lateral dorsal mesencephalic nuclei (MLd). MLd in turn projects to the thalamic *nucleus ovoidalis* (Ov), which in turn projects to Field L in the forebrain. Song processing begins in secondary auditory areas in the caudal mesopallium (cM) and caudomedial nidopallium (NCM), which interact with the anterior forebrain pathway for song plasticity and learning. This pathway includes the hyperstriatum ventrale pars caudalis (HVC) in the dorsal forebrain, which projects to Area X, whose projections form a loop between the dorsolateral thalamus (DLM) and the lateral magnocellular nucleus of the anterior nidopallium (LMAN). Then, both HVC and LMAN project to the robust nucleus of the archipallium (RA), which forms the song motor output pathway (Ball and Balthazart 2010; Ball et al. 2004; Nottebohm et al. 1976).

This system enables birds to process a complex species-specific identification of both self and conspecifics as well as other dynamics in birds' acoustic environments (competitors, prey, predators etc.). These acoustic environments as well as their reproductive and survival relevance are not constant. Territories change hands and the available range of mates fluctuates depending on the time of day or time of year. Auditory sensitivities to these signals must be tuned to appropriate conspecific and perhaps interspecific signals. At the same time, the structure and behavior of song itself must be tuned to reproductively appropriate situations.

Birds from temperate zones in the short days of winter possess small HVC, RA, and depending on the species other structures in the system. When birds are photo-stimulated as photoperiod increases, the song control nuclei grow in parallel with the growth of the testes, and, as they become photorefractory, these structures regress in both size and complexity. Seasonal fluctuations in androgens and estrogens appear to be critical for changes in song control (Ball and Balthazart 2010; Ball et al. 2004; Balthazart et al. 2010). Song control nuclei contain both androgen receptors (AR) and both estrogen receptor subtypes (ERα and ERβ), as well as aromatase (AA), capable of converting androgens into biologically active estrogens. Further, the rate of male song in several songbird species increases when testosterone levels are high, castration decreases this rate, and hormone replacement reestablishes vernal song patterns. Multiple studies have linked the activity of gonadal steroids to both song behavior and regulation of song control nuclei, which have been reviewed extensively (cf. Balthazart et al. 2010).

Even so, a few studies have indicated gonad-independent regulation of song control nuclei under different photoperiods. In both American tree sparrows and house sparrows, HVC and RA increase in size in response to a change in photoperiod from a short day to long day (Bernard et al. 1997a; Gulledge and Deviche 1998; Whitfield-Rucker and Cassone 2000). Castrated birds in these studies also exhibit photostimulated song control nuclei, although the level of induction is not as great as in the sham-operated birds. In the house sparrow study, castrated birds also showed a blunted photorefractory phase. Together, the data indicate that, while the song system certainly responds to the seasonal changes in gonadal steroids, regulation of song control nuclei comprises a gonad-independent as well as a gonad-dependent aspect.

5.10.1 Role of Pineal Melatonin in Song Control

The observation that song control nuclei express melatonin receptors points to a role for melatonin in song behavior and in the growth and regression of song control nuclei (Whitfield-Rucker and Cassone 1996). Melatonin binding and Mel$_{1b}$ receptor mRNA is affected by changing photoperiod, but not by castration (Bentley et al. 2012; Whitfield-Rucker and Cassone 2000). Bentley et al. (1999) have shown that continuous administration exogenous melatonin attenuated the long-day-induced volumetric increase in HVC and also decreased the volume of another song-control nucleus, Area X, in European starlings. This effect was independent of the birds' reproductive state. However, seasonal changes in binding and expression of Mel_{1b} receptor mRNA in Area X was affected by social conditions and laboratory captivity (Bentley et al. 2012). Further, Jansen et al. (2005) have shown that neuronal firing in RA is decreased in vitro by administration of melatonin and that administration of the Mel_{1B} antagonist luzindole decreases song behavior in male zebra finches the next day. However, neither of these studies addresses the rhythmic role of melatonin in the control of song control nuclei and/or song behavior.

We have asked whether PINX of house sparrows and subsequent rhythmic administration of summer-like (short duration) or winter-like (long duration) melatonin affected song control nuclei HVC and RA (Cassone et al. 2008). All birds were PINX and placed in constant darkness (DD), whereupon they all became arrhythmic in locomotor behavior for 30 days. Birds that received no melatonin administration in DD remained arrhythmic, exhibited small, regressed testes and small regressed HVC and RA volumes. Birds that received either the short duration or the long duration melatonin entrained to the melatonin cycle but still exhibited small testes and small song control nuclei. Therefore, we repeated the experiment in intact birds maintained in constant light (LL). In this case, all birds became arrhythmic as above, but all birds exhibited large testes. They also exhibited vocal behavior. Birds that received no melatonin or the short duration melatonin cycle (who entrained to the regime) exhibited large HVC and RA, but birds that received the long duration melatonin entrained to the melatonin regime but showed regressed HVC and RA. This showed that melatonin affects song control nuclei independently of gonadal state, but does melatonin affect song behavior?

As stated above, application of melatonin to brain slices from zebra finches containing RA decreases RA firing rate, and administration of the Mel_{1B} antagonist luzindole affects song duration the next day (Jansen et al. 2005). Derégnaucourt et al. (2012) recently have reported that PINX shortens the duration of song in an LD cycle. Recently, we have asked whether the pineal gland and melatonin cycles influence the daily and circadian pattern of song behavior in zebra finches (Yasuo et al. 2004). In contrast to the Derégnaucourt et al. (2012) study, we did not observe a shortening of the duration of song. However, we did observe effects of both PINX and melatonin under constant environmental conditions. Control birds expressed circadian rhythms of locomotion, song, and call in constant dim light (dimLL) for up to 30 days. PINX birds' locomotion, song, and call in dimLL gradually became arrhythmic, but they became arrhythmic at different rates, such that song became arrhythmic more rapidly than did either locomotor or call rhythms. When PINX birds were placed in a melatonin cycle of 10 h melatonin/24 h, they entrained all three outputs. However, locomotion entrained in 4 days, song in 15 days, and calls took 26 days to entrain. When PINX and control birds were placed in LL, locomotor, song, and call rhythms all became arrhythmic, although PINX birds became arrhythmic more rapidly. In sham birds, song became arrhythmic most rapidly, but in PINX birds locomotor behavior damped most rapidly followed by song and then call behavior. Interestingly, all behaviors in both PINX and sham birds were entrained to melatonin cycles in LL almost immediately. The bottom line for these studies is that locomotor behavior, song production, and call behavior are all under the regulation of the circadian clock and pineal melatonin. However, they are regulated differentially, perhaps via separable circadian clocks.

5.11 Biological Clocks, Migration and Navigation

In migratory birds, circannual and circadian rhythms play integral roles in both the timing of migratory behavior and the capacity for orientation during migration (Gulledge and Deviche 1998; Gwinner 2003; Kramer 1959). Endogenously generated circannual rhythms regulate the initiation of both the vernal and autumnal migrations. Many species of migratory birds, including sparrows, finches, and warblers, maintained in captivity under natural photoperiodic conditions spontaneously exhibit two bouts of migratory behavior in which normally diurnal birds express nocturnal activity called *Zugunruhe* or "migratory restlessness" at the same times of year that coincide with natural migration (Fusani et al. 2009; Stapput et al. 2010). When birds are maintained for more than a year in a constantly equinoctial photoperiod (LD 12:12), they express two bouts of *Zugunruhe* approximately 6 months apart, strongly indicating an internally generated temporal program produces these migratory behaviors.

There is some evidence that a circadian clock and pineal melatonin are important for the expression of *Zugunruhe* (Kumar et al. 2002, 2010). First, when migratory white-throated sparrows, *Zonotrichia albicollis*, *Sylvia* warblers, and other migratory birds are placed in DD or dimLL, *Zugunruhe* is expressed periodically with a stable ψ with circadian patterns of diurnal activity (ψ_{zd}) (Bartell and Gwinner 2005; McMillan 1969, 1970). Secondly, PINX of white-crowned sparrows abolishes both the free-running τ of circadian diurnal activity but also the expression of *Zugunruhe* (McMillan 1969). Thirdly, the pattern of melatonin secretion is altered during *Zugunruhe*. During periods of nocturnal migratory restlessness in the lab, the amplitude of nocturnal melatonin titers is reduced (Fusani and Gwinner 2001; Fusani et al. 2011; Gwinner et al. 1987). If garden warblers, *Sylvia borin*, are disturbed during *Zugunruhe* by feeding, for example, the melatonin amplitude rebounds (Fusani and Gwinner 2004). If melatonin is administered to bramblings, *Fringilla montifringilla*, during *Zugunruhe*, nocturnal activity is suppressed (Fusani and Gwinner 2001).

At this stage, the site for the circadian oscillators responsible for *Zugunruhe* is not known. However, there is evidence that *Zugunruhe* is regulated by circadian oscillators that are separate from, albeit coupled to, the circadian clock system controlling diurnal locomotor behavior. Bartell and Gwinner (2005) have shown that garden warblers placed in a skeleton photoperiod of 11.5D: 1L: 10.55D: 1L entrain their diurnal locomotor behavior with a 24 h period. However, in some birds, *Zugunruhe* free-ran with a long τ, indicating that a separate circadian oscillator must control *Zugunruhe* vs. diurnal locomotor behavior. The authors suggest that the internal coincidence model best explains the data. In this scenario, the circadian clock(s) controlling normal diurnal behavior maintain a stable ψ_{ie} with the photoperiod.

Another clock or population of clocks associated with *Zugunruhe*, the hypothesis goes, is unexpressed unless and until their phase relationship with the circadian clocks controlling diurnal behavior coincide with the solstices. Then, *Zugunruhe* is induced. It is an intriguing idea that awaits the identification of specific structures or molecules associated with *Zugunruhe*.

Perhaps, one of the greatest mysteries associated with birds is their ability to navigate over great distances with remarkable accuracy during semiannual migrations or even in their day to day navigations within their home ranges (Gwinner 2003). There are many theories about avian orientation and navigation, including sensitivities to barometric pressures, olfactory cues, and the Earth's magnetic field (Wiltschko and Wiltschko 2013). These are neither mutually exclusive nor unrelated (Bischof et al. 2011). One of the most enduring ideas is the sun compass or time-compensated navigation. Early work by von Frisch in honeybees has shown that bees learn to visit a food source at a particular time of day and convey that information to the hive via its famous dance (von Frisch 1946/1994). Using this information, bees navigate to the food source by interpolating the position of the sun's azimuth with the bees' internal sense of time. The fact that an internal sense of time (*Zeitgedachnis*) was required for this navigation was demonstrated by shifting the light: dark cycles of the hive by 6 h (or 90° of the 24 h day) and showing that the bees consistently make a 90° error in orientation. This capacity has also been demonstrated in several species of birds, including European starlings and domestic pigeons (Hoffmann 1960; Kramer 1959). One question that has arisen from these studies was whether the internal sense of time was one and the same as the biological circadian clock. In an elegant series of studies, Hoffmann (Hoffmann 1960) showed that starlings maintained in constant dim light exhibited gradual shifting in the birds' orientation with a period equivalent to the period of their free-running rhythms in locomotion. These data suggested that the orientation clock and the circadian clock shared clock properties, but the studies were conducted before any neuroscientific or molecular components of the clock had been identified.

An intriguing link between the circadian clock and avian magnetoreception has been made by several physicists in theoretical quantum coherence properties of cryptochromes. As noted above, cryptochromes are both photopigments in circadian entrainment in insects but a component of the negative element arm of the circadian molecular clockworks (Emery et al. 1998). In birds, *cry1* and *cry2* are expressed rhythmically in the photoreceptor layer of retina of the lateral eyes (Bailey et al. 2004). Cryptochrome photosensitivity derives from light activated transfer of an electron from a pterin cofactor to the flavin moiety on the cryptochrome, but this process has also been shown to be modified by changes in magnetic inflection and amplitude. In insect species such as *Drosophila* and monarch butterflies, *Danaus monarchus*, the ability to perceive changes in magnetic field orientation are both dependent on ultraviolet-A/blue light and the expression of *cry* (Emery et al. 1998; Reppert 2007). Intriguingly, if human *cry2* is expressed in *cry*-knockout *Drosophila*, magnetoreception is restored. Thus, it is possible that even humans have the capacity to detect magnetic fields (Gaston and Menaker 1968; Reppert 2007). In European robins, *Erithracus rubecula*, and garden warblers, magnetic sensitivity is light dependent with maximal sensitivity in the UVA/blue range, consistent with light sensitivity of cryptochromes (Follett et al. 1974; Paulose et al. 2009; Reppert 2007). At this stage, however, it is not clear that cryptochromes actually confer sensitivity to magnetic fields in birds nor is it clear whether the role of cryptochromes derive from their role as components of the molecular clock or whether the molecular clockworks are involved directly in time-compensated navigation Pohl (2000).

5.12 Conclusion: A Chorus of Clocks

This review began with an allegorical discussion of a dawn chorus of birds singing on a spring morning. In this allegory, some birds sing at their own pace, referring directly to their own biological clock entrained independently to the dawn. Others, the chickadees, perhaps, sing and call according to their own circadian clock but are also coupled to each other, responding to a conspecific just over the hedge. Still others, the jays, perhaps, respond to their clocks, their fellow jays and also to some of the other species of the garden, coupled to the community.

This is one way to envision the avian circadian system and perhaps vertebrate circadian clocks in general (Granados-Fuentes and Herzog 2013). At the heart of it, all rhythmic processes are either entrained directly or indirectly to the photoperiod, the rising of the dawn sun. The central core of the avian circadian system, the pineal gland, the SCN and the retina, express robust molecular clockworks, but each component is not capable of self-sustained circadian rhythms. Instead, the pineal gland and retinae maintain SCN rhythmicity by inhibiting SCN output during subjective night through the actions of melatonin. These structures, being oscillators, wane in their outputs as dawn approaches, disinhibiting SCN output. We do not know if SCN activity affects retinal function, but it inhibits pineal melatonin biosynthesis during the day, and the system is maintained in a yin and yang mutually inhibitory relationship. Each of these core oscillators are also pacemakers and influence downstream processes in the brain and periphery via neuroendocrine (pineal and retinal melatonin) and neural (SCN modulation of sympathetic tone) output pathways. In turn, downstream processes possess their own circadian clockworks and entrain their tissue-specific rhythms to one, the other or both outputs of the circadian system. In this way, the system orchestrates coordinated physiological processes that maintain efficient circadian activities (Fig. 5.3). This is the core of the neuroendocrine loop model for avian circadian organization (Cassone and Menaker 1984).

Complementing these processes, other circadian oscillators in the tuberal hypothalamus, song control nuclei, visual system, and undescribed structures associated with migration and navigation maintain stable ψ with the circadian system, responding to its output in differential fashion. Reproductive function appears to be independent, but song and visual system structures are clearly regulated by melatonin, entraining to its nocturnal signal. And thus, it appears to be a cacophony, but it is not. It is a symphony or a chorus at least, coordinated in some processes but independent in others, each tuned to the rising sun and the dawn.

Acknowledgments I'd like to thank Clifford Harpole, Jiffin Paulose, and Ye Li for helpful comments. My lab has been supported by NIH P01 NS39546 and the University of Kentucky.

References

Abraham U, Albrecht U, Gwinner E et al (2002) Spatial and temporal variation of passer Per2 gene expression in two distinct cell groups of the suprachiasmatic hypothalamus in the house sparrow (*Passer domesticus*). Eur J Neurosci 16:429–436

Adachi A, Natesan AK, Whitfield-Rucker MG et al (2002) Functional melatonin receptors and metabolic coupling in cultured chick astrocytes. Glia 39:268–278

Aschoff J (1979) Circadian rhythms: influences of internal and external factors on the period measured in constant conditions. Z Tierpsychol 49:225–249

Bailey MJ, Cassone VM (2005) Melanopsin expression in the chick retina and pineal gland. Brain Res Mol Brain Res 134:345–348

Bailey MJ, Beremand PD, Hammer R et al (2003) Transcriptional profiling of the chick pineal gland, a photoreceptive circadian oscillator and pacemaker. Mol Endocrinol 17:2084–2095

Bailey MJ, Beremand PD, Hammer R et al (2004) Transcriptional profiling of circadian patterns of mRNA expression in the chick retina. J Biol Chem 279:52247–52254

Ball GF, Balthazart J (2010) Seasonal and hormonal modulation of neurotransmitter systems in the song control circuit. J Chem Neuroanat 39:82–95

Ball GF, Auger CJ, Bernard DJ et al (2004) Seasonal plasticity in the song control system: multiple brain sites of steroid hormone action and the importance of variation in song behavior. Ann N Y Acad Sci 1016:586–610

Balthazart J, Charlier TD, Barker JM et al (2010) Sex steroid-induced neuroplasticity and behavioral activation in birds. Eur J Neurosci 32:2116–2132

Bartell PA, Gwinner E (2005) A separate circadian oscillator controls nocturnal migratory restlessness in the songbird *Sylvia borin*. J Biol Rhythms 20:538–549

Bartness TJ, Powers JB, Hastings MH et al (1993) The timed infusion paradigm for melatonin delivery: what has it taught us about the melatonin signal, its reception, and the photoperiodic control of seasonal responses? J Pineal Res 15:161–190

Bell-Pedersen D, Cassone VM, Earnest DJ et al (2005) Circadian rhythms from multiple oscillators: lessons from diverse organisms. Nat Rev Genet 6:544–556

Benoit J, Assenmacher I (1954) Comparative sensitivity of superficial and deep receptors in photosexual reflex in duck. C R Hebd Seances Acad Sci 239:105–107

Bentley GE (2001) Unraveling the enigma: the role of melatonin in seasonal processes in birds. Microsc Res Tech 53(1):63–71

Bentley GE, Van't Hof TJ, Ball GF (1999) Seasonal neuroplasticity in the songbird telencephalon: a role for melatonin. Proc Natl Acad Sci USA 96(8):4674–4679

Bentley GE, Perfito N, Calisi RM (2012) Season and context-dependent sex differences in melatonin receptor activity in a forebrain song control nucleus. Horm Behav 63:829–835. doi:10.1016/j.yhbeh.2012.11.015, pii: S0018-506X(12)00287-5

Bernard DJ, Wilson FE, Ball GF (1997a) Testis-dependent and -independent effects of photoperiod on volumes of song control nuclei in American tree sparrows (Spizella arborea). Brain Res 760:163–169

Bernard M, Iuvone PM, Cassone VM et al (1997b) Avian melatonin synthesis: photic and circadian regulation of serotonin N-acetyltransferase mRNA in the chicken pineal gland and retina. J Neurochem 68:213–224

Binkley S, Riebman JB, Reilly KB (1977a) Timekeeping by the pineal gland. Science 197:1181–1183

Binkley S, Stephens JL, Riebman JB et al (1977b) Regulation of pineal rhythms in chickens: photoperiod and dark-time sensitivity. Gen Comp Endocrinol 32(4):411–416

Bischof HJ, Nießner C, Peichl L et al (2011) Avian ultraviolet/violet cones as magnetoreceptors: the problem of separating visual and magnetic information. Commun Integr Biol 4:713–716

Brandstätter R, Kumar V, Abraham U et al (2000) Photoperiodic information acquired and stored in vivo is retained in vitro by a circadian oscillator, the avian pineal gland. Proc Natl Acad Sci USA 97:12324–12328

Bünning E (1969) Common features of photoperiodism in plants and animals. Photochem Photobiol 9:219–228

Cantwell EL, Cassone VM (2006a) Chicken suprachiasmatic nuclei: I. Efferent and afferent connections. J Comp Neurol 496:97–120

Cantwell EL, Cassone VM (2006b) Chicken suprachiasmatic nuclei: II. Autoradiographic and immunohistochemical analysis. J Comp Neuro 499:442–457

Caprioli M, Ambrosini R, Boncoraglio G et al (2012) Clock gene variation is associated with breeding phenology and maybe under directional selection in the migratory barn swallow. PLoS One 7:e35140. doi:10.1371/journal.pone.0035140

Cassone VM, Menaker M (1983) Sympathetic regulation of chicken pineal rhythms. Brain Res 272:311–317

Cassone VM, Menaker M (1984) Is the avian circadian system a neuroendocrine loop? J Exp Zool 232:539–549

Cassone VM, Moore RY (1987) Retinohypothalamic projection and suprachiasmatic nucleus of the house sparrow, *Passer domesticus*. J Comp Neurol 266:171–182

Cassone VM, Westneat DF (2012) The bird of time: cognition and the avian biological clock. Front Mol Neurosci 5:32

Cassone VM, Takahashi JS, Blaha CD et al (1986) Dynamics of noradrenergic circadian input to the chicken pineal gland. Brain Res 384:334–341

Cassone VM, Brooks DS, Kelm TA (1995) Comparative distribution of 2[^{125}I]iodomelatonin binding in the brains of diurnal birds: out-group analysis with turtles. Brain Behav Evol 45:241–256

Cassone VM, Bartell PA, Earnest BJ et al (2008) Duration of melatonin regulates seasonal changes in song control nuclei of the house sparrow, Passer domesticus: independence from gonads and circadian entrainment. J Biol Rhythms 23:49–58

Chabot CC, Menaker M (1992) Effects of physiological cycles of infused melatonin on circadian rhythmicity in pigeons. J Comp Physiol A 170:615–622

Chaurasia SS, Rollag MD, Jiang G et al (2005) Molecular cloning, localization and circadian expression of chicken melanopsin (Opn4): differential regulation of expression in pineal and retinal cell types. J Neurochem 92:158–170

Chong NW, Bernard M, Klein DC (2000) Characterization of the chicken serotonin N-acetyltransferase gene. Activation via clock gene heterodimer/E box interaction. J Biol Chem 275:32991–32998

Davies WI, Turton M, Peirson SN et al (2012) Vertebrate ancient opsin photopigment spectra and the avian photoperiodic response. Biol Lett 8:291–294

Davies WL, Hankins MW, Foster RG (2010) Vertebrate ancient opsin and melanopsin: divergent irradiance detectors. Photochem Photobiol Sci 9:1444–1457

Dawson A, Goldsmith AR, Nicholls TJ et al (1986) Endocrine changes associated with the termination of photorefractoriness by short daylengths and thyroidectomy in starlings (*Sturnus vulgaris*). J Endocrinol 110:73–79

Derégnaucourt S, Saar S, Gahr M (2012) Melatonin affects the temporal pattern of vocal signatures in birds. J Pineal Res 53:245–258

Dor R, Cooper CB, Lovette IJ et al (2012) Clock gene variation in *Tachycineta swallows*. Ecol Evol 2(1):95–105

Dubocovich ML, Takahashi JS (1987) Use of 2-[125I]iodomelatonin to characterize melatonin binding sites in chicken retina. Proc Natl Acad Sci USA 84:3916–3920

Dubocovich ML, Delagrange P, Krause DN et al (2010) International Union of Basic and Clinical Pharmacology. LXXV. Nomenclature, classification, and pharmacology of G protein-coupled melatonin receptors. Pharmacol Rev 62:343–380

Ebihara S, Uchiyama K, Oshima I (1984) Circadian organization in the pigeon, *Columba livia.*
J Comp Physiol A 154:59–69
Eckel-Mahan K, Sassone-Corsi P (2013) Metabolism and the circadian clock converge. Physiol
Rev 93:107–135
Emery P, So WV, Kaneko M et al (1998) CRY, a *Drosophila* clock and light-regulated cryptochrome,
is a major contributor to circadian rhythm resetting and photosensitivity. Cell 95:669–679
Follett BK, Mattocks PW Jr et al (1974) Circadian function in the photoperiodicinduction of
gonadotropin secretion in the white-crowned sparrow, *Zonotrichialeucophrys gambelii.* Proc
Natl Acad Sci USA 71:1666–1669
Follett BK, Nicholls TJ (1985) Influences of thyroidectomy and thyroxine replacement on photoperiodically controlled reproduction in quail. J Endocrinol 107:211–221
Follett BK, Pearce-Kelly AS (1991) Photoperiodic induction in quail as a function of the period of
the light-dark cycle: implications for models of time measurement. J Biol Rhythms
6:331–341
Follett BK, Kumar V, Juss TS (1992) Circadian nature of the photoperiodic clock in Japanese
quail. J Comp Physiol A 171:533–540
Foster RG, Follett BK, Lythgoe JN (1985) Rhodopsin-like sensitivity of extra-retinal photoreceptors
mediating the photoperiodic response in quail. Nature 313:50–52
Foster RG, Hankins MW (2002) Non-rod, non-cone photoreception in the vertebrates. Prog Retin
Eye Res 21:507–527
Fusani L, Gwinner E (2001) Reduced amplitude of melatonin secretion during migration in the
blackcap (Sylvia atricapilla). In: Goos H et al (eds) Perspective in comparative endocrinology:
unity and diversity. Moduzzi press, Bologna, pp 295–300
Fusani L, Gwinner E (2004) Simulation of migratory flight and sopover affects night levels of
melatonin in a nocturnal migrant. Proc R Soc Lond B271:205–211
Fusani L, Cardinale M, Schwabl I, Goymann W (2011) Food availability but not melatonin affects
nocturnal restlessness in a wild migrating passerine. Horm Behav 59:187–192
Fusani L, Cardinale M, Carere C et al (2009) Stopover decision during migration: physiological
conditions predict nocturnal restlessness in wild passerines. Biol Lett 5:302–305
Gahr M, Kosar E (1996) Identification, distribution, and developmental changes of a melatonin
binding site in the song control system of the zebra finch. J Comp Neurol 367:308–318
Gagné AM, Danilenko KV, Rosolen SG et al (2009) Impact of oral melatonin on the electroretino-
gram cone response. J Circadian Rhythms 7:14
Gaston S, Menaker M (1968) Pineal function: the biological clock in the sparrow? Science
160:1125–1127
Gegear RJ, Foley LE, Casselman A et al (2010) Animal cryptochromes mediate magnetoreception
by an unconventional photochemical mechanism. Nature 463(7282):804–807
Goldman BD (2001) Mammalian photoperiodic system: formal properties and neuroendocrine
mechanisms of photoperiodic time measurement. J Biol Rhythms 16:283–301
Goldsmith AR, Ivings WE, Pearce-Kelly AS et al (1989) Photoperiodic control of the development
of the LHRH neurosecretory system of European starlings (Sturnus vulgaris) during puberty
and the onset of photorefractoriness. J Endocrinol 122:255–268
Granados-Fuentes D, Herzog ED (2013) The clock shop: coupled circadian oscillators. Exp Neurol
243:21–27
Gulledge CC, Deviche P (1998) Photoperiod and testosterone independently affect vocal control
region volumes in adolescent male songbirds. J Neurobiol 36:550–558
Gwinner E (1989) Photoperiod as a modifying and limiting factor in the expression of avian circannual rhythms. J Biol Rhythms 4:237–250
Gwinner E (2003) Circannual rhythms in birds. Curr Opin Neurobiol 13(6):770–778
Gwinner E, Brandstätter R (2001) Complex bird clocks. Philos Trans R Soc Lond B Biol Sci
356:1801–1810
Gwinner E, Subbaraj R, Bluhm CK et al (1987) Differential effects of pinealectomy on circadian
rhythms of feeding and perch hopping in the European starling. J Biol Rhythms 2(2):109–120

Gwinner E, Hau M, Heigl S (1997) Melatonin: generation and modulation of avian circadian rhythms. Brain Res Bull 44:439–444

Hankins MW, Jones RJ, Ruddock KH (1998) Diurnal variation in the b-wave implicittime of the human electroretinogram. Vis Neurosci 15:55–67

Hastings MH, Follett BK (2001) Toward a molecular biological calendar? J Biol Rhythms 16:424–430

Heigl S, Gwinner E (1995) Synchronization of circadian rhythms of house sparrows by oral melatonin: effects of changing period. J Biol Rhythms 10:225–233

Hoffmann K (1960) Experimental manipulation of the orientational clock in birds. Cold Spring Harb Symp Quant Biol 25:379–387

Ikegami K, Yoshimura T (2012) Circadian clocks and the measurement of daylength in seasonal reproduction. Mol Cell Endocrinol 349:76–81

Ikegami K, Katou Y, Higashi K, Yoshimura T (2009) Localization of circadian clock protein BMAL1 in the photoperiodic signal transduction machinery in Japanese quail. J Comp Neurol 517:397–404

Jansen R, Metzdorf R, van der Roest M et al (2005) Melatonin affects the temporal organization of the song of the zebra finch. FASEB J 19:848–850

Kang SW, Thayananuphat A, Bakken T et al (2007) Dopamine-melatonin neurons in the avian hypothalamus controlling seasonal reproduction. Neuroscience 150:223–233

Kang SW, Leclerc B, Kosonsiriluk S et al (2010) Melanopsin expression in dopamine-melatonin neurons of the premammillary nucleus of the hypothalamus and seasonal reproduction in birds. Neuroscience 170:200–213

Karaganis SP, Kumar V, Beremand PD et al (2008) Circadian genomics of the chick pineal gland in vitro. BMC Genomics 9:206. doi:10.1186/1471-2164-9-206

Karaganis SP, Bartell PA, Shende VR et al (2009) Modulation of metabolic and clock gene mRNA rhythms by pineal and retinal circadian oscillators. Gen Comp Endocrinol 161:179–192

Kasal CA, Menaker M, Perez-Polo JR (1979) Circadian clock in culture: N-acetyltransferase activity of chick pineal glands oscillates in vitro. Science 203:656–658

King VM, Follett BK (1997) c-fos expression in the putative avian suprachiasmatic nucleus. J Comp Physiol A 180:541–551

Klein DC, Coon SL, Roseboom PH et al (1997) The melatonin rhythm-generating enzyme: molecular regulation of serotonin N-acetyltransferase in the pineal gland. Recent Prog Horm Res 52:307–357

Konishi H, Foster RG, Follett BK (1987) Evidence for a daily rhythmicity in the acute release of luteinizing hormone in response to electrical stimulation in the Japanese quail. J Comp Physiol A 161:315–319

Kosonsiriluk S, Mauro LJ, Chaiworakul V et al (2013) Photoreceptive oscillators within neurons of the premammillary nucleus (PMM) and seasonal reproduction in temperate zone birds. Gen Comp Endocrinol 190:149–155. doi:10.1016/j.ygcen.2013.02.015, pii: S0016-6480(13)00081-6

Kramer G (1959) Recent experiments of bird navigation. Ibis 101:399–416

Kumar V, Jain N, Follett BK (1996) The photoperiodic clock in blackheaded buntings (*Emberiza melanocephala*) is mediated by a self-sustaining circadian system. J Comp Physiol A 179:59–64

Kumar V, Singh S, Misra M, Malik S et al (2002) Role of melatonin in photoperiodic time measurement in the migratory redheaded bunting (*Emberiza bruniceps*) and the nonmigratory Indian weaver bird (Ploceus philippinus). J Exp Zool 292:277–278

Kumar V, Wingfield JC, Dawson A et al (2010) Biological clocks and regulation of seasonal reproduction and migration in birds. Physiol Biochem Zool 83:827–835

Leclerc B, Kang SW, Mauro LJ et al (2010) Photoperiodic modulation of clock gene expression in the avian premammillary nucleus. J Neuroendocrinol 22:119–128

Li H, Ferrari MB, Kuenzel WJ (2004) Light-induced reduction of cytoplasmic freecalcium in neurons proposed to be encephalic photoreceptors in chick brain. Brain Res Dev Brain Res 153:153–161

Li H, Kuenzel WJ (2008) A possible neural cascade involving the photoneuroendocrine system (PNES) responsible for regulating gonadal development in an avian species, *Gallus gallus*. Brain Res Bull 76:586–596

Liu F, Yuan H, Sugamori KS et al (1995) Molecular and functional characterization of a partial cDNA encoding a novel chicken brain melatonin receptor. FEBS Lett 374(2):273–278

Lu J, Cassone VM (1993) Daily melatonin administration synchronizes circadian patterns of brain metabolism and behavior in pinealectomized house sparrows, Passer domesticus. J Comp Physiol A 173:775–782

Lu J, Zoran MJ, Cassone VM (1995) Daily and circadian variation in the electroretinogram of the domestic fowl: effects of melatonin. J Comp Physiol A 177(3):299–306

Marler P (2004) Bird calls: their potential for behavioral neurobiology. Ann N Y Acad Sci 1016:31–44

Masuda H, Oishi T, Ohtani M, Michinomae M et al (1994) Visual pigments in the pineal complex of the *Japanese quail*, Japanese grass lizard and bullfrog: immunocytochemistry and HPLC analysis. Tissue Cell 26:101–113

Max M, McKinnon PJ, Seidenman KJ, Barrett RK et al (1995) Pineal opsin: a nonvisual opsin expressed in chick pineal. Science 267:1502–1506

McGoogan JM, Cassone VM (1999) Circadian regulation of chick electroretinogram effects of pinealectomy and exogenous melatonin. Am J Physiol 277:R1418–R1427

McMillan JP (1970) Pinealectomy abolishes circadian rhythm of migratory restlessness. J Comp Physiol 79:105–110

McMillan JP (1969) An endogenous rhythm of migratory restlessness in caged birds. Am Zool 9:1065–1072

Meddle SL, Follett BK (1997) Photoperiodically driven changes in Fos expression within the basal tuberal hypothalamus and median eminence of *Japanese quail*. J Neurosci 17(22):8909–8918

Menaker M (1968) Extraretinal light perception in the sparrow. I Entrainment of the biological clock. Proc Natl Acad Sci USA 59:414–421

Menaker M, Eskin A (1967) Circadian clock in photoperiodic time measurement: a test of the Bünning hypothesis. Science 157:1182–1185

Menaker M, Keatts H (1968) Extraretinal light perception in the sparrow. II. Photoperiodic stimulation of testis growth. Proc Natl Acad Sci USA 60:146–151

Menaker M, Underwood H (1976) Extraretinal photoreception in birds. Photophysiology 23:299–306

Menaker M, Roberts R, Elliott J et al (1970) Extraretinal light perception in the sparrow. 3. The eyes do not participate in photoperiodic photoreception. Proc Natl Acad Sci USA 67:320–325

Moore RY, Silver R (1998) Suprachiasmatic nucleus organization. Chronobiol Int 15:475–487

Nakane Y, Yoshimura T (2010) Deep brain photoreceptors and a seasonal signal transduction cascade in birds. Cell Tissue Res 342:341–344

Nakane Y, Ikegami K, Ono H et al (2010) A mammalian neural tissue opsin (Opsin 5) is a deep brain photoreceptor in birds. Proc Natl Acad Sci USA 107:15264–15268

Nakao N, Yasuo S, Nishimura A et al (2007) Circadian clock gene regulation of steroidogenic acute regulatory protein gene expression in preovulatory ovarian follicles. Endocrinology 148:3031–3038

Nakao N, Ono H, Yoshimura T (2008a) Thyroid hormones and seasonal reproductive neuroendocrine interactions. Reproduction 136:1–8

Nakao N, Ono H, Yamamura T et al (2008b) Thyrotrophin in the pars tuberalis triggers photoperiodic response. Nature 452:317–322

Nanda KK, Hamner KC (1958) Studies on the nature of the endogenous rhythm affecting photoperiodic response in the Biloxi soy bean. Bot Gaz 120:14–25

Natesan AK, Cassone VM (2002) Melatonin receptor mRNA localization and rhythmicity in the retina of the domestic chick, *Gallus domesticus*. Vis Neurosci 19:265–274

Nottebohm F, Stokes TM, Leonard CM (1976) Central control of song in the canary. J Comp Neurol 165:457–486

Nozaki S, Wakakura M, Ishikawa S (1983) Circadian rhythm of human electroretinogram. Jpn J Ophthalmol 27:346–352

Okano T, Yoshizawa T, Fukada Y (1994) Pinopsin is a chicken pineal photoreceptive molecule. Nature 372:94–97

Ono H, Nakao N, Yoshimura T (2009) Identification of the photoperiodic signaling pathway regulating seasonal reproduction using the functional genomics approach. Gen Comp Endocrinol 163:2–6

Paulose JK, Peters JL, Karaganis SP et al (2009) Pineal melatonin acts as a circadian zeitgeber and growth factor in chick astrocytes. J Pineal Res 46:286–294

Peters JL, Cassone VM (2005) Melatonin regulates circadian electroretinogram rhythmsin a dose- and time-dependent fashion. J Pineal Res 38:209–215

Pittendrigh CS (1993) Temporal organization: reflections of a Darwinian clock-watcher. Annu Rev Physiol 55:16–54

Pohl H (2000) Circadian control of migratory restlessness and the effects of melatonin in the brambling (*Fringilla montifringila*). Chronobiol Int 17:471–488

Reppert SM (1997) Melatonin receptors: molecular biology of a new family of G protein-coupled receptors. J Biol Rhythms 12(6):528–531

Reppert SM (2007) The ancestral circadian clock of monarch butterflies: role in time-compensated sun compass orientation. Cold Spring Harb Symp Quant Biol 72:113–118

Reppert SM, Weaver DR, Cassone VM et al (1995) Melatonin receptors are for the birds: molecular analysis of two receptor subtypes differentially expressed in chick brain. Neuron 15:1003–1015

Rivkees SA, Cassone VM, Weaver DR et al (1989) Melatonin receptors in chick brain: characterization and localization. Endocrinology 125:363–368

Rowan W (1926) On photoperiodism, reproductive periodicity and the annual migration of birds and certain fishes. Proc Boston Soc Nat Hist 38:147–189

Schleussner G, Gwinner E (1989) Photoperiodic time measurement during the termination of photorefractoriness in the starling (Sturnus vulgaris L.). Gen Comp Endocrinol 75:54–61

Sharp PJ (2005) Photoperiodic regulation of seasonal breeding in birds. Ann N Y Acad Sci 1040:189–199

Siopes TD (1983) Effect of pinealectomy on broodiness of turkey hens. Poult Sci 62:2245–2248

Soni BG, Foster RG (1997) A novel and ancient vertebrate opsin. FEBS Lett 406:279–283

Stapput K, Güntürkün O, Hoffmann KP et al (2010) Magnetoreception of directional information in birds requires nondegraded vision. Curr Biol 20:1259–1262

Steinmeyer C, Kempenaers B, Mueller JC (2012) Testing for associations between candidate genes for circadian rhythms and individual variation in sleep behavior in blue tits. Genetica 140:219–228

Takahashi JS, Menaker M (1982) Role of the suprachiasmatic nuclei in the circadian system of the house sparrow, *Passer domesticus*. J Neurosci 2:815–828

Takahashi JS, Hamm H, Menaker M (1980) Circadian rhythms of melatonin release from individual superfused chicken pineal glands in vitro. Proc Natl Acad Sci USA 77:2319–2322

Thoennissen NH, Thoennissen GB, Abbassi S et al (2012) Transcription factor CCAAT/enhancer-binding protein alpha and critical circadian clock downstream target gene PER2 are highly deregulated in diffuse large B-cell lymphoma. Leuk Lymphoma 53:1577–1585

Tosini G, Baba K, Hwang CK et al (2012) Melatonin: an underappreciated playerin retinal physiology and pathophysiology. Exp Eye Res 103:82–89

Underwood H, Menaker M (1970) Photoreception in sparrows: response to photoperiodic stimuli. Science 169:893

Underwood H, Siopes T (1984) Circadian organization in *Japanese quail*. J Exp Zool 232:557–566

Vakkuri O, Lämsä E, Rahkamaa E et al (1984) Iodinated melatonin: preparation and characterization of the molecular structure by mass and 1H NMR spectroscopy. Anal Biochem 142:284–289

Van Gelder RN (2001) Non-visual ocular photoreception. Ophthalmic Genet 22:195–205

Voisin P, Guerlotté J, Bernard M et al (1992) Molecular cloning and nucleotide sequence of a cDNA encoding hydroxyindole O-methyltransferase from chicken pineal gland. Biochem J 282:571–576 von Frisch K (1946/1994) The "language" of bees and its utilization in agriculture. Experientia 50:406–413

Wainwright SD, Wainwright LK (1979) Chick pineal serotonin acetyltransferase: a diurnal cycle maintained in vitro and its regulation by light. Can J Biochem 57:700–709

Wang G, Wingfield JC (2011) Immunocytochemical study of rhodopsin-containing putative encephalic photoreceptors in house sparrow, *Passer domesticus*. Gen Comp Endocrinol 170:589–596

Wang G, Harpole CE, Trivedi AK, Cassone VM et al (2012) Circadian regulation of bird song call, and locomotor behavior by pineal melatonin in the zebra finch. J Biol Rhythms 27:145–155

Williams H (2004) Birdsong and singing behavior. Ann N Y Acad Sci 1016:1–30

Wiltschko R, Wiltschko W (2013) The magnetite-based receptors in the beak of birds and their role in avian navigation. J Comp Physiol A Neuroethol Sens Neural Behav Physiol 199:89–98

Whitfield-Rucker MG, Cassone VM (1996) Melatonin binding in the house sparrow song control system: sexual dimorphism and the effect of photoperiod. Horm Behav 30:528–537

Whitfield-Rucker M, Cassone VM (2000) Photoperiodic regulation of the male house sparrow song control system: gonadal dependent and independent mechanisms. Gen Comp Endocrinol 118:173–183

Wu WQ, McGoogan JM, Cassone VM (2000) Circadian regulation of visually evokedpotentials in the domestic pigeon, *Columba livia*. J Biol Rhythms 15:317–328

Yasuo S, Ebihara S, Yoshimura T (2004) Circadian expression of clock gene in theoptic tectum of Japanese quail. Brain Res 1005:193–196

Yasuo S, Yoshimura T (2009) Comparative analysis of the molecular basis of photoperiodic signal transduction in vertebrates. Integr Comp Biol 49:507–518

Yasuo S, Yoshimura T, Bartell PA et al (2002) Effect of melatonin administration on qPer2, qPer3, and qClock gene expression in the suprachiasmatic nucleus of *Japanese quail*. Eur J Neurosci 16:1541–1546

Yasuo S, Watanabe M, Iigo M et al (2006) Molecular mechanism of photoperiodic time measurement in the brain of *Japanese quail*. Chronobiol Int 23:307–315

Yoshimura T (2010) Neuroendocrine mechanism of seasonal reproduction in birds and mammals. Anim Sci J 81(4):403–410

Yoshimura T, Yasuo S, Suzuki Y et al (2001) Identification of the suprachiasmatic nucleus in birds. Am J Physiol Regul Integr Comp Physiol 280:R1185–R1189

Yoshimura T, Suzuki Y, Makino E et al (2000) Molecular analysis of avian circadian clock genes. Brain Res Mol Brain Res 78:207–215

Yoshimura T, Yasuo S, Watanabe M et al (2003) Light-induced hormone conversion of T4 to T3 regulates photoperiodic response of gonads in birds. Nature 426:178–181

Zatz M, Mullen DA (1988a) Two mechanisms of photoendocrine transduction in cultured chick pineal cells: pertussis toxin blocks the acute but not the phase-shifting effects of light on the melatonin rhythm. Brain Res 453:63–71

Zatz M, Mullen DA (1988b) Norepinephrine, acting via adenylate cyclase, inhibits melatonin output but does not phase-shift the pacemaker in cultured chick pineal cells. Brain Res 450:137–143

Zimmerman NH, Menaker M (1979) The pineal gland: a pacemaker within the circadian system of the house sparrow. Proc Natl Acad Sci USA 76:999–1003

Chapter 6
Functional Organization of Circadian Timing System of a Diurnal Primate (Marmoset)

Christiane Andressa da Silva, Carolina Carrijo, Kathiane Santana, and John Fontenele Araujo

Abstract Today it is recognized that changes in circadian rhythmicity can cause cardiovascular disorders, digestive disorders, and endocrine disorders, as well as psychiatric disorders. Such evidence of the relevance of circadian rhythmicity and the changes of this rhythmicity to human health is generally derived from temporal challenges that the human organism faces in today's society. Thus, knowledge of functional mechanisms of circadian rhythmicity regulation is important so we can clinically intervene in the intrinsic disorders of circadian rhythmicity as well as for its application in medicine, both for diagnosis and at the therapeutic level. The most studied animal models are the rodents, which have nocturnal habits and polyphasic sleep, indicating the need for an animal model that is diurnal and presents monophasic sleep, more similar to humans, such as the marmoset, *Callithrix jacchus*. In this chapter we describe in detail the recent experimental findings produced in our laboratory and reported in the scientific literature, which allow us to propose the marmoset as a potential model for the study of circadian rhythmicity.

6.1 Introduction

The expression of circadian rhythmicity has been demonstrated in all living beings, including prokaryotes and beings that live in environments with no light and dark variation, as in caves. This observation suggests that the expression of rhythmicity was an adaptation because it allows anticipation of recurring environmental events.

The ubiquity of circadian rhythmicity clearly shows that it is a phenomenon relevant to living beings, including humans. We know that the expression of circadian rhythmicity occurs from the cellular level to the behavioral and social levels. In past years molecular mechanisms involved in the control of circadian rhythms have been discovered, including genes and proteins that are essential for generation of this

C.A. da Silva • C. Carrijo • K. Santana • J.F. Araujo (✉)
Departamento de Fisiologia, Centro de Biociências, Universidade Federal do Rio Grande do Norte, Natal, RN 59078-970, Brazil
e-mail: araujo@cb.ufrn.br

© Springer International Publishing Switzerland 2015
R. Aguilar-Roblero et al. (eds.), *Mechanisms of Circadian Systems in Animals and Their Clinical Relevance*, DOI 10.1007/978-3-319-08945-4_6

rhythmicity, and so these genes and proteins were called clock genes and clock proteins. Additionally, several studies have recently shown that these genes and proteins also act on other biological functions such as the cell division process. This evidence, combined with epidemiological data that show a higher cancer incidence in some shift workers, led the World Health Organization to consider that the "shift work that leads to disturbances of circadian rhythms is probably carcinogenic." Today it is recognized that changes in circadian rhythmicity can cause cardiovascular disorders, digestive disorders, and endocrine disorders (Marcheva et al. 2013), as well as psychiatric disorders (McClung 2013; Chung et al. 2012).

These evidences of the relevance of circadian rhythmicity to human health and the changes of this rhythmicity are generally derived from temporal challenges that the human organism faces in today's society (Erren 2013). Thus, there is an urgent need for construction of social rules for temporal organization, such as at school and at work, respecting circadian rhythmicity. However, as important is the knowledge of the functional mechanisms of circadian rhythmicity regulation (Sahar and Sassone-Corsi 2013), so we can clinically intervene in the intrinsic disorders of circadian rhythmicity as well as its application in medicine, both at diagnosis and at the therapeutic level (Musiek and Fitzgerald 2013; Scheiermann et al. 2013).

Despite evidence of circadian rhythmicity relevance, the most studied animal models are the rodents, which have nocturnal habits and polyphasic sleep, indicating a need to look for an animal model that is diurnal and presents monophasic sleep, more similar to humans. Here we present some evidence of the circadian rhythmicity of the marmoset, a diurnal primate with monophasic sleep.

The marmoset, *Callithrix jacchus*, is a small primate of the Callitrichidae family native to the northeastern region in Brazil (Fig. 6.1). In the Tupi Indians language, it is called "sagui." The marmoset is an animal easy to breed in captivity and has been highlighted as an animal model in biomedical studies (Stellar 1960; Abbott et al. 2003; Ward and Vallender 2012). As a nonhuman primate model, it has achieved a major contributing space, for example, in neuroscience, ethology, pathology, genetics, and pharmacology (Smith et al. 2001; Ludlage and Mansfield 2003; Sasaki et al. 2009). The use of the marmoset in these investigations has been increasing since 1960, with an exponential growth between 1980 and 2000 in the number of publications using *Callithrix jacchus* (Abbott et al. 2003).

Besides its easy maintenance in the laboratory, the marmoset has complex behavioral and vocal repertoires (Stevenson and Rylands 1988; Bezerra and Souto 2008), learning ability in cognitive tasks common to Old World monkeys (Dias et al. 1996; Spinelli et al. 2004; Mendes and Huber 2004), and physiological variables comparable to humans (McNees et al. 1984; Smith et al. 2001; Mansfield 2003), being also used as an animal model for aging studies (Tardif et al. 2011). Additionally, the marmoset behavioral repertoire has been described since the 1970s (Stevenson and Poole 1976), including the behaviors of anxiety and alertness, which are used as study variables in investigations into the administration of anxiogenic and anxiolytic drugs such as diazepam, amphetamine, and modafinil (Barros and Tomaz 2002; van Vliet et al. 2008; Cilia and Piper 1997).

Fig. 6.1 The marmoset, *Callithrix jacchus*, is a small diurnal primate morphologically character-ized by the presence of typical white tufts. (From Christiane A. Silva/LNRB)

We highlight, specifically, its importance in research on internal temporal organization, because in a PubMed search using only the keyword "marmoset" we found 4,166 indexed articles (April 2013), but with the keywords "marmoset and circadian" or "marmoset and sleep," we found only 50 and 19 items, respectively. In this chapter we describe in detail the recent experimental findings produced in our laboratory and reported in the scientific literature, which allow us to propose the marmoset as a potential model for the study of circadian rhythmicity.

6.2 Marmoset Circadian Rhythmicity

The marmoset is a neotropical primate with diurnal habits, and a qualified literature describes the general characteristics of circadian rhythmicity in this species. Studies conducted in the laboratory (Crofts et al. 2001; Philippens et al. 2004; Hoffmann et al. 2012), in seminatural conditions (Menezes et al. 1993), and in the field (Castro et al. 2003) show a diurnal concentration pattern of the marmoset active phase.

In the natural environment the marmosets awake before sunrise, but only come out of their sleeping trees to begin locomotor activity after sunrise, and generally finish the active phase (alpha) about 1 to 2 h before sundown (Erkert 1989; Menezes et al. 1993; Moreira et al. 1991). Under an artificial light–dark cycle (LD 12:12), they present an average alpha of 11.1 ± 0.7 h (Erkert 1989), very similar to that found in studies in natural environmental conditions, 11.6 ± 0.5 h (Menezes et al.

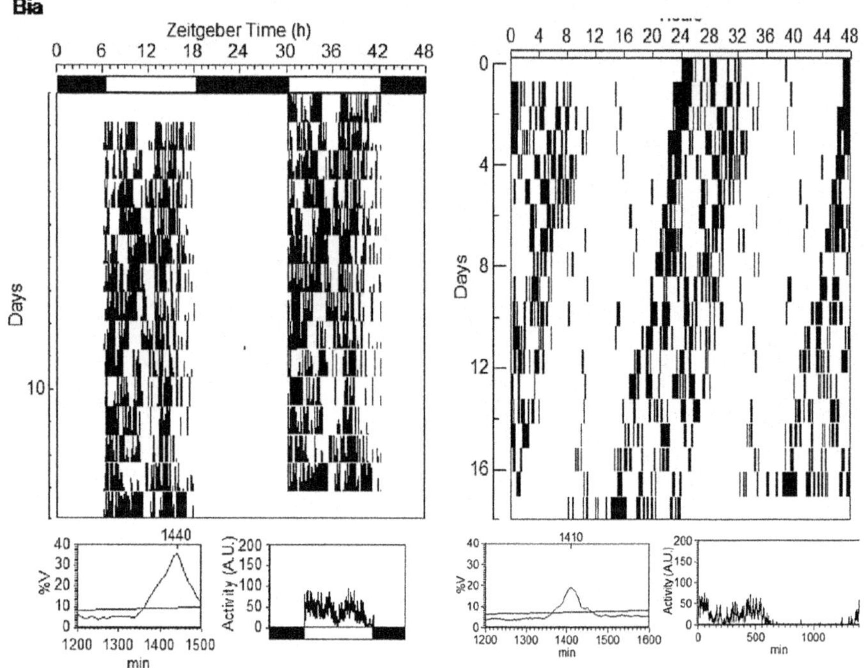

Fig. 6.2 At *left*, 14 days of an adult female under (light–dark) LD 12:12 (≈247:0 lux). At *right*, 18 days of an adult male under LL (≈225 lux). Actograms are in module=24 h. Data are totalized in 5-min intervals

1993). The activity pattern is usually observed as a bimodal profile for locomotor activity, with one peak early in the morning and another one in the afternoon (Fig. 6.2) (Erkert 1989; Menezes et al. 1993; Moreira et al. 1991), whereas other behaviors present a unimodal profile, such as self-grooming, whose peak of duration occurs in the first half of the light phase (Lampert et al. 2011; Menezes et al. 1993).

Under constant environmental conditions, that is, constant light, the marmoset exhibits stable motor activity circadian rhythms (Erkert 1989; Glass et al. 2001) with an endogenous period around 23.3±0.4 h (Erkert 1989; Glass et al. 2001; Silva et al. 2005) and an alpha average duration of 11.3±0.4 h (Fig. 6.2).

According to Erkert (1989) there is no change in the motor activity circadian rhythm of marmosets under different light intensities. However, our findings in marmosets with visual impairment, a congenital blindness caused by degeneration of rods and cones, showed that these animals synchronize the circadian rhythm of locomotor activity to the LD cycle, presented a free-running period different from the control group, and this followed Aschoff's rule (Silva et al. 2005). On the other hand, a finding with *Macaca nemestrina*, an Old World monkey species, shows a positive correlation of period, activity duration, and amount of locomotor activity per period with light intensity (Tokura and Aschoff 1978).

Another study by our group showed that the self-grooming rhythm is also affected by luminous intensity. Under symmetrical LD cycles, the duration and frequency of self-grooming tend to increase when luminous intensity of the light phase varies from 500 to 200 lux but decrease notably under the luminous intensity of 10 lux (Lampert et al. 2011).

The photic phase-response curve of marmosets is characterized by phase delays after pulses in the early subjective night and phase advances after pulses in the late subjective night, similar to that which occurs in mammals in general. The only difference is that the marmosets presented great phase changes in motor activity during the subjective day, when low responsiveness of the circadian timing system to punctual changes in light intensity is expected (Wechselberger and Erkert 1994). According to the authors, the direct influence of light on the circadian system or a nonphotic effect of a behavioral activation induced by light may have caused these phase changes.

Other parameters, besides the observation of locomotor activity, also corroborate the earlier statement, which can be exemplified by temperature variation throughout the 24 hours. Marmosets have elevated temperatures during the day (because of their diurnal habits) that decline during the dark phase (Hoffmann et al. 2012). The levels of cortisol hormone also present circadian variation, with elevated cortisol levels during the beginning of the active phase that significantly decrease during the day (Cross and Rogers 2004). Furthermore, it is observed that when tested for warming level, significant results are allocated in the light phase (Van Vliet et al. 2008; Hoffmann et al. 2012). Therefore, based on all the points raised, there is no doubt that the marmoset is an animal genuinely diurnal.

Although the LD cycle is the main zeitgeber for marmosets, nonphotic cues can also act as zeitgebers for these animals, synchronizing the circadian rhythms in the absence of an external LD cycle or acting in conjunction with photic zeitgebers in the photic resynchronization. Exemplifying these conditions, it was observed that the photic resynchronization to phase delays was accelerated in marmosets by the presence of a conspecific previously synchronized to the new lighting time (Mendes et al. 2008). In two other studies, modulations in motor activity circadian rhythmicity in response to the vocalizations emitted from conspecifics were observed, suggesting synchronization by the sound. These vocalizations were audible in the experiment room and were able to promote positive masking and relative coordination, associated with an increase in activity frequency, in locomotor activity of the animals in one of the studies (Silva et al., unpublished data). In the other study, conducted in a room that allowed the entrance of more vocalizations from conspecifics kept outside than in the previous study, conspecific vocalizations were able to entrain the locomotor activity of three of a total of four marmosets in the experiment, and keeping the same phase relationship that the animals kept outside the room (Gonçalves et al. 2009).

Another important finding that confirms the marmoset diurnality is that, regardless of the latitude where the marmoset lives and seasonality throughout the year, the active phase remains in the light phase (Menezes et al. 1998; Melo et al. 2010). In experiments with short pulses of light of only 20 s and a 1-h induced activity pulse, Glass et al. (2001) were able to construct a phase-response curve to photic stimulus and also a phase-response curve for nonphotic stimulus, respectively. Then, it was observed that the phase-response curves of marmosets are similar to those of rodents, in which the key to the response is the phase of the circadian pacemaker and not that of the activity–rest cycle.

6.3   Photoperiodic Responses in Marmosets

One of the important features for evaluating a circadian system is its response to photoperiod, because the seasonal variations in environmental parameters also promote physiological and behavioral changes and are relevant to the survival of the species. Faced with recurring events, such as the seasons, the organisms can organize their activities to be previously prepared to changes, exhibiting an anticipatory behavior (Golombek and Rosenstein 2010). One of these recurrent events is the photoperiod, that is, the light phase of the day, which passes for different durations and intensities throughout the year and serves as an indicator of seasonality. The photoperiod, food availability, favorable variation in temperature, and rainfall are a set of factors that allow animals to reproduce, migrate, hibernate, dig burrows, and store food when environmental conditions are more favorable, which maximizes an appropriate response to a cyclic environment (Wehr 2001a).

Generally, studies related to seasonality are accomplished with mammalian species that present markedly large changes, resulting from changes in melatonin signaling. As this hormone is secreted at night, the information received by the organism about daylength on these photoperiodic species leads to integrated solutions throughout the system (Wehr 2001a; Nelson et al. 2002). However, not all mammals that are influenced by seasonality have broad answers in their temporal organization.

Among the mammals, seasonality can also be observed in human and nonhuman primates. In human species, studies point to a variation in the birthrate, disease incidence, and immune system responses, in hormonal rhythms (e.g., melatonin, cortisol, prolactin), mood changes, and eating behavior (Wehr et al. 1993; Wehr 2001b; Nelson et al. 2002; Walton et al. 2011). In nonhuman primates, we can mention some changes specifically caused by photoperiod, especially in studies with the nocturnal lemur *Microcebus murinus*, such as changes in activity–rest cycle, temperature rhythm, the steroid hormone DHEA secretion, and metabolic rates (Schilling et al. 1999; Perret and Aujard 2001, 2005; Perret et al. 1998). Changes in locomotor activity rhythm of this nocturnal lemur, which are common in aging, were also observed when the seasonal cycles were accelerated (Perret 1997; Cayetanot et al. 2005). Thus, this nocturnal primate has been pointed out as a model to study the changes in aging of rhythmicity (Aujard et al. 2006; Perret and Aujard 2006).

In our laboratory, we have developed studies of photoperiodic responses in the marmoset. Here we highlight some points along with the studies of other groups.

6.4 Behavioral Aspects of Seasonality in Marmosets

Despite the small variation in photoperiod, animals naturally from equatorial latitudes, such as *C. jacchus*, may have changes in their behavioral repertoire influenced by seasonality. Studies in seminatural conditions, with the time of sunrise varying as much as 40 min, showed a change at the beginning of activity and the bimodal pattern of locomotor circadian rhythm in infants and juveniles (1–10 months old: Menezes et al. 1998; 5–12 months old: Melo et al. 2010).

In artificial conditions of short (LD 8:16) and long (LD 16:08) photoperiods, very different from the marmoset natural environment but found at nonequatorial high-latitude regions, the adult marmosets showed changes in various parameters of circadian rhythmicity, although indicating synchronization of the activity–rest cycle to the photoperiod of long and short days (Carrijo 2013). In short days a decrease in total daily activity and the duration of alpha was observed. On the other side, in the long days there was an increase in the duration of alpha and a decrease in the amplitude and spectral power of the circadian rhythm period.

In our studies we found individual variation in the phase relationship between the onset of light and the beginning of activity, so we suggest that marmosets may use different strategies in response to short and long days, but they all result in synchronization to the tested photoperiods. Additionally, the phase transitions of one LD cycle to another did not occur similarly in all conditions. The decrease of daylength from 12 to 8 h appears to force further the timing system to adjust the onset of locomotor activity, whereas the increase from 12 to 16 h and the decrease from 16 to 12 h promoted the adjustment with greater speed of synchronization.

These differential post-effects during synchronization of the activity–rest rhythm to phase advances or delays were equally observed for the nocturnal species lemur *Microcebus murinus* (Schilling et al. 1999), suggesting that the circadian system functionally works in the same way in diurnal and nocturnal animals. Investigations such as these contribute to the findings on light as a cue of extreme relevance to the primate temporal organization. An example can be observed in humans when undergoing abrupt phase changes in transmeridian flights. There are indications that exposure to light is the best therapeutic strategy to reduce the effects of jet lag (Lee and Galvez 2012), but as the symptoms of jet lag are quite large and complex, further investigations to clarify and establish an adequate prevention strategy in every situation are necessary (Kolla and Auger 2011). We suggest that the marmoset model is an efficient model to conduct these investigations.

Ease or difficulty in adjustment of the circadian timing system to shifts in the environmental light cycle is the result of functional mechanisms related to the interactions among SCN oscillators. This process is determined by the coupling among these oscillators. The physiological and structural mechanisms of coupling among circadian oscillators in the SCN is still an open question, but some evidence suggests that results of several mechanisms such as gap junctions, GABAergic synapses, neuronal adhesion molecules, etc., are involved. Electron microscopy studies are being conducted with the intention of clarifying these mechanisms.

As the marmoset is originally from northeastern Brazil (Rylands et al. 2009), a region of little photoperiodic variation throughout the year (lower latitudes), we could imagine that these animals would not respond distinctly to short and long photoperiods (Fig. 6.3). Thus, the flexibility in temporal adjustment of the marmoset activity–rest cycle to different photoperiods allows these species to have an easy adaptation to different regions, as observed by the presence of *C. jacchus* from north to south in Brazil (Rylands et al. 2008). On the other hand, this flexibility also adds an advantage to the use of this animal as a nonhuman primate model because *C. jacchus* is found at research institutions in several countries (Abbott et al. 2003) with more pronounced photoperiodic variations (considering that the colonies may have a natural LD cycle, beyond the artificial 12:12).

In our studies we also observed that marmosets altered everyday activities in response to the change of a symmetrical photoperiod (12 h light and 12 h dark) to a long photoperiod (16 h light and 8 h dark), simulating the daylength of summer at high-latitude regions. Considering the initial peak of activity usually seen in the marmoset, the animals exhibited a reduction in the rate of vocalization, impoverishment of the vocal repertoire, and decreased exploratory activity stimulated by unfamiliar objects. On the other hand, short days did not affect these parameters. Because in their home region the longer days (indicating summer) are characterized by drought and food shortages, the marmosets are more subject to predation and territory defense. Based on the vocalizations observed in the initial peak of activity in the long days in our study (absence of the long-distance calls, "*loud shrill*," and decrease of the intragroup calls, "*phee*"), we can speculate that this is a possible mechanism of response to this situation in wild life for the marmosets not so exposed. Additionally, as the emission of agonistic and alarm calls (*twitter, tsik, see, seep*) did not change among photoperiods, we suggest that if the daylength is indeed a factor influencing the rate and the type of vocalization, it would not be advantageous to reduce the emission of alarm callings. In this way, the territory and group defense would be maintained when they are most vulnerable to predation and competition in the wild (Carrijo 2013).

Besides the availability of resources, we could imagine that it would not be problematic to allocate activities into the photophase of 8 h because the active phase of the marmoset is less than the daylength of 12 h. Thus, allocation of daily activities in a photophase of 16 h would cause a change in the distribution of activities during the long day. Thus, in relationship to the photoperiods tested, we suggest that increase of photophase is more challenging to marmoset behavioral responses, in contrast to the decrease in photophase duration. This observation is the opposite of what is discussed for species living at high latitudes because they undergo a winter marked by food shortages and low temperatures (Hill et al. 2003; Nelson et al. 2002).

In addition to the laboratory studies just cited, a seasonal pattern in the natural environment in marmoset feeding behavior has also been reported. Castro et al. (2003) observed changes in the pattern of space use by *C. jacchus* between feeding stations with additional behavioral plasticity according to food availability and distribution, even without the marmosets extending the size of the living area. On the other hand, *Callithrix penicillata* can vary the predominant activity pattern,

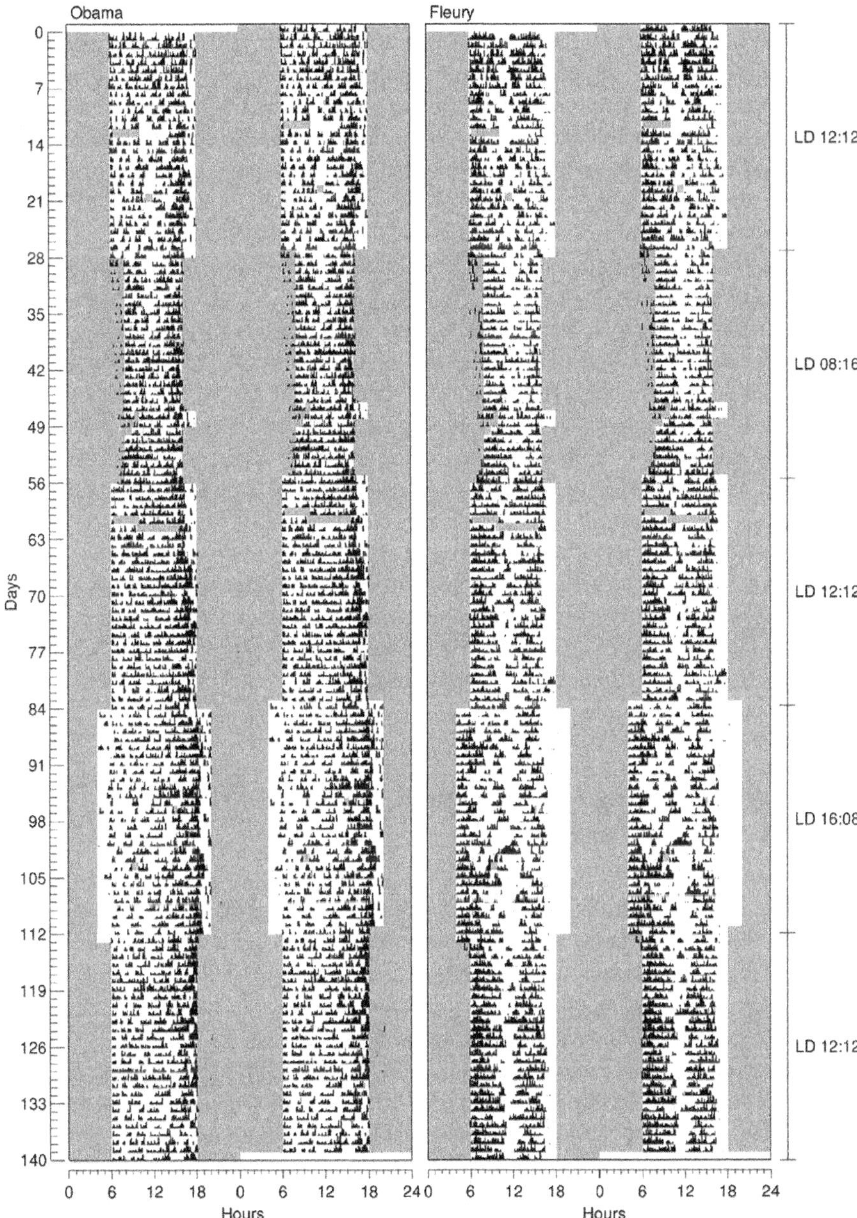

Fig. 6.3 Actograms of two marmosets under symmetrical, short, and long photoperiods: LD 12:12 (day 0), LD 08:16 (day 28), LD 12:12 (day 56), LD 16:08 (day 84), LD 12:12 (day 112). Locomotor activity was recorded by passive infrared sensors positioned over the individual cages. Observing the phase angles between light and activity, it is possible to see the different strategies of adjustment to the light–dark cycle to which the animals were exposed. Under LD 16:08 the animals lengthened the active phase, or by delaying the active phase ending (*left*), or by advancing the beginning of the active phase (*right*) (Carrijo 2013)

alternating between foraging a punctual resource (insects) or displacing in search of other food items, depending on fruit availability and distribution (Vilela and Faria 2004). Similarly, considering other callitrichids, species of the genus *Saguinus* presented alternate responses according to the temporal and spatial distribution of resources over the months, intensively exploring a few food species or being opportunistic and exploring many species. They also experienced the influence of seasonality in resting, grooming, foraging, and displacement behaviors (Egler 2000).

These changes resulting from the seasons and, consequently, from food availability may be related to more propitious moments for breeding throughout the year, which has been observed for both diurnal (*C. jacchus*: Sousa et al. 1999a, b) and nocturnal (*Aotus azarai*: Fernandez-Duque et al. 2002) primates. Regarding *C. jacchus*, there is evidence that progesterone secretion by the corpus luteum can be induced by melatonin administration (Webley and Hearn 1987; Hearn and Webley 1987), the photoperiod biological signal, indicating that marmoset reproductive functions have the potential to be stimulated by photoperiodic variations. This point was suggested by Sousa and collaborators (1999a), who observed a seasonal pattern in the birth of baby marmosets in seminatural captivity, even without variation in food supply. The authors speculated that the observed birth pattern could be related to factors such as the rainy season or photoperiod, even in a region close to Ecuador, that is, with little photoperiodic variation. A discussion of the proximate and ultimate functions of births seasonality observed in several genera of neotropical primates is found in the review of Di Bitetti and Janson (2000).

6.5 Understanding the Rhythm of the Sleep–Wake Cycle in Marmosets

To study circadian rhythmicity is also to study the sleep–wake cycles (SWC). The SWC is a biological rhythm and, given the nature of this oscillatory rhythm, significantly affects the expression of behaviors and functions of living beings. Thus, understanding the SWC rhythm in the marmoset is necessary for a full understanding of the circadian biology in this primate.

As a starting point, we mentioned before that the SWC in marmosets is advanced in relationship to the LD cycle (Erkert 1989; Hoffmann et al. 2012); that is, these animals begin their activities a few minutes before the beginning of the light phase and end the active phase before the beginning of the dark phase. Extrapolating these data to nature, it means these animals awake before sunrise, so marmosets remain alert for activities during the day, such as foraging and social activities of the group, and return to a safe hiding place before sunset.

During times of wakefulness the marmoset interacts socially with other conspecifics, moves around, feeds, and keeps safe from predators, while during the rest period it is predominantly asleep with brief arousals (Sri Kantha and Suzuki 2006). During the night, the marmosets stay on trees in protected conditions, whereas during the day they are moving in search of food, showing alertness to situations considered dangerous, such as a predator attack (Muggleton et al. 2005). One interesting fact is that marmosets have about a 12-h period of monophasic sleep, almost completely concentrated in the dark phase. Another interesting fact is that marmosets have a sleep pattern similar to that of humans. For example, in relationship to its sleep architecture similarity is observed in the number and pattern of non-REM and REM sleep cycles, as well as in switching between one sleep stage and another (Crofts et al. 2001; Philippens et al. 2004). Additionally, these primates also have a high incidence of slow-wave sleep in the first half of the night, which gradually disappears during the second half of the night, giving rise to REM sleep (Crofts et al. 2001).

Given the abundance of studies on the SWC, some already cited and others not mentioned, it is noticeable that several techniques are used to study and evaluate this rhythm in particular. Some techniques evaluate indirectly, such as the infrared sensors positioned above the cage and the actimeters positioned on the back or neck of the marmoset (Sri Kantha and Suzuki 2006; Gonçalves et al. 2009; Melo et al. 2010). In both cases, the main parameter evaluated is the amount of locomotor activity, because by estimating how much the animal remained stationary or moving it is possible to infer the probable sleep and wake episodes. In turn, electrophysiology is considered as a main direct measure of the SWC evaluation. The use of biopotentials to record the electroencephalogram (EEG), electrocorticogram (ECoG), electromyogram (EMG), and electro-oculogram (EOG) serve as faithful parameters that complement each other for the definition of the marmoset SWC state (Crofts et al. 2001; Philippens et al. 2004; Hoffmann et al. 2012). It is worth mentioning that the pioneering studies on this area were made by behavioral observations (Erkert 1989; Menezes et al. 1993). Also, despite the minimal technology employed, they were essential to further development of this area.

6.6 Functional Mechanisms of Circadian Systems in Marmosets

In our laboratory we have investigated the functional mechanisms of the marmoset circadian system using the forced desynchronization model, because, as proposed by Schwartz (2009), this is a noninvasive approach that can investigate the functionality of the circadian system and the consequences of its alterations in memory, mood, and performance. We used a symmetrical LD of 21 h (T21, 10.5:10.5) (Fig. 6.4). Under this condition the marmosets exhibit motor activity with two simultaneous circadian components, one being synchronized to the LD cycle whereas the other results from free-running endogenous rhythmicity of the circadian pacemaker under the influence of masking by the LD cycle (Silva et al., unpublished data).

Physiological rhythms with different simultaneous periods have also been described in *Saimiri sciureus*, another diurnal social primate from South America, also of small size (800–1,200 g) although larger than than the marmoset. Under constant environmental conditions and restriction to a chair (chair-acclimatized),

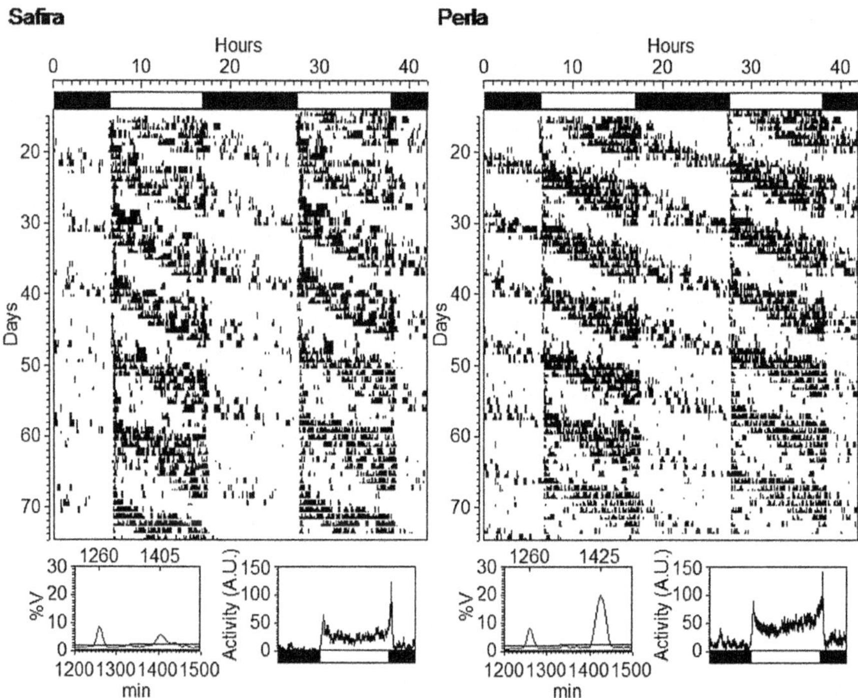

Fig. 6.4 Actograms, periodograms (Sokolove–Bushell), and mean daily profiles of motor activity of two adult females (Safira at *left* and Perla at *right*) under T21 (LD 10.5:10.5 ≈ 247:0 lux), during 60 days. Actograms and mean daily profiles are in module = 21 h. Data are totalized in 5-min intervals

this primate presents spontaneous internal desynchronization between feeding and rectal temperature rhythms ($T = 25$ h) from urine volume and potassium rhythms (Sulzman et al. 1977).

These data of motor activity dissociation of marmosets under T21 (Silva et al., unpublished data) and of internal desynchronization of the squirrel monkey (Sulzman et al. 1977) are evidence that the circadian system of primates is similar to that of rodents, consisting of at least two groups of oscillators coupled together.

An interesting fact is that despite although dissociation of motor activity marmosets continue presenting most of the active phase into the light phase, that is, remain predominantly diurnal in T21, in contrast to *Octodon degus*, a rodent endemic to central Chile, which can switch chronotype according to the period of the LD cycle (Vivanco et al. 2010). Although considered diurnal, this species shows great chronotypic variation in the laboratory, including being able to switch between diurnality and nocturnality depending only on the availability of wheel-running activity in the cage (Otalora et al. 2010; Vivanco et al. 2010). It also presents dissociation of motor activity and temperature under a symmetrical LD cycle of 28 h and of motor activity under a symmetrical LD cycle of 21 h, being the light-dependent component more evident in nocturnal animals (Vivanco et al. 2010).

6.7 Conclusion

From all that has been seen, the relevance of marmosets for biomedical studies is undeniable. Furthermore, that they are in great phylogenetic proximity to humans, because both are in the primate order, is another point that favors their use in research when compared to rodents, the animal models consistently used in research in general. Rodents, in contrast to marmosets and humans, are animals of nocturnal habits with polyphasic sleep that can be allocated into both the dark and light phase. Therefore, the marmoset resembles humans in a number of behavioral and physiological mechanisms, serving as a good experimental subject when the goal is to understand and compare to humans.

References

Abbott DH, Barnett DK, Colman RJ et al (2003) Aspects of common marmoset basic biology and life history important for biomedical research. Comp Med 53:339–350

Aujard F, Séguy M, Terrien J et al (2006) Behavioral thermoregulation in a non human primate: effects of age and photoperiod on temperature selection. Exp Gerontol 41:784–792

Barros M, Tomaz C (2002) Non-human primate models for investigating fear and anxiety. Neurosci Biobehav Rev 26:187–201

Bezerra BM, Souto A (2008) Structure and usage of the vocal repertoire of *Callithrix jacchus*. Int J Primatol 29:671–701

Carrijo C (2013) Influência de fotoperíodo artificial no comportamento de um primata neotropical diurno (Callithrix jacchus). Dissertation, Universidade Federal do Rio Grande do Norte

Castro CSS, Menezes AL, Moreira LFS (2003) Locomotor activity rhythm in free-ranging common marmosets (Callithrix jacchus). Biol Rhythm Res 34:23–30

Cayetanot F, Van Someren EJ, Perret M et al (2005) Shortened seasonal photoperiodic cycles accelerate aging of the diurnal and circadian locomotor activity rhythms in a primate. J Biol Rhythms 20:461–469

Chung JK, Lee KY, Kim SH et al (2012) Circadian rhythm characteristics in mood disorders: comparison among bipolar I disorder, bipolar II disorder and recurrent major depressive disorder. Clin Psychopharmacol Neurosci 10:110–116

Cilia J, Piper DC (1997) Marmoset conspecific confrontation: an ethologically-based model of anxiety. Pharmacol Biochem Behav 58:85–91

Crofts HS, Wilson S, Muggleton NG et al (2001) Investigation of the sleep electrocorticogram of the common marmoset (*Callithrix jacchus*) using radiotelemetry. Clin Neurophysiol 112:2265–2273

Cross N, Rogers LJ (2004) Diurnal cycle in salivary cortisol levels in common marmoset. Dev Psychobiol 45:134–139

Di Bitetti MS, Janson CH (2000) When will the stork arrive? Patterns of birth seasonality in neotropical primates. Am J Primatol 50:109–130

Dias R, Robbins TW, Roberts AC (1996) Primate analogue of the Wisconsin Card Sorting Test: effects of excitotoxic lesions of the prefrontal cortex in the marmoset. Behav Neurosci 110:872–886

Egler SG (2000) Ecologia alimentar e sazonalidade em primatas neotropicais: gênero Marmosetnus. In: Alonso C, Langguth A (eds) A Primatologia no Brasil, vol 7. EDUPB/SBPr, João Pessoa, pp 81–95

Erkert HG (1989) Characteristics of the circadian activity rhythm in common marmosets (*Callithrix J. jacchus*). Am J Primatol 17:271–286

Erren TC (2013) Shift work and cancer research: can chronotype predict susceptibility in night-shift and rotating-shift workers? Occup Environ Med 70:283–284

Fernandez-Duque E, Rotundo M, Ramírez-Llorens P (2002) Environmental determinants of birth seasonality in night monkeys (Aotus azarai) of the Argentinean Chaco. Int J Primatol 23:639–656

Glass JD, Tardif SD, Clements R et al (2001) Photic and nonphotic circadian phase resetting in a diurnal primate, the common marmoset. Am J Physiol Reg Int Comp Physiol 280:191–197

Golombek DA, Rosenstein RE (2010) Physiology of circadian entrainment. Physiol Rev 90: 1063–1102

Gonçalves F, Belísio AS, Azevedo CVM (2009) Effects of nest box availability on the circadian rhythm of common marmosets (Callithrix jacchus). Folia Primatol 80:175–188

Hearn JP, Webley GE (1987) Regulation of the corpus luteum of early pregnancy in the marmoset monkey: local interactions of luteotrophic and luteolytic hormones in vivo and their effects on the secretion of progesterone. J Endocrinol 114:231–239

Hill RA, Barrett L, Gaynor D et al (2003) Day length, latitude and behavioural (in)flexibility in baboons (*Papio cynocephalus ursinus*). Behav Ecol Sociobiol 53:278–286

Hoffmann K, Coolen A, Schlumbohm C et al (2012) Remote long-term registrations of sleep-wake rhythms, core body temperature and activity in marmoset monkeys. Behav Brain Res 235: 123–133

Kolla BP, Auger RR (2011) Jet lag and shift work sleep disorders: how to help reset the internal clock. Cleve Clin J Med 78:675–684

Lampert RMS, Azevedo CVM, Menezes AAL (2011) Influence of different light intensities on the daily grooming distribution of common marmosets *Callithrix jacchus*. Folia Primatol 82: 131–142

Lee A, Galvez JC (2012) Jet lag in athletes. Sports Health 4:211–216

Ludlage E, Mansfield K (2003) Clinical care and disease of the common marmosets (*Callithrix jacchus*). Comp Med 54:369–382

Mansfield K (2003) Marmoset models commonly used in biomedical research. Comp Med 53:383–392

Marcheva B, Ramsey KM, Peek CB et al (2013) Circadian clocks and metabolism. Handb Exp Pharmacol 217:127–155

McClung CA (2013) How might circadian rhythms control mood? Let me count the ways…. Biol Psychiatry. doi:10.1016/j.biopsych.2013.02.019

McNees DW, Lewis RW, Ponzio BJ et al (1984) Blood chemistry of the common marmoset (Callithrix jacchus) maintained in an indoor-outdoor environment: primate comparisons. Primates 25:103–109

Melo PR, Belísio AS, Menezes AAL et al (2010) Influence of seasonality on circadian motor activity rhythm in common marmosets during puberty. Chronobiol Int 27:1420–1437

Mendes N, Huber L (2004) Object permanence in common marmosets (*Callithrix jacchus*). J Comp Psychol 118:103–112

Mendes ALB, Menezes AAL, Azevedo CVM (2008) The influence of social cues on circadian activity rhythm resynchronization to the light-dark cycle in common marmosets *Callithrix jacchus*. Biol Rhythm Res 39:469–479

Menezes AAL, Moreira LFS, Azevedo CVM et al (1993) Behavioral rhythms in the captive common marmoset (*Callithrix jacchus*) under natural environmental conditions. Braz J Med Biol Res 26:741–745

Menezes AAL, Moreira LFS, Menna-Barreto LS (1998) Annual variation in an ultradian component in the locomotor activity rhythm of the common marmoset (*Callithrix jacchus*). Biol Rhythm Res 29:556–562

Moreira LFS, Sousa MBC, Menezes AAL et al (1991) Ritmo circadiano da atividade motora do sagüi comum (*Callithrix jacchus*). A Primatologia no Brasil 3:25–33

Muggleton NG, Smith AJ, Scott EA et al (2005) A long-term study of the effects of diazinon on sleep, the electrocorticogram and cognitive behaviour in common marmosets. J Psychopharmacol 19:455–466

Musiek ES, Fitzgerald GA (2013) Molecular clocks in pharmacology. Handb Exp Pharmacol 217:243–260

Nelson RJ, Demas GE, Klein SL et al (2002) Seasonal patterns of stress, immune function and disease. Cambridge University Press, New York

Otalora BB, Vivanco P, Madariaga AM et al (2010) Internal temporal order in the circadian system of a dual-phasing rodent, the *Octodon degus*. Chronobiol Int 27:1564–1579

Perret M (1997) Change in photoperiodic cycle affects life span in a prosimian primate (*Microcebus murinus*). J Biol Rhythms 12:136–145

Perret M, Aujard F (2001) Daily hypothermia and torpor in a tropical primate: synchronization by 24-h light-dark cycle. Am J Physiol Reg Int Comp Physiol 281:R1925–R1933

Perret M, Aujard F (2005) Aging and season affect plasma dehydroepiandrosterone sulfate (DHEA-S) levels in a primate. Exp Gerontol 40:582–587

Perret M, Aujard F (2006) Vieillissement et rythmes biologiques chez les primates. Med Sci (Paris) 22:279–283

Perret M, Aujard F, Vannier G (1998) Influence of daylength on metabolic rate and daily water loss in the male prosimian primate *Microcebus murinus*. Comp Biochem Physiol A Mol Integr Physiol 119:981–989

Philippens IH, Kersten CJ, Vanwersc RA et al (2004) Sleep and sleep EEG spectra in marmoset monkeys. Sleep Wake Res Neth 15:49–51

Rylands AB, Mittermeier RA, de Oliveira MM et al (2008) *Callithrix jacchus*. In: IUCN 2012. IUCN Red List of Threatened Species. Version 2012.2. www.iucnredlist.org

Rylands AB, Coimbra-Filho AF, Mittermeier RA (2009) The systematics and distribution of the marmosets (*Callithrix, Calibella, Cebuella, and Mico*) and Callimico (*Callimico*) (Callitrichidae, Primates). In: Ford SM, Porter LM, Davis LLC (eds) The smallest anthropoids: the marmoset/callimico radiation. Springer, Nova Iorque, pp 25–61

Sahar S, Sassone-Corsi P (2013) The epigenetic language of circadian clocks. Handb Exp Pharmacol 217:29–44

Sasaki E, Suemizu H, Shimada A et al (2009) Generation of transgenic non-human primates with germline transmission. Nature 459:523–527

Scheiermann C, Kunisaki Y, Frenette PS (2013) Circadian control of the immune system. Nat Rev Immunol 13:190–198

Schilling A, Richard J, Servière J (1999) Duration of activity and period of circadian activity–rest rhythm in a photoperiod-dependent primate, *Microcebus murinus*. Comptes Rendus de l'Académie des Sciences - Series III - Sciences de la Vie 322:759–770

Schwartz WJ (2009) Circadian rhythms: a tale of two nuclei. Curr Biol 19:460–462

Silva MMA, Albuquerque AM, Araujo JF (2005) Light-dark cycle synchronization of circadian rhythm in blind primates. J Circad Rhythm 3:10–16

Smith D, Trennery P, Farningham D et al (2001) The selection of marmoset monkeys (Callithrix jacchus) in pharmaceutical toxicology. Lab Anim 35:117–130

Sousa MBC, Peregrino HPA, Cirne MFC et al (1999a) Reproductive patterns and birth seasonality in a South-American breeding colony of common marmosets, Callithrix jacchus. Primates 40:327–336

Sousa MBC, Silva HPA, Vidal JF (1999b) Litter size does not interfere with fertility in common marmoset, Callithrix jacchus. Folia Primatol 70:41–46

Spinelli S, Pennanen L, Dettling AC et al (2004) Performance of the marmoset monkey on computerized tasks of attention and working memory. Cogn Brain Res 19:123–137

Sri Kantha S, Suzuki J (2006) Sleep quantitation in common marmoset, cotton top tamarin and squirrel monkey by non-invasive actigraphy. Comp Biochem Physiol 144:203–210

Stellar E (1960) The marmoset as a laboratory animal: maintenance, general observations of behavior, and simple learning. J Comp Physiol Psychol 53:1–10

Stevenson MF, Poole TB (1976) An ethogram of the common marmoset (*Callithrix jacchus*): general behavioural repertoire. Anim Behav 24:428–451

Stevenson MF, Rylands AB (1988) The marmosets, genus *Callithrix*. In: Mittermeier AB, Rylands AT, Coimbra-Filho AF, Fonseca GAB (eds) Ecology and behavior of neotropical primates. World Wildlife Foundation, New York, pp 131–221

Sulzman FM, Fuller CA, Moore-ede MC (1977) Spontaneous internal desynchronization of circadian rhythms in the squirrel monkey. Comp Biochem Physiol 58:63–67

Tardif SD, Mansfield KG, Ratnam R et al (2011) The marmoset as a model of aging and age-related diseases. ILAR J 52:54–65

Tokura H, Aschoff J (1978) Circadian activity rhythms of the pig-tailed macaque, *Macaca nemestrina*, under constant illumination. Pflugers Arch 376:241–243

Van Vliet SAM, Jongsma MJ, Vanwersch RAP et al (2008) Efficacy of caffeine and modafinil in counteracting sleep deprivation in the marmoset monkey. Psychopharmacology 197:59–66

Vilela SL, Faria DS (2004) Seasonality of the activity pattern of *Callithrix penicillata* (Primates, *Callitrichidae*) in the cerrado (scrub savanna vegetation). Braz J Biol 64:363–370

Vivanco P, Otalora BB, Rol MA et al (2010) Dissociation of the circadian system of *Octodon degus* by T28 and T21 light-dark cycles. Chronobiol Int 27:1580–1595

Walton JC, Weil ZM, Nelson RJ (2011) Influence of photoperiod on hormones, behavior, and immune function. Front Neuroendocrinol 32:303–319

Ward JM, Vallender EJ (2012) The resurgence and genetic implications of New World primates in biomedical research. Trends Genet 28:586–591

Webley GE, Hearn JP (1987) Local production of progesterone by the corpus luteum of the marmoset monkey in response to perfusion with chorionic gonadotrophin and melatonin in vivo. J Endocrinol 112:449–457

Wechselberger E, Erkert HG (1994) Characteristics of the light-induced phase response of circadian activity rhythms in common marmosets, Callithrix J. jacchus (Primates-Cebidae). Int Soc Chronobiol 11:275–284

Wehr TA (2001a) Seasonal photoperiodic responses of the human circadian system. In: Takahashi JS, Turek FW, Moore RY (eds) Handbook of behavioral neurobiology: circadian clocks, vol 12. Kluwer/Plenum, New York, pp 715–744

Wehr TA (2001b) Photoperiodism in humans and other primates: evidence and implications. J Biol Rhythms 16:348–364

Wehr TA, Moul DE, Barbato G et al (1993) Conservation of photoperiod-responsive mechanisms in humans. Am J Physiol 265:R846–R857

Part II
Mechanisms of Circadian Oscillation

Chapter 7
Intracellular Calcium as a Clock Output from SCN Neurons

Raúl Aguilar-Roblero, Mauricio Díaz-Muñoz, Adrian Báez-Ruíz,
Daniel Quinto-Muñoz, Gabriella Lundkvist, and Stephan Michel

Abstract In mammals, the major circadian clock is located in the suprachiasmatic nuclei (SCN). The molecular oscillator in these neurons is driven by transcriptional–translational feedback loops (TTL) among clock genes that generate a circadian periodicity. To fulfill its role as pacemaker, the molecular oscillation must be translated to an electrical signal in SCN neurons, which will be transmitted to the rest of the brain and eventually the organism. The mechanisms involved in this process remain mostly unknown, but some information is already available. Among the ion channels in SCN neurons which are regulated by the circadian clock, only the manipulations of the fast delayed rectifier (fDR) and large-conductance (BK) K^+ currents have shown to affect circadian rhythmicity either in neuronal firing pattern or behavior. On the other hand, data from rat and mouse clearly indicate that intracellular Ca^{2+} channels sensitive to ryanodine (RyR) are part of an output pathway of the clock in SCN neurons. Intracellular Ca^{2+} signals mediate between the molecular circadian clock and the neuronal plasma membrane of SCN neurons and thus can modulate the excitability and firing frequency according to the time of day. Intracellular Ca^{2+} mobilization through RyRs may affect neuronal excitability directly through Ca^{2+}-modulated plasma membrane channels and indirectly as a second messenger activating protein kinases regulating a variety of cellular processes converging at the cell membrane.

R. Aguilar-Roblero (✉) • D. Quinto-Muñoz
División de Neurociencias, Instituto de Fisiología Celular, Universidad Nacional
Autónoma de México, Circuito exterior S/N, Ciudad Universitaria, Coyoacán,
México, D.F. 04510, México
e-mail: raguilar@ifc.unam.mx

M. Díaz-Muñoz • A. Báez-Ruíz
Departamento de Neurobiología Molecular y Celular, Instituto de Neurobiología,
Universidad Nacional Autónoma de México, Querétaro 76230, QRO, México

G. Lundkvist
Department of Neuroscience, Karolinska Institutet, Stockholm, Sweden

S. Michel
Department of Molecular Cell Biology, Leiden University Medical Center,
2300 RC Leiden, The Netherlands

© Springer International Publishing Switzerland 2015   115
R. Aguilar-Roblero et al. (eds.), *Mechanisms of Circadian Systems
in Animals and Their Clinical Relevance*, DOI 10.1007/978-3-319-08945-4_7

7.1   The Circadian Clock in Mammals

Circadian rhythms are daily fluctuations in biochemical, physiological, and behavioral parameters. These rhythms are generated by specialized cells in the organisms, which function as biological clocks (Pittendrigh 1993). In mammals, the major circadian clock is located in the suprachiasmatic nuclei (SCN), located in the brain, at the basal portion of the anterior hypothalamus (Klein et al. 1991) (Fig. 7.1).

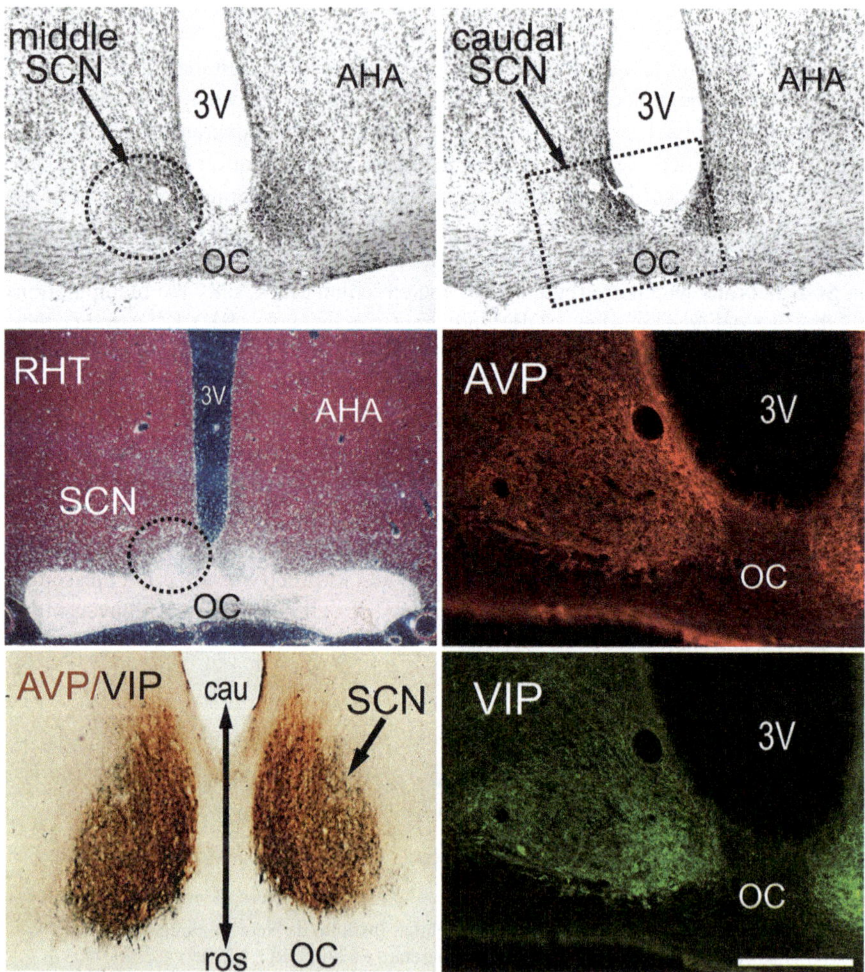

Fig. 7.1 The SCN from the rat. *Top row*: Nissl staining at two coronal levels of the suprachiasmatic nuclei (SCN). *Middle left*: The retino-hypothalamic tract (RHT) as seen by autoradiography to 3H-proline injected to the posterior chamber of the eye. *Middle* and *bottom right*: Coronal sections showing immunofluorescence to vasopressin (AVP, rhodamine) and vasoactive intestinal polypeptide (VIP, fluorescein). *Bottom left*: Double immunohistochemistry to AVP (*brown*) and VIP (*dark blue*) from a horizontal slice, caudal (cau) and rostral (ros) are indicated. Other abbreviations: *OC* optic chiasm, *3V* third ventricle, *AHA* anterior hypothalamic area. *Calibration bar*: 500 μm *top row*; 800 μm, *middle left*, 250 μm, *middle* and *bottom right*; 900 μm, *bottom left arrow*

The SCN is considered the central circadian clock because it is the only identified cell group in the body to fulfill the criteria to be considered as a self-sustained pacemaker (Menaker et al. 1978) controlling daily rhythms in behavior and physiology in mammals. These criteria include the following: (1) when it is lesioned or ablated from the organism, circadian rhythms in behavior are eliminated (Moore and Eichler 1972; Stephan and Zucker 1972); (2) circadian rhythms are not recovered unless the same tissue is transplanted back into the organism (Drucker-Colín et al. 1984; Lehman et al. 1987; Ralph et al. 1990; Aguilar-Roblero et al. 1994) (Fig. 7.2); (3) the SCN expresses circadian rhythms in metabolic and electrical activity, even when this tissue is isolated from the rest of the organism and kept in vitro as acute slices (Fig. 7.3) or cultured as isolated cells or organotypic slices (Schwartz and Gainer 1977; Inouye and Kawamura 1979; Shibata et al. 1982, 1983; Green and Gillette 1982; Welsh et al. 1995; Wilsbacher et al. 2002); (4) finally, elcetrical stimulation of the SCN, in situ or in vitro, phase shifts circadian rhythms in the locomotor activity of free moving organism (Rusak and Groos 1982) or in the neuronal activity rhythm in the isolated SCN tissue, respectively (Ding et al. 1994; Hurst et al. 2002).

7.2 The Molecular Circadian Clock

The molecular clocks in eukaryotes are constituted by transcription–translation feedback loops (TTL) in which dynamics lead them to oscillate with a circadian ($\approx$24 h) periodicity. In *Drosophila* and *rodents*, the core genes of the TTL share a promoter region called E-Box in which helix–loop–helix transcription factors bind and initiate gene transcription; in turn, the proteins from these genes accumulate in the cytoplasm and return dimerized to the nucleus, where they dislocate the binding of the helix–loop–helix transcription factors to the E-Box and suppress their own transcription. In rodents the canonical model of molecular circadian clock involves the following processes: the transcription factors CLOCK and BMAL-1 dimerize in the nuclei and bind to E-Boxes in the promoters to induce transcription of the following genes, *per1*, *per2*, *cry1*, and *cry2*. The proteins translated from the corresponding mRNAs (PER1, PER2, CRY1, and CRY2) are phosphorylated in the cytoplasm and form heterodimers among themselves to enter back into the nucleus. Once in the nucleus, the PER/CRY dimers inhibit their own transcription by removing CLOCK/BMAL1 complex from the E-Box. PER/CRY dimers are eventually removed from the nucleus and hydrolyzed, thus allowing the CLOCK/BMAL complex to bind again to the E-Box, and the cycle reinitiates with a periodicity close to 24 h (Lowrey and Takahashi 2004; Lee et al. 2011). For a recent discussion of the TTL model, see Brown et al. (2012).

The molecular oscillator described above has been demonstrated not only in SCN neurons but also in other brain regions and peripheral tissues (Yamazaki et al. 2000; Balsalobre et al. 2000; Wilsbacher et al. 2002; Reppert and Weaver 2002;

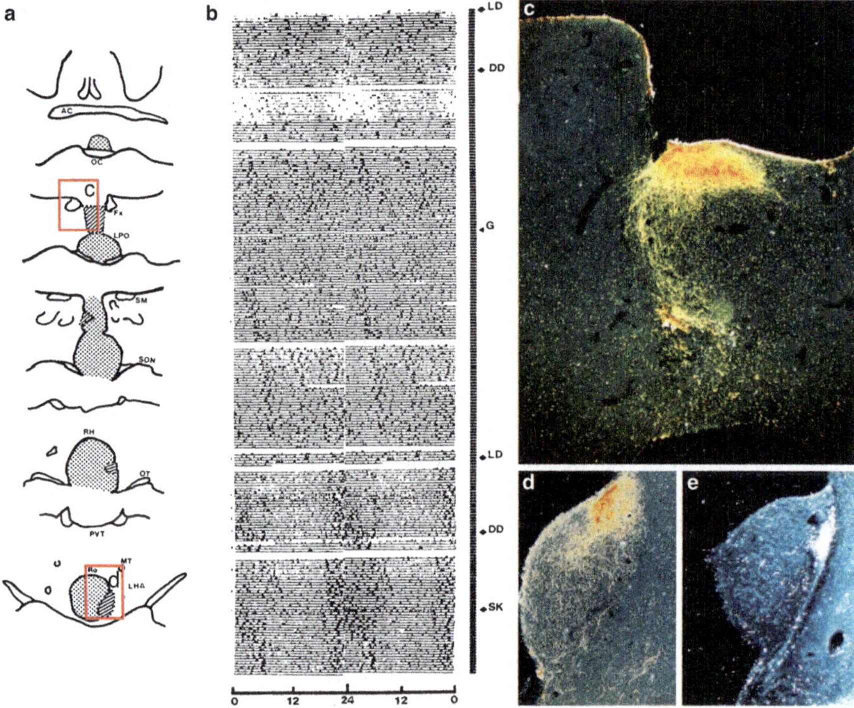

Fig. 7.2 SCN lesion in hamster abolishes wheel-running circadian rhythmicity, while grafting fetal hypothalamic area containing SCN restores rhythmicity. (**a**) Reconstruction of electrolytic lesion (*dotted area*) and location of the graft (*diagonal lines*) from the same animal showed in **b**. (**b**) Actogram of wheel running of a hamster sustaining an SCN lesion 2 weeks earlier. *LD* light–dark (14:10 h), *DD* constant darkness, G the day the animal received the fetal SCN graft. *SK* skeleton photoperiod (1 h lights on, 13:9 h *dark intervals*). (**c**) Dark field micrograph from VIP immunostaining of the graft placed in rostral third ventricle (shown in the *red rectangle C* in **a**). (**d**) Dark field micrograph from AVP immunostaining of the graft located caudally in the third ventricle (shown in the *red rectangle d* in **a**). (**e**) Dark field micrograph from adjacent section to **d** showing retinal ganglion projections to the graft labeled by TMB reaction to cholera toxin injected to the posterior eye chamber. Modified from Aguilar-Roblero et al. (1994)

Abe et al. 2002; Nishide et al. 2006), which raised the question whether cells other than SCN neurons should be considered also as circadian clocks. These cells are commonly referred to as peripheral oscillators. Transgenic rats and mice that transcribe luciferase—the enzyme responsible to the light emission in fireflies—every time the promoter of the per1 gene is transcribed, have been used as real-time reporters of

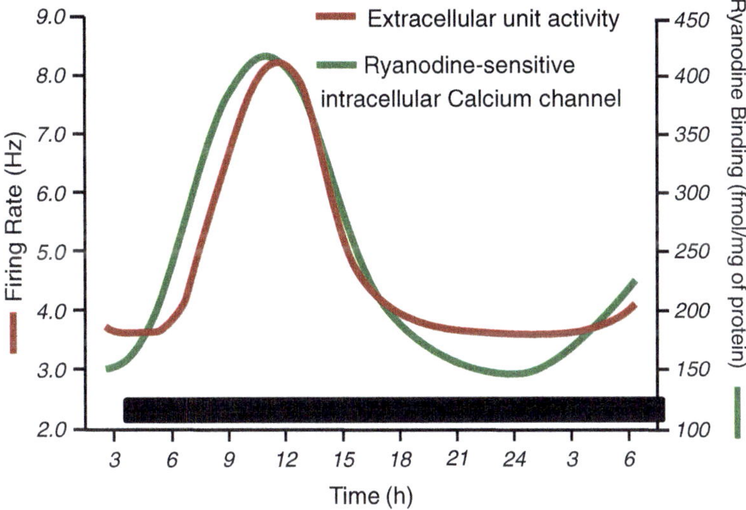

Fig. 7.3 Circadian rhythms in acute SCN slices kept in vitro. Firing rate and ryanodine binding to SCN were obtained in independent experiments. The time correlation of the opening of the ryanodine-receptor intracellular Ca^{2+} channel (RyR) indicated by increased binding to the rise in electrical activity leads us to suggest a causal relation

circadian dynamics. Cultured SCN explants from these animals are able to sustain circadian rhythms in *per*1-driven luminescence for prolonged intervals of time, whereas in explants from peripheral oscillators, the luminescence rhythms dampen after two to seven cycles (Yamazaki et al. 2000; Wilsbacher et al. 2002; Izumo et al. 2003). These results led to suggest that peripheral oscillators are driven by the central pacemaker in the SCN. This view was challenged by the finding of long-term (up to 20 days) luminescence circadian rhythms in explants from a knockin mouse expressing Period2::Luciferase fusion protein, in both the SCN and peripheral oscillators. These results demonstrate that peripheral tissues also contain self-sustained circadian oscillators that do not depend on the SCN and shift the role of the SCN to a phase coordinator rather than a central pacemaker driving peripheral oscillators (Yoo et al. 2003). Furthermore, the dampening of rhythms found at peripheral tissue level in per1-luc transgenic animals could be at least partially due to the uncoupling of oscillators at cellular level. This is further supported by the observation that bioluminescence rhythms of peripheral oscillators are restored by a serum shock to the cells in culture (Balsalobre et al. 2000; Jung et al. 2003; Izumo et al. 2003). Thus, peripheral oscillators could be phase coordinated (as suggested by Yoo et al. 2003) by the SCN outputs in vivo and probably regulate circadian rhythms at specific tissue and cellular level (Prolo et al. 2005; Loh et al. 2011).

7.3 Beyond the Transcription–Translation Feedback Loop

Previous findings underline the relevance of intercellular communication in the coupling processes among oscillators, which has led to a reevaluation of the role of the SCN neuronal network to maintain coherent phases as part of the clock mechanism. Cultured explants from the SCN present long-term luminescence rhythms with relatively stable phase relations as long as the network communication is intact. Although even in this conditions different regions within the SCN are not completely synchronous to each other, a wave of activation spreading throughout the nuclei with earliest peaking populations of neurons was observed in the medial and dorsal SCN (Yan and Okamura 2002; Yamaguchi et al. 2003; Evans et al. 2011; Foley et al. 2011; Nakamura et al. 2005). In contrast, in cultures from dissociated and dispersed SCN neurons studied at the individual cell level, the rhythms persist for long periods of time, but the phases drift apart from each other. These results indicate that although each SCN neuron has the molecular clock machinery, the SCN network plays a relevant role maintaining a coherent circadian signal by coupling individual oscillators. Further suggested roles of the SCN network are to provide stability and robustness to the circadian clock (Mohawk and Takahashi 2011; see Chap. 9 by Morgado et al.).

Recent studies indicate that the clock mechanism may involve other elements besides the molecular oscillator previously described. The inhibition of adenylyl cyclase by the administration of MDL-12,330A has been shown, in a dose-dependent manner, to decrease the amplitude of circadian transcription and protein synthesis in SCN explants from m*Per*1::*luciferase* and mPER2::LUC mice. This clearly indicates that intracellular cAMP signaling is necessary to sustain the progression of the transcriptional rhythms, probably by the activation of CRE sequences (O'Neill et al. 2008). There is also evidence that neuronal firing feedback amplifies the molecular oscillator. In *Drosophila* electrical silencing of pacemaker neurons disrupts circadian rhythms in locomotion and clock gene expression (Nitabach et al. 2002). In mammals, it has also been shown that action potentials are crucial for a robust expression of molecular rhythms. The blockade of generation of action potentials in the SCN neurons by tetrodotoxin (TTX), a toxin that blocks fast voltage-gated Na^+ channels, results in a decay of expression of circadian luminescence in SCN explants from the m*Per1-Luc* mouse; when analyzed at the cellular level, the amplitude of luminescence rhythmicity decreases and in some neurons disappears, while the individual acrophases among those cells that remain rhythmic become desynchronized (Yamaguchi et al. 2003). The observed effects of TTX on circadian per1-driven luminescence rhythms may depend on Ca^{2+} influx that results from membrane depolarization during action potentials, since Ca^{2+} influx also contributes to sustain the molecular circadian oscillation in SCN neurons. Per1-luc rhythmicity from SCN rat explants and PER2::LUC mouse explants is abolished by decreasing Ca^{2+} concentration in the extracellular medium below 300 mM or by the administration of increasing doses of the intracellular Ca^{2+} chelator BAPTA-AM (Lundkvist et al. 2005). This hypothesis that Ca^{2+} influx is necessary to sustain the molecular circadian oscillation is also supported by the elimination of Per2 and Bmal1 circadian expression in cadmium-treated SCN2.2 cells (Sang-Soep et al. 2005).

7.4 Ionic Currents Involved in Neuronal Firing in the SCN

In order to be useful to the rest of the organism, the circadian time signal generated by the molecular clock must be translated into a neuronal firing pattern.

SCN neurons show circadian oscillations in firing frequency in vivo (Inouye and Kawamura 1979) and in vitro (Shibata et al. 1982; Green and Gillette 1982; Welsh et al. 1995); SCN neurons produce action potentials at a higher frequency and show a higher input resistance during the subjective day than during the subjective night. Neuronal firing rate and membrane excitability in SCN neurons depend on the interaction of a number of ionic currents through the plasma membrane (Schaap et al. 2003; Kuhlman and McMahon 2006; Brown and Piggins 2007; Aguilar-Roblero et al. 2009; Colwell 2011). The action potential is triggered by a combination of currents including a slow-inactivating Na^+ leak current sensitive to riluzole (Kononenko et al. 2004a, b), a persistent Na^+ current insensitive to TTX (Pennartz et al. 1997; Jackson et al. 2004), a hyperpolarization-activated conductance (I_h) (Akasu et al. 1993; de Jeu and Pennartz 1997), and a hyperpolarization-activated conductance gated by cAMP (HCN) (Atkinson et al. 2011). The progressive activation of such inward currents drives the membrane potential from the afterhyperpolarization at the end of an action potential to the threshold of activation for the voltage-gated fast Na^+ channels, as well as the L- and T-type Ca^{2+} channels responsible for the next action potential (Huang 1993; Pennartz et al. 2002; Nahm et al. 2005). Repolarization of the membrane is driven by outward K^+ currents such as the fast delayed rectifier (fDR) channels and the transient A-type K^+ current (I_A), which have been demonstrated to modulate the duration of the action potential and thus to participate in the modulation of SCN firing frequency (Bouskila and Dudek 1995; Itri et al. 2005; Kudo et al. 2011). Finally, the afterhyperpolarization is mainly driven by several Ca^{2+}-modulated K^+ currents such as the large-conductance current (BK) iberiotoxin sensitive (Pitts et al. 2006; Vandael et al. 2010), the small-conductance current (SK) apamin sensitive (Teshima et al. 2003), an iberiotoxin- and apamin-insensitive $K^+_{(Ca)}$ current (Cloues and Sather 2003), and a barium-sensitive K^+ current ($K_{(Ba)}$) (De Jeu et al. 2002).

7.5 Molecular Clock Control of SCN Neuronal Firing Rate

The circadian modulation of SCN neuronal firing rate is essential for the expression of circadian rhythms. At the time, the mechanisms for the control of membrane excitability by the molecular circadian clock components are not completely identified. Ion channel activity can be either regulated by differential expression of genes coding for membrane channels or, as often found, by posttranslational modifications of these proteins. Outside the SCN, numerous studies have demonstrated the importance of kinase/phosphatase activity in mediating short-term changes in channel function that alter electrical excitability (Smart 1997; Mathie 2007; Okamura 2007; Ebner-Bennatan et al. 2012; He et al. 2013; Kyle et al. 2013).

Circadian modulation of ionic currents has been reported in L-type Ca^{2+} and K^+ conductances, but its contribution to the expression of circadian SCN firing has only been clearly shown for K^+ (Colwell 2011). High firing rate from SCN neurons during the day is regulated by fDR K^+ currents (Itri et al. 2005). The fast kinetics of fDR K^+ currents allows quick repolarization of the neuronal membrane to end the action potential without altering action potential threshold or amplitude (Rudy and McBain 2001). Acute blockade of the fDR with 0.5 mM 4-aminopyridine (AP) or 1 mM tetraethylammonium (TEA) during the day decreases the firing rate of SCN neurons, thus blunting the day/night difference in spontaneous activity, while long-term application of 4-AP (0.5 mM) prevents the expression of multiunit electrical activity rhythm from the SCN (Itri et al. 2005). Lower firing rate of SCN neurons during the night involves the large-conductance Ca^{2+}-activated K^+ channel (BK). The activity and mRNA levels of this channel is increased during the night as compared to the day (Pitts et al. 2006). Interestingly, pharmacological block of BK in this study led to a significant reduction of spontaneous spike frequency during the day. This suggests a dual role of BK in regulating spike frequency, which was also found after genetic activation of BK in SCN neurons (Montgomery and Meredith 2012). Kcnma1−/− knockout mice, which lack BK channels, show disrupted behavioral circadian rhythms and a high SCN neuronal action potential frequency during the night, close to the action potential frequency characteristic from daytime recorded SCN neurons, in spite of circadian expression of clock genes such as Bmal-1 and mPer2 (Meredith et al. 2006; Kent and Meredith 2008).

7.6 Intracellular Ca^{2+} as an Output from the Molecular Clock in SCN Neurons

Neuronal calcium dynamics embodies a code in which the frequency and the amplitude of calcium transients regulate the transcriptional and metabolic status of the cell (Pralong et al. 1994; Spitzer 2002; Neher and Sakaba 2008). Ca^{2+} homeostasis in neurons involves (1) its influx through voltage- or ligand-gated ionic channels or (2) its release from the endoplasmic reticulum, Golgi apparatus, and nucleus (Finkbeiner and Greenberg 1998), (3) Ca^{2+} ATPases located in the plasma membrane and endoplasmic reticulum, and (4) several Ca^{2+}-binding proteins (Carafoli et al. 2001). Thus different Ca^{2+} pools have specific roles in regulating the spatio-temporal neuronal responses to Ca^{2+} signaling (Rizzuto 2001). Intracellular Ca^{2+} stores in the endoplasmic reticulum are a key component in the dynamic of $[Ca^{2+}]_i$. Thapsigargin-sensitive Ca^{2+}-ATPase is responsible for the Ca^{2+} uptake into the stores, while Ca^{2+} release depends mainly on high-conductance Ca^{2+} channels termed ryanodine and inositol (1,4,5)-triphosphate receptors, RyRs and IP3Rs, respectively (Meldolesi and Pozzan 1998; Verkhratsky 2005).

In the SCN there are reports on the role of intracellular Ca^{2+} as a signal involved in resetting the clock, maintaining the molecular oscillation and transmitting the rhythmicity from the clock itself to the efferents involved in the expression of the overt rhythmicity. The manipulation of calmodulin levels phase shifts metabolic and electrical rhythms in SCN neurons (Shibata and Moore 1994), while RyRs mediate light-induced phase delays during early subjective night (Ding et al. 1998). The role of Ca^{2+} influx in sustaining the circadian oscillation of the per1 and per2 clock gene has been shown in SCN neurons (Lundkvist et al. 2005; Sang-Soep et al. 2005).

A circadian rhythm in free cytoplasmic Ca^{2+}, originating from intracellular Ca^{2+} storages, may be one of the first transmission elements linking the molecular oscillator to the circadian modulation of SCN activity, particularly the modulation of SCN firing rate (Fig. 7.3). In the SCN intracellular Ca^{2+} regulation shows a clear circadian rhythm. The SCN expresses both types of intracellular Ca^{2+} channels, inositol triphosphate (IP3) and ryanodine sensitive (RyR). Binding experiments with radioactive ligands demonstrate that RyR but not the IP3 presents circadian regulation in constant darkness, while in other areas of the brain outside the SCN, both IP3 and RyR activity remain constant throughout the day. The peak expression of the RyR binding circadian rhythm in the SCN occurs during the middle of the subjective day at circadian time (CT) 07 and is due to an increase in the protein expression of the neuronal RyR type 2 (RyR2), present in neurons but not astrocytes (Díaz-Muñoz et al. 1999) (Fig. 7.4a-c). Circadian rhythmicity in mRNA expression of RyR is under direct control of bmal1, one of the positive regulators of the clock genes, as bmal1 knockout mice show decreased expression of RyR mRNA and protein (Pfeffer et al. 2009).

Concentrations of free intracellular Ca^{2+} also show a clear circadian fluctuation as measured with fura2 AM. During the subjective day, (ZT 1–6) $[Ca^{2+}]_i$ is higher as compared with the subjective night (ZT 13–18), and this circadian fluctuation occurs even in animals housed in constant darkness (Colwell 2000). This rhythm was abolished with the administration of TTX, which blocks Na^+ channels, or methoxyverapamil that blocks L-type Ca^{2+} channels, concluding that membrane events are associated with this rhythm. In contrast, in SCN organotypic cultures, using a chameleon Ca^{2+} reporter, circadian rhythms in $[Ca^{2+}]_i$ and neuronal firing rate can be dissociated (Ikeda et al. 2003). In this study TTX abolished the rhythm in SCN neuronal firing but did not affect the rhythm in $[Ca^{2+}]_i$. However, closing the intracellular Ca^{2+} channel sensible to ryanodine (RyR) with high doses of ryanodine abolished both the rhythmicity in intracellular Ca^{2+} concentration and electrical activity in SCN neurons. In contrast, the inhibition of voltage-gated Ca^{2+} channels by nifedipine did not affect any of the rhythms.

Ryanodine at a dose of 0.1 µM activates the RyR and causes release of Ca^{2+} from the endoplasmic reticulum, while ryanodine at a dose of 100 µM has the opposite effect and prevents the release of Ca^{2+} (Chu et al. 1990). In rats recorded in constant dim red light, chronic administration of different doses of ryanodine into the SCN via a cannula connected to an osmotic minipump has an effect on behavioral circadian rhythms (Mercado et al. 2009). Intracellular Ca^{2+} release by the activation of RyRs by 0.1 µM ryanodine induced a significant shortening of the endogenous period (Fig. 7.4d), whereas the inhibition of the Ca^{2+} channels by 100 µM ryanodine disrupted the circadian rhythmicity. After withdrawing the pharmacological treatments, the period and phase of rhythmicity returned to basal and expected values.

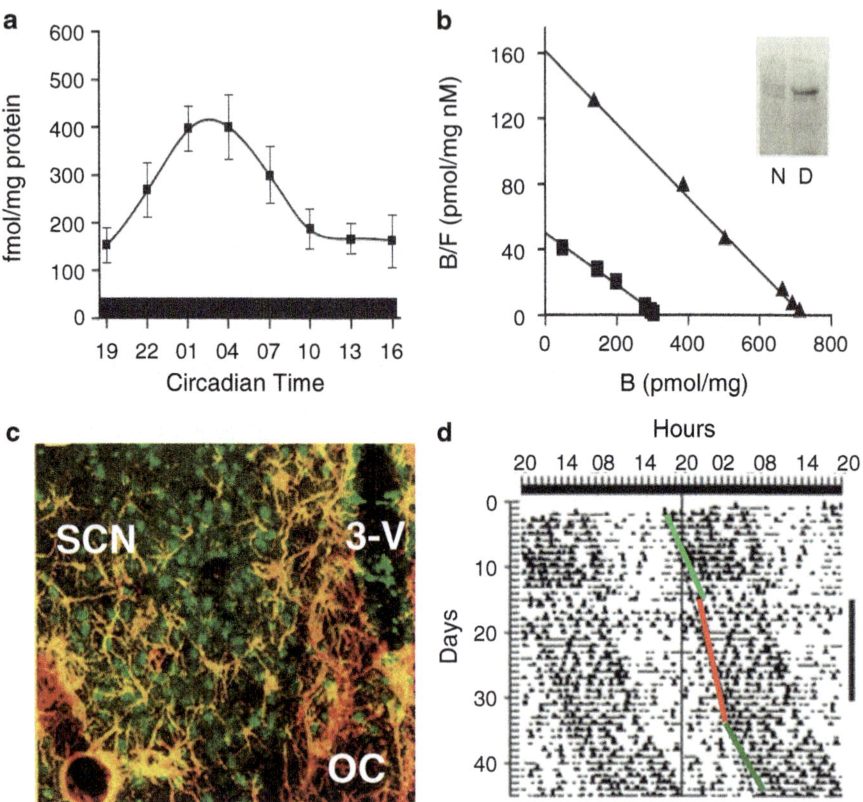

Fig. 7.4 The circadian rhythm in the RyR binding to ryanodine (**a**) corresponds to changes in Bmax as indicated from Scatchard analysis (**b**) comparing day (*triangles*) vs. night (*squares*); Western blot to RyR type 2 confirmed higher levels of protein during the day (*inset*). (**c**) Confocal microscopy showed that RyR-2 immunofluorescence was located in neurons but not in astrocytes (labeled to GFAP). Abbreviations: *SCN* suprachiasmatic nucleus, *3V* third ventricle, *OC* optic chiasm. (**d**) Actogram of drinking behavior from a rat receiving 2 weeks of 100 nM ryanodine into the SCN. The drug was continuously delivered (*vertical black bar* on the *right*) with an ALZET miniosmotic pump. The period of rhythmicity shortened during the administration (*red line*) in comparison with before and after the administration (*green lines*). Since neither the period nor the phase was affected but only ryanodine administration, we conclude the effects were in the output pathway from the clock to the effectors of rhythmicity

These results suggest that changes in overt rhythms induced by both doses of ryanodine did not directly involve the clock mechanism.

Pharmacological manipulation of ryanodine-sensitive intracellular Ca^{2+} channels modifies SCN neuronal firing rate, recorded by perforated patch from rat brain slices (Aguilar-Roblero et al. 2007). Thus, the increase of $[Ca^{2+}]_i$ by opening the RyRs by the administration of either 1 mM caffeine or 100 nM ryanodine increases the firing frequency in 58 % of SCN neurons, without having any effect on the firing threshold. Closure of RyRs by the administration of 10 μM dantrolene decreases firing rate in 64 % of the tested neurons, whereas the administration of 80 μM ryanodine decreases firing rate in 100 % of neurons selected by their irregular firing rate and clear hyperpolarizing after potential during basal conditions. Synaptic activity in the SCN can be suppressed by pharmacological blockade of both GABAergic and glutaminergic receptors. Thus, the administration of a cocktail containing 10 μM bicuculline ($GABA_A$ receptors), 50 μM AP-V (NMDA receptors), and 10 μM DNQX (AMPA receptors) inhibits synaptic transmission but does not prevent ryanodine effects on firing frequency (increase by 100 nM or decrease by 80 μM), clearly indicating that ryanodine modulation of SCN firing rate is due to a direct action on intracellular Ca^{2+} mobilization through RyRs and not to indirect effects on the synaptic network (Fig. 7.5). Since the effects of both doses of ryanodine were observed both during daytime (ZT4–10) and nighttime (ZT17–22), the pharmacological gating of the RyRs is able to override the circadian clock control of SCN neuronal firing rate.

Similar results were found in SCN slices from C57Bl/6 mice. Activating RyRs with 100 nM ryanodine increased $[Ca^{2+}]_i$ in about 40 % of neurons, as measured with the Ca^{2+} indicator dye Fura-2 (Fig. 7.6a). Application of the RyR antagonist dantrolene (10 μM) decreased free cytoplasmic Ca^{2+} in 54 % of the recorded neurons. The responses to the RyR agonist or antagonist were dependent on the basal $[Ca^{2+}]_i$ of the cell (Fig. 7.6b) which may in turn be correlated to the activation state of the RyR. Neither activation of nor blocking of the RyR had major effects on the period or phase of PER2 expression assessed by luminescence in SCN slices from PER2::LUC transgenic mice (Fig. 7.6c). This indicates that acute intracellular Ca^{2+} release from RyR in SCN neurons does not affect the clock angular velocity (unpublished data). The data are consistent with the behavioral study in rats described above that concluded no impact of RyR activation on molecular clock mechanisms (Mercado et al. 2009). The effects of manipulating the RyR on the neuronal firing, measured in cell-attached configuration, were similar to those previously reported in rats, in which the firing rate increased after 100 nM ryanodine and decreased after 10 μM dantrolene. The proportion of SCN neurons showing ryanodine-induced increase in $[Ca^{2+}]_i$ is similar to the percentage of cells increasing neuronal activity after RyR activation. This relationship also holds true for the reduction of $[Ca^{2+}]_i$ and spike frequency after blocking the RyR by dantrolene. A potential link between RyR-induced changes in $[Ca^{2+}]_i$ and neuronal excitably could be implemented at least in part by the Ca^{2+}-dependent K^+ channel BK, which seems to contribute to the high spike frequency during the day (Pitts et al. 2006).

7.7 Concluding Remarks

Free intracellular Ca^{2+} is a versatile intracellular signal involved in numerous functions in virtually every cell. The complex morphology and compartmentalization of neurons adds to Ca^{2+} versatility as intracellular signal in the nervous system. Thus, the specific cellular compartment, the amplitude of cytoplasmic calcium rise, and its

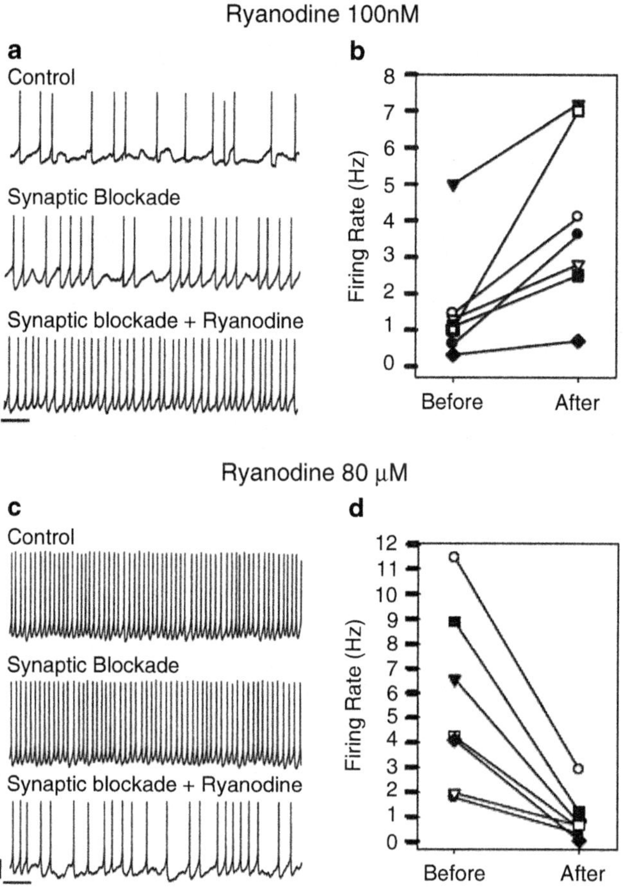

Fig. 7.5 Pharmacological manipulation of the RyRs modulates the firing rate from SCN neurons recorded in perforated patch configuration. Low dose of ryanodine (100 nM) opens the RyR and releases Ca^{2+} from the endoplasmic reticulum to the cytoplasm, while a high dose (80 µM) closes the RyR and leads to a decrease in free Ca^{2+} in the cytoplasm. **a** and **c** show typical traces from two neurons receiving either 100 nM or 80 µM ryanodine. Synaptic blockade to isolate the recorded neuron from the network was accomplished by a cocktail of bicuculline (10 µM), AP-V (50 µM), and DNQX (10 µM). **b** and **d** show the summary of neurons receiving either 100 nM or 80 µM ryanodine. Modified from Aguilar-Roblero et al. (2007)

time course determine the target of the Ca^{2+} signaling process. Depending on the compartment where the $[Ca^{2+}]_i$ increase occurs, it could induce neurotransmitter exocytosis, calcium-induced calcium release, ionic channel gating, specific protein phosphorylation, or gene transcription. Furthermore, all these processes span over a wide time range, from neurotransmitter release in microseconds to gene transcription lasting for minutes or even hours (Berridge et al. 2003).

It is not surprising that Ca^{2+} signaling in SCN neurons participates in a variety of processes related to its role as a circadian clock, which includes resetting the clock,

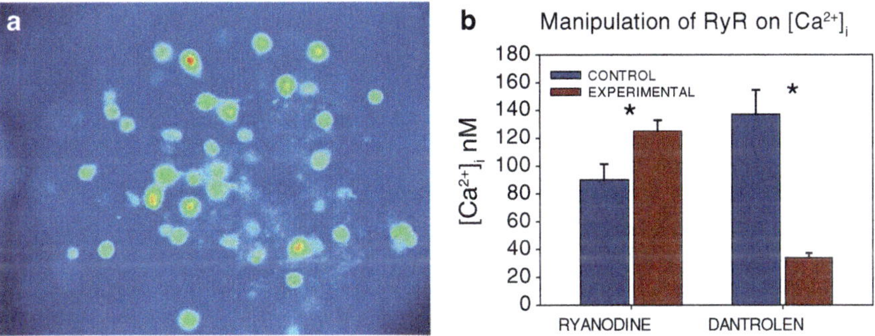

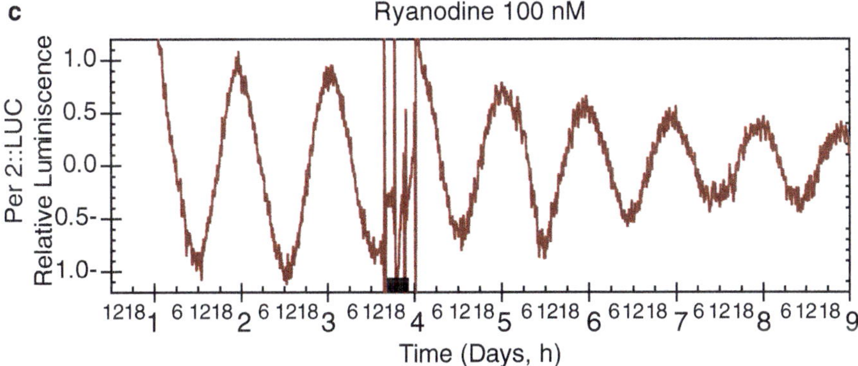

Fig. 7.6 Ryanodine-mediated changes of $[Ca^{2+}]_i$ do not affect PER2 rhythm in mouse SCN neurons. (**a**) Fluorescent image of SCN neurons loaded with the Ca^{2+} indicator dye Fura2. (**b**) Ryanodine (100 nM) application significantly increased $[Ca^{2+}]_i$ in a subset of neurons with low baseline $[Ca^{2+}]_i$ levels, and blocking RyR with dantrolene (10 µM) results in a decrease of $[Ca^{2+}]_i$ in cells with high baseline $[Ca^{2+}]_i$ levels ($n = 3$; $*P < 0.05$). (**c**) Administration of 100 nM ryanodine for 6 h dld not affect the phase or the period of the PER2::LUC rhythm. This finding suggests that Ca^{2+} release from the endoplasmic reticulum via RyRs is an output from the molecular oscillator to the neuronal membrane

sustaining the molecular circadian oscillation, and transmitting clock signal to its outputs to express the overt rhythmicity. Nevertheless, the complexity of Ca^{2+} signaling makes difficult to experimentally dissect these processes. In this chapter, we have focused on the participation of Ca^{2+} signaling in the clock output. The first step of the clock output requires that the oscillation of clock genes be translated to a code produced by action potentials in SCN neurons, which in turn could be transmitted to the rest of the brain and eventually the rest of the organism. Here we have reviewed evidence that the RyR is under direct control of the molecular clock and its manipulation modulates $[Ca^{2+}]_i$ and the excitability and firing frequency from SCN neurons, but does not affect the period or phase of PER2::LUC luminescence rhythm or behavioral overt rhythmicity. The evidence obtained so far clearly suggests that Ca^{2+} signaling is one of the output pathways from the molecular circadian clock to the neuronal plasma membrane.

A pending question on the role of Ca^{2+} signaling as a clock output is to clearly identify those channels modulated by this signaling pathway. It is worth noticing that the BK channel is directly gated by $[Ca^{2+}]_i$, and during the day, when $[Ca^{2+}]_i$ reaches its maximum, the firing rate of SCN neurons is decreased after blocking the BK channel. Aside from direct modulation of neuronal excitability by Ca^{2+}-gated channels, indirect channel modulation could involve protein kinases regulating a variety of cellular processes converging at the cell membrane. It is thus clear that further studies are needed to unravel the complex participation of Ca^{2+} signaling as a clock output and to identify all membrane ionic channels contributing to the circadian rhythm in neuronal firing.

Acknowledgments We thank José Luis Chávez, Ana María Escalante, and Francisco Pérez for skillful technical assistance. This work was partially supported by grants from CONACyT 128528, PAPIIT IN204811, and FONCICYT 91984.

References

Abe M, Herzog ED, Yamazaki S et al (2002) Circadian rhythms in isolated brain regions. J Neurosci 22(1):350–356

Aguilar-Roblero R, Morin LP, Moore RY (1994) Morphological correlates of circadian rhythm restoration induced by transplantation of the suprachiasmatic nucleus in hamsters. Exp Neurol 136:1–11

Aguilar-Roblero R, Mercado C, Alamilla J et al (2007) Ryanodine receptor Ca2+-release channels are an output pathway for the circadian clock in the rat suprachiasmatic nuclei. Eur J Neurosci 26(3):575–582

Aguilar-Roblero R, Alamilla J, Mercado C et al (2009) Neuronal activity in the suprachiasmatic nuclei: cellular and molecular mechanisms. In: Fanjul-Moles ML, Aguilar-Roblero R (eds) Comparative aspects of circadian rhythms. Research Signpost, Kerala, India, pp 185–203

Akasu T, Shoji S, Hasuo H (1993) Inward rectifier and low-threshold calcium currents contribute to the spontaneous firing mechanism in neurons of the rat suprachiasmatic nucleus. Pflugers Arch 425:109–116

Atkinson SE, Maywood ES, Chesham JE et al (2011) Cyclic AMP signalling controls action potential firing rate and molecular circadian pacemaking in the suprachiasmatic nucleus. J Biol Rhythms 26(3):210–220

Balsalobre A, Brown SA, Marcacci L et al (2000) Resetting of circadian time in peripheral tissues by glucocorticoid signaling. Science 289(5488):2344–2347

Berridge MJ, Bootman MD, Roderick HL (2003) Calcium signalling: dynamics, homeostasis and remodelling. Nat Rev Mol Cell Biol 4:517–529

Bouskila Y, Dudek FE (1995) A rapidly activating type of outward rectifier K+ current and A-current in rat suprachiasmatic nucleus neurons. J Physiol 488(2):339–350

Brown TM, Piggins HD (2007) Electrophysiology of the suprachiasmatic circadian clock. Prog Neurobiol 82(5):229–255

Brown SA, Kowalska E, Dallham R (2012) (Re)inventing the circadian feedback loop. Dev Cell 22:477–487

Carafoli E, Santella L, Branca D et al (2001) Generation, control, and processing of cellular calcium signals. Crit Rev Biochem Mol Biol 36(2):107–260

Chu A, Díaz-Muñoz M, Hawkes MJ et al (1990) Ryanodine as a probe for the functional state of the skeletal muscle reticulum calcium release channel. Mol Pharmacol 37:735–741

Cloues RK, Sather WA (2003) Afterhyperpolarization regulates firing rate in neurons of the suprachiasmatic nucleus. J Neurosci 23(5):1593–1604

Colwell CS (2000) Circadian modulation of calcium levels in cells in the suprachiasmatic nucleus. Eur J Neurosci 12:571–576

Colwell CS (2011) Linking neural activity and molecular oscillations in the SCN. Nat Rev Neurosci 12:553–569

de Jeu MT, Pennartz CM (1997) Functional characterization of the H-current in SCN neurons in subjective day and night: a whole-cell patch-clamp study in acutely prepared brain slices. Brain Res 767:72–80

de Jeu M, Geurtsen A, Pennartz C (2002) A Ba(2+)-sensitive K(+) current contributes to the resting membrane potential of neurons in rat suprachiasmatic nucleus. J Neurophysiol 88(2):869–878

Díaz-Muñoz M, Dent A, Granados-Fuentes D et al (1999) Circadian modulation of the ryanodine receptor type 2 in the SCN of rodents. Neuroreport 10:481–486

Ding JM, Chen D, Weber ET et al (1994) Resetting the biological clock: mediation of nocturnal circadian shifts by glutamate and NO. Science 266(5191):1713–1717

Ding JM, Buchanan GF, Tischkau SA et al (1998) A neuronal ryanodine receptor mediates light-induced phase delays of the circadian clock. Nature 394:381–384

Drucker-Colín R, Aguilar-Roblero R, García-Hernández F et al (1984) Fetal suprachiasmatic nucleus transplants: diurnal rhythm recovery of lesioned rats. Brain Res 311:353–357

Ebner-Bennatan S, Patrich E, Peretz A et al (2012) Multifaceted modulation of K+ channels by protein-tyrosine phosphatase ε tunes neuronal excitability. J Biol Chem 287(33):27614–27628

Evans JA, Leise TL, Castanon-Cervantes O et al (2011) Intrinsic regulation of spatiotemporal organization within the suprachiasmatic nucleus. PLoS One 6:e15869. doi:10.1371/journal.pone.0015869

Finkbeiner S, Greenberg ME (1998) Ca2+ channel-regulated neuronal gene expression. J Neurobiol 37(1):171–189

Foley NC, Tong TY, Foley D et al (2011) Characterization of orderly spatiotemporal patterns of clock gene activation in mammalian suprachiasmatic nucleus. Eur J Neurosci 33:1851–1865

Green DJ, Gillette R (1982) Circadian rhythm of firing rate recorded from single cells in the rat suprachiasmatic brain slice. Brain Res 245(1):198–200

He C, Chen F, Li B et al (2013) Neurophysiology of HCN channels: from cellular functions to multiple regulations. Prog Neurobiol. doi:10.1016/j.pneurobio.2013.10.001, pii: S0301-0082(13)00101-9 [Epub ahead of print]

Huang RC (1993) Sodium and calcium currents in acutely dissociated neurons from rat suprachiasmatic nucleus. J Neurophysiol 70(4):1692–1703

Hurst WJ, Mitchell JW, Gillette MU (2002) Synchronization and phase-resetting by glutamate of an immortalized SCN cell line. Biochem Biophys Res Commun 298(1):133–143

Ikeda M, Sugiyama T, Wallace CS (2003) Circadian dynamics of cytosolic and nuclear Ca2+ in single suprachiasmatic nucleus neurons. Neuron 38:253–263

Inouye ST, Kawamura H (1979) Persistence of circadian rhythmicity in a mammalian hypothalamic "island" containing the suprachiasmatic nucleus. Proc Natl Acad Sci U S A 76(11):5962–5966

Itri JN, Michel S, Vansteensel MJ et al (2005) Fast delayed rectifier potassium current is required for circadian neural activity. Nat Neurosci 8:650–856

Izumo M, Johnson CH, Yamazaki S (2003) Circadian gene expression in mammalian fibroblasts revealed by real-time luminescence reporting: temperature compensation and damping. Proc Natl Acad Sci U S A 100(26):16089–16094

Jackson AC, Yao GL, Bean BP (2004) Mechanism of spontaneous firing in dorsomedial suprachiasmatic nucleus neurons. J Neurosci 24:7985–7998

Jung H, Choe Y, Kim H et al (2003) Involvement of CLOCK:BMAL1 heterodimer in serum-responsive mPer1 induction. Neuroreport 14(1):15–19

Kent J, Meredith AL (2008) BK channels regulate spontaneous action potential rhythmicity in the suprachiasmatic nucleus. PLoS One 3(12):e3884. doi:10.1371/journal.pone.0003884

Klein DC, Moore RY, Reepert S (eds) (1991) Suprachiasmatic nucleus. The mind's clock. Oxford University Press, New York

Kononenko I, Medina I, Dudek FE (2004a) Persistent subthreshold voltage-dependent cation single channels in suprachiasmatic nucleus neurons. Neuroscience 129:85–92

Kononenko NI, Shao LR, Dudek FE (2004b) Riluzole-sensitive slowly inactivating sodium current in rat suprachiasmatic nucleus neurons. J Neurophysiol 91:710–718

Kudo T, Loh DH, Kuljis D et al (2011) Fast delayed rectifier potassium current: critical for input and output of the circadian system. J Neurosci 31(8):2746–2755

Kuhlman SJ, McMahon DG (2006) Encoding the ins and outs of circadian pacemaking. J Biol Rhythms 21(6):470–481

Kyle BD, Hurst S, Swayze RD et al (2013) Specific phosphorylation sites underlie the stimulation of a large conductance, Ca(2+)-activated K(+) channel by cGMP-dependent protein kinase. FASEB J 27(5):2027–2038

Lee H-M, Chen R, Kim H et al (2011) The period of the circadian oscillator is primarily determined by the balance between casein kinase 1 and protein phosphatase 1. Proc Natl Acad Sci U S A 108(39):16451–16456

Lehman MN, Silver R, Gladstone WR et al (1987) Circadian rhythmicity restored by neural transplant. Immunocytochemical characterization of the graft and its integration with the host brain. J Neurosci 7(6):1626–1638

Loh DH, Dragich JM, Kudo T et al (2011) Effects of vasoactive intestinal peptide genotype on circadian gene expression in the suprachiasmatic nucleus and peripheral organs. J Biol Rhythms 26(3):200–209

Lowrey PL, Takahashi JS (2004) Mammalian circadian biology: elucidating genome-wide levels of temporal organization. Annu Rev Genomics Hum Genet 5:407–441

Lundkvist GB, Kwak Y, Davis EK et al (2005) A calcium flux is required for circadian rhythm generation in mammalian pacemaker neurons. J Neurosci 25:7682–7686

Mathie A (2007) Neuronal two-pore-domain potassium channels and their regulation by G protein-coupled receptors. J Physiol 578(Pt 2):377–385

Meldolesi J, Pozzan T (1998) The heterogeneity of ER Ca2+ stores has a key role in nonmuscle cell signaling and function. J Cell Biol 142(6):1395–1398

Menaker M, Takahashi JS, Eskin A (1978) The physiology of circadian pacemakers. Annu Rev Physiol 40:501–526

Mercado C, Díaz-Muñoz M, Alamilla J et al (2009) Ryanodine-sensitive intracellular Ca2+ channels in rat suprachiasmatic nuclei are required for circadian clock control of behavior. J Biol Rhythms 24(3):203–210

Meredith AL, Wiler SW, Miller BH et al (2006) BK calcium-activated potassium channels regulate circadian behavioral rhythms and pacemaker output. Nat Neurosci 9(8):1041–1049

Mohawk JA, Takahashi JS (2011) Cell autonomy and synchrony of suprachiasmatic nucleus circadian oscillators. Trends Neurosci 34(7):349–358

Montgomery JR, Meredith AL (2012) Genetic activation of BK currents in vivo generates bidirectional effect on neuronal excitability. Proc Natl Acad Sci U S A 109(46):18997–19002

Moore RY, Eichler VB (1972) Loss of a circadian adrenal corticosterone rhythm following suprachiasmatic lesions in the rat. Brain Res 42:201–206

Nahm SS, Farnell YZ, Griffith W et al (2005) Circadian regulation and function of voltage-dependent calcium channels in the suprachiasmatic nucleus. J Neurosci 25(40):9304–9308

Nakamura W, Yamazaki S, Takasu NN et al (2005) Differential response of Period 1 expression within the suprachiasmatic nucleus. J Neurosci 25(23):5481–5487

Neher E, Sakaba T (2008) Multiple roles of calcium ions in the regulation of neurotransmitter release. Neuron 59(6):861–872

Nishide SY, Honma S, Nakajima Y et al (2006) New reporter system for Per1 and Bmal1 expressions revealed self-sustained circadian rhythms in peripheral tissues. Genes Cells 11(10):1173–1182

Nitabach MN, Blau J, Holmes TC (2002) Electrical silencing of Drosophila pacemaker neurons stops the free-running circadian clock. Cell 109:485–495

O'Neill JS, Maywood ES, Chesham JE et al (2008) cAMP-dependent signaling as a core component of the mammalian circadian pacemaker. Science 320:949–953

Okamura Y (2007) Biodiversity of voltage sensor domain proteins. Pflugers Arch 454(3):361–371

Pennartz CM, Bierlaagh MA, Geurtsen AM (1997) Cellular mechanisms underlying spontaneous firing in rat suprachiasmatic nucleus: involvement of a slowly inactivating component of sodium current. J Neurophysiol 78:1811–1825

Pennartz CM, de Jeu MT, Bos NP et al (2002) Diurnal modulation of pacemaker potentials and calcium current in the mammalian circadian clock. Nature 416(6878):286–290

Pfeffer M, Müller CM, Mordel J et al (2009) The mammalian molecular clockwork controls rhythmic expression of its own input pathway components. J Neurosci 29(19):6114–6123

Pittendrigh CS (1993) Temporal organization: reflections of a Darwinian clock-watcher. Annu Rev Physiol 55:16–54

Pitts GR, Ohta H, McMahon DG (2006) Daily rhythmicity of large-conductance Ca2+-activated K+ currents in suprachiasmatic nucleus neurons. Brain Res 1071:54–62

Pralong WF, Spät A, Wollheim CB (1994) Dynamic pacing of cell metabolism by intracellular Ca2+ transients. J Biol Chem 269(44):27310–27314

Prolo LM, Takahashi JS, Herzog ED (2005) Circadian rhythm generation and entrainment in astrocytes. J Neurosci 25(2):404–408

Ralph MR, Foster RG, Davis FC et al (1990) Transplanted suprachiasmatic nucleus determines circadian period. Science 247(4945):975–978

Reppert SM, Weaver DR (2002) Coordination of circadian timing in mammals. Nature 418:935–941

Rizzuto R (2001) Intracellular Ca(2+) pools in neuronal signalling. Curr Opin Neurobiol 11(3):306–311

Rudy B, McBain CJ (2001) Kv3 channels: voltage-gated K+ channels designed for high-frequency repetitive firing. Trends Neurosci 24(9):517–526

Rusak B, Groos G (1982) Suprachiasmatic stimulation phase shifts rodent circadian rhythms. Science 215(4538):1407–1409

Sang-Soep N, Farnell YZ, Griffith W et al (2005) Circadian regulation and function of voltage-dependent calcium channels in the suprachiasmatic nucleus. J Neurosci 25:9304–9308

Schaap J, Pennartz C, Meijer JH (2003) Electrophysiology of the circadian pacemaker in mammals. Chonobiol Int 20(2):171–188

Schwartz WJ, Gainer H (1977) Suprachiasmatic nucleus: use of 14C-labeled deoxyglucose uptake as a functional marker. Science 197(4308):1089–1091

Shibata S, Moore RY (1994) Calmodulin inhibitors produce phase shifts of circadian rhythms in vivo and in vitro. J Biol Rhythms 9(1):27–41

Shibata S, Oomura Y, Kita H et al (1982) Circadian rhythmic changes of neuronal activity in the suprachiasmatic nucleus of the rat hypothalamic slice. Brain Res 247(1):154–158

Shibata S, Liou SY, Ueki S (1983) Development of the circadian rhythm of neuronal activity in suprachiasmatic nucleus of rat hypothalamic slices. Neurosci Lett 43(2–3):231–234

Smart TG (1997) Regulation of excitatory and inhibitory neurotransmitter-gated ion channels by protein phosphorylation. Curr Opin Neurobiol 3:358–367

Spitzer NC (2002) Activity-dependent neuronal differentiation prior to synapse formation: the functions of calcium transients. J Physiol Paris 96(1–2):73–80

Stephan FK, Zucker I (1972) Circadian rhythms in drinking behavior and locomotor activity of rats are eliminated by hypothalamic lesions. Proc Natl Acad Sci U S A 69:1583–1586

Teshima K, Kim SH, Allen CN (2003) Characterization of an apamin-sensitive potassium current in suprachiasmatic nucleus neurons. Neuroscience 120:65–73

Vandael DH, Marcantoni A, Mahapatra S et al (2010) Ca(v)1.3 and BK channels for timing and regulating cell firing. Mol Neurobiol 42:185–198

Verkhratsky A (2005) Physiology and pathophysiology of the calcium store in the endoplasmic reticulum of neurons. Physiol Rev 85(1):201–279

Welsh DK, Logothetis DE, Meister M et al (1995) Individual neurons dissociated from rat suprachiasmatic nucleus express independently phased circadian firing rhythms. Neuron 14(4):697–706

Wilsbacher LD, Yamazaki S, Herzog ED et al (2002) Photic and circadian expression of luciferase in mPeriod1-luc transgenic mice in vivo. Proc Natl Acad Sci U S A 99(1):489–494

Yamaguchi S, Isejima H, Matsuo T et al (2003) Synchronization of cellular clocks in the suprachiasmatic nucleus. Science 302(5649):1408–1412

Yamazaki S, Numano R, Abe M et al (2000) Resetting central and peripheral circadian oscillators in transgenic rats. Science 288(5466):682–685

Yan L, Okamura H (2002) Gradients in the circadian expression of Per1 and Per2 genes in the rat suprachiasmatic nucleus. Eur J Neurosci 15:1153–1162

Yoo SH, Yamasaky S, Lowry PL et al (2003) PERIOD2::LUCIFERASE real-time reporting of circadian dynamics reveals persistent circadian oscillations in mouse peripheral tissues. Proc Natl Acad Sci U S A 101(15):5339–5346

Chapter 8
GABA$_A$ Receptor-Mediated Neurotransmission in the Suprachiasmatic Nucleus

Charles N. Allen, Nathan J. Klett, Robert P. Irwin, and Mykhaylo G. Moldavan

Abstract GABAergic neurotransmission is a fundamental component of the suprachiasmatic nucleus (SCN) neural network, and virtually all SCN neurons communicate using GABA as a neurotransmitter. GABAergic neurotransmission plays a critical role in light-induced phase shifts, synchronization of the dorsal and ventral SCN, and, although controversial, synchronization of the circadian phase of individual SCN neurons. The circadian clock regulates the strength of GABA$_A$ receptor-mediated neurotransmission although the signaling mechanisms mediating this regulation are not known. GABA released from axon terminals acts on synaptic GABA$_A$ receptors producing postsynaptic currents that have a rapid onset and offset and desensitize in the continued presence of GABA. In the SCN, the postsynaptic GABA$_A$ receptor-mediated currents may be excitatory or inhibitory depending on the time of day. Once released GABA is removed from the synaptic cleft by specific sodium–chloride-dependent transporters (GAT). Some GABA can diffuse out of the synaptic cleft and act on extrasynaptic GABA$_A$ receptors. These extrasynaptic GABA$_A$ receptors have high affinity for GABA and show little or no desensitization. They mediate a "tonic" GABA$_A$ current that could modulate the input–output characteristics of individual SCN neurons. While significant scientific questions remain about the roles of GABAergic neurotransmission in the circadian timing signals, recent findings have yielded important advances in our understanding of GABAergic neurotransmission in the SCN.

8.1 General

The circadian system in mammals consists of a central circadian oscillator located in the suprachiasmatic nucleus (SCN) that aligns the phases of circadian clocks located throughout the body (Yamazaki et al. 2000, 2009; Yoo et al. 2004). The SCN, a

C.N. Allen (✉) • N.J. Klett • R.P. Irwin • M.G. Moldavan
Oregon Institute of Occupational Health Sciences, Oregon
Health & Science University, Portland, OR 97239, USA
e-mail: allenc@ohsu.edu

© Springer International Publishing Switzerland 2015
R. Aguilar-Roblero et al. (eds.), *Mechanisms of Circadian Systems in Animals and Their Clinical Relevance*, DOI 10.1007/978-3-319-08945-4_8

small bilateral hypothalamic nucleus, contains neurons expressing a molecular clock based on a transcription–translation feedback loop that generates a 24-h timing signal (Reppert and Weaver 2001; Welsh et al. 2010). Individual SCN neurons are imprecise single cellular oscillators that have a range of circadian periods. Intercellular communication provided by the SCN neural network synchronizes these cellular oscillators to a common period (Welsh et al. 1995, 2010; Michel and Colwell 2001; vanderLeest et al. 2009; Freeman et al. 2013). An animal's behavioral phenotype is determined by the integrated activity of this network of SCN neurons (Low-Zeddies and Takahashi 2001; Liu et al. 2007; Ko et al. 2010). The SCN network is critical for the generation of precise circadian timing signals and the stabilization of the circadian clock. In fact, the neural network can "rescue" circadian timing in the presence of circadian clock gene mutations (Liu et al. 2007; vanderLeest et al. 2009; Ko et al. 2010). GABAergic neurotransmission is a fundamental component of the SCN neural network, and virtually all SCN neurons communicate using GABA as a neurotransmitter (Moore and Speh 1993; Cardinali and Golombek 1998; Jobst et al. 2004). GABAergic neurotransmission (Fig. 8.1) plays a critical role in light-induced phase shifts, nonphotic phase changes, and synchronization of the dorsal and ventral SCN. In turn, the circadian clock regulates the strength of the GABA neurotransmission and the effect of GABA on the activity of SCN neurons (Wagner et al. 1997, 2001; De Jeu and Pennartz 2002; Gompf and Allen 2004; Itri et al. 2004; Choi et al. 2008).

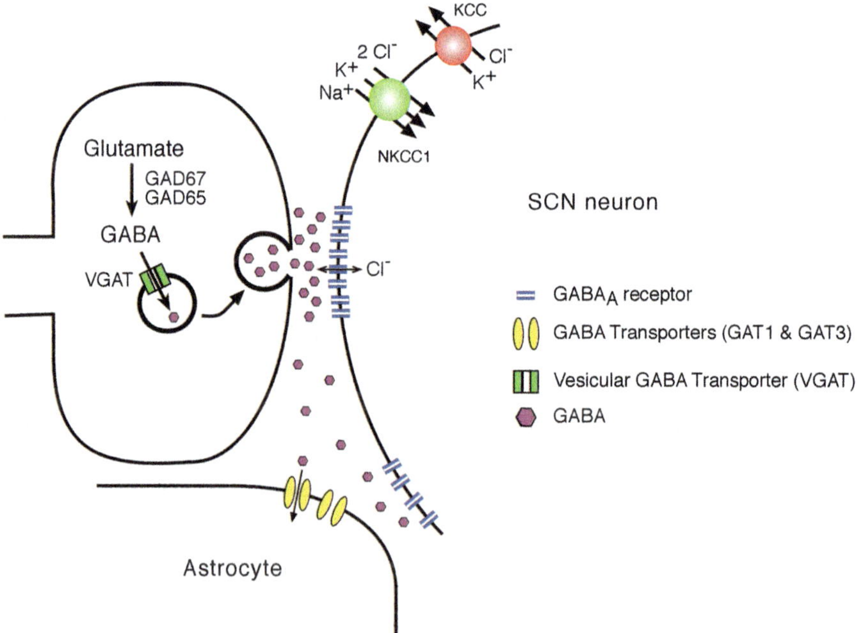

Fig. 8.1 Diagram of the key elements of a GABAergic synapse

8.2 Introduction to GABA$_A$ Receptor-Mediated Neurotransmission

8.2.1 GABA Synthesis, Release, Transport, and Degradation

GABA is synthesized from glutamate by two isoforms of glutamate decarboxylase (GAD), GAD67 and GAD65, named for the molecular size of the proteins, 67 and 65 kDa, respectively. Nearly all neurons in the SCN of the mammalian species examined contain GAD (Van den Pol 1980; Card and Moore 1984; Moore and Speh 1993; Jansen et al. 1994; O'Hara et al. 1995; Gao and Moore 1996; Belenky et al. 2003; Jobst et al. 2004). GAD67 and GAD65 genes and proteins have only 65 % identity, and these unique structural features underlie the differential functional patterns of GAD67 and GAD65 activity (Fenalti et al. 2007). In addition, GAD67 and GAD65 genes contain distinctive regulatory elements suggesting that different intracellular signaling pathways regulate the transcription of the two enzymes.

GAD67 and GAD65 are members of the pyridoxal-5'-phosphate-dependent enzyme superfamily. The enzyme is active when pyridoxal-5'-phosphate is bound and inactive when it is absent (Battaglioli et al. 2003). When glutamate is decarboxylated to produce GABA, the pyridoxal-5'-phosphate may remain bound in which case the GAD enzyme remains active. Alternately, the GAD is inactive if the pyridoxal-5'-phosphate unbinds. GAD67 is constitutively active while a significant fraction of GAD65 is inactive due to structural differences in the pyridoxal-5'-phosphate regulatory site, which is a result of the unbinding reaction occurring at a significantly higher rate in GAD65 compared to GAD67 (Fenalti et al. 2007). The different levels of constitutive activity and the localization of GAD67 to the soma and dendrites and GAD65 to membranes of axon terminals and synaptic vesicles suggest unique roles for GAD67 and GAD56 in GABAergic neurotransmission. For example, inactive GAD65 may provide a reservoir of synthetic capacity to rapidly increase the amount of GABA available for synaptic release.

GAD67 and GAD65 have different patterns of expression in the SCN. GAD67 mRNA showed no 24-h rhythm of expression in rats although one study noted higher levels at CT0 and CT 14 (Cagampang et al. 1996; Huhman et al. 1996, 1999). Panda et al. (2002) observed a 24-h rhythm of GAD67 expression in the mouse SCN. The disparate observations may reflect differences in the sensitivity and signal-to-noise ratio of the autoradiographic in situ hybridization method compared to high-density microarrays. No rhythm in GAD65 mRNA was observed in mice maintained in constant darkness (DD) (Panda et al. 2002). In rats, GAD65 expression has a diurnal rhythm with a peak at ZT0 and a trough at ZT20 (rats; light:dark (LD) 12:12 h). The increase was larger in the dorsomedial portion of the SCN (Huhman et al. 1996). A similar pattern of expression was observed in rats (LD 12:12 h) on the first day after release into DD (Cagampang et al. 1996). However, the rhythm of GAD65 mRNA expression disappeared in rats after 9 days in DD suggesting that the light input pathways regulate GAD65 expression (Huhman et al. 1999).

Similarly, GAD activity showed a diurnal rhythm with levels peaking late in the day and early in the night (Aguilar-Roblero et al. 1993). Together these data indicate that GAD65 expression and activity increase in parallel to the increased frequency of GABA-mediated synaptic currents observed during the subjective day (Itri and Colwell 2003; Itri et al. 2004).

GABA is loaded into synaptic vesicles by the vesicular GABA transporter (VGAT). VGAT is also called the vesicular inhibitory amino acid transporter (VIAAT) and will also transport glycine into synaptic vesicles (McIntire et al. 1997; Gasnier 2004). The VGAT moves GABA into synaptic vesicles utilizing the H^+ gradient, an electrochemical gradient established by an H^+-ATPase (Hsu et al. 1999). Trafficking signals target the VGAT to specific vesicles allowing specific pools to be filled more rapidly or more fully (Santos et al. 2013). In addition, the regulation of the VGAT activity by phosphorylation or other biochemical modifications will regulate the GABA concentration in vesicles (Hsu et al. 1999). Reducing the amount of GABA in synaptic vesicles reduces the amplitude and frequency of $GABA_A$ receptor-mediated currents (Riazanski et al. 2011). However, the importance of these mechanisms in regulating SCN neural network activity has not been studied. Neither VGAT mRNA nor vesicular VGAT expression was found to be rhythmic in constant darkness or in a light–dark cycle (Panda et al. 2002; Darna et al. 2009). Interestingly, mice housed in a LD (12:12) cycle had higher levels of VGAT expression compared to animals housed in DD (Darna et al. 2009). It would be interesting to know if animals housed in LD have higher vesicular GABA concentrations and larger and more frequent $GABA_A$ receptor-mediated currents than those housed in DD.

The depolarization of a presynaptic terminal releases GABA into the synaptic cleft where it is estimated to reach a concentration of 1.5–3 mM. High affinity GABA transporters (GAT) rapidly remove GABA from the synaptic cleft with an estimated time constant of 75–125 μs (Jones et al. 2001; Mozrzymas et al. 2003). Four high affinity GABA transporters (GAT-1, GAT-2, GAT-3, BGT-1) have been identified with GAT-1 and GAT-3 being responsible for terminating GABA signaling in the brain (Dalby 2003). GATs are secondary active transporters that utilize the Na^+ and Cl^- electrochemical gradients to move 2 Na^+ and 1 Cl^- ion for each GABA molecule transported (Kavanaugh et al. 1992). This stoichiometry results in a lower limit (~400 nM) below which GAT cannot reduce extracellular GABA (Attwell and Mobbs 1994; Cavelier et al. 2005). During prolonged or high-frequency GABA release, the transporter's capacity is exceeded, and GABA diffuses out of the synapse and activates extrasynaptic receptors (Isaacson et al. 1993; Dalby 2003; Semyanov et al. 2003; Glykys and Mody 2007). GABA that "spills over" from the synapse may also activate $GABA_B$ receptors and inhibit the release of glutamate from terminals of the retinohypothalamic tract (Moldavan and Allen 2013). Under appropriate conditions, GATs may act in reverse to move GABA from the cell cytoplasm into the extracellular space (Richerson and Wu 2003). Evidence for GAT-mediated GABA release has not been reported in the SCN.

8.2.2 GABA$_A$ Receptors

Neurotransmission at GABAergic synapses is important in the SCN neural network. GABA is the major neurotransmitter at intra-SCN synapses, and nearly all neurons in the SCN contain glutamic acid decarboxylase, the GABA-synthesizing enzyme (Van den Pol 1980; Belenky et al. 2003). The SCN is estimated to contain between 8,000 and 12,000 neurons. Each SCN neuron receives approximately 1,400 synaptic contacts located on the soma and dendrites (Guldner 1984). Approximately half of these synaptic connections stain positive for markers of GABAergic neurotransmission (Decavel and Van den Pol 1990; Belenky et al. 2008). SCN neurons have short axons that project to other SCN neurons as well as to neighboring hypothalamic nuclei (Van den Pol 1980; Strecker et al. 1997). SCN neurons communicate with each other via GABA$_A$ receptor-mediated neurotransmission (Strecker et al. 1997).

GABA activates ionotropic GABA$_A$ receptors and metabotropic GABA$_B$ receptors. GABA$_A$ receptors have a pentameric structure that forms an agonist gated Cl⁻ channel. The activation of GABA$_A$ receptors in the adult brain has traditionally been thought to be inhibitory due to the activation of a Cl⁻ conductance that hyperpolarizes the membrane. The activation of GABA$_A$ receptors may also produce shunting inhibition that results from an increase in membrane conductance that shunts depolarizing currents (Staley and Mody 1992). SCN neurons are a rare example of adult neurons in which GABA acts as both an inhibitory and excitatory neurotransmitter (Wagner et al. 1997, 2001; De Jeu and Pennartz 2002; Choi et al. 2008; Irwin and Allen 2009). In contrast, GABA$_B$ receptors are composed of seven transmembrane spanning subunits that signal through Gi-type G-proteins and in the SCN regulate transmitter release from presynaptic retinohypothalamic tract terminals by inhibiting voltage-gated Ca^{2+} channels and activating potassium currents in SCN neurons (Jiang et al. 1995; Moldavan et al. 2006; Moldavan and Allen 2013).

8.2.2.1 GABA$_A$ Receptor Structure

GABA$_A$ receptors have diverse physiological properties and sensitivity to pharmacological agents—properties conferred by the specific subunits that make up the GABA$_A$ receptor complex (Rudolph and Mohler 2004). Individual neurons express multiple types of GABA$_A$ receptors, and this diversity of GABA-mediated signaling is essential for rhythmicity and synchrony in neural networks (Aradi et al. 2002; Aradi and Soltesz 2002). Eighteen GABA$_A$ receptor subunits have been identified including α(1–6), β(1–3), γ(1–3), ρ(1–3), δ, ϵ, and π subunits (Barnard et al. 1998). GABA$_A$ receptors are composed of five subunits—two α subunits, two β subunits, and a γ subunit (Fig. 8.2). In some GABA$_A$ receptors, the γ subunit is replaced with a δ, ϵ, ρ (1–3), or π subunit (Barnard et al. 1998).

Fig. 8.2 Putative subunit composition of GABA$_A$ receptors in the SCN. The structure of GABA$_A$ receptors in the SCN can be predicted based on the subunit composition of GABA$_A$ receptors in other parts of the brain. (**a**) Synaptic receptors. (**b**) Extrasynaptic receptors. (**c & d**) Receptors located both synaptically and extrasynaptically

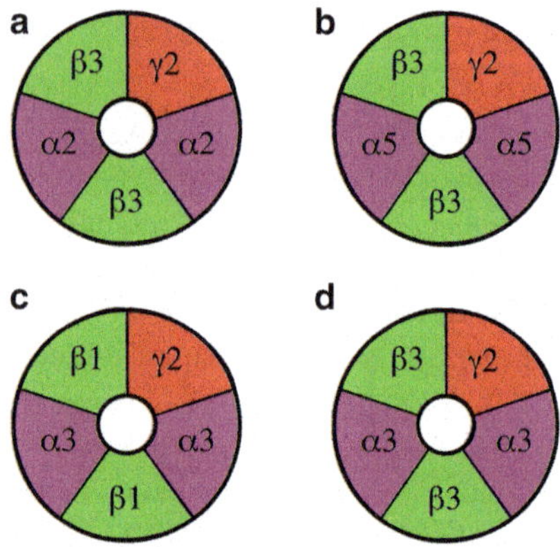

In the SCN, three α (α2, α3, and α5), two β (β1 and β3), and two γ (γ1 and γ2) GABA$_A$ receptor subunits have been identified using in situ hybridization and immunocytochemical techniques (Gao et al. 1995; O'Hara et al. 1995; Naum et al. 2001; Belenky et al. 2003; Jobst et al. 2004). β2/3 immunoreactivity was observed in the ventrolateral SCN of the hamster but not the rat (Gao et al. 1995; Naum et al. 2001; Jobst et al. 2004). β2/3 subunit messenger RNA was detected in the SCN of mice, but a concern was raised that this signal may result from the inclusion of additional hypothalamic tissue in the sample (Gao et al. 1995; O'Hara et al. 1995). Therefore, the difference in β2/3 expression may be a species difference (hamster and mouse versus rat). Messenger RNA for β1 subunit was identified in the mouse and hamster SCN (O'Hara et al. 1995; Naum et al. 2001). The β1 subunit in the SCN of Syrian hamsters housed in a light–dark cycle (14:10) had a circadian and diurnal rhythm with a peak amplitude at midnight and a nadir at midday. In contrast, the α2, α5, and β5 subunits were not rhythmic (Naum et al. 2001).

While there are several hundred possible GABA$_A$ receptor subunit combinations, only a limited number of subunit combinations are actually expressed in neurons (Barnard et al. 1998; Fritschy and Brunig 2003; Mody and Pearce 2004; Mohler 2006; Olsen and Sieghart 2008). Based on the pattern of GABA$_A$ receptor expression in other brain areas, we can surmise the subunit composition of GABA$_A$ receptors expressed by SCN neurons. First, GABA$_A$ receptors consisting of α2β3γ2 subunits are probably expressed at GABAergic synapses since these receptors are mainly synaptic in the hippocampus (O'Hara et al. 1995; Naum et al. 2001). Second, a likely GABA$_A$ subunit combination is α3β1γ2 or α3β3γ2, which may be located synaptically or extrasynaptically (O'Hara et al. 1995). Third, the GABA$_A$ receptor combination α5β3γ2 is likely to be located extrasynaptically since α5-containing receptors are found at extrasynaptic locations. Interestingly, the predominant GABA$_A$ receptor in the brain, α1β2γ2, does not appear to be present in the SCN since the α1 subunit is not expressed in hamster, rat, or mouse SCN (Gao et al. 1995; O'Hara et al. 1995; Jobst et al. 2004). Although δ subunits have not been identified with neurochemical techniques, there are pharmacological data that suggest that δ-containing GABA receptors are located in the SCN (Ehlen and Paul 2009; McElroy et al. 2009).

The subunit composition of GABA$_A$ receptors determines their activation, inactivation, and desensitization kinetics, as well as their affinity for GABA and consequently the strength of GABA$_A$-mediated synaptic currents (Schofield and Huguenard 2007). Synaptic receptors typically have lower agonist affinities and desensitize more rapidly than extrasynaptic receptors (Mody and Pearce 2004; Mody 2005; Mohler 2006). For example, α2-containing GABA$_A$ receptors, which are generally localized to synapses, have a lower affinity for GABA and desensitize faster than extrasynaptic GABA$_A$ receptors containing the α5 subunit (Mody 2001; Semyanov et al. 2004). The high GABA affinity allows extrasynaptic GABA$_A$ receptors to respond to the low GABA concentrations (0.3–1 μM) present in the extrasynaptic space (Lerma et al. 1986; Tossman et al. 1986). The extremely slow rate of desensitization displayed by these extrasynaptic receptors underlies the continuous activation necessary for the GABA-mediated tonic current, which is observed in many neurons (Stell and Mody 2002).

The subunit composition of the GABA$_A$ receptors also determines receptor's sensitivity to specific pharmacological agents. Several pharmacological agents that affect circadian phase also interact with GABA$_A$ receptors in a subunit-specific manner. The benzodiazepine, diazepam, which alters circadian period and blocks light-induced phase shifts, increases GABA$_A$-mediated currents through receptor channels composed of γ2 subunits together with α1, α2, α3, or α5 subunits (Ralph and Menaker 1986; Subramanian and Subbaraj 1996). The imidazobenzodiazepine L655,708 has an affinity for GABA$_A$ receptors containing α5 subunits 50 times higher than GABA$_A$ receptors containing other α subunits (Ing and Poulter 2007). L655,708 will potentiate, at low concentrations, currents mediated by GABA$_A$ receptors containing the α5 subunit (Ing and Poulter 2007). L655,708 acting in the SCN to increase a tonic GABA current could be used to study the role of extrasynaptic GABA$_A$ receptors in the SCN network.

8.2.2.2 GABA$_A$ Receptor-Mediated Neurotransmission

GABA binding to the GABA$_A$ receptor results in a conformational change that opens an anion selective channel. GABA$_A$ receptors in the SCN are primarily permeable to Cl$^-$ (~80 %) although there is also a significant HCO$_3^-$ conductance (~20 %) (Wagner et al. 2001). For this reason, GABA$_A$ receptor-mediated current amplitude and direction depend on the chemical and electrical gradients of the Cl$^-$ ion. In most adult neurons, the intracellular Cl$^-$ concentration is low compared to the extracellular Cl$^-$ concentration. Therefore, the equilibrium potential for Cl$^-$ (E_{Cl}) is more negative than the resting membrane potential, and GABA$_A$ receptor-mediated currents are consequently hyperpolarizing (inhibitory). However, if the intracellular Cl$^-$ concentration is high, E_{Cl} may became more positive than the resting membrane potential, and GABA efflux will depolarize the cell membrane causing neuronal excitation (Woodin et al. 2003; Fiumelli et al. 2005).

Depending on the circadian time, GABA$_A$ receptor-mediated neurotransmission in the adult SCN can be both excitatory and inhibitory. GABA was first reported to increase the firing rate of SCN neurons during the day and inhibit SCN neurons during the night (Wagner et al. 1997). This observation was not absolute—the majority (~75 %) of SCN neurons were activated during the day compared to only 32 % at night. This circadian control of excitability was attributed to a circadian regulation of the intracellular Cl$^-$ concentration (Wagner et al. 1997). Subsequent research confirmed GABA's dualistic role as both an inhibitory and excitatory neurotransmitter in the SCN; however, in these studies GABA was more likely to be an excitatory neurotransmitter during the night than during the day (De Jeu and Pennartz 2002; Choi et al. 2008; Irwin and Allen 2009). These inconsistencies have not been explained but may be due to differences in methodology: whole cell vs. perforated patch recording. Also, when extracellular recording electrodes are used to record the action potential firing rate, a shunting effect of GABA$_A$ receptor activation could mask a relatively depolarized E_{Cl}. Further, all of these interpretations are complicated by differences in solutions and the fact that traumatized, damaged, and ischemic neurons have all been shown to display depolarized E_{Cl}. Furthermore, whether or not GABA is excitatory or inhibitory or evokes no response ultimately depends on the cell's resting membrane potential. Regardless of the differences in experimental observations, it is clear that a substantial proportion of SCN neurons respond to GABA in an excitatory manner and that the proportion of SCN neurons that are excited by GABA shows a circadian or diurnal rhythm.

In neurons, the basal intracellular Cl$^-$ concentration is set by the combined activity of several chloride channels and transporters (Fig. 8.3, however see Glykys et al. 2014). Of these, the sodium–potassium–chloride cotransporter 1 (NKCC1) has

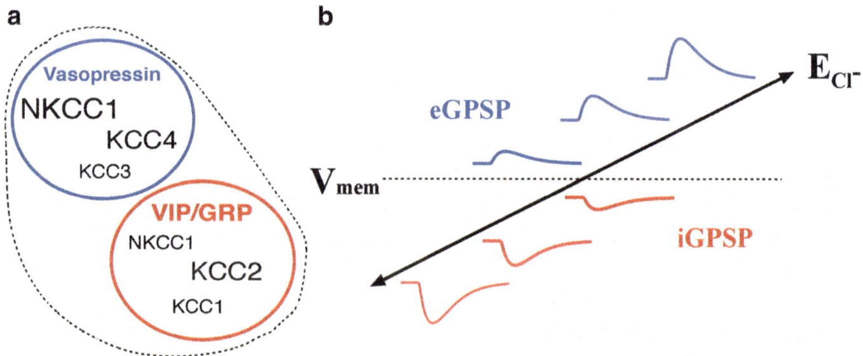

Fig. 8.3 Regulation of the intracellular Cl$^-$ gradient in SCN neurons. (**a**) Diagram showing the relative expression of Cl$^-$ cotransporters in SCN neurons expressing different neuropeptides (Belenky et al. 2010). (**b**) Schematic displaying the relationship between the Cl$^-$ equilibrium potential (E_{Cl}), membrane potential (V_{mem}), and the magnitude and the direction of GABA-induced changes in membrane potential. An increase in Cl$^-$ influx will increase [Cl$^-$]i and make E_{Cl} more positive, while Cl$^-$ efflux will decrease [Cl$^-$]i and produce a more negative E_{Cl}. When E_{Cl} is more negative than V_{mem}, GABA hyperpolarizes the membrane potential (inhibitory GABA-mediated postsynaptic potential, iGPSP). An E_{Cl} more positive than V_{mem} depolarizes the membrane potential (excitatory GABA-mediated postsynaptic potential, eGPSP)

received the most attention for maintaining Cl$^-$ influx. Under normal neuronal conditions, NKCC1 moves one Na$^+$ and one K$^+$ ion into the neuron for every two Cl$^-$ ions. NKCC1 was found in the soma and dendrites of virtually every SCN neuron examined (Belenky et al. 2010). Further, a circadian rhythm of NKCC1 protein level was observed in the dorsal SCN, but not in the ventral SCN (Choi et al. 2008). Blocking NKCC1 with bumetanide reduces or eliminates depolarizing GABA responses and increases the magnitude of inhibitory GABA responses (Irwin and Allen 2009). These data demonstrate that in the absence of NKCC1-mediated Cl$^-$ influx, the intracellular Cl$^-$ concentration will decrease, resulting in larger inhibitory GABA responses (Choi et al. 2008; Irwin and Allen 2009).

Chloride efflux is mediated by the members of the potassium–chloride cotransporter family (KCC1–4), which move one K$^+$ and one Cl$^-$ out of the neuron for each cycle. The four members of the KCC family are regionally and phenotypically distributed in the SCN. KCC3 and KCC4 are expressed in vasopressin-expressing SCN neurons, while KCC1 and KCC2 are expressed in vasoactive intestinal peptide (VIP)- and gastrin-releasing peptide (GRP)-expressing SCN neurons (Belenky et al. 2008; Belenky et al. 2010). KCC1, KCC3, and KCC4 all have a lower affinity (15–20 mM) for Cl$^-$ than KCC2 (5–7 mM). Due to its higher affinity for Cl$^-$, KCC2 is expected to transport Cl$^-$ out of a neuron at lower Cl$^-$ concentrations compared to the other KCCs. Given the different Cl$^-$ affinities of the various KCC cotransporters, it is expected that the intracellular Cl$^-$ concentration should reach lower concentrations in cells expressing KCC2 than in the neurons expressing KCC3 and KCC4. Consequently, the lower intracellular Cl$^-$ concentration in KCC2-expressing neurons would engender more hyperpolarizing GABA responses. The overall balance between NKCC1 (Cl$^-$ influx) and KCC2 (Cl$^-$ efflux) expression and activity in SCN neurons helps determine the intracellular Cl$^-$ concentration and thereby enable both depolarizing and hyperpolarizing GABA responses.

8.3 Circadian Regulation of GABAergic Neurotransmission

Functionally, GABAergic neurotransmission in the SCN has been proposed to modulate retinal input to the SCN (Jiang et al. 1995; Moldavan et al. 2006; Moldavan and Allen 2013), synchronize short-term firing patterns between SCN neurons (Kononenko and Dudek 2004), synchronize circadian firing patterns (Liu and Reppert 2000; Shirakawa et al. 2000), and regulate the amplitude of circadian rhythms (Wagner et al. 2001; Aton et al. 2006; Choi et al. 2008). Spontaneous and miniature GABA$_A$ receptor-mediated currents are observed in the majority of SCN neurons (Kim and Dudek 1992; Jiang et al. 1997; Strecker et al. 1997; Itri and Colwell 2003; Gompf and Allen 2004; Itri et al. 2004; Gompf et al. 2006). The frequency of the spontaneous fast synaptic GABA currents shows a day–night difference in the dorsal but not the ventral SCN (Itri and Colwell 2003; Itri et al. 2004). The frequency of the GABA$_A$ receptor-mediated currents peaks late in the subjective day and early in the subjective night (Itri and Colwell 2003; Itri et al. 2004).

The increased GABAergic activity precedes the peak in extracellular GABA concentration, which occurs early in the night (Aguilar-Roblero et al. 1993). The application of GABA or the $GABA_A$ agonist muscimol can elevate or reduce the intracellular calcium concentration in different populations of SCN neurons, while the $GABA_B$ agonist baclofen lowers intracellular calcium in both groups (Irwin and Allen 2009). Further, the application of gabazine, a $GABA_A$ receptor antagonist, altered the resting intracellular calcium concentration inversely to the direction induced by GABA (Irwin and Allen 2009). The strength of $GABA_A$ receptor-mediated synaptic transmission is regulated in a diurnal manner (Gompf and Allen 2004). Short-term synaptic depression exhibits a diurnal rhythm, with a majority of synapses during the day and few synapses at night showing synaptic depression, implying that neurotransmission within this local circuitry is under the control of the circadian clock. The spontaneous daytime action potential firing frequencies observed in the SCN correspond to the frequencies that elicit synaptic depression at synapses between SCN neurons (Gompf and Allen 2004). Therefore, both the frequency and synaptic strength of intra-SCN GABAergic neurotransmission may signal a cell's relative circadian phase to neighboring neurons.

Albus et al. (2005) found two peaks in the ensemble firing pattern of the SCN on the first day following a 6-h phase delay. They demonstrated that the peaks could be separated into an early component intrinsic to the dorsal SCN and a late component intrinsic to the ventral SCN. Combined with findings from previous studies, these results were interpreted as evidence that (1) the two peaks reflect two pacemakers, the ventral SCN which shifts more rapidly than the dorsal SCN, (2) the ventral SCN has a strong phase-shifting effect on the dorsal SCN, and (3) GABA acting through $GABA_A$ receptors plays a critical role in synchronizing these regional oscillators, perhaps by exciting the dorsal SCN at night and inhibiting the ventral SCN during the day. Aton et al. (2006) found that blockade of $GABA_A$ receptors for up to 10 days increased the cell-to-cell synchrony and individual cell amplitude of circadian rhythms in cultured SCN. These results were interpreted as GABA is (1) not required in synchronizing daily rhythms in the SCN and (2) restricts the amplitude of circadian rhythms in firing rate.

One possible union of these data recognizes the antagonistic role of GABA in the SCN as it adjusts to large changes in ambient light cycles (e.g., transmeridian travel or perhaps seasonal) or to natural variability in the cycle-to-cycle synchrony of SCN neurons. Following a large delay in the light cycle as in Albus et al. (2005), GABA slows the shifting of neurons to local time. Addition of bicuculline (or cutting a large portion of the GABA projections) would release this antagonism so that cells in the ventral SCN would shift rapidly to the new local time. It could be that neurons in the dorsal SCN shift slowly under these conditions because GABA is excitatory instead of inhibitory, but single-cell data are needed. During stable entrainment or free-running conditions, the loss of GABA's antagonism would lead to a tighter phase distribution of cells within the SCN and less cycle-to-cycle variability. Further studies are also required to distinguish the role of fast synaptic vs. tonic $GABA_A$ receptor-mediated neurotransmission.

8.4 GABA Effects on the Circadian Clock

GABAergic neurotransmission acts as a nonphotic entrainment signal. Injections of the GABA$_A$ receptor agonist muscimol into the golden hamster SCN produced phase advances during the day and phase delays at night (Smith et al. 1989). Interestingly, the response of the diurnal Nile grass rat (*Arvicanthis niloticus*) was dramatically different; muscimol during the day produced phase delays, while nighttime injections produced no phase changes (Novak and Albers 2004). Muscimol injections reduce the light induction of *Per1* during the early and late night, while *Per2* mRNA induction is blocked during the late night. The authors postulated that light and GABAergic signaling converged on *Per* expression to regulate the circadian clock activity (Ehlen et al. 2006, 2008). The activation of GABA$_A$ receptors with muscimol changes the expression of the clock genes *Per1* and *Per2* in a manner that is dependent on the time of day. During the day (CT6), muscimol decreases *Per1* levels while not altering *Per2* levels. Early in the night (CT13.5), muscimol inhibits light-induced increases in *Per1* but not *Per2*. Late in the night (CT19), muscimol will inhibit the light-induced increase in both *Per1* and *Per2* (Ehlen et al. 2006).

In vivo, the inhibition of GABAergic neurotransmission with bicuculline increases the magnitude of light-induced phase delays but has no effect on light-induced phase advances (Gillespie et al. 1996). A similar effect was observed when the phase delay was produced by a cocktail of neuropeptides (Gillespie et al. 1996). 4,5,6,7-tetrahydroisoxazolo[5,4-c]pyridin-3-ol (THIP), a GABA$_A$ receptor agonist that preferentially activates extrasynaptic GABA$_A$ receptors containing the δ or $\alpha4$ subunit, injected into the hamster SCN inhibited both light-induced phase advances and delays (Ehlen and Paul 2009). THIP alone had no direct effect on *Per1* or *Per2* expression but inhibited the induction of *Per1* and *Per2* observed with a light pulse at circadian time 13.5 (CT13.5). These data provide pharmacological evidence for the presence of δ-containing GABA receptors in the SCN and point to the importance of extrasynaptic GABA$_A$ receptors in the activity of the SCN neural network activity (Ehlen and Paul 2009).

The final question is what role GABAergic neurotransmission plays in the synchronization of individual SCN neuronal oscillators. GABA synchronized the action potential firing pattern of SCN neurons grown in dispersed cell cultures (Liu and Reppert 2000). Similarly, bicuculline, a GABA$_A$ receptor antagonist, desynchronized Per2::LUC rhythms in organotypic SCN cultures (Ko et al. 2010). In contrast, Aton et al. (2006) concluded based on the chronic application of the GABA antagonist bicuculline to organotypic SCN cultures of Per2:LUC mice that GABAergic neurotransmission was not involved in synchronizing the circadian clocks of individual SCN neurons (Aton and Herzog 2005; Aton et al. 2006). The experimental differences that produce these disparate data are not known. Recently, Freeman et al. (2013) demonstrated that the GABAergic neurotransmission reduced the precision of circadian rhythms in individual neurons. They concluded that GABA$_A$ receptor-mediated neurotransmission introduces "jitter" to the circuit that opposes the synchronizing effects of VIP signaling (Freeman et al. 2013). Interestingly, a separate study found that GABAergic neurotransmission reduced the regularity of the action potential firing frequency of SCN neurons (Kononenko and Dudek 2004). It will be interesting to determine whether this irregularity of action potential firing frequency underlies the GABA-mediated network "jitter."

The crucial question is what is the role or roles that GABAergic neurotransmission plays in the SCN? The evolving role of GABA in the SCN presents a highly complex system varying over circadian time, with a myriad of receptor forms, different levels of control, and both excitatory and inhibitory effects. Though not fully understood, recent findings have yielded important advances in our understanding of the role of GABAergic neurotransmission in the SCN.

Acknowledgments The work was supported by grants from NINDS (NS036607) and NIGMS (GM096972) to CNA.

References

Aguilar-Roblero R, Verduzco-Carbajal L, Rodríguez C et al (1993) Circadian rhythmicity in the GABAergic system in the suprachiasmatic nuclei of the rat. Neurosci Lett 157:199–202

Albus H, Vansteensel MJ, Michel S et al (2005) A gabaergic mechanism is necessary for coupling dissociable ventral and dorsal regional oscillators within the circadian clock. Curr Biol 15:886–893

Aradi I, Soltesz I (2002) Modulation of network behaviour by changes in variance in interneuronal properties. J Physiol (Lond) 538:227–251

Aradi I, Santhakumar V, Chen K et al (2002) Postsynaptic effects of GABAergic synaptic diversity: regulation of neuronal excitability by changes in IPSC variance. Neuropharmacology 43:511–522

Aton SJ, Herzog ED (2005) Come together, right…now: synchronization of rhythms in a mammalian circadian clock. Neuron 48:531–534

Aton SJ, Huettner JE, Straume M et al (2006) GABA and Gi/o differentially control circadian rhythms and synchrony in clock neurons. Proc Natl Acad Sci U S A 103:19188–19193

Attwell D, Mobbs P (1994) Neurotransmitter transporters. Curr Opin Neurobiol 4:353–359

Barnard EA, Skolnick P, Olsen RW et al (1998) International Union of Pharmacology. XV. Subtypes of gamma-aminobutyric acidA receptors: classification on the basis of subunit structure and receptor function. Pharmacol Rev 50:291–313

Battaglioli G, Liu H, Martin DL (2003) Kinetic differences between the isoforms of glutamate decarboxylase: implications for the regulation of GABA synthesis. J Neurochem 86:879–887

Belenky MA, Sagiv N, Fritschy J et al (2003) Presynaptic and postsynaptic GABA(A) receptors in rat suprachiasmatic nucleus. Neuroscience 118:909–923

Belenky MA, Yarom Y, Pickard GE (2008) Heterogeneous expression of gamma-aminobutyric acid and gamma-aminobutyric acid-associated receptors and transporters in the rat suprachiasmatic nucleus. J Comp Neurol 506:708–732

Belenky MA, Sollars PJ, Mount DB et al (2010) Cell-type specific distribution of chloride transporters in the rat suprachiasmatic nucleus. Neuroscience 165:1519–1537

Cagampang FRA, Rattray M, Powell JF et al (1996) Circadian changes of glutamate decarboxylase 65 and 67 mRNA in the rat suprachiasmatic nuclei. Neuroreport 7:1925–1928

Card JP, Moore RY (1984) The suprachiasmatic nucleus of the golden hamster: immunohistochemical analysis of cell and fiber distribution. Neuroscience 13:415–431

Cardinali DP, Golombek DA (1998) The rhythmic GABAergic system. Neurochem Res 23:607–614

Cavelier P, Hamann M, Rossi D et al (2005) Tonic excitation and inhibition of neurons: ambient transmitter sources and computational consequences. Prog Biophys Mol Biol 87:3–16

Choi HJ, Lee CJ, Schroeder A et al (2008) Excitatory actions of GABA in the suprachiasmatic nucleus. J Neurosci 28:5450–5459

Dalby NO (2003) Inhibition of gamma-aminobutyric acid uptake: anatomy, physiology and effects against epileptic seizures. Eur J Pharmacol 479:127–137

Darna M, Schmutz I, Richter K et al (2009) Time of day-dependent sorting of the vesicular glutamate transporter to the plasma membrane. J Biol Chem 284:4300–4307

De Jeu M, Pennartz CMA (2002) Circadian modulation of GABA function in the rat suprachiasmatic nucleus: excitatory effects during the night phase. J Neurophys 87:834–844

Decavel C, Van den Pol AN (1990) GABA: a dominant neurotransmitter in the hypothalamus. J Comp Neurol 302:1019–1037

Ehlen JC, Paul KN (2009) Regulation of light's action in the mammalian circadian clock: role of the extrasynaptic GABAA receptor. Am J Physiol Regul Integr Comp Physiol 296:R1606–R1612

Ehlen JC, Novak CM, Karom MC et al (2006) GABAA receptor activation suppresses Period 1 mRNA and Period 2 mRNA in the suprachiasmatic nucleus during the mid-subjective day. Eur J Neurosci 23:3328–3336

Ehlen JC, Novak CM, Karom MC et al (2008) Interactions of GABA A receptor activation and light on period mRNA expression in the suprachiasmatic nucleus. J Biol Rhythms 23:16–25

Fenalti G, Law RH, Buckle AM et al (2007) GABA production by glutamic acid decarboxylase is regulated by a dynamic catalytic loop. Nat Struct Mol Biol 14:280–286

Fiumelli H, Cancedda L, Poo MM (2005) Modulation of GABAergic transmission by activity via postsynaptic Ca2+-dependent regulation of KCC2 function. Neuron 48:773–786

Freeman GM Jr, Krock RM, Aton SJ et al (2013) GABA networks destabilize genetic oscillations in the circadian pacemaker. Neuron 78:799–806

Fritschy JM, Brunig I (2003) Formation and plasticity of GABAergic synapses: physiological mechanisms and pathophysiological implications. Pharmacol Ther 98:299–323

Gao B, Moore RY (1996) Glutamic acid decarboxylase message isoforms in human suprachiasmatic nucleus. J Biol Rhythms 11:172–179

Gao B, Fritschy JM, Moore RY (1995) GABAA-receptor subunit composition in the circadian timing system. Brain Res 700:142–156

Gasnier B (2004) The SLC32 transporter, a key protein for the synaptic release of inhibitory amino acids. Pflugers Arch 447:756–759

Gillespie CF, Huhman KL, Babagbemi TO et al (1996) Bicuculline increases and muscimol reduces the phase-delaying effects of light and VIP/PHI/GRP in the suprachiasmatic region. J Biol Rhythms 11:137–144

Glykys J, Dzhala V, Egawa K et al (2014) Local impermeant anions establish the neuronal chloride concentration. Science 343:670–675

Glykys J, Mody I (2007) The main source of ambient GABA responsible for tonic inhibition in the mouse hippocampus. J Physiol 582:1163–1178

Gompf HS, Allen CN (2004) GABAergic synapses of the suprachiasmatic nucleus exhibit a diurnal rhythm of short-term synaptic plasticity. Eur J Neurosci 19:2791–2798

Gompf HS, Irwin RP, Allen CN (2006) Retrograde suppression of GABAergic currents in a subset of SCN neurons. Eur J Neurosci 23:3209–3216

Guldner FH (1984) Suprachiasmatic nucleus: numbers of synaptic appositions and various types of synapses. A morphometric study on male and female rats. Cell Tissue Res 235:449–452

Hsu CC, Thomas C, Chen W et al (1999) Role of synaptic vesicle proton gradient and protein phosphorylation on ATP-mediated activation of membrane-associated brain glutamate decarboxylase. J Biol Chem 274:24366–24371

Huhman KL, Hennessey AC, Albers HE (1996) Rhythms of glutamic acid decarboxylase mRNA in the suprachiasmatic nucleus. J Biol Rhythms 11:311–316

Huhman KL, Jasnow AM, Sisitsky AK et al (1999) Glutamic acid decarboxylase mRNA in the suprachiasmatic nucleus of rats housed in constant darkness. Brain Res 851:266–269

Ing T, Poulter MO (2007) Diversity of GABA(A) receptor synaptic currents on individual pyramidal cortical neurons. Eur J Neurosci 25:723–734

Irwin RP, Allen CN (2009) GABAergic signaling induces divergent neuronal Ca(2+) responses in the suprachiasmatic nucleus network. Eur J Neurosci 30:1462–1475

Isaacson JS, Solís JM, Nicoll RA (1993) Local and diffuse synaptic actions of GABA in the hippocampus. Neuron 10:165–175

Itri J, Colwell CS (2003) Regulation of inhibitory synaptic transmission by vasoactive intestinal peptide (VIP) in the mouse suprachiasmatic nucleus. J Neurophysiol 90:1589–1597

Itri J, Michel S, Waschek JA et al (2004) Circadian rhythm in inhibitory synaptic transmission in the mouse suprachiasmatic nucleus. J Neurophysiol 92:311–319

Jansen HT, Gong Q, Norgren RB Jr et al (1994) Single- and double-label immunocytochemical study of the ovine suprachiasmatic nucleus (SCN): GABAergic and peptidergic relationships. Brain Res Bull 34:499–506

Jiang Z-G, Allen CN, North RA (1995) Presynaptic inhibition by baclofen of retinohypothalamic excitatory synaptic transmission in rat suprachiasmatic nucleus. Neuroscience 64:813–819

Jiang Z-G, Yang Y-Q, Liu Z-P et al (1997) Membrane properties and synaptic inputs of suprachiasmatic nucleus neurons in rat brain slices. J Physiol (Lond) 499:141–159

Jobst EE, Robinson DW, Allen CN (2004) Potential pathways for intercellular communication within the calbindin subnucleus of the hamster suprachiasmatic nucleus. Neuroscience 123:87–99

Jones MV, Jonas P, Sahara Y et al (2001) Microscopic kinetics and energetics distinguish GABA(A) receptor agonists from antagonists. Biophys J 81:2660–2670

Kavanaugh MP, Arriza JL, North RA et al (1992) Electrogenic uptake of gamma-aminobutyric acid by a cloned transporter expressed in Xenopus oocytes. J Biol Chem 267:22007–22009

Kim YI, Dudek FE (1992) Intracellular electrophysiological study of suprachiasmatic nucleus neurons in rodents: inhibitory synaptic mechanisms. J Physiol (Lond) 458:247–260

Ko CH, Yamada YR, Welsh DK et al (2010) Emergence of noise-induced oscillations in the central circadian pacemaker. PLoS Biol 8:e1000513

Kononenko NI, Dudek FE (2004) Mechanism of irregular firing of suprachiasmatic nucleus neurons in rat hypothalamic slices. J Neurophysiol 91:267–273

Lerma J, Herranz AS, Herreras O et al (1986) In vivo determination of extracellular concentration of amino acids in the rat hippocampus. A method based on brain dialysis and computerized analysis. Brain Res 384:145–155

Liu C, Reppert SM (2000) GABA synchronizes clock cells within the suprachiasmatic circadian clock. Neuron 25:123–128

Liu AC, Welsh DK, Ko CH et al (2007) Intercellular coupling confers robustness against mutations in the SCN circadian clock network. Cell 129:605–616

Low-Zeddies SS, Takahashi JS (2001) Chimera analysis of the Clock mutation in mice shows that complex cellular integration determines circadian behavior. Cell 105:25–42

McElroy B, Zakaria A, Glass JD et al (2009) Ethanol modulates mammalian circadian clock phase resetting through extrasynaptic gaba receptor activation. Neuroscience 164:842–848

McIntire SL, Reimer RJ, Schuske K et al (1997) Identification and characterization of the vesicular GABA transporter. Nature 389:870–876

Michel S, Colwell CS (2001) Cellular communication and coupling within the suprachiasmatic nucleus. Chronobiol Int 18:579–600

Mody I (2001) Distinguishing between GABAA receptors responsible for tonic and phasic conductances. Neurochem Res 26:907–913

Mody I (2005) Aspects of the homeostaic plasticity of GABAA receptor-mediated inhibition. J Physiol 562:37–46

Mody I, Pearce RA (2004) Diversity of inhibitory neurotransmission through GABA(A) receptors. Trends Neurosci 27:569–575

Mohler H (2006) GABA(A) receptor diversity and pharmacology. Cell Tissue Res 326:505–516

Moldavan MG, Allen CN (2013) GABAB receptor-mediated frequency-dependent and circadian changes in synaptic plasticity modulate retinal input to the suprachiasmatic nucleus. J Physiol (Lond) 591:2475–2490

Moldavan MG, Irwin RP, Allen CN (2006) Presynaptic GABAB receptors regulate retinohypothalamic tract synaptic transmission by inhibiting voltage-gated Ca2+ channels. J Neurophysiol 95:3727–3741

Moore RY, Speh JC (1993) GABA is the principal neurotransmitter of the circadian system. Neurosci Lett 150:112–116

Mozrzymas JW, Zarnowska ED, Pytel M et al (2003) Modulation of GABA(A) receptors by hydrogen ions reveals synaptic GABA transient and a crucial role of the desensitization process. J Neurosci 23:7981–7992

Naum OG, Fernanda Rubio M, Golombek DA (2001) Rhythmic variation in gamma-aminobutyric acid(A)-receptor subunit composition in the circadian system and median eminence of Syrian hamsters. Neurosci Lett 310:178–182

Novak CM, Albers HE (2004) Novel phase-shifting effects of GABAA receptor activation in the suprachiasmatic nucleus of a diurnal rodent. Am J Physiol Regul Integr Comp Physiol 286:R820–R825

O'Hara BF, Andretic R, Heller HC et al (1995) GABAA, GABAC, and NMDA receptor subunit expression in the suprachiasmatic nucleus and other brain regions. Mol Brain Res 28:239–250

Olsen RW, Sieghart W (2008) International Union of Pharmacology. LXX. Subtypes of gamma-aminobutyric acid(A) receptors: classification on the basis of subunit composition, pharmacology, and function. Update. Pharmacol Rev 60:243–260

Panda S, Antoch MP, Miller BH et al (2002) Coordinated transcription of key pathways in the mouse by the circadian clock. Cell 109:307–320

Ralph MR, Menaker M (1986) Effects of diazepam on circadian phase advances and delays. Brain Res 372:405–408

Reppert SM, Weaver DR (2001) Molecular analysis of mammalian circadian rhythms. Annu Rev Physiol 63:647–676

Riazanski V, Deriy LV, Shevchenko PD et al (2011) Presynaptic CLC-3 determines quantal size of inhibitory transmission in the hippocampus. Nat Neurosci 14:487–494

Richerson GB, Wu Y (2003) Dynamic equilibrium of neurotransmitter transporters: not just for reuptake anymore. J Neurophysiol 90:1363–1374

Rudolph U, Mohler H (2004) Analysis of GABAA receptor function and dissection of the pharmacology of benzodiazepines and general anesthetics through mouse genetics. Annu Rev Pharmacol Toxicol 44:475–498

Santos MS, Park CK, Foss SM et al (2013) Sorting of the vesicular GABA transporter to functional vesicle pools by an atypical dileucine-like motif. J Neurosci 33:10634–10646

Schofield CM, Huguenard JR (2007) GABA affinity shapes IPSCs in thalamic nuclei. J Neurosci 27:7954–7962

Semyanov A, Walker MC, Kullmann DM (2003) GABA uptake regulates cortical excitability via cell type-specific tonic inhibition. Nat Neurosci 6:484–490

Semyanov A, Walker MC, Kullmann DM et al (2004) Tonically active GABA(A) receptors: modulating gain and maintaining the tone. Trends Neurosci 27:262–269

Shirakawa T, Honma S, Katsuno Y et al (2000) Synchronization of circadian firing rhythms in cultured rat suprachiasmatic neurons. Eur J Neurosci 12:2833–2838

Smith RD, Inouye S, Turek FW (1989) Central administration of muscimol phase-shifts the mammalian circadian clock. J Comp Physiol 164:805–814

Staley KJ, Mody I (1992) Shunting of excitatory input to dentate gyrus granule cells by depolarizing GABAA receptor-mediated postsynaptic conductances. J Neurophysiol 68:197–212

Stell BM, Mody I (2002) Receptors with different affinities mediate phasic and tonic GABA(A) conductances in hippocampal neurons. J Neurosci 22:RC223

Strecker GJ, Wuarin JP, Dudek FE (1997) GABAA-mediated local synaptic pathways connect neurons in the rat suprachiasmatic nucleus. J Neurophysiol 78:2217–2220

Subramanian P, Subbaraj R (1996) Diazepam modulates the period of locomotor rhythm in mice (Mus booduga) and attenuates light-induced phase advances. Pharmacol Biochem Behav 54:393–398

Tossman U, Jonsson G, Ungerstedt U (1986) Regional distribution and extracellular levels of amino acids in rat central nervous system. Acta Physiol Scand 127:533–545

Van den Pol AN (1980) The hypothalamic suprachiasmatic nucleus of rat: intrinsic anatomy. J Comp Neurol 191:661–702

vanderLeest HT, Rohling JH, Michel S et al (2009) Phase shifting capacity of the circadian pacemaker determined by the SCN neuronal network organization. PLoS One 4:e4976

Wagner S, Castel M, Gainer H et al (1997) GABA in the mammalian suprachiasmatic nucleus and its role in diurnal rhythmicity. Nature 387:598–603

Wagner S, Sagiv N, Yarom Y (2001) GABA-induced current and circadian regulation of chloride in neurones of the rat suprachiasmatic nucleus. J Physiol (Lond) 537:853–869

Welsh DK, Logothetis DE, Meister M et al (1995) Individual neurons dissociated from rat suprachiasmatic nucleus express independently phased circadian firing rhythms. Neuron 14:697–706

Welsh DK, Takahashi JS, Kay SA (2010) Suprachiasmatic nucleus: cell autonomy and network properties. Annu Rev Physiol 72:551–577

Woodin MA, Ganguly K, Poo MM (2003) Coincident pre- and postsynaptic activity modifies GABAergic synapses by postsynaptic changes in Cl- transporter activity. Neuron 39:807–820

Yamazaki S, Numano R, Abe M et al (2000) Resetting central and peripheral circadian oscillators in transgenic rats. Science 288:682–685

Yamazaki S, Yoshikawa T, Biscoe EW et al (2009) Ontogeny of circadian organization in the rat. J Biol Rhythms 24:55–63

Yoo SH, Yamazaki S, Lowrey PL et al (2004) PERIOD2::LUCIFERASE real-time reporting of circadian dynamics reveals persistent circadian oscillations in mouse peripheral tissues. Proc Natl Acad Sci U S A 101:5339–5346

Chapter 9
Relevance of Network Organization in SCN Clock Function

Elvira Morgado, Claudia Juárez-Portilla, Ann-Judith Silverman, and Rae Silver

Abstract The bilateral suprachiasmatic nuclei (SCN) of the hypothalamus are the loci of a circadian pacemaker that coordinates rhythms throughout the body. The ~10,000 cells of *each* nucleus are organized into distinct core and shell divisions, each of which is comprised of clusters of various cell types. In dispersed cell preparations, individual SCN neurons express a wide range of circadian periods. In a slice preparation, there is a characteristic peak and trough time of clock gene expression in specific subregions, indicating that the networks into which cells are organized are an important aspect of SCN function. Understanding the relationship of individual cells to that of larger neuronal groupings and to brain circuits in which they participate is a problem of general interest in neuroscience. The SCN is an especially tractable system for such queries as the function of this nucleus is known, and circadian rhythms in activity of individual cells and in the SCN as a whole can each be tracked and associated with the overall behavior and physiology of the intact animal.

E. Morgado
Department of Psychology, Barnard College, New York, NY 10027, USA

Facultad de Biología, Universidad Veracruzana, Xalapa, Mexico

C. Juárez-Portilla
Department of Psychology, Barnard College, New York, NY 10027, USA

Centro de Investigaciones Biomédicas, Universidad Veracruzana, Xalapa, Mexico

A.-J. Silverman
Department of Pathology and Cell Biology, Columbia University Medical Center,
New York, NY 10032, USA

R. Silver (⊠)
Department of Psychology, Barnard College, New York, NY 10027, USA

Department of Psychology, Columbia University, 1190 Amsterdam Avenue, Room 406
Schermerhorn Hall Mail Code 5501, New York, NY 10027, USA

Department of Pathology and Cell Biology, Columbia University Medical Center,
New York, NY 10032, USA
e-mail: QR@columbia.edu

© Springer International Publishing Switzerland 2015
R. Aguilar-Roblero et al. (eds.), *Mechanisms of Circadian Systems
in Animals and Their Clinical Relevance*, DOI 10.1007/978-3-319-08945-4_9

This chapter focuses on the cellular and circuit organization of the SCN to understand how the SCN encodes circadian time, the ways in which it is synchronized to the local environment and how it adjusts its activity in the presence of mutations in key regulatory genes. While the clock analogy is useful, it can be extended to better capture aspects of SCN function that are not "clock-like".

9.1 Introduction

Daily oscillations in physiology and behavior are ubiquitous in the animal and plant kingdoms. They function to keep organisms in synchrony with regularly occurring changes in their local environment. One aspect that makes these oscillations supremely interesting is the fact that most rhythms in metabolism, physiology, and behavior are manifestations of intrinsic timekeeping machinery: they do not depend on driving signals from the environment for their regular recurrence. The internally organized circadian rhythms serve to coordinate the timing of many diverse responses, including sleep–wake cycle, hormonal secretion, locomotor activity, heartbeat, renal blood flow, immune system responses to antigens, among others. In mammals, the phase of bodily circadian rhythms is controlled by an endogenous timing system, a brain clock located in the suprachiasmatic nucleus (SCN) of the hypothalamus. Two major properties characterize this central clock. First, it is capable of generating autonomous, synchronized circadian rhythms. Second, the SCN clock can be reset by cues from the environment. The most salient of these cues is the solar light–dark (LD) cycle, but also effective are nonphotic temporal cues such as temperature, timed food shortage (or availability) or treatment with chronobiotic drugs (Challet and Pévet 2003). While the most detailed analyses of this nucleus are based on studies of mice, rats, and hamsters, the general organization appears to be highly conserved among mammalian species.

Evidence for the existence and function of this brain clock has been accumulating ever since it was first discovered in lesion studies in 1972 (Moore and Eichler 1972; Stephan and Zucker 1972). In that work, ablation of the SCN resulted in loss of circadian rhythms in adrenal corticosterone secretion and in locomotor activity, respectively. Subsequent work based on monitoring of electrical activity of the SCN, in a slice preparation in vitro, established that the tissue functions as a clock, tracking time of day in the absence of input from the rest of the brain. Finally, grafts of fetal SCN tissue transplanted into the third ventricle of SCN lesioned host animals can reestablish overt rhythmicity in locomotor activity, with the period of the donor rather than the host (rev in Weaver 1998). These findings establish not only that the mammalian biological clock is endogenous, but also that it is sufficient to restore rhythmicity with the clock's intrinsic period, with no evident contribution of timing properties by the host.

9.1.1 Clock Components, Inputs and Outputs

We generally conceptualize the brain clock as having three main components (1) input signals and pathways, (2) a circadian pacemaker or the clock itself, and (3) its output pathways. On the input side, in mammals, photic information from the eye reaches the SCN via a direct retinohypothalamic tract (RHT). Information about the internal state of the animal is also communicated to the SCN. For example, the gonads secrete steroid hormones into the circulation on a seasonal basis, and testosterone binds to androgen receptors in the SCN (Karatsoreos et al. 2007). Both the retinal input and the cells bearing receptors for androgens lie in an area that has been designated the "core" of the SCN (Moore 1996a; Leak et al. 1999; Moore et al. 2002). Cells of the core send their output to the shell region. The organization of core and shell is discussed further in the sections that follow.

Via its output pathways, the clock's oscillatory rhythm is transmitted to the body, thereby modulating the expression of numerous behaviors, among which the sleep/wake cycle is the most salient. That the SCN may send out more than one type of signal is suggested by transplant studies. Consistent with such a view, there is evidence of both neural efferents (van den Pol 1991) and paracrine signaling (Silver et al. 1990; Maywood et al. 2011) from the SCN neurons to targets within and outside of the nucleus.

9.1.2 Molecular Clocks

The field of circadian rhythms was transformed with the discovery of mechanisms of oscillation at the cellular/molecular level. A brief overview is presented here, as this topic has been amply reviewed (Zhang and Kay 2010; Buhr and Takahashi 2013). The core of the circadian molecular clock is an autoregulatory transcriptional and translational feedback loop (Fig. 9.1). The transcription factors CLOCK and BMAL1 dimerize and directly and indirectly activate transcription of a family of *Period* (*Per1*, *Per2*, *Per3*) and *Cryptochrome* (*Cry1*, *Cry2*) genes through E-box elements (5′-CACGTG-3′). The PER and CRY proteins accumulate in the cytosol, and following phosphorylation, they are translocated into the nucleus, where they inhibit the activity of CLOCK and BMAL1. The turnover of the inhibitory PER and CRY proteins leads to a new cycle of activation by CLOCK and BMAL1 via E-box elements.

In addition to this core loop, there is a secondary negative feedback loop, in which REV-ERBα and ROR (both nuclear receptors) act to stabilize *Bmal1* transcription, thereby contributing to clock precision and robustness (Reppert and Weaver 2001, 2002; Ueda et al. 2005; Ukai and Ueda 2010). These transcription–translation feedback loops, present within individual cells, provide the basic mechanism of circadian oscillation. Entrainment to local environmental time is thought to result from adjustments of various steps in these oscillating feedback loops.

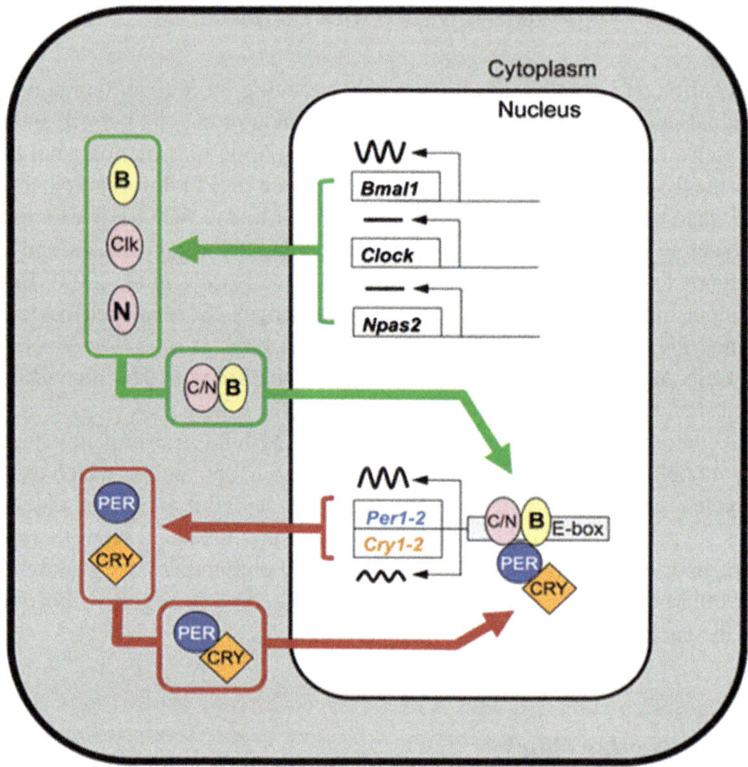

Fig. 9.1 Molecular clock of the suprachiasmatic nucleus (SCN). A simplified view of the molecular circadian clock. The mammalian clock is composed of a negative feedback loop involving clock genes clock, bmal1, per1, per2, cry1, and cry2. Transcription factors CLOCK and BMAL1 interact and activate per and cry gene expression by binding to an E-box in their promoter regions. PER and CRY proteins heterodimerize, translocate to the nucleus, and inhibit Clock/Bmal1-induced gene expression. After the PER/CRY complex is degraded, the cycle starts again. The alternation between the activation and the repression has a cycle length of about 24 h. (From Yu and Weaver 2011)

An important property of the SCN as a network is the synchronization of the individual oscillator cells. Loss or mutation of specific clock genes alters period length (*Cry1, Cry2, Per1, Per3*) and decreases (*Clock, Per2*) or causes immediate loss (*Bmal1*) of circadian rhythmicity under constant conditions. *Cry1/Cry2* and *Per1/Per2* double mutant animals lose circadian rhythmicity immediately under constant conditions, confirming the importance of these genes in the clock mechanism (Vitaterna et al. 1999; Bae et al. 2001; Zheng et al. 2001). CRY1 and CRY2 appear to be necessary for the development of intercellular networks that subserve coherent rhythm expression in adult SCN. There are developmental changes in SCN circadian rhythms in Cry1 and Cry2 double deficient (Cry-null mice). Rhythms in electrical activity and PER expression were detected in both neonatal and adult SCN individual cells; however, at the level of the SCN as a whole, circadian rhythms are found in only neonatal not adult SCN. Importantly synchronized circadian rhythms in adult Cry-null SCN were restored by coculture with neonatal, but not juvenile, SCN (Ono et al. 2013).

Present Goal The molecular revolution enabled substantial exploration of the mechanisms underlying SCN clock function. Our goal is to review current understanding of how the individual "clock" cells of this small nucleus in the hypothalamus encode circadian time and how they work together to set the schedule of daily behavioral and physiological responses throughout the entire body. We discuss the importance of the interconnections among the varied cells and clusters of cells of the SCN, to explore how SCN circuitry underlies its master clock function. We also point to some of the consequences of failures in brain clock function. More specifically, we consider circadian oscillation when molecular elements of the feedback cycle are missing, when key subregions of the SCN are separated from each other, and the consequences of circadian disruptions associated with jet lag.

9.2   SCN Intrinsic Anatomy

The SCN is a bilaterally symmetrical structure located at the base of the brain, in the anterio-ventral hypothalamus situated dorsal to, and extending into, the optic chiasm (Fig. 9.2). The SCN arises from the ventral diencephalic germinal epithelium as a component of the periventricular cell groups (Altman and Bayer 1986). It has a rostrocaudal extent of approximately 600 μm in the mouse and the rat (van den Pol 1980; Abrahamson and Moore 2001), 650 μm in the hamster (Card and Moore 1984) and 4.7 mm in humans (Moore 1996b). It is comprised of ~10,000 neurons on each side in rodents, and ~45,000 in humans (Abrahamson and Moore 2001). The neurons in the SCN are small, about 8–10 μm of diameter, densely packed, and able to generate self-sustained oscillations of electrical activity (Welsh et al. 1995). In addition to neurons, the SCN contains glial cells of many shapes and sizes. They provide structural support to surrounding neurons, regulate ionic and neurotransmitter levels in extracellular fluid, and secrete neurotrophic factors (rev in Middeldorp and Hol 2011; Paratcha and Ledda 2008; Verkhratsky and Steinhäuser 2000 for further detail). Glial cells appear to contribute to the pacemaker function of the SCN, as mice that have a mutation in glial fibrilliary acidic protein (GFAP) have altered circadian activity rhythms in a constant light environment (Moriya et al. 2000). Most work on SCN organization has been done on material examined in the coronal plane and focused on anatomy and function in the area of its greatest extent, specifically, a caudal aspect that includes the core and shell regions. A sagittal view (Fig. 9.2) of the SCN provides a window into its overall extent.

9.2.1   Afferents

The core SCN receives photic input from the retina through two different pathways. Photic input reaches the nucleus via the monosynaptic RHT (Moore and Lenn 1972), and indirectly through the geniculo-hypothalamic tract (GHT), which originates in NPY-containing cells of the thalamic intergeniculate leaflet (IGL) and anterior portions

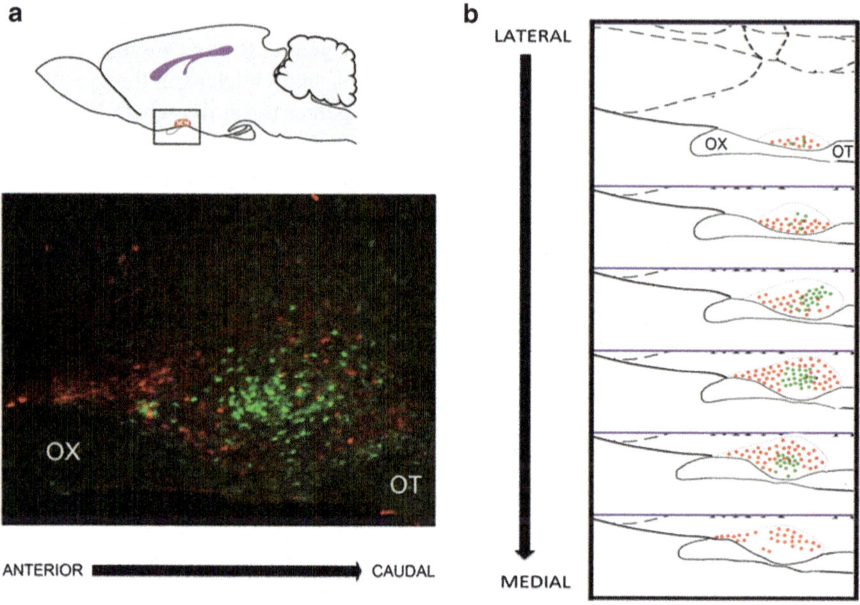

Fig. 9.2 The suprachiasmatic nucleus (SCN). (**a**) Sagittal view of a rodent brain (*upper panel*) illustrates the approximate location of the SCN in the anterior hypothalamus (*red oval*). The *lower panel* shows a photomicrograph through the mid-SCN region, in the same orientation as the cartoon with GFP:GRP cells in *green* and AVP in *red*. (**b**) Schematic of the SCN, in a sagittal plane, from the lateral to the medial aspect. The core region is delimited by the *red dots*, and the shell region is indicated by *green dots*. *OX*, optic chiasm, *OT*, optic tract

of the ventral lateral geniculate nucleus (vLGN) (Morin et al. 1992; Harrington 1996). In addition, the core SCN receives nonphotic input, which arrives from the serotonergic neurons of the dorsal and median raphe nuclei pathway (Moga and Moore 1997; Hay-Schmidt et al. 2003) (Fig. 9.3). The IGL and the dorsal raphe communicate information about locomotor activity (Janik and Mrosovsky 1994; Kuroda et al. 1997), and IGL may also provide metabolic information (Saderi et al. 2013).

The shell region receives dense input from hypothalamus and limbic areas such as the ventral subiculum and limbic cortex (Leak et al. 1999; Moga and Moore 1997; Krout et al. 2002 for further details). Finally, afferents from the paraventricular nucleus of the midline thalamus are distributed throughout the SCN (Moga et al. 1995; Moga and Moore 1997). This specialization in input pathways to regionally distinct populations of cells points to functional heterogeneity within the nucleus.

9.2.2 Endocrine Signals to SCN

While the majority of studies focus on the neural afferent input to the SCN, endogenous humoral signals also act on the SCN, enabling integration in internal and external temporal information. Gonadectomy results in marked alteration of

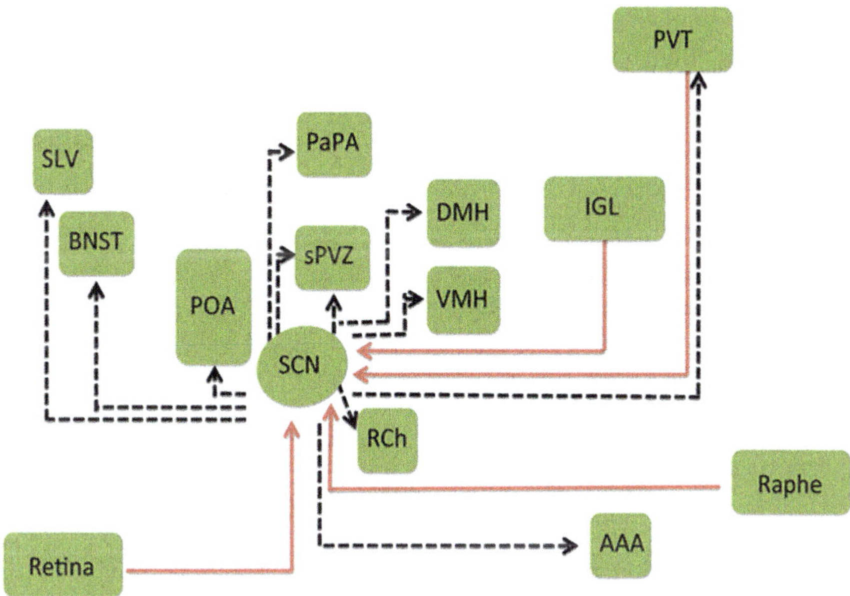

Fig. 9.3 Organization of neural inputs to and outputs from the suprachiasmatic nucleus (SCN). In the schematic illustration the SCN receives direct photic input through the retinohypothalamic tract, indirect photic input from the intergeniculate leaflet (IGL) of the thalamus, and serotonergic input from the raphe. Afferents from other hypothalamic nuclei and parts of the limbic system, among other afferents, including the hippocampus and brainstem. The SCN efferents reach the dorsomedial nucleus of the hypothalamus (DMH) and medial subparaventricular zone (sPVZ) and the lateral sPVZ. *LS* lateral septum, *ILC* infralimbic cortex, *VS* ventral subiculum, *VMH* ventromedial nucleus of the hypothalamus, *ARC* arcuate nucleus, *LHA* lateral hypothalamic area. (Modified from Morin 2013)

circadian behaviors in mice, including lengthened free-running period, decreased precision of daily onset of running, and elimination of early-evening but not late-night activity bouts (Karatsoreos et al. 2007; Daan et al. 1975). Androgen replacement restores these responses. The evidence suggests that the brain clock is an important site of these androgen actions. As already noted, the rodent SCN is composed of a core and a shell region. Androgen receptors (AR) are localized to the core SCN (Karatsoreos et al. 2007). Analysis of a transgenic mouse bearing two reporter molecules driven by the AR targeted to both membrane and nucleus indicates that projections of AR-containing cells form a dense plexus in the core, with their fibers exiting the SCN dorsally. Dual label immunochemistry indicates that gastrin-releasing peptide cells contain AR and that these cells express FOS after a light pulse. As is well known, increasing intensities of constant light parametrically increase circadian period, and this effect was potentiated at all intensities by gonadectomy (Butler et al. 2012). Gonadectomy reduces the FOS response after a phase-shifting light pulse, whereas androgen replacement restores the response to that seen in intact animals. Gonadectomy also increased glial fibrilliary acidic protein expression and decreased expression of the synaptic proteins synaptophysin and postsynaptic density 95 (Karatsoreos et al. 2011). Treatment of gonadectomized animals with the nonaromatizable androgen dihydrotestosterone restored glial fibrilliary acidic protein, postsynaptic density 95, and synaptophysin in the SCN and reinstated the normal pattern of responses to light. Together, the results reveal a role for androgens in regulating circuitry in the mouse SCN, with functional consequences for photic phase resetting stimuli.

AR expression is sexually dimorphic in humans and rodents, with greater expression in males than in females (reviewed in Karatsoreos and Silver 2007). Both Western blots and immunochemistry confirm that ARs are more highly expressed in male than in female mice. Gonadectomy eliminates and androgen treatment restores these sex differences, indicating that the response is dependent on circulating hormones rather than on genetic factors (Iwahana et al. 2008). Interestingly, while there is little ER alpha in the SCN, ERbeta-IR neurons are localized to the shell subdivision of the nucleus, with significantly greater numbers in females than in males (Vida et al. 2008). Treatment of either gonadectomized mice of either sex with estradiol benzoate reduced the number of ERbeta-IR neurons, with no effect of ERalpha-IR. The sexually differentiated expression and distribution of ARs and ERs in various cell populations of the SCN suggest multiple modes of hormone action within the SCN.

9.2.3 SCN Efferent Connections

Monosynaptic SCN efferents primarily reach nearby hypothalamic targets (Moore 1996a; Abrahamson and Moore 2001). The densest projections to the subparaventricular zone (sPVZ), the preoptic area (POA), bed nucleus of the stria terminalis (BNST), the lateral septum (LS), retrochiasmatic area (RCA), the arcuate nucleus (ARC), and the dorsomedial hypothalamus (DMH) come from the shell, whereas the lateral sPVZ receives most of its input from the core (Watts et al. 1987; Leak and Moore 2001) (Fig. 9.3). Limited projections to the forebrain, midline thalamus, IGL, and periaqueductal gray are also present. These direct SCN recipient areas in turn signal many other brain regions with the result that most of the brain gets rhythmic input and has daily rhythms of activity (Harbour et al. 2013; Morin 2013). Circadian rhythms are also expressed in virtually all peripheral bodily tissues. Time signals from the SCN travel to peripheral organs by the sympathetic and parasympathetic nervous systems and by circadian fluctuations in adrenal glucocorticoids (Ota et al. 2012).

9.2.4 SCN Diffusible Signals

In addition to these neural efferents, the SCN also produces diffusible signals, which represent another SCN output. When SCN donor tissue graft is encapsulated within a semipermeable membrane, circadian locomotor activity is restored

(Silver et al. 1996; LeSauter and Silver 1998). The polymer capsule allows chemicals to diffuse through but does not permit fibers to exit the graft. Several diffusible signals have been studied, including prokineticin 2 (PK2) (Cheng et al. 2002; Prosser et al. 2007), transforming growth factor-alpha (TGF-α) (Kramer et al. 2001), and cardiotrophin-like cytokine (Kraves and Weitz 2006). Both PK2 and TGF-α are secreted in a circadian fashion. When infused into the third ventricle of the rat or hamster brain, they suppress locomotor activity. This may be a mechanism for supporting rest–activity cycles. While it has been suggested that diffusible signals produced by the SCN may travel to distal tissues located widely in the brain and even in the gut (Schibler and Sassone-Corsi 2002), the distance over which such signals can travel is actually not known.

Diffusible signals from the SCN serve to communicate not only with nearby extra-SCN regions, but also within the nucleus itself. Diffusible interneuronal signals have been examined in an SCN slice coculture using wild-type "donor graft" SCN to drive pacemaking (monitored by PER2::LUCIFERASE bioluminescence) in "host target" SCN deficient either in elements of neuropeptidergic signaling or in elements of the core feedback loop (Maywood et al. 2011). A hierarchy of neuropeptidergic signals underpins paracrine regulation by vasoactive intestinal polypeptide (VIP) augmented by contributions from arginine vasopressin (AVP) and gastrin-releasing peptide (GRP). Also paracrine signaling maintains circadian pacemaking in arrhythmic Cry-null SCN, deficient in essential elements of the transcriptional negative feedback loops (Ono et al. 2013). These studies indicate that paracrine neuropeptidergic signals determine cell- and circuit-level circadian pacemaking within the SCN. In summary, the pacemaking function in the in vitro coculture preparation, and in whole animal transplant experiments, is specified by the genotype of the graft, confirming graft-derived factors as determinants of the host rhythm. The next question is how the pacemaking function of the brain's master clock's many individual neurons is achieved, given the heterogeneity of its inputs and outputs. This can be addressed by examining the functional and spatial organization of the SCN.

9.3 Functional Organization of SCN Neurons

9.3.1 Properties and Functions of Core and Shell SCN

Early studies suggested that rhythmicity was lost if the entire nucleus was lesioned, but was sustained if any part of the nucleus survived ablation. With the advent of markers enabling identification of specific subregions of the nucleus, it became clear that the core and shell have markedly different properties. Period genes Per1, Per2, and Per3 mRNA expression are differentially regulated in the two compartments. The shell exhibits endogenous rhythmicity in Per1, Per2, and Per3 mRNA. The core compartment does not have high-amplitude rhythmic mRNA expression but has temporally gated light-induced Per1 and Per2 and high levels of endogenous non-rhythmic Per3 mRNA expression (see below for discussion of gating). These results reveal differential regulation of clock genes in the two distinct SCN regions (Hamada et al. 2001). Furthermore, lesions of the SCN core eliminate all circadian rhythms, including locomotor activity, drinking, gnawing, body temperature, and hormone secretion. Loss of these rhythms occurs even when substantial numbers of cells in the SCN shell survive the lesion (LeSauter and Silver 1999; Kriegsfeld et al. 2004). This finding is further confirmed in work on SCN slices in vitro where the ventral SCN (containing the core) is separated from dorsal SCN. In slice studies, when core and shell of mouse SCN were separated, the shell was desynchronized while the core was not, pointing to the importance of connections from the core in synchronizing neurons in the shell (Yamaguchi et al. 2003). Such findings lead to the hypothesis that cells of the core, which themselves do not appear to have high-amplitude oscillation of clock genes, are necessary to coordinate oscillation of network of cells in the shell.

9.3.2 Function of the Shell SCN

Disrupting the intercellular coupling among neurons in the SCN indicated the existence of regional differences in the periods of the oscillating small-latticed regions of the SCN in an imaging study using a transgenic rat carrying a luciferase reporter gene for $Per2$ ($Per2$::dluc) (Koinuma et al. 2013). Here, the results show that the SCN can be divided into two regions—one with periods shorter than 24 h and another with periods longer than 24 h. It appears that the topographically specific and multiphase oscillations of clock gene expression are inherent properties of the SCN. The foregoing studies point to properties of SCN networks. The robust oscillation of the SCN rests on a specific linkage of oscillations that bear a characteristic pattern of changes in spatial extent of the nucleus and changes within subregions that occur in specific temporal niches for each region: time spatial and temporal organization of the SCN. Why such heterogeneity is necessary for SCN function is discussed further below.

Such findings draw our attention to the properties of individual cells versus that of SCN networks.

9.3.3 Relation of Cells to Networks

When individual cells are dispersed, the period of the individual oscillators is much broader than that seen in cells within the intact network (Welsh et al. 1995). Within cultured explants, individual neurons express rhythms with a period of about 24 h, while dissociated neurons that are plated at low density oscillate with a much broader range of periods (Welsh et al. 1995; Liu et al. 1997; Herzog et al. 1998; Honma et al. 1998; Nakamura et al. 2001). This indicates that the period of the SCN tissue as a whole requires an interconnected networks to generate its typical coherent oscillation. Also, while dispersed SCN cells sustain rhythms, they are unstable oscillators when dispersed, indicating that they require network interactions for robust cycling (Aton et al. 2005).

9.3.4 Gating Mechanisms

As we have seen, a fundamental question with regard to the organization of circadian timing is how the SCN achieves a coherent oscillation while their constituent independent cellular oscillators express a wide range of periods. The SCN tissue shows specific modes of coupling mediated by specialized mechanisms, not seen in other tissues bearing oscillators. Furthermore, the amplitude of the SCN intracellular clock is highly responsive to and dependent on intercellular signaling. The most prominent network property of the SCN is the coupling of its cellular oscillators which then produce a coherent circadian oscillation at the level of the tissue. One might imagine that the individual SCN cells might be coupled and that the consensus output of individual oscillators is a summation of weak coupling among cells. Instead, it is clear that in each circadian cycle, activation of PER expression begins in dorsomedial SCN, and proceeds ventrally and laterally, followed by marked activation of an inner area or "cap" just dorsal to the core (Foley et al. 2011; Nakamura et al. 2001; Yan and Okamura 2002). This evidence argues strongly that SCN coupling is mediated by specific neural circuits.

One model, attractive in its simplicity, is neutral with regard to inter-neuronal coupling mechanisms and rests on the differences between core and shell properties. Specifically, the differences in oscillation between core and shell cells suggest a model that incorporates non-rhythmic "gate" cells and rhythmic oscillator cells with a wide range of periods. In this model, the gate provides daily input to oscillator cells and is in turn regulated (directly or indirectly) by the oscillator cells. Individual oscillators with initial random phases use the phase-setting information from the gate cells to self-assemble so as to maintain cohesive rhythmic output. The model explains how individual SCN cells oscillate independently and yet work together to produce a coherent rhythm (Antle et al. 2003). An extension of this model considers how the system responds when the gate cells are activated by an external stimulus, such as an entraining or phase-setting signal. Here, the model shows that exogenous triggering of the gate over a number of days can organize a completely arrhythmic system, simulating the light cue-dependent reappearance of rhythmicity in a population of disorganized, independent oscillators. A significant merit of this model is that a single mechanism (i.e., the output of gate cells) can account for not only free-running and entrained rhythmicity but also limits of entrainment, a PRC with both delay and advance zones, and the light-dependent reappearance of rhythmicity in an arrhythmic animal (Antle et al. 2007). (To intuit this scheme, imagine a row of children on swings. Even though they may start out swinging along separately, if their parents coordinate the time at which they push their offspring, they will start to swing in phase with each other). In summary, individual cells are oscillators, but the oscillation of the SCN requires network properties. While the foregoing model focuses on the relation of core to shell cells, it is also interesting to consider the relation of shell cells to each other.

Clearly the contributions of the chemical and electrical interconnections among neurons are essential to circuit-level organization; how the ensemble of oscillators work together is not known. By applying quantitative methods it is possible to deconstruct the interactions between the spatial and the temporal organization of circadian oscillations in organotypic slices from mice with circadian abnormalities. Clues to SCN network organization come from studies of mice lacking Cryptochrome genes (Cry1, Cry2), which are essential for cell-autonomous oscillation, and in other SCN of mice lacking the VPAC2 receptor, which is necessary for circuit-level integration (Pauls et al. 2014). The SCN of wild-type mice show a strong link between the temporal rhythm of the bioluminescence profiles of PER2::LUC and spatial coherence in portions of tissue. Cry-null SCN have stable spatial organization but lack temporal organization, while some VPAC2-null specimens can exhibit temporal organization in the absence of spatial organization. The results indicate that spatial and temporal organization are separable, that they may have different mechanistic origins (cell-autonomous vs. interneuronal signaling), and that both are necessary to maintain robust and organized circadian rhythms throughout the SCN. This work provides evidence that the coherent properties of the neuronal circuitry, revealed in the spatially organized clusters, are essential to the pacemaking function of the SCN.

9.4   Peptides of the SCN

While the foregoing consideration of networks indicates that diffusible signals cannot account for the phase dispersion of SCN oscillators, it is nevertheless the case that SCN cells can release peptidergic signals. The consideration of SCN peptides points to another level of organization of neurons. It is well established that SCN neurons are comprised of many different peptidergic cell types. These are not dispersed through the nucleus, but instead are organized within clusters of similar cells (Fig. 9.4). However, there is evidence that regionally clustered peptidergic cells have different free-running periods. For example, the period of AVP-releasing rhythm in the dorsal SCN is shorter than that in the ventral SCN (Noguchi et al. 2004; Noguchi and Watanabe 2008). Such findings point to the potential role of paracrine/autocrine signals.

Typically, the core and shell are characterized by different peptidergic cell types. In most species, AVP cells are located in the shell and VIP cells in the core. That said, cells of the core, and those of regions intercalated between distinctly core and shell areas are somewhat more variable among species—likely associated with adaptive species differences. A sense of the relative numbers of SCN

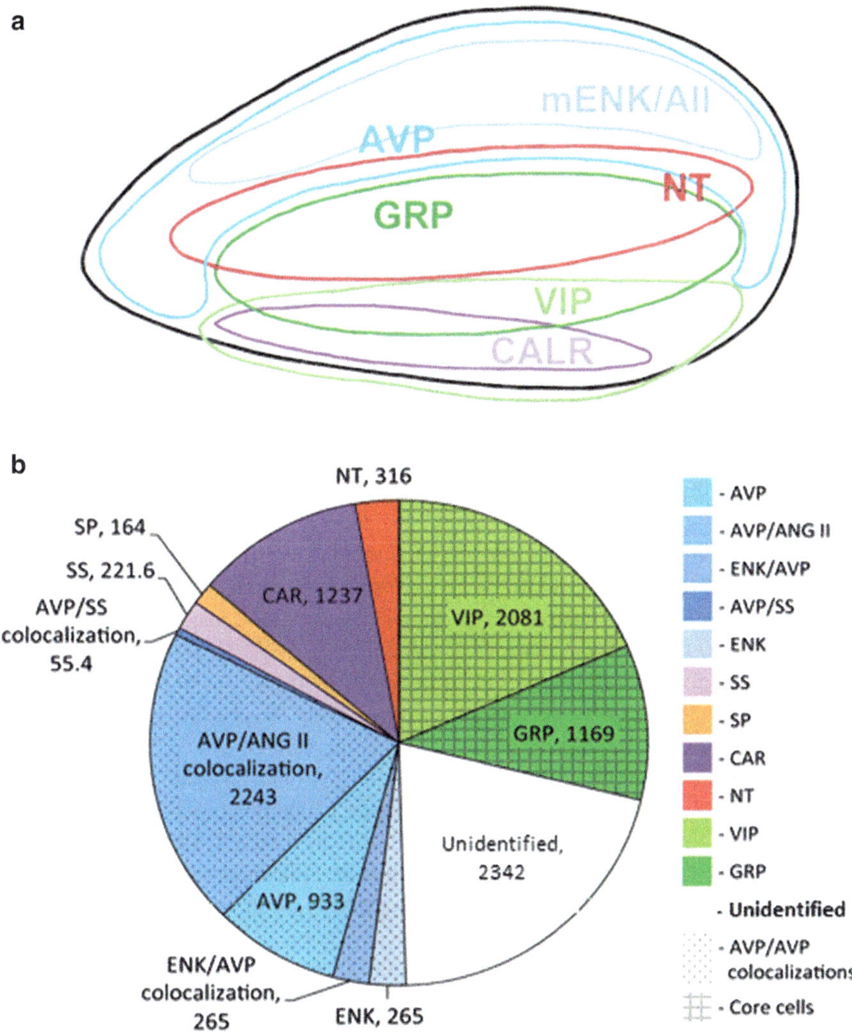

Fig. 9.4 Chemoarchitecture of the suprachiasmatic nucleus. (**a**) The approximate location of clusters of cells bearing peptides of the core and shell regions is seen in sagittal section view. Cells in the core contain VIP, GRP, and CALR and small populations of NT and SP (not shown). Cells in the shell bear AVP and ENK among other peptides. (Modified from Yan et al. 2007). (**b**) The approximate numbers of peptidergic cells of each type are depicted in the "color wheel" based on single- and double-label studies. Note that many peptides are co-expressed in a given cell type and that approximately 20 % of cells remain to be identified (modified from Moore et al. 2002)

cells bearing reasonably well-studied peptidergic phenotypes is provided in Fig. 9.4b and highlights that fact that a substantial fraction of cells remain unidentified (based on data in Moore et al. 2002). In fact, somewhat surprisingly, a study using ultrahigh performance tandem mass spectrometry produced a list of 102 endogenous peptides in the rat SCN, including 33 previously unidentified peptides, and 12 peptides that were post-translationally modified (Lee et al. 2010). To date, research has focused on a few SCN peptides. We present an overview of this work as the peptidergic phenotypes of the SCN and their specialized properties have been amply reviewed (Antle and Silver 2005; Mohawk and Takahashi 2011; Slat et al. 2013).

9.4.1 AVP

AVP-containing cells, lying in the shell region of the SCN, represent about a third of the cell population in the SCN in the rat (Abrahamson and Moore 2001; Moore et al. 2002). AVP release from SCN neurons is under circadian control and Clock mutant mice lack these AVP rhythms (Jin et al. 1999; Silver et al. 1999). In the cerebrospinal fluid, there are SCN-dependent circadian fluctuations of AVP levels with a morning peak and evening trough (Noguchi et al. 2004). AVP serves in the amplification and synchronization of the endogenous rhythmicity of the clock itself. Electrical activity of the SCN is highest during the day, when AVP release is high. Consistent with these findings, SCN neurons generally increase their firing rate following the application of AVP (Ingram et al. 1996). This clock-controlled neuropeptide is thought to play an important role in transmitting the output of the circadian clock to targets outside the SCN (Buijs and Kalsbeek 2001). For example, in the common vole, vasopressin in the SCN is correlated with the expression of circadian locomotor behavior (Jansen et al. 2007). Animals deficient for AVP or its receptor (V1a receptor) have rhythms with attenuated amplitudes (Li et al. 2009). AVP-deficient Brattleboro rats display low-amplitude daily rhythms in sleep–wake, body temperature, plasma melatonin, and SCN firing rates (Ingram et al. 1996). Mice lacking the V1aR show diminished rhythms (Robinson et al. 1988).

Coculture experiments in which diffusible signals are effective but synaptic connections are absent suggested another role for AVP in the SCN. Blocking AVP receptors does not abolish SCN rhythms in vitro (just as Brattleboro rats continue to display circadian rhythms). But in the absence of AVP receptors, wild-type SCN tissue in cocultures fails to restore rhythms to VIP-deficient SCN (Maywood et al. 2011). It is possible that AVP normally amplifies SCN rhythms and, when VIP signaling has been compromised, AVP can act as a weak synchronizing agent to coordinate the rhythms among the circadian cells of the SCN, perhaps acting through other receptors. It had been shown in early studies of the rat SCN that the oscillators in ventral and dorsal AVP cells contribute differently to period length; dorsal cells have shorter period length while those with longer periods are in the ventral aspect (Noguchi et al. 2004). Because the period of activity in the ventral SCN is approximately that of the SCN as a whole, it has been suggested that the oscillation of the shell cells is entrained by cells in the core area.

9.4.2 VIP

VIP is one of the best understood SCN peptides—because studies of mutant mice in which VIP is absent fit well with work on mice lacking the VIP receptor (VPAC2R) (Piggins and Cutler 2003; Vosko et al. 2007). VIP cells constitute about 24 % of the cells in the nucleus (Abrahamson and Moore 2001; Moore et al. 2002; Atkins et al. 2010), and is the most prevalent neuropeptide within the core area: its receptor is widely distributed in the SCN (Usdin et al. 1994; Kalamatianos et al. 2004; Kallo et al. 2004). VIP acts at VPAC2Rs to depolarize SCN neurons by closing potassium channels. VIP is released with circadian rhythmicity under LD conditions, but that rhythm is abolished in constant darkness suggesting a role for VIP in the maintenance of circadian rhythms with respect to environmental lighting (Francl et al. 2010; Shinohara et al. 1998). VIP is inhibited by serotonergic activation, suggesting a possible mode for the attenuating action of 5-hydroxytryptamine (5-HT) receptors in SCN photic signaling (Francl et al. 2010). In addition, VIP can phase-shift circadian rhythms of locomotor behavior and of firing rate of SCN neurons as well as clock gene expression, similar to the action of environmental light. Mutant mice lacking VIP or the $VPAC_2$ receptor have weak behavioral rhythms with multiple period components. In SCN slices from these mutant mice, neuronal firing and clock gene rhythms are suppressed, largely due to desynchronization among cells (Maywood et al. 2011). However, daily application of a VIP agonist to mutant SCN cultures restores synchrony (Aton et al. 2005).

In the intact animal, there is a dose-dependent effect of VIP that is specific to time of day. VIP maximally delays circadian rhythms in the SCN around subjective dusk (Reed et al. 2001; An et al. 2011). Around the transition to dawn, VIP produces small-phase advances. When applied daily, VIP entrains the SCN in a slice preparation (An et al. 2011). Using a coculture preparation Maywood et al. (2011) showed that a wild-type SCN slice could restore circadian oscillation in a VIP-deficient SCN slice. Thus it appears that VIP can diffuse out of the SCN and is both necessary and sufficient for sustained SCN rhythms. However, wild-type SCN also restores circadian rhythms to VPAC2R-deficient SCN, though slowly. However, VIP is not the only factor capable of producing such synchronization by paracrine signals.

9.4.3 GRP

GRP is produced by approximately 10 % of SCN neurons of the core region (Abrahamson and Moore 2001; Antle et al. 2005; Atkins et al. 2010; Moore et al. 2002). The GRP receptor (BBR2) mRNA appears throughout the SCN, with more in the dorsal SCN (Aida et al. 2002; Karatsoreos et al. 2006). In some ways, GRP function in the SCN overlaps with VIP. In fact, about some VIP cells in SCN also express GRP mRNA in the rat (Kawamoto et al. 2003).

GRP has been implicated in the transmission of photic information to the SCN. Anatomical evidence indicates that the cell somata of GRP neurons of the core SCN extend fibers into the shell (Silver et al. 1999; Karatsoreos et al. 2004; Drouyer et al. 2010), where they communicate with other types of SCN neurons. GRP application induces coordinated circadian rhythms in SCN slices from mice deficient in VPAC2R (Brown et al. 2005; Maywood et al. 2006). GRP, acting through BB2 receptors, induces *Per1*, *Per2*, and c-*fos* expression in the shell SCN (Aida et al. 2002). Like VIP neurons, GRP neurons receive retinal input and respond to nocturnal light with increased transcription of c-fos and the Period genes (Bryant et al. 2000; Karatsoreos et al. 2004). Again like VIP, GRP signals through increases in cAMP (Gamble et al. 2007), and GRP application *in vivo* and *in vitro* can shift SCN rhythms (Piggins et al. 1995; Antle et al. 2005; Kallingal and Mintz 2006; Gamble et al. 2007). Unlike VIP and VPAC2Rs, blocking GRP receptors (BBR2) does not abolish SCN rhythms, but it does prevent SCN cocultures from restoring rhythms to VIP-deficient SCN (Maywood et al. 2011). Taken together, the data suggest that both VIP and GRP participate in entrainment of SCN circadian rhythms and are important in synchronizing SCN neurons, but that GRP is less powerful in this regard.

9.4.4 CalR, mENK/AII, and NT and Unidentified Cell Populations

Neurons of the SCN shell contain angiotensin II (AII) and met-enkephalin (mENK1) whereas neurons of the SCN core synthesize calretinin (CalR, a calcium-binding protein) and neurotensin (NT). CalR cells are found in the ventrolateral SCN, where the majority of RHT fibers terminate, with a distinct cluster of cells in the dorsolateral region (Silver et al. 1999). CalR neurons represent the 14 % of the nucleus and are distributed throughout most of the SCN, but are generally absent from the central SCN (Moore et al. 2002). CalR-producing neurons colocalize GABA, and receive input from nonvisual cortical and subcortical regions (Moore et al. 2002). CalR may act as a Ca^{2+}-buffering protein in mouse retinorecipient SCN neurons which have an important role in daily rhythms of intracellular Ca^{2+} mobilization (Ikeda and Allen 2003).

NT, which is present in 4 % of SCN neurons (Moore et al. 2002), has phase-shifting effects on the firing rate in SCN slices and this effect may be mediated by NTS1 and NTS2 receptors. These receptors signal via phospholipase C (PLC; Coogan et al. 2000; Reed and Piggins 2002).

mENK1 neurons are located primarily dorsomedially in the shell subdivision of the SCN; these neurons project to the ipsilateral and contralateral shell and to ipsilateral extrinsic projection areas (Abrahamson and Moore 2001). mENK1 is colocalized with AVP in approximately 50 % of the neurons (Moore et al. 2002).

9.5   Circuits and Spatiotemporal Communication of Clock Gene Activity

9.5.1   Intercellular Signaling Maintains SCN Circadian Oscillation

Inactivation of the receptor for VIP and pituitary adenylate cyclase activating polypeptide (PACAP) leads to a loss of normal circadian activity and a failure to express clock genes such as *Per1*, *Per2*, and *Cry1* in a circadian fashion, as well as a failure of circadian expression of AVP in the SCN (Harmar et al. 2002). Synchrony requires that individual neurons (on average) share the same period but does not necessarily require that they share the same phase. During a circadian cycle *in vivo*, the clock genes *Per1* and *Per2* are first expressed in cells lying in the dorsomedial part of the SCN (near the third ventricle) and then spread out to the central region and finally to the ventral region of rat and mouse (Yan and Okamura 2002; Quintero et al. 2003). These results were amplified by using the Per1: luciferase model (transgenic mice express the gene for luciferase, a bioluminescent enzyme, under the Per1 promoter) where the rhythmicity of Per1 promoter activity can be visualized by the bioluminescence of luciferase (Yamaguchi et al. 2003). In this work, dorsal cells were no longer synchronous when isolated from the ventral core SCN by a knife cut. Furthermore, application of cycloheximide (a protein synthesis inhibitor) stopped the molecular clock and reset SCN neurons to a common phase. Once the cycloheximide was removed, the cells gradually reestablished their original phase relationships (Yamaguchi et al. 2003). These results confirm that differentially phased neuronal clocks are topographically arranged across the SCN and that the phase relationships among the cellular oscillators are an intrinsic property of SCN circuitry.

Bioluminescence imaging of coronal SCN slices cultured from PER2::LUC mice produces real-time longitudinal monitoring of rhythmic expression of the fusion protein in various subregions of the SCN (Yan et al. 2007). Bioluminescence was initially elevated in the dorsal, central, and dorsomedial SCN and then expanded to the rest of the SCN slice before regressing back. SCN slices that contained both core and shell showed strong PER2::LUC rhythms in all subregions with relative phases that remained consistent over several circadian cycles and across all slices (Yan et al. 2007). These descriptive characterizations have been extended using cluster analysis to assess the spatiotemporal activation patterns of PER2 bioluminescence in acute brain slices from PER2::LUC knockin mice. The results indicate that circadian oscillation is characterized by a stable daily cycle of PER2 expression involving orderly serial activation of specific SCN subregions, followed by a silent interval (Foley et al. 2011). It appears that the precise timing of PER2 expression within individual neurons is dependent on their location within the nucleus, and that small groups of neurons within the SCN give rise to distinctive and identifiable subregions (Fig. 9.5). In contrast, in SCN lacking the VPAC2 receptor, which is necessary for circuit-level

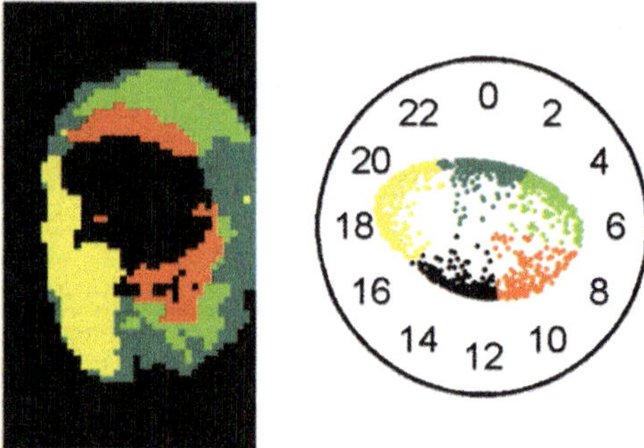

Fig. 9.5 The spatial and temporal organization of the suprachiasmatic nucleus (SCN). To understand the relationship of changes in individual cells and the circuits of which they are a part, it is necessary to visualize changes that occur at each of these levels of analysis over time, as can be done using quantitative analytic tools. The left side of the figure depicts cell groupings based on cluster analysis of phase relationships. The right side of the image points to times of day, based on spectral clustering, when each cluster peaks in its activity—assessed by PER::LUC expression (From Pauls et al. 2014). This analysis contrasts the results in wild-type and mutant SCN (CRY-null and VIP receptor null), and indicates that spatial and temporal organizations are separable and may have different mechanistic origins. Moreover, coupling of the two is necessary to maintain robust and organized circadian rhythm throughout the SCN

integration, specimens can exhibit temporal organization in the absence of spatial organization (Pauls et al. 2014). The results indicate that spatial and temporal organizations are separable, pointing to their basis in different mechanistic origins (cell-autonomous vs. interneuronal signaling). The results also indicate that robust oscillation requires an orderly link, identified in the cluster analysis, between temporal and spatial changes in the nucleus. The foregoing studies suggest that the robust rhythmicity of the SCN rests on a specific linkage of oscillations that bear a characteristic pattern of changes across the spatial extent of the nucleus; changes within subregions occur in specific temporal niches for each region.

It has been suggested that phase dispersal among different cell populations within the SCN may be functionally important for output signals. Possibly, it allows for multiple, variously phased output signals to targets, enabling temporally specialized and plastic regulation of peripheral rhythms (Kalsbeek et al. 2006). On the other hand, evidence from flies suggests that clock neuron network consists of multiple independent oscillators that are *each* capable of supporting activity bouts (Yao and Shafer 2014). In summary, it seems that behavioral rhythms in the fly, as in mammals, emerge from the interactions of many independent oscillators.

9.6 Jet Lag and Seasonality

Understanding the organization of the SCN sets the stage for exploring the basic adaptive functions of the circadian system. These rely on functional plasticity in the SCN's neuronal network organization, and involve changes in phase relations among oscillatory neurons. This is well illustrated in adjustments to shifted environmental cycle such as occurs during jet lag, and in seasonal adaptations to environmental changes. Both daily photic entrainment and seasonal adaptation to changes in day length arise from a plastic reorganization of SCN oscillatory activity.

9.6.1 Jet Lag

Jet lag results from rapid flight across several time zones, and is characterized by temporary symptoms including fatigue, insomnia, and other physiological, psychological, and cognitive disruption. The rapid change in time zones desynchronizes circadian rhythms among the various tissues and organs of the body. Of interest here is the role of the SCN, and the evidence suggests that a dissociation and slow resynchronization of endogenous oscillators within the SCN after a shift in the light–dark cycle underlies the behavioral and physiological symptoms of jet lag.

Following a phase shift in the light–dark cycle of several hours, distinct region-specific changes occur in the SCN. For example, after a 6 h delay of the light schedule, there was a bimodal pattern of electrical activity in the SCN overall. This is attributable to fast resetting in the ventral SCN and distinct, slow resetting in the dorsal SCN (Albus et al. 2005). A similar effect is seen when SCN molecular expression profiles are monitored. The phases of the RNA expression of Per1 and Per2 and Cry1 shifted rapidly in the core SCN, but were relatively slow to shift in the shell region. After a 10 h delay, molecular resynchrony required approximately 6 days (Nagano et al. 2003; Davidson et al. 2009). Monitoring *Per1-luc* expression in vitro in the rat SCN and peripheral tissues to evaluate re-entrainment revealed that peripheral tissues took much longer to complete a shift than did the SCN, and that each tissue readjusted at a different rate (Yamazaki et al. 2000). Similarly, mPER2::LUC mice subjected to a 6-h advance of the light cycle showed significant differences in shifting kinetics among organs such as the SCN, thymus, spleen, esophagus, and lung (Davidson et al. 2009). The 6 h shift to a new phase was achieved at different rates by each tissue. Although the thymus and lung had completed the shift by day 3, the esophagus was only partially shifted and the spleen had barely shifted at all. Resynchronization appears to become complete for thymus, lung, spleen, and esophagus around day 5. All of these methods converge to indicate that adjustments of the clock to a new time require several days, and point to jet lag as the consequence of processes associated with entrainment to a new LD cycle.

9.6.2 Seasonality

Many organisms adapt to seasonal environmental changes, with species-typical adjustments in activity, reproductive behavior, reproductive physiology and metabolism, fur and feather color, among other changes (Hazlerigg and Loudon 2008).

Of interest here is the evidence that the SCN is a key component in measuring day length and that the SCN network adjusts to changes in day length, thereby contributing to seasonal cyclicity. The activity of the SCN, whether measured by electrical activity, FOS expression, or clock gene expression indicates that the duration of the response reflects the light–dark cycle. In long days the duration of SCN clock gene expression (Sumová et al. 2004) and electrical activity (Brown and Piggins 2009) is extended, compared to short days. Conversely, these responses are compressed in short days (Mrugala et al. 2000).

The response of the SCN subregions to day length is different for the core and shell (Yan and Silver 2008). This was seen by using c-FOS and the clock protein PER1 as markers for neuronal activity in Syrian hamsters housed in long (LD 16:8 h light:dark) vs. short (SD 8:16 h light:dark) days. During SD, there was no detectable phase dispersion across the rostrocaudal extent of the nucleus. In contrast, during LD, rhythms in the caudal SCN phase led those in the mid- and rostral SCN by 4–8 h and 8–12 h, respectively. The dark-phase FOS expression depends on tonic light exposure rather than on intrinsic clock properties. The results point to the occurrence of two separate populations of SCN cells, one responding to acute and the other to tonic light exposure. The results suggest that the seasonal encoding of day length by the SCN entails reorganization of its constituent oscillators by a subgroup of neurons in the SCN core that respond to tonic photic cues. The suggestion is that activation of this population of neurons may alter the strength of the intercellular connections in specific components of the SCN network, thereby changing the phase dispersion among SCN oscillators.

Recordings of single-cell electrical activity patterns in acutely prepared SCN brain slices from animals kept in 12:12 light–dark cycles showed that individual SCN cells are active for only short periods of time—about 4–5 h (Schaap et al. 2003; Brown et al. 2006). In contrast, the population pattern shows a near sinusoidal pattern of electrical activity (Gillette 1986). Furthermore, single cells may be active at different times of day, at either the beginning, middle, or end of the day (van der Leest et al. 2009). This result suggests the very interesting hypothesis that changes in the waveform of the SCN overall are the result of alterations in the phase relationship across many individual oscillators (see Meijer et al. 2010 for further discussion).

9.7 Overview and Summary: SCN as a Brain Clock and More

The concept of a brain clock is appealing and intuitively easy to grasp. A clock keeps time, keeps track of its count, and displays what it has counted. Keeping time requires counting cycles or units of time. The earth and the sun constitute our oldest clock, and are the basis of all other clocks. The brain clock relies on this ancient timekeeping system, and more.

Since its discovery in 1972, the SCN has been viewed as the locus of the master daily clock of the body. We know now that the SCN has essential clock properties in that it counts units of time. The SCN counts out days: it also tracks seasons and counts out annual cycles. But the SCN also incorporates non-clock-like features, and acts more like an analog input device, integrating information from many sources. While the SCN uses the photic information provided by the rotation of earth around sun, it is also sensitive to nonphotic cues. For example, methamphetamine changes the period of the clock (Honma et al. 1986). This is no earth-sun clock. The SCN clock, unlike the earth-sun clock, may function well or poorly. Hormones influence the period of the clock, depending on the photic conditions (Butler et al. 2012). The amplitude and period of the clock change with age (rev Hofman and Swaab 2006) and with genetic mutations, as discussed above. Finally, the SCN does not simply keep track of its count and display time. It dictates the timing of events throughout the body. While we know of this function through studies of efferent monosynaptic connections of SCN neurons and diffusible signals from SCN to targets within and outside of the nucleus, we know barely anything of the timing SCN signal(s) to targets.

One of the most useful tools of neuroscience research is to link specific brain regions to specific functions. For mammalian brains, this can be very difficult to implement, as individual brain regions participate in so many different functions, and most functions are represented in many loci rather than resting in a single center. In contrast, the SCN is an outstanding example of the benefit of this strategy because decades of research consistently show that the SCN of the anterior hypothalamus are the site of important integration of information from the internal and external environment and communication of that information to the rest of the body.

Acknowledgments The research from Rae Silver lab reported here was supported by NIH grants NS37919 MH 075045 and NSF grant 21-02-0530-120045 (to R.S.) and postdoctoral fellowship awards from CONACYT 186901 (to E.M.) and CONACYT 186902 (to C.J.P).

References

Abrahamson EE, Moore RY (2001) Suprachiasmatic nucleus in the mouse: retinal innervation, intrinsic organization and efferent projections. Brain Res 916(1–2):172–191

Aida R, Moriya T, Araki M et al (2002) Gastrin-releasing peptide mediates photic entrainable signals to dorsal subsets of suprachiasmatic nucleus via induction of *Period* gene in mice. Mol Pharmacol 61:26–34

Albus H, Vansteensel MJ, Michel S et al (2005) A GABAergic mechanism is necessary for coupling dissociable ventral and dorsal regional oscillators within the circadian clock. Curr Biol 15:886–893

Altman J, Bayer SA (1986) The development of the rat hypothalamus. Adv Anat Embryol Cell Biol 100:1–178

An S, Irwin RP, Allen CN et al (2011) Vasoactive intestinal polypeptide requires parallel changes in adenylate cyclase and phospholipase C to entrain circadian rhythms to a predictable phase. J Neurophysiol 105(5):2289–2296

Antle MC, Silver R (2005) Orchestrating time: arrangements of the brain circadian clock. Trends Neurosci 28(3):145–151

Antle MC, Foley DK, Foley NC et al (2003) Gates and oscillators: a network model of the brain clock. J Biol Rhythms 18(4):339–350

Antle MC, Kriegsfeld LJ, Silver R (2005) Signaling within the master clock of the brain: localized activation of mitogen-activated protein kinase by gastrin-releasing peptide. J Neurosci 25(10):2447–2454

Antle MC, Foley NC, Foley DN et al (2007) Gates and oscillators II: Zeitgebers and the network model of the brain clock. J Biol Rhythms 22:14–25

Atkins N Jr, Mitchell JW, Romanova EV et al (2010) Circadian integration of glutamatergic signals by little SAAS in novel suprachiasmatic circuits. PLoS One 5(9):e12612

Aton SJ, Colwell CS, Harmar AJ et al (2005) Vasoactive intestinal polypeptide mediates circadian rhythmicity and synchrony in mammalian clock neurons. Nat Neurosci 8:476–483

Bae K, Jin X, Maywood ES et al (2001) Differential functions of *mPer1*, *mPer2*, and *mPer3* in the SCN circadian clock. Neuron 30:525–536

Brown TM, Piggins HD (2009) Spatiotemporal heterogeneity in the electrical activity of suprachiasmatic nuclei neurons and their response to photoperiod. J Biol Rhythms 24:44–54

Brown TM, Hughes AT, Piggins HD (2005) Gastrin-releasing peptide promotes suprachiasmatic nuclei cellular rhythmicity in the absence of vasoactive intestinal polypeptide-VPAC2 receptor signaling. J Neurosci 25(48):11155–11164

Brown TM, Banks JR, Piggins HD (2006) A novel suction electrode recording technique for monitoring circadian rhythms in single and multiunit discharge from brain slices. J Neurosci Methods 156(1):173–181

Bryant DN, LeSauter J, Silver R et al (2000) Retinal innervation of calbindin-D28K cells in the hamster suprachiasmatic nucleus: ultrastructural characterization. J Biol Rhythms 15(2): 103–111

Buhr ED, Takahashi JS (2013) Molecular components of the mammalian circadian clock. Handb Exp Pharmacol 217:3–27

Buijs RM, Kalsbeek A (2001) Hypothalamic integration of central and peripheral clocks. Nat Rev Neurosci 2:521–526

Butler MP, Karatsoreos IN, LeSauter J et al (2012) Dose-dependent effects of androgens on the circadian timing system and its response to light. Endocrinology 153(5):2344–2352

Card JP, Moore RY (1984) The suprachiasmatic nucleus of the golden hamster: Immunohistochemical analysis of cell and fiber distribution. Neuroscience 13:415–431

Challet E, Pévet P (2003) Interactions between photic and nonphotic stimuli to synchronize the master circadian clock in mammals. Front Biosci 8:246–257

Cheng MY, Bullock CM, Li C et al (2002) Prokineticin 2 transmits the behavioural circadian rhythm of the suprachiasmatic nucleus. Nature 417(6887):405–410

Coogan AN, Rawlings N, Luckman SM et al (2000) Effects of neurotensin on discharge rates of rat suprachiasmatic nucleus neurons in vitro. Neuroscience 103(3):663–672

Daan S, Damassa D, Pittendrigh CS et al (1975) An effect of castration and testosterone replacement on a circadian pacemaker in mice (Mus musculus). Proc Natl Acad Sci U S A 72:3744–3777

Davidson AJ, Castanon-Cervantes O, Leise TL et al (2009) Visualizing jet lag in the mouse suprachiasmatic nucleus and peripheral circadian timing system. Eur J Neurosci 29:171–180

Drouyer E, LeSauter J, Hernandez AL et al (2010) Specializations of gastrin-releasing peptide cells of the mouse suprachiasmatic nucleus. J Comp Neurol 518(8):1249–1263

Foley NC, Tong TY, Foley D et al (2011) Characterization of orderly spatiotemporal patterns of clock gene activation in mammalian suprachiasmatic nucleus. Eur J Neurosci 33(10):1851–1865

Francl JM, Kaur G, Glass JD (2010) Regulation of vasoactive intestinal polypeptide release in the SCN circadian clock. Neuroreport 21(16):1055–1059

Gamble KL, Allen GC, Zhou T et al (2007) Gastrin-releasing peptide mediates light-like resetting of the suprachiasmatic nucleus circadian pacemaker through cAMP response element-binding protein and Per1 activation. J Neurosci 27(44):12078–12087

Gillette MU (1986) The suprachiasmatic nuclei: circadian phase-shifts induced at the time of hypothalamic slice preparation are preserved in vitro. Brain Research 379:176–181

Hamada T, LeSauter J, Venuti JM et al (2001) Expression of period genes: rhythmic and nonrhythmic compartments of the suprachiasmatic nucleus pacemaker. Neuroscience 21(19):7742–7750

Harbour VL, Weigl Y, Robinson B et al (2013) Comprehensive mapping of regional expression of the clock protein PERIOD2 in rat forebrain across the 24-h day. PLoS One 8:e76391

Harmar AJ, Marston HM, Shen S et al (2002) The VPAC2 receptor is essential for circadian function in the mouse suprachiasmatic nuclei. Cell 109:497–508

Harrington ME (1996) The ventral lateral geniculate nucleus and the intergeniculate leaflet: Interrelated structures in the visual and circadian systems. Neuroscience Biobehav Rev 21(5):705–727

Hay-Schmidt A, Vrang N, Larsen PH et al (2003) Projections from the raphe nuclei to the suprachiasmatic nucleus of the rat. J Chem Neuroanat 25:293–310

Hazlerigg D, Loudon A (2008) New insights into ancient seasonal life timers. Curr Biol 18:R795–R804

Herzog ED, Takahashi JS, Block GD (1998) Clock controls circadian period in isolated suprachiasmatic nucleus neurons. Nature Neurosci 1:708–713

Hofman MA, Swaab DF (2006) Living by the clock: the circadian pacemaker in older people. Ageing Res Rev 5(1):33–51

Honma KI, Honma S, Hiroshige T (1986) Disorganization of the rat activity rhythm by chronic treatment with methamphetamine. Physiol Behav 38:687–695

Honma S, Shirakawa T, Katsuno Y et al (1998) Circadian periods of single suprachiasmatic neurons in rats. Neurosci Lett 250:157–160

Ikeda M, Allen CN (2003) Developmental changes in calbindin-Dk28 and calretinin expression in the mouse suprachiasmatic nucleus. Eur J Neurosci 17:1111–1118

Ingram CD, Snowball RK, Mihai R (1996) Circadian rhythm of neuronal activity in suprachiasmatic nucleus slices from the vasopressin-deficient Brattleboro rat. Neuroscience 75(2): 635–641

Iwahana E, Karatsoreos I, Shibata S et al (2008) Gonadectomy reveals sex differences in circadian rhythms and suprachiasmatic nucleus androgen receptors in mice. Horm Behav 53:422–430

Janik D, Mrosovsky N (1994) Intergeniculate leaflet lesions and behaviorally-induced shifts of circadian rhythms. Brain Res 651:174–182

Jansen K, Van der Zee EA, Gerkema MP (2007) Vasopressin immunoreactivity, but not vasoactive intestinal polypeptide, correlates with expression of circadian rhythmicity in the suprachiasmatic nucleus of voles. Neuropeptides 41(4):207–216

Jin X, Shearman LP, Weaver DR et al (1999) A molecular mechanism regulating rhythmic output from the suprachiasmatic circadian clock. Cell 96(1):57–68

Kalamatianos T, Kalló I, Piggins HD et al (2004) Expression of VIP and/or PACAP receptor mRNA in peptide synthesizing cells within the suprachiasmatic nucleus of the rat and in its efferent target sites. J Comp Neurol 475(1):19–35

Kallingal GJ, Mintz EM (2006) Glutamatergic activity modulates the phase-shifting effects of gastrin-releasing peptide and light. Eur J Neurosci 24(10):2853–2858

Kallo I, Kalamatianos T, Wiltshire N et al (2004) Transgenic approach reveals expression of the VPAC2 receptor in phenotypically defined neurons in the mouse suprachiasmatic nucleus and in its efferent target sites. Eur J Neurosci 19(8):2201–2211

Kalsbeek A, Palm IF, La Fleur SE et al (2006) SCN outputs and the hypothalamic balance of life. J Biol Rhythms 21:458–469

Karatsoreos IN, Silver R (2007) Minireview: The neuroendocrinology of the suprachiasmatic nucleus as a conductor of body time in mammals. Endocrinology 148:5640–5647

Karatsoreos IN, Yan L, LeSauter J et al (2004) Phenotype matters: identification of light-responsive cells in the mouse suprachiasmatic nucleus. J Neurosci 24:68–75

Karatsoreos IN, Romeo RD, McEwen BS et al (2006) Diurnal regulation of the gastrin-releasing peptide receptor in the mouse circadian clock. Eur J Neurosci 23(4):1047–1053

Karatsoreos IN, Wang A, Sasanian J et al (2007) A role for androgens in regulating circadian behavior and the suprachiasmatic nucleus. Endocrinology 148(11):5487–5495

Karatsoreos IN, Butler MP, Lesauter J, Silver R (2011) Androgens modulate structure and function of the suprachiasmatic nucleus brain clock. Endocrinology 152:1970–1978

Kawamoto K, Nagano M, Kanda F et al (2003) Two types of VIP neuronal components in rat suprachiasmatic nucleus. J Neurosci Res 74(6):852–857

Koinuma S, Asakawa T, Nagano M et al (2013) Regional circadian period difference in the suprachiasmatic nucleus of the mammalian circadian center. Eur J Neurosci 38:2832–2841

Kramer A, Yang FC, Snodgrass P et al (2001) Regulation of daily locomotor activity and sleep by hypothalamic EGF receptor signaling. Science 294:2511–2515

Kraves S, Weitz CJ (2006) A role for cardiotrophin-like cytokine in the circadian control of mammalian locomotor activity. Nature Neurosci 9:212–219

Kriegsfeld LJ, LeSauter J, Silver R (2004) Targeted microlesions reveal novel organization of the hamster suprachiasmatic nucleus. J Neurosci 24:2449–2457

Krout KE, Kawano J, Mettenleiter TC et al (2002) CNS input to the suprachiasmatic nucleus of the rat. Neuroscience 110(1):73–92

Kuroda H, Fukushima M, Nakai M et al (1997) Daily wheel running activity modifies the period of free-running rhythm in rats via intergeniculate leaflet. Physiol Behav 61:633–637

Leak RK, Moore RY (2001) Topographic organization of suprachiasmatic nucleus projection neurons. J Comp Neurol 433(3):312–334

Leak RK, Card JP, Moore RY (1999) Suprachiasmatic pacemaker organization analyzed by viral transynaptic transport. Brain Res 819:23–32

Lee JE, Atkins N, Hatcher NG et al (2010) Endogenous peptide discovery of the rat circadian clock: a focused study of the suprachiasmatic nucleus by ultrahigh performance tandem mass spectrometry. Mol Cell Proteomics 9:285–297

LeSauter J, Silver R (1998) Output signals of the SCN. Chronobiol Int 15(5):535–550

LeSauter J, Silver R (1999) Localization of a suprachiasmatic nucleus subregion regulating locomotor rhythmicity. J Neurosci 19:5574–5585

Li JD, Burton KJ, Zhang C et al (2009) Vasopressin receptor V1a regulates circadian rhythms of locomotor activity and expression of clock-controlled genes in the suprachiasmatic nuclei. Am J Physiol Regul Integr Comp Physiol 296:R824–R830

Liu C, Weaver DR, Jin X et al (1997) Molecular dissection of two distinct actions of melatonin on the suprachiasmatic circadian clock. Neuron 19:91–102

Maywood ES, Reddy AB, Wong GK et al (2006) Synchronization and maintenance of timekeeping in suprachiasmatic circadian clock cells by neuropeptidergic signaling. Curr Biol 16(6):599–605

Maywood ES, Chesham JE, O'Brien JA et al (2011) A diversity of paracrine signals sustains molecular circadian cycling in suprachiasmatic nucleus circuits. Proc Natl Acad Sci USA 108(34):14306–14311

Meijer JH, Michel S, van der Leest HT et al (2010) Daily and seasonal adaptation of the circadian clock requires plasticity of the SCN neuronal network. Eur J Neurosci 32:2143–2151

Middeldorp J, Hol EM (2011) GFAP in health and disease. Progress Neurobiol 93(3):421–443

Moga MM, Moore RY (1997) Organization of neural inputs to the suprachiasmatic nucleus in the rat. J Comp Neurol 389:508–534

Moga MM, Weis RP, Moore RY (1995) Efferent projections of the paraventricular thalamic nucleus in the rat. J Comp Neurol 359(2):221–238

Mohawk JA, Takahashi JS (2011) Cell autonomy of suprachiasmatic nucleus circadian oscillators. Trends Neurosci 34(7):349–358

Moore RY (1996a) Entrainment pathways and the functional organization of the circadian system. In: Buijs R, Romijn HJ, Pennartz CMA, Mirmiran M (eds) Hypothalamic integration of circadian rhythms. Elsevier, Amsterdam, pp 103–119

Moore RY (1996b). Organization of the human circadian system. Air Force Office Scientific research. Final report, pp: 1–10

Moore RY, Eichler VB (1972) Loss of circadian adrenal corticosterone rhythm following suprachiasmatic nucleus lesions in the rat. Brain Res 42:201–206

Moore RY, Lenn NJ (1972) A retinohypothalamic projection in the rat. J Comp Neurol 146:1–14

Moore RY, Speh JC, Leak RK (2002) Suprachiasmatic nucleus organization. Cell Tissue Res 309:89–98

Morin LP (2013) Neuroanatomy of the extended circadian rhythm system. Experimental Neurol 243:4–20

Morin LP, Blanchard JH, Moore RY (1992) Intergeniculate leaflet and suprachiasmatic nucleus organization and connections in the hamster. Vis Neurosci 8:219–230

Moriya T, Yoshinobu Y, Kouzu Y et al (2000) Involvement of glial fibrillary acidic protein (GFAP) expressed in astroglial cells in circadian rhythm under constant lighting conditions in mice. J Neurosci Res 60:212–218

Mrugala M, Zlomanczuk P, Jagota A et al (2000) Rhythmic multiunit neural activity in slices of hamster suprachiasmatic nucleus reflects prior photoperiod. Am J Physiol 278:R987–R994

Nagano M, Adachi A, Nakahama K et al (2003) An abrupt shift in the day/night cycle causes desynchrony in the mammalian circadian center. J Neurosci 23:6141–6151

Nakamura W, Honma S, Shirakawa T et al (2001) Regional pacemakers composed of multiple oscillator neurons in the rat suprachiasmatic nucleus. Eur J Neurosci 14(4):666–674

Noguchi T, Watanabe K (2008) Regional differences in circadian period within the suprachiasmatic nucleus. Brain Res 1239:119–126

Noguchi T, Watanabe K, Ogura A et al (2004) The clock in the dorsal suprachiasmatic nucleus runs faster than that in the ventral. Eur J Neurosci 20:3199–3202

Ono D, Honma S, Honma KI (2013) Cryptochromes are critical for the development of coherent circadian rhythms in the mouse suprachiasmatic nucleus. Nat Commun 4:1666

Ota T, Fustin JM, Yamada H et al (2012) Circadian clock signals in the adrenal cortex. Mol Cell Endocrinology 349(1):30–37

Paratcha G, Ledda F (2008) GDNF and GFRa: a versatile molecular complex for developing neurons. Trends Neurosci 31(8):384–391

Pauls S, Foley N, Foley DK, LeSauter J et al (2014) Differential contributions of intra- and intercellular mechanisms to spatial and temporal architecture of the suprachiasmatic nucleus circadian circuitry in wild-type, CRY- and VPAC2–null mutant mice. Eur J Neurosci 40:2528–2540

Piggins HD, Cutler DJ (2003) The roles of vasoactive intestinal polypeptide in the mammalian circadian clock. J Endocrinology 177:7–15

Piggins HD, Antle MC, Rusak B (1995) Neuropeptides phase shift the mammalian circadian pacemaker. J Neurosci 15(8):5612–5622

Prosser HM, Bradley A, Chesham JE et al (2007) Prokineticin receptor 2 (Prokr2) is essential for the regulation of circadian behavior by the suprachiasmatic nuclei. Proc Natl Acad Sci USA 104:648–653

Quintero JE, Kuhlman SJ, McMahon DG (2003) The biological clock nucleus: A multiphasic oscillator network regulated by light. J Neurosci 23(22):8070–8076

Reed HE, Piggins HD (2002) Neurotensin phase-shifts the firing rate rhythm of neurons in the rat suprachiasmatic nuclei in vitro. Eur J Neurosci 16(2):339–344

Reed HE, Cutler DJ, Coen CW et al (2001) Vasoactive intestinal polypeptide (VIP) phase-shifts the rat suprachiasmatic nucleus clock in vitro. Eur J Neurosci 13(4):839–843

Reppert SM, Weaver DR (2001) Molecular analysis of mammalian circadian rhythms. Ann Rev Physiol 63(1):647–676

Reppert SM, Weaver DR (2002) Coordination of circadian timing in mammals. Nature 418:935–941

Robinson BG, Frim DM, Schwartz WJ et al (1988) Vasopressin mRNA in the suprachiasmatic nuclei: daily regulation of polyadenylate tail length. Science 241(4863):342–344

Saderi N, Cazarez-Márquez F, Buijs FN et al (2013) The NPY intergeniculate leaflet projections to the suprachiasmatic nucleus transmits metabolic conditions. Neuroscience 246:291–300

Schaap J, Albus H, van der Leest TH et al (2003) Heterogeneity of rhythmic suprachiasmatic nucleus neurons: Implications for circadian waveform and photoperiodic encoding. Proc Natl Acad Sci USA 100:15994–15999

Schibler U, Sassone-Corsi P (2002) A web of circadian pacemakers. Cell 111:919–922

Shinohara K, Tominaga K, Inouye SIT (1998) Luminance-dependent decrease in vasoactive intestinal polypeptide in the rat suprachiasmatic nucleus. Neuroscience Lett 251(1):21–24

Silver R, Lehman MN, Gibson M et al (1990) Dispersed cell suspensions of fetal SCN restore circadian rhythmicity in SCN-lesioned adult hamsters. Brain Res 525(1):45–58

Silver R, LeSauter J, Tresco PA et al (1996) A diffusible coupling signal from the transplanted suprachiasmatic nucleus controlling circadian locomotor rhythms. Nature 382(6594):810–813

Silver R, Sookhoo AI, LeSauter J et al (1999) Multiple regulatory elements result in regional specificity in circadian rhythms of neuropeptide expression in mouse SCN. Neuroreport 10: 3165–3174

Slat E, Freeman GM Jr, Herzog ED (2013) The clock in the brain: neurons, glia and networks in daily rhythms. In: Kramer A, Merrow M (eds) Circadian clocks. Handbook of Experimental pharmacology, vol 217. Springer, Heidelberg, pp 104–123

Stephan FK, Zucker I (1972) Circadian rhythms in drinking behavior and locomotor activity of rats are eliminated by hipothalamic lesions. Proc Nat Acad Sci USA 69:1583–1586

Sumová A, Bendová Z, Sládek M et al (2004) Seasonal molecular timekeeping within the rat circadian clock. Physiol Res 53:S167–176

Ueda HR, Hayashi S, Chen W et al (2005) System-level identification of transcriptional circuits underlying mammalian circadian clocks. Nat Genet 37:187–192

Ukai H, Ueda HR (2010) Systems biology of mammalian circadian clocks. Ann Rev Physiol 72:579–603

Usdin TB, Bonner TI, Mezey E (1994) Two receptors for vasoactive intestinal polypeptide with similar specificity and complementary distributions. Endocrinology 135(6):2662–2680

van den Pol AN (1980) The hypothalamic suprachiasmatic nucleus of rat: intrinsic anatomy. J Comp Neurol 191(4):661–702

van den Pol AN (1991) The suprachiasmatic nucleus: morphological and cytochemical substrates for cellular interaction. In: Klein DC, Moore RY, Reppert SM (eds) Suprachiasmatic nucleus: the mind's clock. Oxford University Press, New York, pp 17–50

van der Leest HT, Rohling JH, Michel S et al (2009) Phase shifting capacity of the circadian pacemaker determined by the SCN neuronal network organization. PLoS One 4(3):e4976

Verkhratsky A, Steinhäuser C (2000) Ion channels in glial cells. Brain Res Rev 32(2–3):380–412

Vida B, Hrabovszky E, Kalamatianos T et al (2008) Oestrogen receptor alpha and beta immunoreactive cells in the suprachiasmatic nucleus of mice: distribution, sex differences and regulation by gonadal hormones. J Neuroendocrinol 20:1270–1277

Vitaterna MH, Selby CP, Todo T et al (1999) Differential regulation of mammalian Period genes and circadian rhythmicity by cryptochromes 1 and 2. Proc Natl Acad Sci USA 96(21): 12114–12119

Vosko AM, Schroeder A, Loh DH et al (2007) Vasoactive intestinal peptide and the mammalian circadian system. Gen Comp Endocrinol 152:165–175

Watts AG, Swanson LW, Sanchez-Watts G (1987) Efferent projections of the suprachiasmatic nucleus I. Studies using anterograde transport of Phaseolus vulgaris- Leucoagglutinin in the rat. J Comp Neurol 258:204–229

Weaver DR (1998) The suprachiasmatic nucleus: a 25-year retrospective. J Biol Rhythms 13(2): 100–112

Welsh DK, Logothetis DE, Meister M et al (1995) Individual neurons dissociated from rat suprachiasmatic nucleus express independently phased circadian firing rhythms. Neuron 14(4):697–706

Yamaguchi S, Isejima H, Matsuo T et al (2003) Synchronization of cellular clocks in the suprachiasmatic nucleus. Science 302:1408–1412

Yamazaki S, Numano R, Abe M et al (2000) Resetting central and peripheral circadian oscillators in transgenic rats. Science 288:682–685

Yan L, Okamura H (2002) Gradients in the circadian expression of Per1 and Per2 genes in the rat suprachiasmatic nucleus. Eur J Neurosci 15:1153–1162

Yan L, Silver R (2008) Day-length encoding through tonic photic effects in the retinorecipient SCN region. Eur J Neurosci 28:2108–2115

Yan L, Karatsoreos I, Lesauter J et al (2007) Exploring spatiotemporal organization of SCN circuits. Cold Spring Harb Symp Quant Biol 72:527–541

Yao Z, Shafer OT (2014) The Drosophila circadian clock is a variably coupled network of multiple peptidergic units. Science 343(6178):1516–1520

Yu EA, Weaver DR (2011) Disrupting the circadian clock: gene-specific effects on aging, cancer, and other phenotypes. Aging 13(5):479–493

Zhang EE, Kay SA (2010) Clocks not winding down: unraveling circadian networks. Nat Rev Mol Cell Biol 11:764–776

Zheng B, Albrecht U, Kaasik K et al (2001) Nonredundant roles of the *mPer1* and *mPer2* genes in the mammalian circadian clock. Cell 105:683–694

Chapter 10
Behavioral, Physiological, and Neuroendocrine Circadian Rhythms During Lactation

Mario Caba, Stefan Waliszewski, and Enrique Meza

Abstract Lactation in mammals implies great challenges for the mothers, who must adapt their behavior and physiology to fulfill the costly demands of motherhood. Lactation is the main distinctive feature of mammals, and it is expressed in different modalities depending on the particular evolutionary characteristics of each species. Intense research, using mainly rodents and ungulates, is contributing to understand the complex physiological changes and their interaction with the nervous system during this period. Although this is a field of intense research, very little is known about the circadian rhythms of the mother during lactation. In the present chapter, we explore the importance of circadian rhythms during this period, focusing mainly in the rabbit as a model. There is a major reason for emphasizing attention on this species. The mother rabbit nurses her pups just once a day, for a period of less than 5 min. Although this behavior has long been known, only in recent years has it been firmly established that this event occurs around every 24 h, that is, with circadian periodicity. This is a unique characteristic among mammals, and perhaps for this reason the circadian control of lactation had been underestimated in considering that most of the research during lactation had been conducted in other groups of mammals. Then, in this chapter we present behavioral, physiological, and neuroendocrine circadian rhythms during lactation in the rabbit, and whenever possible we include available literature in other mammal species, including humans.

10.1 Introduction

Lactation causes, in mothers, great challenges that necessitate adaptations in their behavior and physiology so that they can fulfill the costly demands of motherhood. Lactation, the main distinctive feature of mammals, is expressed in different modalities, depending on the particular evolutionary characteristics of each species. Intense research, using mainly rodents and ungulates, contributes to understanding

M. Caba (✉) • S. Waliszewski • E. Meza
Centro de Investigaciones Biomédicas, Universidad Veracruzana,
Xalapa 91000, VER, México
e-mail: mcaba@uv.mx

© Springer International Publishing Switzerland 2015
R. Aguilar-Roblero et al. (eds.), *Mechanisms of Circadian Systems in Animals and Their Clinical Relevance*, DOI 10.1007/978-3-319-08945-4_10

the complex physiological changes occurring during this period as well as their interaction with the nervous system. Even though the physiological and neuroendocrine adaptation to lactation is a field of intense research, the interactions between circadian rhythms of the mother and lactation remain largely unknown. In the present chapter we explore the importance of circadian rhythms during this period, focusing mainly on the rabbit as a model. The major reason for emphasizing this species is that rabbit mothers nurse their pups only once a day and for a period of less than 5 min. Although this behavior has been documented for a long time, it is only in the last years that it was firmly established that this event occurs around every 24 h, that is, with circadian periodicity. This characteristic, which is unique among mammals, has probably contributed to underestimating the circadian control of lactation because most of the research during lactation had been conducted in mammalian species that do not present such a marked circadian periodicity of nursing. Hence, in the present chapter we present behavioral, physiological, and neuroendocrine circadian rhythms during lactation in the rabbit, and whenever possible we refer to other mammal species, including humans.

10.2 Molecular Basis of Circadian Rhythms and Entrainment

10.2.1 The Circadian System

Circadian rhythms are physiological and behavioral changes with a periodicity of around 24 h that allow subjects to anticipate daily environmental fluctuations. In mammals, the circadian system is composed of various circadian clocks organized hierarchically in several peripheral and central tissues. Within these various structures, the suprachiasmatic nucleus (SCN), located in the anterior hypothalamus, is the master circadian clock (Moore and Eichler 1972; Stephan and Zucker 1972). Indeed, lesions of the SCN abolish circadian rhythmicity and behavior whereas transplanting donor SCN tissue into arrhythmic hosts restores it (Drucker-Colín et al. 1984; Lehman et al. 1987). Moreover, circadian rhythms in neural firing rate persist in isolated SCN tissue maintained in culture (Green and Gillette 1982), which indicates that inputs from extra-SCN brain regions are not necessary to maintain rhythmicity in this nucleus. As a whole, these studies confirm that the SCN is the main biological pacemaker.

10.2.2 Clock Genes in Central and Peripheral Tissues

At the molecular level, circadian clocks are driven by an autoregulatory transcriptional/translational negative feedback loop of clock genes and their protein products, with a period of approximately 24 h (Reppert and Weaver 2001). In mammals, the core feedback loop includes the genes *Clock*, *Bmal1*, *Period* (*Per*) *1*, *Per2*, *Cry1*, *Cry2*, and their proteins, among others. Proteins PER1 and CRY are the negative, and CLOCK and BMAL1, the positive, limbs of the transcriptional/translational loop. Also, posttranscriptional and posttranslational events are important for circadian coordination (Ueda et al. 2005; Ripperger and Schibler 2006). The discovery of clock genes has represented a milestone for the study of circadian rhythms. As expected, these genes and their proteins are expressed in the SCN with a circadian rhythm under both light–dark and constant light conditions (reviewed in Reppert and Weaver 2001). Further, several structures in the brain express daily or circadian oscillations of clock genes and their protein products, for example, the olfactory bulbs, hippocampus, cerebral cortex, and pituitary gland, as well as paraventricular, supraoptic, retrochiasmatic, and arcuate nuclei in the hypothalamus, among others (Abe et al. 2002; Granados-Fuentes et al. 2006; Kriegsfeld et al. 2003; Wakamatsu et al. 2001). However, many peripheral tissues also contain these genes, and they also show circadian oscillations (Dibner et al. 2010). In fact, various studies have indicated that in some cases the circadian oscillations of particular tissues are maintained in the absence of the SCN. Specifically, the olfactory bulb (Granados-Fuentes et al. 2006) and retina (Tosini and Menaker 1996) are considered autonomous self-sustaining oscillators. Nonetheless, very little is known about the functional significance of circadian oscillations of clock genes in extra-SCN tissue (Granados-Fuentes et al. 2006; Nolasco et al. 2012), including neuroendocrine cells (Kriegsfeld et al. 2003; see following).

10.2.3   Photic and Nonphotic Entrainment

The SCN needs entraining signals to synchronize its self-sustaining oscillations in clock genes and their proteins with the geographical environmental conditions. This connection is very important to ensure that behavioral and physiological functions of individuals are coordinated among members of the same species and with their environment. Photic information is conveyed to the SCN through the retinohypothalamic tract, projecting from ganglion cells in the eye. Photic stimulation is so powerful that a light pulse delivered at either early or late subjective night produces phase advances or phase delays in behavioral activity in nocturnal rodents (Takahashi et al. 1984).

In humans, consequences of disrupting the normal light–dark cycle are clearly illustrated by the jet lag disturbances in adult subjects (Sack et al. 2007), but this problem has been explored also in preterm infants. Preterm neonates are usually maintained in an intensive care environment with continuous bright light. However, when they are exposed to a light–dark cycle, they gain weight faster, have improved formula tolerance, and higher oxygen saturation, and they are released earlier from hospital stay compared to infants maintained in constant bright light (Vásquez et al. 2011). These results emphasize the importance of changes in lighting conditions through the day for the proper development of infants. Furthermore, proper lighting conditions are prominent in neonatal synchronization; the circadian rhythms of the mother play an essential role through the differential composition of breast milk between the light versus the obscure phases of the day (Cubero et al. 2007).

Besides light, there are also other stimuli that specifically can entrain the SCN, such as novelty-induced wheel-running or administration of chronobiotic drugs, which are termed nonphotic cues (Challet and Pevet 2003). In addition to the SCN, both clock genes and their proteins in peripheral and central regions are also reset, and shift their oscillations, by stimuli such as stress or periodic feeding (Takahashi et al. 2001; Morgado et al. 2011). For this reason, assessment of rhythms of clock genes and their proteins in specific areas or nuclei in central and peripheral tissues is commonly used as an index of synchronization by a specific stimulus (Feillet et al. 2008). Using this strategy it has been found that the mammary gland and some nuclei in the brain are entrained during lactation, not only in the rabbit but also in other mammals, including humans.

10.3 Behavioral and Physiological Adaptations During Lactation

10.3.1 Locomotor Behavior During Lactation

Under normal conditions, light is the main entraining signal for the SCN, which determines the diurnal or nocturnal pattern of activity of a given species. In the rabbit, there are contradictory reports, some indicating that this species is diurnal (Van Hof-van Duin 1971), whereas others indicate that it is nocturnal (Mykytowycz and Rowley 1958), and yet others consider that they are crepuscularly active (Nuboer et al. 1983). However, chronobiological controlled experiments unequivocally demonstrated that the rabbit is nocturnal. Under a 12-h dark/12-h light schedule, most of their activity was displayed during the dark period (Jilge 1980). Moreover, under the same conditions, locomotor activity, hard feces excretion, food intake, water intake, and urine excretion were significantly higher during the dark period and free-ran with a circadian period length greater than 24 h, typical of nocturnally active animals (Jilge 1991). Nevertheless, in non-sound-isolated conditions, subjects exhibit the same aforementioned parameters, mostly during the hours of elevated external noise. If this elevated noise occurs during the day, then most rabbits behave as diurnal (Jilge 1991). Recently we observed that rabbits do not express a typical response to light in the SCN, as do other nocturnal animals, which confirms flexibility in the activity pattern of this species (Juárez et al. 2013).

This circadian flexibility was explored in relationship to locomotor behavior in lactating does. Intact female rabbits in non-sound-isolated conditions exhibit a circadian rhythm (periodogram, X^2 $P<0.001$) with increased activity around the first hours of light and lower activity during the evening and most of the night

(Meza et al. 2008). However, this activity pattern is not maintained during lactation. Circadian recordings indicate that during early lactation, the maximal locomotor behavior of does shifts to around the hours coinciding with the nursing of pups (Fig. 10.1).

Whether nursing is scheduled in the day or the night, locomotor behavior is maximal at around the same hours and consequently the does behave mostly as diurnal or as nocturnal, depending on the time at which nursing is imposed (Meza et al. 2008). This shift in locomotor behavior suggests an alteration in the circadian oscillations of the SCN. Indeed, as shown in Fig. 10.2, in intact control females, PER1 in SCN has a rhythm that reaches a peak at ZT11 (ZT0 = 7 a.m., lights on), but timing of nursing decreases PER1 at times of their maximal expression when compared to intact does (Meza et al. 2008).

In contrast, nursing deprivation for 48 h significantly increases PER1 levels at peak time, in comparison to nursed females (Meza et al. 2011; Fig. 10.3). In the Syrian hamster, exposure to a novel running wheel downregulates clock genes in the SCN (Maywood et al. 1999). This stimulus is considered a nonphotic cue that induces a decrease, contrary to the upregulation of clock genes by a light pulse (Yan and Silver 2004). On the basis of this evidence we propose that suckling stimulus is a nonphotic stimulus for the rabbit SCN.

In other species, such as the Djungarian (*Phodopus campbelli*) and Siberian (*Phodopus sungorus*) hamsters, during lactation there is a significant decrease in amplitude of the circadian activity and wheel-running rhythms as the result of an abrupt decrease in mean dark-phase activity during around the first two thirds of lactation (Scribner and Wynne-Edwards 1994). Also in the rat, free-running activity decreased during lactation (Rosenwasser et al. 1987). This decrease in activity is correlated with an increase in the time spent in the nest, but does not implicate a phase shift as in the rabbit doe.

10.3.2 Body Temperature

In rodents, during lactation the dam shows hyperthermia, in comparison to cycling and pregnant subjects (Woodside and Leon 1980; Scribner and Wynne-Edwards 1994). Perhaps this hyperthermia occurs because the mother has frequent contact bouts with her pups. Indeed, she nurses her young every 24–54 min for a duration of 20–30 min each bout, without an apparent 24-h rhythm (Lincoln et al. 1973). However, even though dams have longer contact bouts during the day, their temperature continues to reach a peak during the night (Leon et al. 1984; Scribner and Wynne-Edwards 1994). In contrast, in the rabbit the peak of temperature does not occur in a phase-dependent manner but depends on the timing of lactation. During pregnancy, temperature has a circadian rhythm with a peak that free-runs with a period length > 24 h (Jilge et al. 2001). After parturition, however, the temperature peak coincides with the only once-daily lactation bout. Temperature starts to increase around 3 h before nursing, reaching highest values at the time of nursing

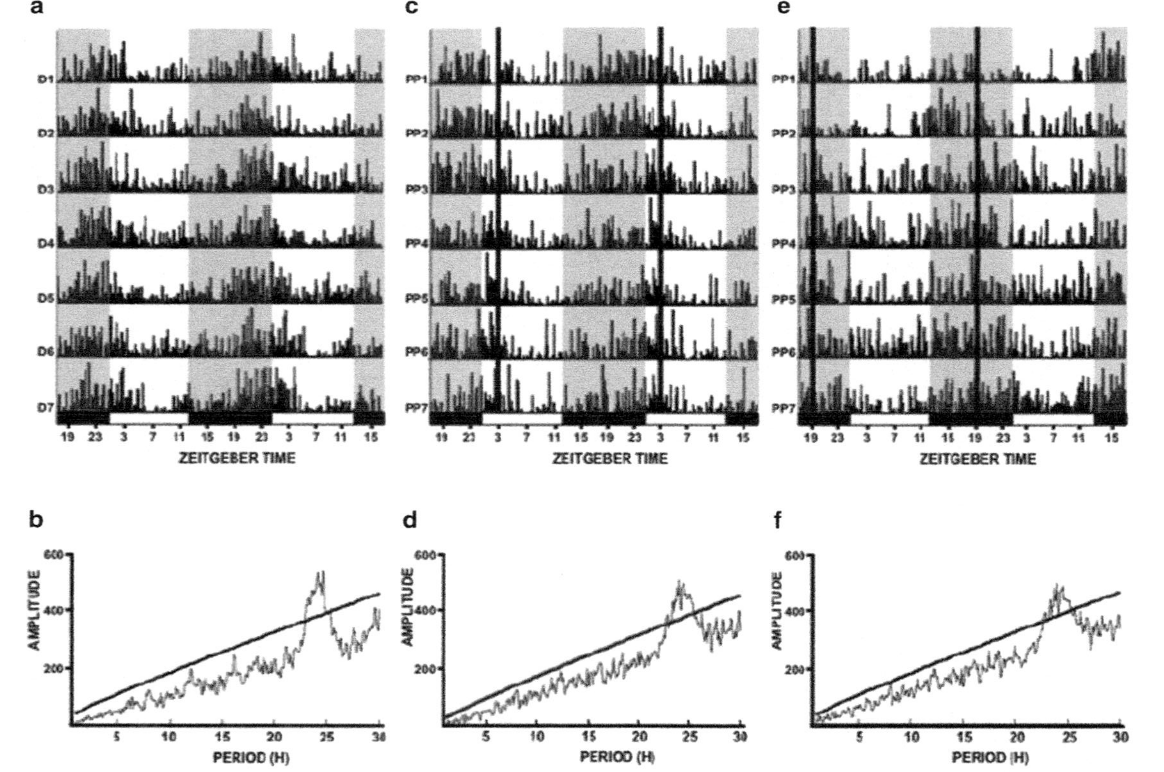

Fig. 10.1 Actograms and periodograms of representative Intact, Nursing-ZT03, and Nursing-ZT19 groups. (**a**, **c**, **e**) Double-plotted actograms of Intact (**a**), Nursing-ZT03 (**c**), and Nursing-ZT19 (**e**) groups. *Black* and *white bars* at *bottom* represent light–dark (L/D) condition. *Shaded areas* represent geographical night. *Black vertical line* represents time of nursing. (**b**, **d**, **f**) Periodograms of locomotor activity of actograms of Intact (**b**), Nursing-ZT03 (**d**), and Nursing-ZT19 (**f**) females indicate a rhythm of 24 h (X^2, $P < 0.001$) (Meza et al. 2008)

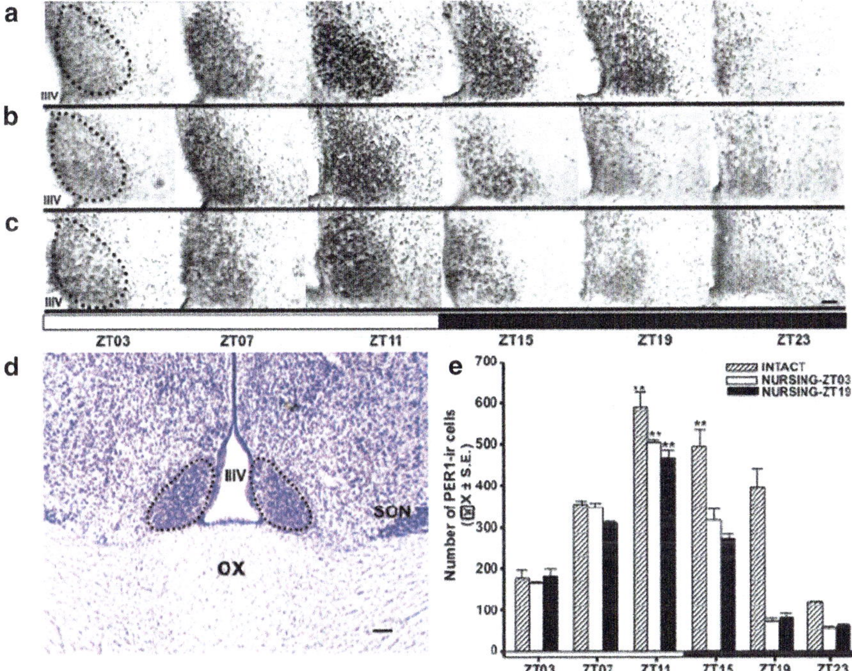

Fig. 10.2 Rhythmic expression of PER1-ir in the suprachiasmatic nucleus (SCN) in intact and nursing does. (**a–c**) Micrographs of representative sections illustrating the expression of PER1 protein at the level of the middle portion of the SCN in Intact (**a**), Nursing-ZT03 (**b**), and Nursing-ZT19 (**c**) groups at six different time points throughout a complete 24-h cycle. (**d**) Micrograph (thionine stain) denotes the level at which analyses were performed. *Dotted line* delimits SCN. *IIIV* third ventricle, *OX* optic chiasm. (**e**) PER1-ir cells in Intact, Nursing-ZT03, and Nursing-ZT19 groups. **$P < 0.001$ for differences between the higher and the lower within the same group. Values are mean ± SEM. *Black* and *white bar* at *bottom* of **c** and **e** represents L/D condition. *Bar* = 100 µm; *ZT* zeitgeber time, ZT0 = light on at 07:00 (Meza et al. 2008)

(Jilge et al. 2001). This result supports the proposal that rabbit mothers have a circadian clock mechanism that "anticipates" nursing time, which seems to affect behavioral, neural, and physiological parameters (Caba and González-Mariscal 2009). Then, it is possible that daily suckling modulates the circadian temperature rhythm, although this had not been explored.

10.3.3 Mammary Gland

Recently, studies in several species have led to the discovery that the mammary gland has a circadian clock. In mice, clock genes and their proteins change their expression in the mammary gland according to the development and reproductive

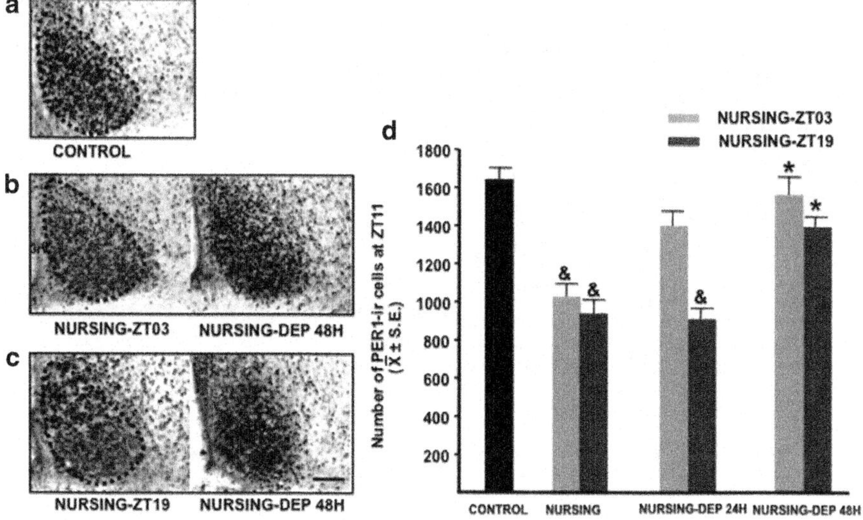

Fig. 10.3 Expression of PER1 in the suprachiasmatic nucleus (SCN). *Left*: Micrographs of representative sections illustrating the expression of PER1 at the level of the middle portion of the time of their maximal expression (ZT11) in control (**a**), Nursing-ZT03 (**b**), and Nursing-ZT19 (**c**) does and in corresponding nursing-deprived subjects for 48 h. *Right*: (**d**) Quantitative analysis of PER1 in control, Nursing-ZT03, Nursing-ZT19, and nursing-deprived subjects for 24 and 48 h. *$P < 0.05$, significant difference between nursing group and respective nursing-deprived group. &$P < 0.05$, significant difference versus the control group. *3V* third ventricle. Values are given as the mean ± SE. *Bar* = 100 μm, *ZT* zeitgeber time, ZT0 = light on at 07:00 (Meza et al. 2011)

cycle of females. For instance, PER1 protein was not detected in the mammary glands of virgin mice but was robustly expressed in ductal epithelial cells during lactation (Metz et al. 2006). Moreover, the changes of clock genes in the mammary epithelial cell lines are associated with expression of specific differentiation markers. In virgin and pregnant females, the mammary glands show a high expression of *Per2* mRNA, which coincides with high levels of c-Myc and Cyclin D1 mRNA. These genes are downstream of the circadian-controlled genes that regulate cell-cycle progression (Nagoshi et al. 2004) and indicate growth and proliferation in undifferentiated cells of the mammary gland (Metz et al. 2006). Furthermore, *Per1* and *Bmal1* mRNA levels were elevated in late pregnant and lactating mammary tissues, and this was correlated with high levels of β-casein mRNA, which indicates differentiation. Also, alteration of the amplitude, but not of the phase, in the circadian expression of clock genes was found in the mammary gland during lactation when compared with that of virgin females (Metz et al. 2006).

Similar changes have been also found in other species. In rats, during lactation there is a coordinated change in the expression of core clock genes and their regulators, with an increase of the proteins of the positive limb in relationship to those of the negative limb during lactation in comparison to pregnant females (Casey et al. 2009). In cows at mid-lactation, there is also a circadian pattern expression of core clock genes mRNA isolated from milk fat globules, which correlates with the percentage of milk fat (Plaut and Casey 2012). In humans, genes specific to the mammary epithelial cells from milk fat globules show a circadian expression during lactation. Approximately 7 % of genes, including circadian core and clock-controlled genes involved in cell development, growth, proliferation, and cell morphology, showed highest expression at different hours through the day (Maningat et al. 2009). As a whole, the evidence indicates that expression of clock genes is important for the observed changes in morphology and metabolic capacity of the mammary gland according to the reproductive state of the females. As in the mammary gland, the transition from pregnancy to lactation implicates parallel changes in other peripheral tissues such as the liver. During this transition, an upregulation of genes related to P450 pathways catalyzes many reactions involved in synthesis of cholesterol, steroids, and other lipids, while at the same time a downregulation of genes related to the breakdown of fatty acids allows fats to be spared for milk synthesis (Casey et al. 2009).

The circadian oscillation of clock genes in the mammary gland suggests that this is important for the regulation of diurnal variations in milk composition in mammals (Nielsen et al. 2003; Cubero et al. 2007; Lubetzky et al. 2007), and this has been confirmed in human neonates. Analyses of proteins, carbohydrates, nucleotides, and amino acids indicate that the composition of breast milk follows a circadian pattern (Cubero et al. 2007). The consequences of these differences had been tested in relationship to consolidation of the sleep–wake rhythm. About 30 % of bottle-fed infants present difficulties in sleeping during the night, but enrichment of formula milk according to its differential day-versus-night components significantly improves the development and consolidation of the normal sleep–wake rhythm (Aparicio et al. 2007). It had been proposed that a key component related to this effect is the amino acid tryptophan, which is related to the synthesis of the hormone melatonin, which is secreted only during the night (Aparicio et al. 2007).

Besides these changes in peripheral tissues, there is also evidence of changes in clock genes and their protein products at brain level. The rabbit doe offers a unique opportunity to explore the physiological significance of clock genes in neuroendocrine cells. As mentioned, lactation in the rabbit is circadian; hence it is possible that neuroendocrine cells associated with the production of hormones related to milk production and yield could be affected by the periodic suckling of pups.

10.4 Entrainment of Neuroendocrine Systems During Lactation

10.4.1 Entrainment of Neuroendocrine Oxytocinergic System

In the rabbit, oxytocinergic cells are located in the supraoptic (SON) and paraventricular hypothalamic (PVN) nuclei and in the lateral hypothalamic area (LHA), as in other mammals (Caba et al. 1996; Hou-Yu et al. 1986). In the rabbit, oxytocinergic cells undergo dramatic changes between estrus and the transition period from pregnancy to lactation. The number of oxytocin (OT) cells increases significantly between estrus and late pregnancy/postpartum day (PD) 1 in the LHA, with a twofold increase in the PVN between estrus and PD1 (Caba et al. 1996). Moreover, there also is a significant increase in the OT somal area at PD1 compared with both estrus and late pregnancy (Caba et al. 1996). Also, in the rat, the oxytocinergic system undergoes morphological changes in the number of cells and their processes from pregnancy to lactation (Jirikowski et al. 1989), sustained by a significant increase in the synthesis of OT mRNA (Caldwell et al. 1987). These morphological changes around and after parturition are related to the behavioral and physiological changes induced by OT during these periods, as the onset of maternal behavior and mother–pup bonding (Lévy et al. 1992; Pedersen et al. 1994; Febo et al. 2005). After parturition, there are profound changes in the mother as lactation implicates a complex reorganization of the mother's physiology to support the energetic demand of pups, and as maternal behavior implicates profound organizational changes in the brain.

In the rabbit doe the brief suckling stimulus of less than 5 min daily is enough to maintain the amount and rate of milk secretion, in contrast to rodent species, which need a more frequent suckling stimulus through the day (Grosvenor and Mena 1974). This suckling stimulus elicits a massive release of both OT and PRL into the blood every 24 h (Fuchs et al. 1984), and pups ingest around 30 % of their body weight in milk in just less than 5 min at PD7 (Caba et al. 2003). In considering that, in contrast to rodents, the rabbit lactation is circadian, we aimed to explore first whether SON and PVN express PER1 and, if so, then to evidence a possible rhythm and its possible synchronization in these two structures by periodic nursing. To this aim we explored the effect of suckling in neuroendocrine cells by immunohistochemistry of PER1 and OT, by scheduling nursing at two different times: one group nursed during the night and another nursed during the day.

Under light–dark (L/D) conditions, both SON and PVN in control does (nonpregnant and nonlactating females) expressed a rhythm of PER1 protein with a peak at the early night at ZT15 in the PVN and at ZT07–11 in the SON. This rhythm is affected by nursing. Whether nursing was scheduled in the day or in the night, this rhythm shifted and PER1 peaked 4–8 h after suckling by pups. This effect was specific to these nuclei as PER1 rhythm in the SCN did not shift in any condition (Meza et al. 2008; Fig. 10.2). Moreover, lactation also synchronizes PER1 rhythm in oxytocinergic cells. In both SON and PVN, PER1 in OT cells had a rhythm that reached a peak at ZT15, but again this peak shifted to 4–8 h after nursing (Meza et al. 2008; Fig. 10.4).

10.4.2 Entrainment of Neuroendocrine Dopaminergic System

Prolactin (PRL) secretion is under an inhibitory control by dopamine, although some releasing factors, as oxytocin, have been also identified (Grattan and Kokay 2008). In the rat, three dopaminergic populations release dopamine (DA) into the

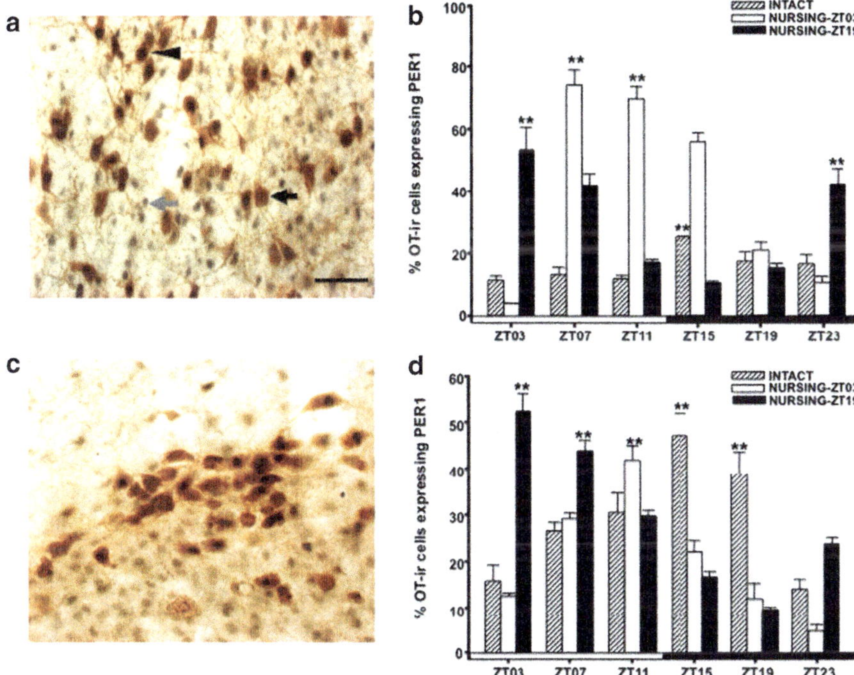

Fig. 10.4 Induction of PER1-/OT in the paraventricular nucleus (PVN) and supraoptic nucleus (SON). Micrographs of representative sections of PVN (**a**) and SON (**c**) in nursing does. *Gray arrow*, PER1; *black arrow*, OT; *arrowhead*, PER1/OT. (**b, d**) Percentages of PER1-/OT in (**b**) PVN and (**d**) SON of intact, Nursing-ZT03 and Nursing-ZT19 groups. ***P*<0.01 for differences between higher and lower values within the same group. Values are mean ± SEM. *Black* and *white bar* at *bottom* of **b** and **d** represents L/D condition of female rabbits. *ZT* zeitgeber time (ZT0=light on at 07:00). *Bar* = 50 μm (Meza et al. 2008)

hypothalamo-hypophyseal portal blood: they are tuberoinfundibular (TIDA) and tuberohypophyseal neurons with cell bodies in the arcuate nucleus (AR), and periventricular hypophysial (PHDA) dopaminergic population with cell bodies in the periventricular region, to the sides of the third ventricle (Ben-Jonathan and Hnasko 2001). PRL secretion has a robust circadian rhythm that persists under constant lighting conditions (Mai et al. 1994), perhaps supported by clock genes oscillations. In fact, dopaminergic cells in the arcuate nucleus express *Per1* and their proteins differentially during the day versus the night (Kriegsfeld et al. 2003; Sellix et al. 2006). The *Per1* rhythm is in phase with that of the light–dark cycle and shifts accordingly to advances or delays of the light schedule (Abe et al. 2002), suggesting that this rhythm is controlled by the SCN. Indeed, neuroanatomical evidence indicates SCN projections to the AR, and destruction of the SCN abolishes circadian rhythms of PRL (Bethea and Neill 1980).

The presence of clock genes in dopaminergic cells, however, also suggests that besides the SCN, these cells could be entrained by a specific timed stimulus other than the L/D cycle. To explore this possibility we evaluated the effect of suckling stimulus on PER1 expression in three dopaminergic populations: TIDA, PHDA, and incertohypothalamic dopaminergic (IHDA) populations. Both TIDA and PHDA, but not IHDA, have been found to be involved in the control of PRL. Then, we explored two populations that project to the adenohypophysis and one population, IHDA, that projects to other regions of the brain (Cheung et al. 1998). Dopaminergic populations were identified by tyrosine hydroxylase (TH), the rate-limiting enzyme of catecholamine synthesis, in considering that DA is the only catecholamine produced in these three cell groups (Björklund and Lindvall 1984). Similar to the foregoing experiment with OT, we explored the effect of suckling on dopaminergic cells by immunohistochemistry of PER1 and TH, scheduling suckling at two different times, with one group nursing during the night and another nursing during the day.

Double-labeled PER1/TH cells were identified in all three nuclei in all conditions. However, their number changes according to the physiological condition of the subjects and timing of nursing. In control nonlactating females, PER1/TH cells reach a peak at ZT15, but this maximal expression shifts to 4–8 h after timing of nursing in TIDA and PHDA neurons; data on TIDA neurons are presented in Fig. 10.5. In contrast, in IHDA the number of PER1/TH cells was scarce and did not change significantly in any of the conditions explored (Meza et al. 2011).

10.4.3 Which Is the Entraining Signal to Neuroendocrine Cells?

We hypothesized that suckling is the signal for entraining of the oxytocin and dopaminergic neuroendocrine cells. Concerning OT, PER1 in OT cells significantly decreased at peak time in both SON and PVN in lactating females that did not receive this stimulus during 48 h (Meza et al. 2008). Regarding TH, we observed a similar effect in PHDA but not in TIDA neurons, which suggests a differential effect of suckling on TH populations (Meza et al. 2011). In contrast, at PD7 the suckling stimulus induces FOS, a protein considered an index of neural activation, in the SON and PVN in comparison to nonnursing females (González-Mariscal et al. 2009). In the rat, suckling by pups also induces FOS in OT cells in both SON and PVN and OT mRNA in PVN cells (Eriksson et al. 1996). Moreover, in the rat, absence of suckling induces an upregulation of TH mRNA in arcuate neurons, observed as early as 1.5 h after the pup's removal. Furthermore, prevention of suckling by covering one side of the dam's teats induces an upregulation of TH mRNA in the contralateral side of the arcuate (Berghorn et al. 2001).

At behavioral level, experiments in the rabbit support the importance of suckling stimulus on the doe. It was found that there is a critical threshold, that is, amount of nipple stimulation, that determines the "turn-off" of the nursing bout and consequently the time in the nest box (González-Mariscal et al. 2012). Importantly, the time that the mother spends inside the nest is normal when provided with 6–8 pups

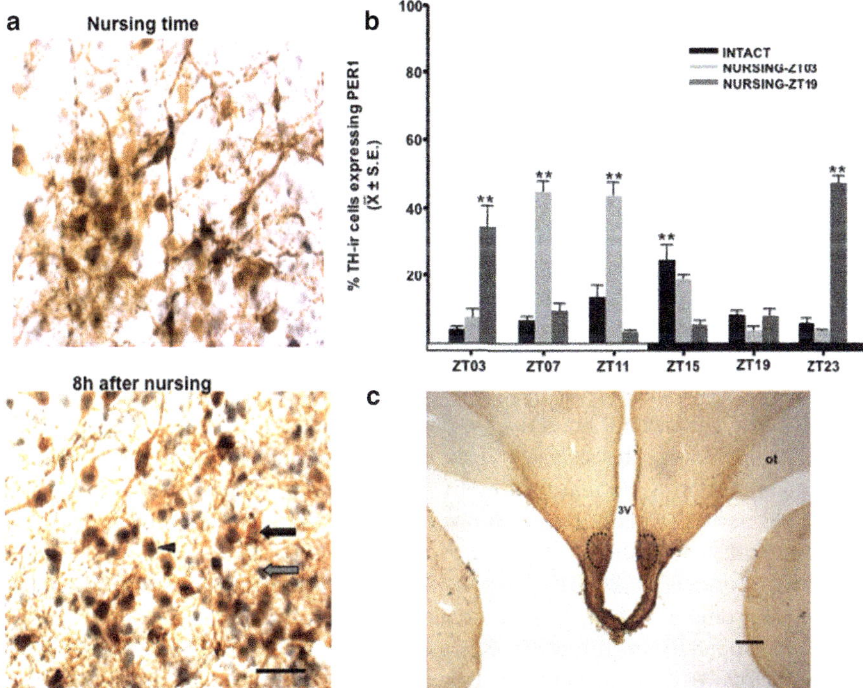

Fig. 10.5 Coexpression of tyrosine hydroxylase (TH) cells with PER1 protein (PER/TH) in the female rabbit brain. (**a**) Micrographs of representative sections of tuberoinfundibular dopaminergic (TIDA) populations at the time of nursing and 8 h later in does nursing at ZT03. *Grey arrow*, PER1 staining; *black arrow*, TH staining; *arrowhead*, PER1/TH double-labeled cells. *Bar*=50 μm. (**b**) Percentages of PER1/TH cells in TIDA in Nursing-ZT03 and Nursing-ZT19 groups compared to intact subjects. ***P*<0.01 for differences between higher and lower values within the same group. Values are mean + SEM. *Black* and *white bar* at the *bottom* represents light/dark (L/D) condition. *ZT* zeitgeber time (ZT0=light on at 07:00). (**c**) Micrograph of representative section of tuberoinfundibular dopaminergic (TIDA) population *Dotted line* delimits the area where analysis was performed. *3V* third ventricle, *ot* optic tract. *Bar*=1 mm (Modified from Meza et al. 2011)

and occurs just once a day with circadian periodicity, but when fewer pups are provided she shows multiple entrances to the nest (González-Mariscal et al. 2013). This experiment confirms that not only suckling, but also its intensity, are critical to maintain the circadian rhythmicity of lactation.

10.4.4 What Is the Significance of the Shift of the PER1 Rhythm in Neuroendocrine Cells in the Lactating Rabbit?

Evidence from several species indicates that the suckling stimulus is transmitted through the spinal cord and brainstem to the hypothalamus, to both PVN and SON (Wakerly et al. 1994). Also, FOS is induced in the PVN as a consequence of suckling (Tsingotjidou and Papadopoulos 1996). It has also been proposed that afferent stimulus from the nipples affects TIDA neurons, regulating TH mRNA (Berghorn et al. 2001). We proposed previously that the suckling stimulus shifts clock gene central core oscillations in neuroendocrine cells in the rabbit during lactation (Meza et al. 2008, 2011). Several lines of evidence support this proposal. It is well known that oscillations of these genes are reset by a photic stimulus (Yan and Silver 2004). However, experiments in vitro and in vivo indicate that some stimuli, other than light, can also affect oscillations of these genes. A serum shock, administration of various compounds such as cAMP, protein kinase C, glucocorticoid hormones, or Ca^{2+} to a cultured rat-1 fibroblast cell line, can all trigger a surge of *Per1* transcription and elicit and shift circadian expression of clock genes and other genes (Balsalobre et al. 2000). In adult rats, exposure to stressors (Takahashi et al. 2001) or scheduled feeding (Caba et al. 2008; Feillet et al. 2008) also induces a similar effect in several brain nuclei. Overall we consider that suckling seems to be the main stimulus that entrains clock gene oscillations in neuroendocrine cells in the rabbit. During normal conditions, the SCN sends signals to the brain and peripheral tissues to maintain synchronization, and the clock gene oscillations are in phase with that of the master clock (Abe et al. 2002; Granados-Fuentes et al. 2006; Kriegsfeld et al. 2003). However, it had been proposed that, during special conditions, some local cells in specific nuclei uncouple from the SCN oscillations and now respond to specific needs (Kriegsfeld and Silver 2006). These local clocks then shift their oscillations and drive the expression of particular clock-controlled genes that act as outputs to regulate particular physiological conditions. By using the rabbit doe we confirmed that dopaminergic cells contain the molecular machinery necessary to oscillate independently and also that clock genes oscillations seem to uncouple from the SCN and now are entrained by a different stimulus, specifically suckling.

Suckling stimulus seems to be a timing signal for neuroendocrine dopaminergic cell populations and oxytocinergic cells in the SON and PVN, because these cells also shift their clock gene rhythm according to nursing schedule. Future experiments should explore which intracellular signals are driven by this shift of clock gene oscillations and how they are related to the cyclic synthesis, storage, and release of both OT and DA, to support circadian lactation in the rabbit.

10.5 Changes in Other Neuronal Systems

Lactation imposes a significant impact on metabolic homeostasis, mainly as the result of the increased energy demands of milk production that induce a general negative energy balance. To fulfill this, the mother experiences a three- to fourfold increase in food consumption (Malabu et al. 1994), and as a consequence there are significant differences in several metabolites, neurotransmitters, and hormones in lactating, compared with pregnant and virgin, females (Malabu et al. 1994; Smith et al. 2010). However, some compounds have particular relevance in considering their circadian implications, for example, the orexinergic system. Orexin cells (ORX) are located in the lateral hypothalamus and in nonlactating rodents these cells show a higher activation, as indicated by protein FOS, during the active phase of subjects, which is consistent with the role of ORX in arousal (Martínez et al. 2002). In contrast, during lactation levels of FOS in these cells are elevated during the day in nocturnal rodents (España et al. 2004). This observation is interesting as these subjects exhibit high levels of attention to their litter, as well as nursing bouts, during the day; concomitantly lactating females exhibit a disruption of normal day-time sleeping (Grota and Ader 1974). It will be important to explore possible alterations during lactation of clock genes in ORX cells in other brain regions than the lateral hypothalamus, for example, in the medial preoptic area, which contains abundant ORX terminals and receptors (España et al. 2001; Zhang et al. 2005). Analysis of this area in the context of circadian rhythms during lactation is important for two additional reasons. It is related to the control of temperature, which changes during lactation, as already mentioned, and plays a key role in maternal behavior (González-Mariscal and Kinsley 2008; Numan and Insel 2003).

Finally, it is well established that a negative energy balance is associated with a suppression of reproductive function and ovarian cyclicity (Hill et al. 2008), a rhythm controlled by the SCN (Everett and Sawyer 1950; Goldman 1999). It had been proposed that the metabolic changes of hyperphagia during lactation, besides acting directly on GnRH cells, also affect the kisspeptin system in the arcuate nucleus, which contributes to the disruption of pulsatile GnRH release during lactation (Smith et al. 2010). The mechanisms by which SCN is involved in the suppression of GnRH release during lactation deserve further attention.

10.6 Conclusions and Perspectives

Circadian rhythms during lactation have been scarcely explored. Given the importance of circadian rhythms for coordinating homeostatic drive, the changes from pregnancy through lactation imply a profound reorganization of behavioral and physiological parameters that affects both peripheral and central structures. Clock genes are a very useful tool to explore such changes, as evidenced in the mammary gland and in neuroendocrine cells in the brain. The rabbit doe offers an excellent opportunity to explore in detail the interrelationship between lactation and circadian rhythms, considering their unique characteristic of circadian lactation among mammals. The translational impact of this field is clearly evidenced by the circadian changes in breast milk composition and the control of day–light regime for the proper development of human infants.

Acknowledgments We gratefully thank Dr. Pascal Poindron for corrections and helpful comments on this contribution.

References

Abe M, Herzog ED, Yamazaki S et al (2002) Circadian rhythms in isolated brain regions. J Neurosci 22:350–356

Aparicio S, Garau C, Estevan S et al (2007) Chrononutrition: use of dissociated day/night infant milk formulas to improve the development of the wake-sleep rhythms. Effects of tryptophan. Nutr Neurosci 10:137–143

Balsalobre A, Marcacci L, Schibler U (2000) Multiple signaling pathways elicit circadian gene expression in cultured Rat-1 fibroblasts. Curr Biol 10:1291–1294

Ben-Jonathan N, Hnasko R (2001) Dopamine as a prolactin (PRL) inhibitor. Endocr Rev 22: 724–763

Berghorn KA, Le W-W, Sherman TH, Hoffman GE (2001) Suckling stimulus suppresses messenger RNA for tyrosine hydroxylase in arcuate neurons during lactation. J Comp Neurol 438: 423–432

Bethea CL, Neill JD (1980) Lesions of the suprachiasmatic nuclei abolish the cervically stimulated prolactin surges in the rat. Endocrinology 107:1–5

Björklund A, Lindvall O (1984) Dopamine-containing systems in the CNS. In: Björklund A, Hökfelt T (eds) Handbook of chemical neuroanatomy, vol. 2, Classical transmitters in the CNS, Part 1. Elsevier, Amsterdam, pp 55–122

Caba M, González-Mariscal G (2009) The rabbit pup, a natural model of nursing-anticipatory activity. Eur J Neurosci 30:1697–1706

Caba M, Silver R, González-Mariscal G et al (1996) Oxytocin and vasopressin immunoreactivity in rabbit hypothalamus during estrus, late pregnancy, and postpartum. Brain Res 720:7–16

Caba M, Rovirosa MJ, Silver R (2003) Suckling and genital stroking induces Fos expression in hypothalamic oxytocinergic neurons of rabbit pups. Dev Brain Res 143:119–128

Caba M, Tovar A, Silver R et al (2008) Nature's food anticipatory experiment: entrainment of locomotor behavior, suprachiasmatic and dorsomedial hypothalamic nuclei by suckling in rabbit pups. Eur J Neurosci 27:432–443

Caldwell JD, Greer RE, Johnson MF et al (1987) Oxytocin and vasopressin immunoreactivity in hypothalamic and extrahypothalamic sites in late pregnant and postpartum rats. Neuroendocrinology 46:39–47

Casey T, Patel O, Dykema K et al (2009) Molecular signatures reveal circadian clocks may orchestrate the homeorhetic response to lactation. PLoS One 4:e7395

Challet E, Pevet P (2003) Interactions between photic and nonphotic stimuli to synchronize the master circadian clock in mammals. Front Biosci 8:246–257

Cheung S, Ballew JR, Moore KE et al (1998) Contribution of dopamine neurons in the medial zona incerta to the innervations of the central nucleus of the amygdala, horizontal diagonal band of Broca and hypothalamic paraventricular nucleus. Brain Res 808:174–181

Cubero J, Narciso D, Terrón P et al (2007) Chrononutrition applied to formula milks to consolidate infants' sleep/wake cycle. Neuro Endocrinol Lett 28:360–366

Dibner C, Schibler U, Albrecht U (2010) The mammalian circadian timing system: organization and coordination of central and peripheral clocks. Annu Rev Physiol 72:517–549

Drucker-Colín R, Aguilar-Roblero R, García-Hernández F et al (1984) Fetal suprachiasmatic nucleus transplants: diurnal rhythm recovery of lesioned rats. Brain Res 311:353–357

Eriksson M, Ceccatelli S, Uvnäs-Moberg K et al (1996) Expression of Fos-related antigens, oxytocin, dynorphin and galanin in the paraventricular and supraoptic nuclei of lactating rats. Neuroendocrinology 63:356–367

España RA, Baldo BA, Kelley AE et al (2001) Wake-promoting and sleep-suppressing actions of hypocretins (orexin): basal forebrain sites of action. Neuroscience 106:699–715

España RA, Berridge CW, Gammie SC (2004) Diurnal levels of Fos immunoreactivity are elevated within hypocretins neurons in lactating mice. Peptides 25:1927–1934

Everett JW, Sawyer CH (1950) A 24-hour periodicity in the "LH-release apparatus" of female rats, disclosed by barbiturate sedation. Endocrinology 47:198–218

Febo M, Numan M, Ferris CF (2005) Functional magnetic resonance imaging shows oxytocin activates brain regions associated with mother-pup bonding during suckling. J Neurosci 25: 11637–11644

Feillet CA, Mendoza J, Albrecht U et al (2008) Forebrain oscillators ticking with different clock hands. Mol Cell Neurosci 37:209–221

Fuchs A-R, Cubile L, Dawood MY et al (1984) Release of oxytocin and prolactin by suckling in rabbits throughout lactation. Endocrinology 114:462–469

Goldman BD (1999) The circadian timing system and reproduction in mammals. Steroids 64: 679–685

González-Mariscal G, Kinsley CH (2008) From indifference to ardor: the onset, maintenance, and meaning of the maternal brain. In: Pfaff D, Arnold A, Edgen A, Fahrbach S, Rubin R (eds) Hormones, brain and behavior. Academic, San Diego, pp 333–367

González-Mariscal G, Jiménez A, Chirino R et al (2009) Motherhood and nursing stimulate c-FOS expression in the rabbit forebrain. Behav Neurosci 123:731–739

González-Mariscal G, Toribio A, Gallegos JA et al (2012) The characteristics of suckling stimulation determine the daily duration of mother-young contact and milk output in rabbits. Dev Psychobiol. doi:10.1002/dev.21071

González-Mariscal G, Lemus A, Vega-González A et al (2013) Litter size determines periodicity of nursing in rabbits. Chronobiol Int 30:711–718

Granados-Fuentes D, Tseng A, Herzog ED (2006) A circadian clock in the olfactory bulb controls olfactory responsivity. J Neurosci 26:12219–12225

Grattan DR, Kokay IC (2008) Prolactin: a pleiotropic neuroendocrine hormone. J Neuroendocrinol 20:752–763

Green DJ, Gillette R (1982) Circadian rhythm of firing rate recorded from single cells in the rat suprachiasmatic brain slice. Brain Res 245:198–200

Grosvenor CE, Mena F (1974) Neural and hormonal control of milk secretion and milk ejection. In: Larson BL, Smith VR (eds) Lactation: a comprehensive treatise. Academic, New York, pp 227–276

Grota LJ, Ader R (1974) Behavior of lactating rats in a dual-chambered maternity cage. Horm Behav 5:275–282

Hill JW, Elmquist JK, Elias CF (2008) Hypothalamic pathways linking energy balance and reproduction. Am J Physiol Endocrinol Metab 294:E827–E832

Hou-Yu A, Lamme AT, Zimmerman EA et al (1986) Comparative distribution of vasopressin and oxytocin neurons in the rat brain using a double label procedure. Neuroendocrinology 44:235–246

Jilge B (1980) The effect of two different light intensities on spontaneous period and phase angle difference of caecotrophy-rhythm in the rabbit. J Interdiscipl Cycle Res 11:41–54

Jilge B (1991) The rabbit: a diurnal or a nocturnal animal? J Exp Anim Sci 34:170–183

Jilge B, Kuhnt B, Landerer W et al (2001) Circadian temperature rhythms in rabbit pups and in their does. Lab Anim 35:364–373

Jirikowski GF, Caldwell JD, Pilgrim C et al (1989) Changes in immunostaining for oxytocin in the forebrain of the female rat during late pregnancy, parturition and early lactation. Cell Tissue Res 256:411–417

Juárez C, Morgado E, Meza E et al (2013) Development of retinal projections and response to photic input in the suprachiasmatic nucleus of New Zealand white rabbits. Brain Res 1499: 21–28

Kriegsfeld LJ, Silver R (2006) The regulation of neuroendocrine function: timing is everything. Horm Behav 49:557–574

Kriegsfeld LJ, Korets R, Silver R (2003) Expression of the circadian clock gene Period1 in neuroendocrine cells: an investigation using mice with a Per1::GFP transgene. Eur J Neurosci 17:212–220

Lehman MN, Silver R, Gladstone WR et al (1987) Circadian rhythmicity restored by neural transplant. Immunocytochemical characterization of the graft and its integration with the host brain. J Neurosci 7(6):1626–1638

Leon M, Adels L, Coopersmith R et al (1984) Diurnal cycle of mother-young contact in Norway rats. Physiol Behav 32:999–1003

Lévy F, Kendrick KM, Keverne EB et al (1992) Intracerebral oxytocin is important for the onset of maternal behavior in inexperienced ewes delivered under peridural anesthesia. Behav Neurosci 106:427–432

Lincoln DW, Hill A, Wakerley JD (1973) The milk-ejection reflex of the rat: an intermittent function not abolished by surgical levels of anesthesia. J Endocrinol 57:459–476

Lubetzky R, Mimouni FB, Dollberg S et al (2007) Consistent circadian variations in creamatocrit over the first 7 weeks of lactation: a longitudinal study. Breastfeed Med 2:15–18

Mai LM, Shieh KR, Pan JT (1994) Circadian changes of serum prolactin levels and tuberoinfundibular dopaminergic neuron activities in ovariectomized rats treated with or without estrogen: the role of the suprachiasmatic nuclei. Neuroendocrinology 60:520–526

Malabu UH, Kilpatrick A, Ware M et al (1994) Increased neuropeptide Y concentrations in specific hypothalamic regions of lactating rats: possible relationship to hyperphagia and adaptive changes in energy balance. Peptides 15:83–87

Maningat PD, Sen P, Rijnkels M et al (2009) Gene expression in the human mammary epithelium during lactation: the milk fat globule transcriptome. Physiol Genomics 37:12–22

Martínez GS, Smale L, Nunez AA (2002) Differential expression of orexin receptors 1 and 2 in the rat brain. Brain Res 955:1–7

Maywood ES, Mrosovsky N, Field MD et al (1999) Rapid down-regulation of mammalian period genes during behavioral resetting of the circadian clock. Proc Natl Acad Sci U S A 96: 15211–15216

Metz RP, Qu X, Laffin B et al (2006) Circadian clock and cell cycle gene expression in mouse mammary epithelial cells and in the developing mouse mammary gland. Dev Dyn 235: 263–271

Meza E, Juárez C, Morgado E et al (2008) Brief daily suckling shifts locomotor behavior and induces PER1 protein in paraventricular and supraoptic nuclei, but not in the suprachiasmatic nucleus, of rabbit does. Eur J Neurosci 28:1394–1403

Meza E, Waliszewski SM, Caba M (2011) Circadian nursing induces PER1 protein in neuroendocrine tyrosine hydroxylase neurons in the rabbit doe. J Neuroendocrinol 23:472–480

Moore RY, Eichler VB (1972) Loss of a circadian adrenal corticosterone rhythm following suprachiasmatic lesions in the rat. Brain Res 42:201–216

Morgado E, Juárez C, Melo AI et al (2011) Artificial feeding synchronizes behavioral, hormonal, metabolic and neural parameters in mother-deprived neonatal rabbit pups. Eur J Neurosci 34: 1807–1816

Mykytowycz R, Rowley I (1958) Continuous observations of the activity of the wild rabbit, Oryctolagus cuniculus (L), during 24-hr periods. CSIRO Wild Res 3:26–31

Nagoshi E, Saini C, Bauer C et al (2004) Circadian gene expression in individual fibroblasts: cell autonomous and self-sustained oscillators pass time to daughter cells. Cell 119:693–705

Nielsen NI, Ingvartsen KL, Larsen T (2003) Diurnal variation and the effect of feed restriction on plasma and milk metabolites in TMR-fed dairy cows. J Vet Med A Physiol Pathol Clin Med 50:88–97

Nolasco N, Juárez C, Morgado E et al (2012) A circadian clock in the olfactory bulb anticipates feeding during food anticipatory activity. PLoS One 7(19):e47779

Nuboer JFW, Nuys WM, Van Steenbergen JC (1983) Colour changes in a light regimen as synchronizers of circadian activity. J Comp Physiol A 151:359–366

Numan M, Insel TR (2003) The neurobiology of parental behavior. Springer, New York

Pedersen CA, Caldwell JD, Walker C et al (1994) Oxytocin activates the postpartum onset of rat maternal behavior in the ventral tegmental and medial preoptic areas. Behav Neurosci 108: 1163–1171

Plaut K, Casey T (2012) Does the circadian system regulate lactation? Animal 6:394–402

Reppert SM, Weaver DR (2001) Molecular analysis of mammalian circadian rhythms. Annu Rev Physiol 63:647–676

Ripperger JA, Schibler U (2006) Rhythmic CLOCK-BMAL1 binding to multiple E-box motifs drives circadian Dbp transcription and chromatin transitions. Nat Genet 38:369–374

Rosenwasser AM, Hollander SJ, Adler NT (1987) Effects of pregnancy and parturition on free-running circadian activity rhythms in the rat. Chronobiol Int 4:183–187

Sack RL, Auckley D, Auger RR et al (2007) Circadian rhythm sleep disorders: Part I, Basic principles, shift work and jet lag disorders. An American Academy of Sleep Medicine Review. Sleep 30:1460–1483

Scribner SJ, Wynne-Edwards KE (1994) Disruption of body temperature and behavior rhythms during reproduction in dwarf hamsters (Phodopus). Physiol Behav 55:361–369

Sellix MT, Egli M, Poletini MO et al (2006) Anatomical and functional characterization of clock gene expression in neuroendocrine dopaminergic neurons. Am J Physiol Regul Integr Comp Physiol 290:R1309–R1323

Smith SM, True C, Grove KL (2010) The neuroendocrine basis of lactation-induced suppression of GnRH: role of kisspeptin and leptin. Brain Res 1364:139–152

Stephan FK, Zucker I (1972) Circadian rhythms in drinking behavior and locomotor activity of rats are eliminated by hypothalamic lesions. Proc Natl Acad Sci U S A 69:1583–1586

Takahashi JS, DeCoursey PJ, Bauman L et al (1984) Spectral sensitivity of a novel photoreceptive system mediating entrainment of mammalian circadian rhythms. Nature 308:186–188

Takahashi S, Yokota SI, Hara R et al (2001) Physical and inflammatory stressors elevate circadian clock gene mPer1 levels in the paraventricular nucleus of the mouse. Endocrinology 142: 4910–4917

Tosini G, Menaker M (1996) Circadian rhythms in cultured mammalian retina. Science 272: 419–421

Tsingotjidou A, Papadopoulos GC (1996) Neuronal expression of Fos-like protein along the afferent pathway of the milk-ejection reflex in the sheep. Brain Res 741:309–313

Ueda HR, Hayashi S, Chen W et al (2005) System-level identification of transcriptional circuits underlying mammalian circadian clocks. Nat Genet 37:187–192

Van Hof-Van Duin J (1971) Locomotor activity in normal and dark-reared rabbits. Doc Ophthalmol 30:327–330

Vásquez S, Ángeles-Castellanos M, Escobar C (2011) Preterm infants have improved growth in light/dark cycle compared with continuous bright light. In: Third world congress of chronobiology, P043, Puebla, México, pp 162

Wakamatsu H, Yoshinobu Y, Aida R et al (2001) Restricted-feeding-induced anticipatory activity rhythm is associated with a phase-shift of the expression of mPer1 and mPer2 mRNA in the cerebral cortex and hippocampus but not in the suprachiasmatic nucleus of mice. Eur J Neurosci 13:1190–1196

Wakerly JB, Clarke G, Summerlee AJS (1994) Milk ejection and its control. In: Knovil E, Neill JD (eds) The physiology of reproduction, 2nd edn. Raven, New York, pp 1131–1177

Woodside B, Leon M (1980) Thermoendocrine influences on maternal nesting behavior in rats. J Comp Physiol Psychol 94:41–60

Yan L, Silver R (2004) Resetting the brain clock: time course and localization of mPER1 and mPER2 protein expression in suprachiasmatic nuclei during phase shifts. Eur J Neurosci 19:1105–1109

Zhang S, Blache D, Vercoe PE et al (2005) Expression of orexin receptors in the brain and peripheral issues of the male sheep. Regul Pept 124:81–87

Chapter 11
A Time to Wake, a Time to Sleep

Horacio O. de la Iglesia and Michael L. Lee

Abstract Sleep is a defining trait in animals. Although the function(s) of sleep remain elusive, it is clear that sleep is necessary for survival and its timing precisely regulated. Two processes operate to determine when sleep will occur and how intense it will be. The homeostatic process determines the buildup of sleep pressure that results from extended wakefulness, and the circadian process regulates sleep pressure according to a circadian clock located in the suprachiasmatic nucleus of the hypothalamus. This chapter discusses the experimental evidence for these regulatory processes. In mammals, the interplay between the homeostatic and circadian regulatory limbs not only determines the timing and intensity of sleep but also leads to a stereotypic temporal organization of sleep stages. Whereas increased homeostatic drive favors slow-wave sleep, increased circadian drive favors rapid eye movement (REM) sleep. Environmental challenges that disrupt the timing of sleep affect its amount as well as its quality.

11.1 Sleep–Wake Cycle

The sleep–wake cycle represents the most remarkable alternation of behavioral and physiological states. We transition from freely behaving active individuals during waking, to a state of unconsciousness characterized by low cortical activity, to a paradoxical state of mind in which our cortex is highly active displaying dreams yet our muscles paralyzed preventing us from acting them out. Strikingly, we repeat this cycle every 24 h until our very last day. Why animals have evolved to expose themselves daily to such a vulnerable and seemingly underachieving behavioral state is

H.O. de la Iglesia (✉)
Department of Biology, University of Washington, Seattle, WA, USA

Program in Neurobiology and Behavior, University of Washington, Seattle, WA, USA
e-mail: horaciod@uw.edu

M.L. Lee
Division of Sleep and Circadian Disorders, Brigham and Women's Hospital, Boston, MA, USA

Division of Sleep Medicine, Harvard Medical School, Boston, MA, USA

© Springer International Publishing Switzerland 2015
R. Aguilar-Roblero et al. (eds.), *Mechanisms of Circadian Systems in Animals and Their Clinical Relevance*, DOI 10.1007/978-3-319-08945-4_11

one of the biggest mysteries in biology; yet, all animals need sleep for survival. Sleep loss is associated with acute and chronic adverse health outcomes and even death. The need to sleep has led to accurate mechanisms to time sleep and wake to when they are most adaptive. At sunset, while most anthropoid monkeys are starting their long nighttime sleep bout, nocturnal rodents and bats are waking up to start foraging. While all animals sleep, some species only sleep a couple of hours a day, whereas others sleep up to 16 h a day. Sleep and sleep loss have intrigued humans from early times, as it is clear by the appearance of sleep and insomnia as central themes in the mythology and literature of virtually all civilizations (Borges 1999; Summers-Brenmer 2008). How sleep and its associated behavioral states are timed within the 24-h cycle is still a field of active research.

Far from the misleading appearance of sleep as an inanimate state, the brain can express relatively high electrical activity during sleep. Sleep transitions between different stages represented by distinct electrical brain activity in a stereotypic pattern, creating a conserved sleep "architecture." In this chapter, we will cover the basic mechanisms by which sleep and specific sleep stages are temporally organized throughout the 24-h day, focusing specifically on the sleep–wake cycle as an output of the circadian system.

11.2 Measuring Sleep

Sleep represents a reversible state—as opposed to certain types of coma or hibernation—that is characterized by reduced or absent consciousness, and decreased muscle activity, metabolism, and sensitivity to the environment. The reduced muscle and metabolic activity may be part of an energy conservation strategy to allow the organism to conserve resources at strategic times of day.

While reduced behavioral activity has historically been a key indicator of sleep, the development of the electroencephalogram (EEG) to measure brain activity represented the dawn of sleep research. EEG-defined sleep stages were described in the mid-1930s (Loomis et al. 1935; Davis et al. 1937), and Aserinsky and Kleitman (1953) identified rapid eye movement (REM) sleep. Since then, our understanding of sleep has expanded, but these early descriptions of sleep stages have remained largely unchanged. In mammals, the EEG is the gold standard for describing and recording sleep. EEG, through electrodes placed on the scalp, monitors electrical brain activity generated by cortical neurons in the superficial layer and enables accurate detection of sleep and wake.

During wakefulness mammals exhibit high-frequency low-voltage electrical activity. At the transition from wake to sleep, the voltage of the recording increases, representing synchronous activity, while the frequency decreases.

Sleep is categorized into two broad types: non-rapid eye movement (NREM) sleep and REM sleep. NREM sleep is divided further into three stages: stage 1, stage 2, and stage 3 (Iber et al. 2007) (Fig. 11.1). The NREM sleep stages are typically characterized by the presence of specific frequencies in the EEG wave, which parallel the depth of sleep with progressively increased arousal thresholds. Stage 1

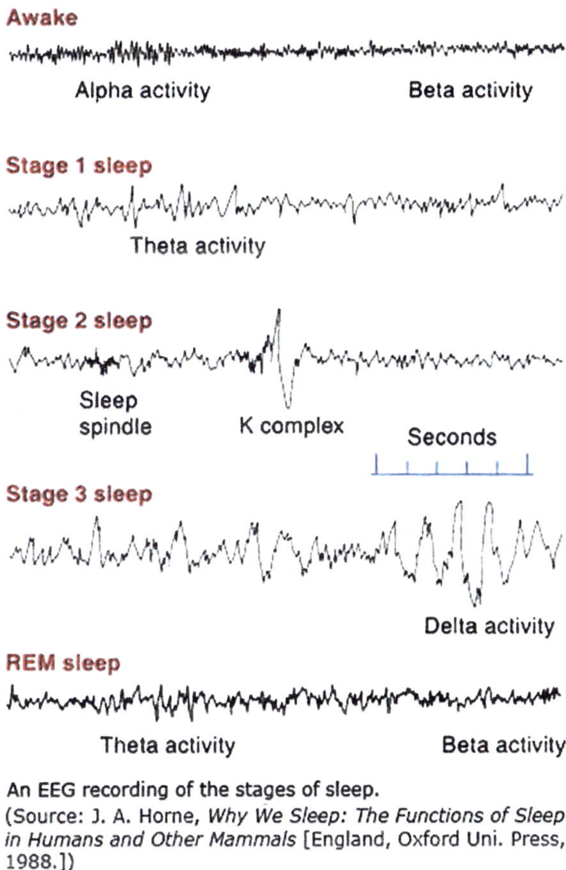

An EEG recording of the stages of sleep.
(Source: J. A. Horne, *Why We Sleep: The Functions of Sleep in Humans and Other Mammals* [England, Oxford Uni. Press, 1988.])

Fig. 11.1 Recording of sleep stages through EEG. Each trace is a representative recording for the indicated stage. Note the increase in voltage amplitude as NREM progresses from stage 1 to stage 3, in which SWS with typical high-amplitude/low-frequency delta waves appear. Note also the similarity between the wake and REM sleep EEG. *Source*: Horne (1988)

NREM sleep is characterized by reduced electromyogram (EMG)-measured muscular activity, lower frequency, and higher amplitude synchronous EEG signals, with increased theta (4.5–8 Hz) activity and lower alpha (8–12 Hz) activity. During this stage, slow rolling eye movements replace the rapid saccadic eye movements of wakefulness. Stage 2 NREM sleep contains brief high-voltage K-complexes and bursts of sleep spindles (12–16 Hz). Muscle and eye activity remain low. Stage 3 NREM sleep is known as deep or slow-wave sleep (SWS) due to the presence of slow waves or delta wave activity, which are 0.5–4 Hz, large amplitude electrical waves. REM sleep, characterized in many species by rapid oculomotor activity, is also known as paradoxical sleep because the EEG resembles a wake state with beta (12–30 Hz) and theta (4–7 Hz) activity. During REM sleep, voluntary muscle activity is inhibited creating a state of paralysis, also known as atonia, that affects all muscles except oculomotor and respiratory muscles. Dreams typically take place during

REM sleep. The rapid eye movements and atonia of REM sleep are often measured via electrooculography and EMG recordings, respectively.

While EEG has remained the definitive method of measuring sleep for nearly half a century, other methods of sleep measurement capitalize on physiological correlates of sleep. The reduction or near absence of locomotor activity provides an easily measurable marker of sleep. Locomotor activity can be measured using a range of techniques. Accelerometers and global positioning satellite monitoring can be attached to animals by collars around the neck or legs, and they monitor locomotor activity and range of movement over time in an open environment. Locomotor activity in animals in their natural environment can be monitored for long durations across far distances. For example, in nocturnal owl monkeys, long-term locomotor activity was recorded for up to 18 months (Fernandez-Duque et al. 2010). In a confined environment, locomotor activity can be measured more precisely using a non-invasive infrared grid that counts beam breaks. Recently, video tracking of animals has become attractive, as hardware and software advances have enabled the accurate automated scoring of not only wake and sleep but the characterization of NREM and REM sleep stages in rodents (Fisher et al. 2012; McShane et al. 2012). Video monitoring allows the tracking of multiple animals in the same living environment and for the screening of many animals to search for interesting sleep phenotypes. In *Drosophila*, dozens of animals can be simultaneously tracked in test tubelike actimeters. The high-throughput measurement of sleep combined with the powerful fruit fly molecular genetics greatly accelerates the identification of genes involved in the regulation of sleep (Cirelli 2009; Sehgal and Mignot 2011).

In humans, actigraphy is frequently used to monitor activity levels, and advanced algorithms have been developed to identify periods of sleep (Sadeh et al. 1994; de Souza et al. 2003). Wristwatch-like activity monitors can record up to weeks of activity. As opposed to EEG, determining sleep from actigraphy is less accurate, as movement does not always correlate with the absence of sleep and lack of movement with its presence. However, actigraphy has the benefit of being a simple, inexpensive method for measuring sleep for long durations with no maintenance.

11.3 Timing of Sleep

Under natural conditions each species has a typical 24-h temporal distribution of wake and sleep, which determines a "24-h temporal niche" for that species. Humans are diurnal and display nocturnal sleep, whereas nocturnal species typically sleep during the day. What determines the temporal organization of sleep and wake? In other words, why does a songbird sleep during the night and an owl during the day?

We have all experienced how the pressure to fall asleep increases as we stay awake. Sleep pressure increases as a function of wake time; in contrast, sleep pressure decreases with sleep time, and the longer sleep lasts, the more difficult it is to sustain it. This regulatory process is known as the homeostatic or sleep–wake-dependent regulatory process of sleep (Borbely 1982; Borbely et al. 2001).

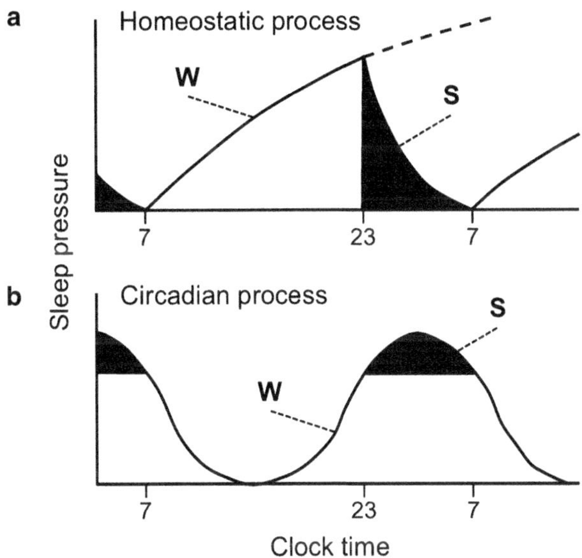

Fig. 11.2 Sleep pressure in normal healthy humans is regulated by homeostatic (**a**) and circadian (**b**) sleep regulatory processes. *W*, wake; *S*, sleep. *Black regions* indicate sleep time, which occurs between 23:00 and 7:00. The *dotted line* on (**a**) depicts how sleep pressure increases through the homeostatic mechanism if there is sleep deprivation

Figure 11.2b schematizes the dependence of sleep pressure on wake and sleep times. On a typical day a human adult wakes up in the morning (7:00) with little or no sleep pressure. As the individual progresses through the day, sleep pressure increases steadily with the duration of wake, reaching a daily wake-dependent threshold at the usual bedtime (23:00). When he/she falls asleep and as sleep takes place during the night (black areas in the plot), sleep pressure decreases to once again reach a sleep-dependent minimum. Figure 11.2a also shows that interfering with the subject's habitual bedtime—working late into the night or a Saturday night party—will allow the "sleep homeostat" to increase sleep pressure beyond normal daily values (dotted line). Thus, the homeostatic process of sleep keeps track of both sleep and wake hours and will prevent excessive sleep deprivation when the sleep pressure is so high that sleep must occur; conversely, it will allow wake after a given duration of sufficient sleep.

The homeostatic regulatory process determines the need for sleep and its intensity when sleep begins (see below), but it does not determine the phase of sleep within the 24-h day, that is, at what time of the day sleep and wake should take place. How do diurnal and nocturnal animals determine their respective nighttime and daytime sleep bouts? The second regulatory process of sleep, known as the circadian process, regulates the time of the day sleep will likely occur, regardless of the sleep or wake history. In other words, sleep pressure oscillates with a 24-h period and this oscillation, with a peak of sleep propensity at a specific time of the day, represents an output of the circadian system. Figure 11.2b illustrates the circadian process of sleep as an oscillation of sleep pressure. In humans whose circadian system is entrained to the solar light–dark (LD) cycle, the circadian peak of sleep propensity occurs around 3 AM, which explains the high sleep propensity we feel at this time when we try to stay up all night. Interestingly, the peak of circadian sleep pressure does not occur at the same time as the peak of homeostatic sleep pressure; instead, as the homeostatic drive to sleep wanes, the circadian drive to sleep reaches its peak. Because overall propensity to stay asleep depends on the cumulative contributions of the homeostatic and circadian processes, this paradoxical mismatch between the homeostatic and circadian peaks of sleep pressure allows us to have an extended and consolidated bout of nocturnal sleep (Dijk and Czeisler 1994).

11.4   Uncovering the Circadian Nature of Sleep

What is the evidence that this peak of sleep pressure in the second half of the night represents a circadian rhythm? The golden standard for the demonstration of the circadian nature of a rhythm is its persistence under constant laboratory conditions, which reveals the existence of the circadian clock that generates the rhythm. Studies in the 1960s showed that indeed the human rest–activity cycle persists with a period of circa 24 h in humans (Aschoff 1965), and free-running rhythms of EEG-detected sleep stages were later confirmed (Czeisler et al. 1980b). Furthermore, several laboratories have reported clear circadian rhythms of polysomnographic sleep in nocturnal rodents and nonhuman primates under free-running conditions (see Mistlberger 2005 for an in-depth review). In fact, the circadian regulation of sleep appears to be a common feature in virtually all species studied so far, including invertebrates (Sehgal and Mignot 2011).

Interestingly, the temporal organization of human sleep under constant laboratory conditions is rather different from that under a normal 24-h LD cycle. First, the self-selected circadian time to go to bed coincides with the minimum of core body temperature (CBT) (Czeisler et al. 1980b; Zulley et al. 1981) whereas, sleep onset under a 24-h LD cycle occurs typically at the end of the evening (2300 in Fig. 11.2), after several hours of waking due to exposure to daytime light and several hours before the temperature minimum. Second, humans under constant environmental conditions often adopt rest–activity—and associated sleep–wake—cycles with periods that are much longer than 24 h. Under these conditions, the regulation of the rest–activity cycle is out of synchrony with the circadian regulation of several physiological processes, leading to a state known as spontaneous internal desynchrony. Although at first glance this appears to challenge the notion of a robust circadian regulation of sleep, careful analysis of the timing of sleep under spontaneous internal desynchrony has revealed some of the key features of its circadian regulation. First, most of the spontaneous sleep episodes in internally desynchronized subjects are initiated during the circadian trough of temperature. In other words, although they freely choose their bedtime, there is a clear propensity for them to fall asleep at a very specific circadian time. Second, the duration of each sleep episode depends on the circadian time at which sleep started. The longest sleep bouts are those started during the peak of rhythmic CBT, and sleep episodes that start at the rising phase of the CBT rhythm are shorter. This rhythm of sleep propensity is also manifested during the wake episodes as increased alertness at circadian times when the CBT is at its peak (Czeisler et al. 1980b).

Other experimental protocols have revealed circadian rhythmicity in sleep propensity. One of these protocols displaces sleep to times in which sleep typically does not take place. By extending the habitual wake time, the sleep onset can be forced to occur at different circadian times. These sleep displacement protocols have demonstrated that sleep can be started at any circadian time as long as the wake time before is sufficiently long. Similar to sleep under spontaneous internal desynchrony, the circadian regulation of sleep in displacement protocols is evidenced by longer sleep episodes when they are started at the circadian peak of CBT as well as by the increased self-rated sleepiness at specific circadian times (Akerstedt and Gillberg 1981; Dijk et al. 1991 and see review by Borbely et al. 2001).

A third experimental approach to uncover the circadian regulation of sleep is the forced desynchrony (FD) protocol. Nathaniel Kleitman pioneered this experimental approach in studies in which he subjected himself to rest–activity cycles with periods different from 24 h (Kleitman 1939; Kleitman and Kleitman 1953). He discovered that despite the fact that he could schedule himself to these rest–activity cycles, his rhythm of body temperature did not seem to synchronize to the artificially short or long days. Kleitman's experiments led to the development of FD protocols as the most useful approach to unmask the circadian process of sleep regulation. In some of these studies, subjects are exposed to short sleep–wake cycles (for instance, 30 min of sleep scheduled every 90 min) for 1 or more days. Sleepiness just before and after these short sleep episodes as well as the amount and type of sleep within each scheduled sleep period are assessed (Moses et al. 1975, 1978 and see review by Borbely et al. 2001). Other studies use longer rest–activity periods that are closer to 24 h but still outside of the limits of entrainment of the circadian system. FD protocols offer several advantages to free-running and sleep displacement studies. Because the amount of sleep in a 24-h period can be maintained, FD protocols do not have the confounding effect of long-term sleep deprivation present in displaced-sleep studies. Furthermore, the strictly imposed non-24-h sleep–wake cycle assures that the circadian rhythms that do not entrain to this schedule, such as the CBT rhythm, come predictably in and out of phase with scheduled sleep. This represents an important advantage to spontaneous internal desynchrony, in which the experimenter cannot predict the misalignment between the self-selected sleep–wake cycle and physiological circadian rhythms. FD protocols have confirmed circadian rhythms of alertness and sleep propensity with peaks respectively aligned with the peak and trough of the CBT rhythm (see reviews by Borbely et al. 2001; Czeisler and Dijk 2001). Importantly, FD systematically changes the phase relationship between the homeostatic regulatory process—tracking the imposed rest–activity cycle—and the circadian regulatory process, following the endogenous circadian clock. This feature of the protocol has had remarkable value in discriminating how each of the two regulatory processes modulates specific sleep stages (see below).

11.5 Architecture of Sleep

11.5.1 NREM/REM Cycle

Young healthy adults sleep for about 8 h a night and cycle between NREM and REM sleep with a stereotypic temporal distribution. The NREM–REM cycle lasts approximately 90–100 min, with a total of 4–6 cycles per night. In general, in the healthy young adult, NREM sleep accounts for 75–90 % of sleep time (3–5 % stage I, 50–60 % stage II, and 10–20 % stage III). REM sleep accounts for 10–25 % of sleep time (Fig. 11.3). In nocturnal rodents, the sleep cycle is shorter, approximately 15 min, and sleep is typically much more fragmented, namely, short periods of sleep are interrupted by wake episodes (Borbely et al. 2001; Siegel 2005).

The transition from wake to sleep begins with stage 1 NREM sleep. Following stage 1 is stage 2 NREM sleep, characterized by the presence of K-complexes and sleep spindles (Fig. 11.1). Throughout stage 2, high-voltage slow-wave activity appears and increases until meeting the criteria for stage 3 NREM sleep, also known as deep sleep or SWS. From stage 3, the individual will typically transition to stage 2 before moving into REM sleep, approximately 80–90 min after sleep onset.

The proportion of NREM (especially SWS) and REM sleep changes across a night of sleep. REM sleep duration is quite short in the first cycle, approximately 1–5 min, and increases towards the latter cycles. SWS, on the other hand, has an opposite trend and its duration decreases with successive sleep cycles. This highly

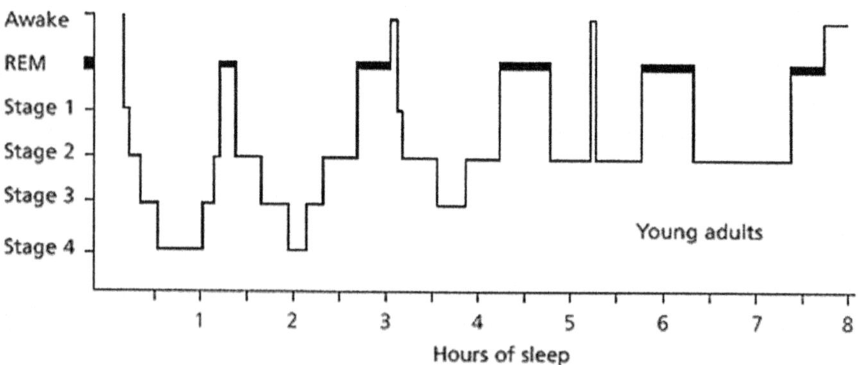

Fig. 11.3 Hypnogram of sleep architecture of a healthy young adult across an 8-h sleep period. See text for details. *Source*: Neubauer (1999). *Note*: NREM Stage 4 was redefined in 2007 by the American Association of Sleep Medicine to be included within NREM stage 3

conserved temporal landscape of sleep stages is known as sleep architecture. Normal sleep architecture is the result of a complex interaction between the homeostatic and circadian regulatory processes of sleep. Interestingly, although the rodent sleep cycle is shorter and sleep is much more fragmented, the regulation of sleep stages by the dual regulatory process of sleep is similar. Normal sleep architecture can be disrupted by sleep deprivation and challenges to the circadian system as well as by environmental changes that can destabilize the normal phase relationship between the sleep homeostat and the circadian system.

11.5.2   Homeostatic Sleep Pressure

Figure 11.2b clearly shows that sleep pressure increases with extended sleep deprivation. In all species studied so far, the onset of sleep after sleep deprivation shows increased SWS duration per sleep cycle and increased slow-wave activity, and the duration and intensity of SWS are proportional to the previous wake duration (reviewed by Borbely et al. 2001; Achermann and Borbely 2003). This conserved feature of the response to sleep deprivation has made of SWS a signature of the homeostatic regulatory process. Under normal conditions, homeostatic pressure is higher at the beginning of a sleep bout and decreases until the animal wakes up. SWS is thus concentrated at the beginning of the sleep bout and decreases in duration over the course of the night, matching the time course of homeostatic sleep pressure.

11.5.3   Circadian Regulation of Sleep

The circadian regulatory process times sleep to a specific phase of the LD cycle (Fig. 11.2b). Whereas SWS depends strongly on homeostatic regulation, REM sleep and the tightly associated regulation of CBT are under robust regulation by the circadian system. As the nocturnal bout of sleep progresses, homeostatic pressure, and its characteristic increased SWS, decreases. In contrast, circadian pressure increases and peaks towards 3 AM, leading to sleep cycles that get poorer in SWS but richer in REM sleep (Fig. 11.3).

The same protocols that have been key to understand the interplay between the homeostatic and circadian sleep regulatory processes in modulating sleep pressure have also been critical to unmask their influence on specific sleep states. Spontaneous internal desynchrony (Czeisler et al. 1980a, b), sleep displacement (Akerstedt and Gillberg 1981), and FD protocols (Dijk and Czeisler 1994, 1995; Dijk et al. 1999; Wyatt et al. 1999) have all shown that sleep onsets are characterized by increased SWS regardless of the circadian time at which they occur. In contrast, REM sleep propensity peaks at a specific circadian time, which coincides with the trough of the CBT rhythm, and this propensity is rather independent from previous wake time (Fig. 11.4).

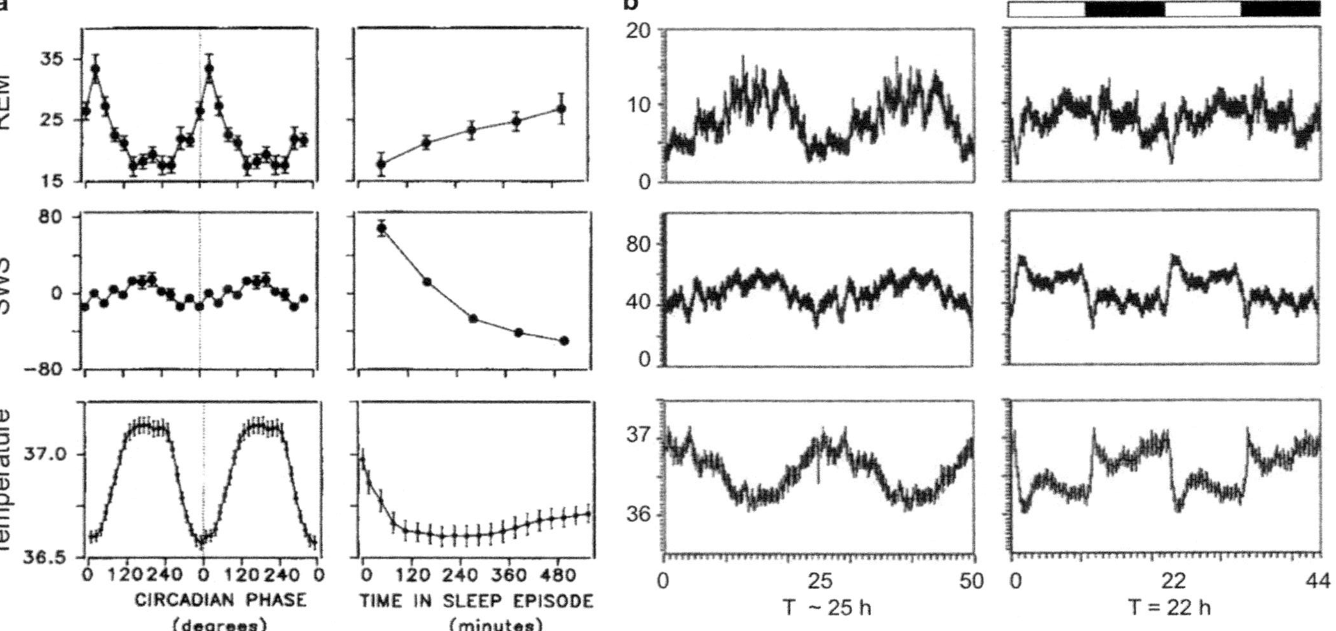

Fig. 11.4 Sleep stages are regulated by the homeostatic and circadian regulatory processes of sleep. (**a**) *Left*: in human adults under a 28-h FD protocol mean waveforms (double plotted) for REM sleep, SWS and CBT are plotted relative to the circadian period (in degrees using CBT rhythm as a phase marker). *Right*: the same data is plotted relative to the scheduled sleep time during the FD protocol, which lasted 9.33 h per cycle. (**b**) Mean waveforms for the same variables as in (**a**) but for a rat under a FD protocol induced by exposure to 22-h LD cycle. *Left*: waveforms for the three variables are plotted relative to the period of the temperature rhythm that is dissociated from the LD cycle. *Right*: the same data is plotted relative to the LD cycle (shown on *top* as *black* and *white bars*). Note that both in **a** and **b**, REM sleep has a very strong circadian regulation, independent of schedule sleep, whereas SWS depends strongly on scheduled sleep. In humans sleep is scheduled to 9.33 h every cycle and only the sleep episodes are used to generate the waveform. In rats, sleep cannot be scheduled, but the artificially short day imposes a sleep–wake cycle, and the whole cycle is used to generate the waveform. Data represent % of time in each state except for SWS in (**a**) which represents % slow-wave activity as deviations from the mean. Temperature is expressed in °C. *Sources*: Dijk and Czeisler (1995) for (**a**) and Cambras et al. (2007) for (**b**)

Evidence in the rat indicates that the key features of sleep-stage regulation by the circadian and homeostatic processes of sleep are similar to those in humans. Rats are nocturnal and concentrate most of their sleep during the light phase. The first half of this diurnal sleep is richer in NREM sleep and SWS, whereas REM sleep becomes more abundant towards the second half of the light phase. FD can be induced in the rat by exposure to artificially short (22 h) LD cycles. Under these conditions, two rest–activity rhythms are present within the same individual. The first rhythm is entrained to the LD cycle, whereas the second is dissociated from the LD cycle and has a period longer than 24 h (~25 h) (Campuzano et al. 1998; de la Iglesia et al. 2004). Under this FD protocol, SWS oscillates in synchrony with both the 22- and 25-h bouts of rest, but REM sleep only oscillates in synchrony with the 25-h bout. As in humans under FD protocols, the peak of REM sleep propensity in the forced desynchronized rat coincides with the minimum of the CBT rhythm (Cambras et al. 2007) (Fig 11.4b).

11.6   Disruption of Sleep Architecture

Normal healthy sleep architecture can become disrupted with perturbations to the circadian system. The desynchronization of the homeostatic and circadian regulation of sleep stages, a salient feature in humans exhibiting internal circadian desynchrony, disrupts sleep architecture. Internal circadian desynchrony can result from environmental "temporal challenges" such as shift work and jet lag, but is also associated with disorders including all forms of depression, other mental disorders, and total blindness.

Shift work can include night shifts, extended work hours, or rotating shifts (variable work schedule that includes day and night shifts). Poor sleep quality is self-reported among rotating shift workers (Fido and Ghali 2008; Zvcrcv and Misiri 2009) and night shift workers (Boivin et al. 2012; Juda et al. 2013; Rahman et al. 2013). Under laboratory settings, humans on a FD protocol experience less sleep efficiency and lower subjective sleep quality (O'Donnell et al. 2009). While these studies suggest circadian disruptions should result in abnormal sleep architecture in shift workers, this has yet to be investigated directly. Studies examining the rates of sleep–wake transitions in elderly (Klerman et al. 2004) and patients with obstructive sleep apnea (Bianchi et al. 2010) have documented fragmented sleep architecture in those populations.

Most psychiatric disorders show severely disrupted circadian rhythms (see reviews by Boivin 2000; Wirz-Justice 2006; McCarthy and Welsh 2012). This is particularly the case for all forms of depression, including manic-depressive illness (also known as bipolar disease) and winter depression, which are characterized by a misalignment between the timing of sleep onset and the timing of circadian rhythms such as rhythmic melatonin release (Lewy et al. 2006, 2007). REM sleep timing is among the circadian rhythms that are misaligned with sleep onset in depressive patients, which typically show an advanced REM sleep onset in their night sleep. Together, these findings have led to the "phase-advanced hypothesis," which postulates that in depressive patients the central circadian clock is advanced relative to the onset of sleep (Wehr et al. 1979). Support for a causal role of circadian misalignment in triggering depression comes from the fact that sleep deprivation or strictly scheduled sleep has antidepressive effects (Boivin 2000; Riemann et al. 2002; Wirz-Justice 2006).

Internal desynchrony of circadian rhythms represents a chronic ailment in totally blind patients. Czeisler and colleagues showed that some blind patients retain their ability to synchronize their circadian rhythms to the LD cycle, whereas others show free-running circadian rhythms despite their self-imposed 24-h sleep–wake cycle (Czeisler et al. 1995). Later studies in animal models showed that retinal degeneration that spares intrinsically photoreceptive retinal ganglion cells preserves the ability of the master circadian pacemaker to be synchronized by the LD cycle (Schmidt et al. 2011). In totally blind patients, however, the central clock is not entrained to the LD cycle, and they typically show either an abnormal phase relationship between their circadian rhythms and the LD cycle or a free-running circadian system. Studies of their circadian rhythms in melatonin release and their temporal distribution of sleep and alertness have indicated that they behave similarly to sighted subjects under free-running conditions, namely, that increased sleepiness and decreased alertness is present during the middle of the circadian night (Lockley et al. 1999, 2007, 2008). Although EEG has not been performed in these studies, one would predict that the misalignment between the circadian and homeostatic processes in totally blind patients would lead to severe disruption of sleep architecture.

Desynchronization of sleep regulators occurs with jetlag or an abrupt phase shift of the circadian system, which disrupts the timing between the homeostatic and circadian sleep processes. In rats, an abrupt 6-h phase delay rapidly shifts the timing of the sleep–wake cycle; however, the circadian system requires several days to adjust to the phase delay. During this adjustment, the timing of locomotor activity and delta power of NREM sleep is able to quickly shift to the new schedule. The timing of REM sleep and CBT, instead, requires several days to entrain to the new schedule. As a consequence, on the days immediately after the phase shift, the peaks of REM sleep and CBT occur earlier—relative to delta power of NREM and locomotor activity—than under an unaltered 24-h LD cycle (Lee et al. 2009) (Fig. 11.5). The mistiming of REM and NREM sleep pressure results in abnormal sleep architecture, with longer durations of REM occurring in the early sleep period. One outcome of disrupted sleep regulation is the presence of REM sleep stages following wake without the typical intermediate NREM sleep. REM after wake rarely occurs in normal healthy sleep and is a characteristic of narcolepsy, a sleep disorder in which individuals are unable to maintain stable wakefulness and fall directly into REM sleep with muscle paralysis.

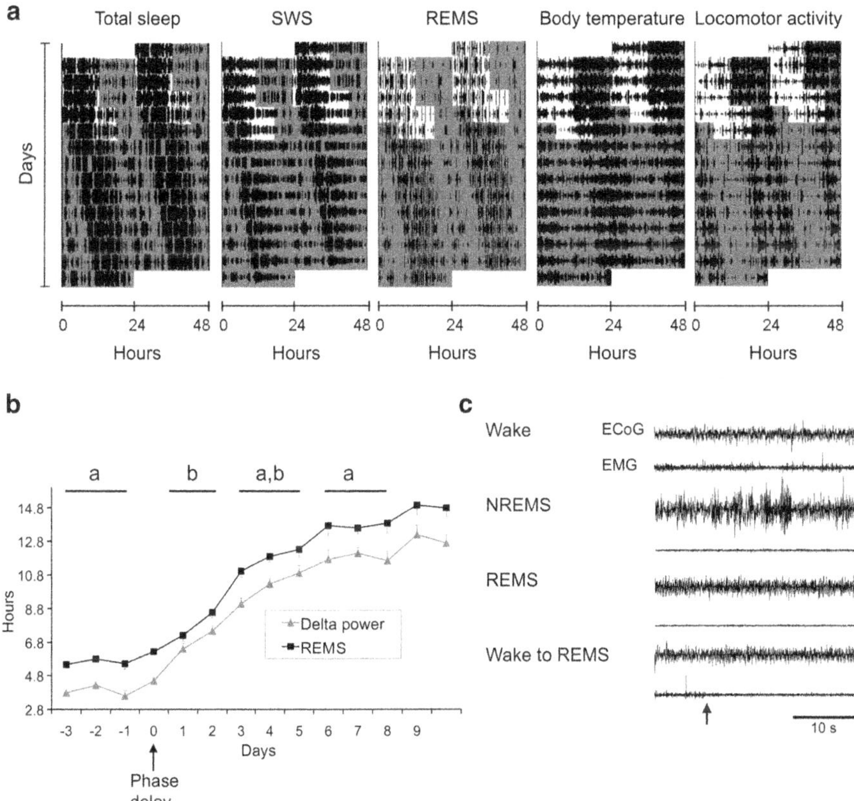

Fig. 11.5 (**a**) Actogram of daily total sleep, REM sleep, delta power of NREM, CBT, and locomotor activity of a rat exposed to an abrupt 6-h phase delay in the LD cycle. *White* and *gray areas* represent lights ON and OFF, respectively. (**b**) The daily time course of the peak of REM and delta power before and after the 6-h delay in the LD cycle (day 0) with standard error bars ($n = 5$). Blocks of days marked with *different letters* denote statistically significant differences in the peak-time difference between the REM and NREM delta power. (**c**) Spontaneous transition from wake to REM sleep in animal immediately after the abrupt phase delay. Representative EEG and EMG recordings are shown for each vigilance state; the wake-to-REM sleep transition (*arrow*) is shown in the bottom traces

11.7   Molecular Genetics of Sleep

The sleep–wake cycle and sleep architecture are directly influenced by environmental factors that can affect both homeostatic and circadian regulation. Genetic factors, on the other hand, are equally important, as they can directly affect not only the generation of specific sleep stages but also the homeostatic and circadian regulation of sleep. A review of the genes involved in sleep regulation is beyond the scope of this chapter (for recent reviews in this topic, see Jones et al. 2013; Andretic et al.

2008; Cirelli 2009; Sehgal and Mignot 2011); we will specifically focus on genes that have been directly implicated in the regulation of sleep timing that results from the homeostatic and circadian regulatory processes.

The genetics and molecular neurobiology behind the homeostatic regulation of sleep is poorly understood. In other words, genes whose products are essential for the buildup of sleep pressure during wake or of its decrease during sleep have not been identified. Sleep homeostasis is such a conserved trait across the animal phylogenetic tree that it likely serves an essential function for survival. Indeed, extended sleep deprivation leads to death in rats (Rechtschaffen et al. 1983). This may in part be why there are no single-point mutations known to abolish the homeostatic regulation of sleep. Although the evolutionarily ancestral function of sleep is unknown, it is likely that the process of sleep was co-opted for additional functions as animals evolved into more complex organisms. This notion is supported by the fact that sleep deprivation negatively impacts a wide range of biological processes, including immune response, metabolic balance, memory consolidation, and thermoregulation. Thus, sleep homeostasis may be the result of the effects of extended wake and sleep on several systems and therefore rely on the combined effects of several gene products. Nevertheless, some transmitter systems represent compelling candidates for "sleep homeostats." Among these, adenosine, the product of the degradation of ATP and cAMP, can be produced in or released into the extracellular space and interact with specific neuronal receptors. Adenosine agonists or treatments that increase endogenous adenosine levels increase sleep and specifically stimulate SWS. Adenosine receptor antagonists, such as caffeine, inhibit sleep and SWS (Bjorness and Greene 2009). This evidence, together with the fact that adenosine levels increase with extended wake in specific brain areas, has led to the notion that adenosine levels may reflect the state of the sleep homeostat; nevertheless, several experiments have challenged this notion and it is clear that there may be other transmitter systems involved in the homeostatic buildup of sleep pressure during wake (Bjorness and Greene 2009; Sehgal and Mignot 2011).

In contrast to the limited success of the molecular genetics approach in uncovering the mechanisms underlying the homeostatic process of sleep regulation, the strategy has had remarkable success in identifying several genes involved in the generation of circadian rhythmicity and the interaction between their gene products (Zhang and Kay 2010; Mohawk et al. 2012). Interestingly, whereas no known mutations can abolish the homeostatic regulation of sleep, the mutation or knockout of specific clock genes can lead to total circadian arrhythmicity in the rest activity cycle. Most of these mutations, however, do not induce dramatic changes in the amount of daily sleep and wake; they simply abolish the animal's ability to time them.

The circadian clockwork relies on transcriptional and translational feedback loops that result in the rhythmic expression of clock genes. In mammals, CLOCK and BMAL1 proteins dimerize and induce the transcription of genes of the PER and CRY families, whose protein products form complexes that inhibit the CLOCK/BMAL1 induction of transcription. This feedback loop results in the inhibition of the *PER* and *CRY* genes by their own protein products, resulting in circadian rhythmic levels of their mRNAs and proteins. The length of this circadian cycle is in part regulated by casein kinase I (CKI) through the phosphorylation of the PER proteins, which changes their half-life. This transcription–translation feedback loop is remarkably conserved among animals, and the high homology among genes in species ranging from *Drosophila* to humans has led to the identification of clock genes whose polymorphisms in both coding and noncoding regions are predictive of specific sleep–wake timing phenotypes.

The first hint that clock gene polymorphisms were associated with sleep–wake timing in humans came from a study in 410 adults, in which a single-nucleotide polymorphism in the 3′ untranslated region of the CLOCK gene, 3111C, was associated with evening-activity preference (Katzenberg et al. 1998). A second study found a polymorphism in the number of repeats in a repeat region of the PER3 protein, which was associated with morningness (5 repeat) or eveningness (4 repeats); the short repeat also showed strong association with a delayed sleep-phase syndrome, a pathological extreme evening-activity preference (Archer et al. 2003). The same group later identified a polymorphism in the untranslated region of the *PER2* gene that is predictive of morning preference. Whereas these studies used questionnaires to assess morning/evening preference, a recent study used actimetry-derived daily activity profiles to confirm the association of a common polymorphism (rs7221412) in the noncoding region of the *PER1* gene with delayed timing of activity. The authors found that homozygotes for the polymorphism had a 67-min delayed activity pattern compared to the genotypes with no copies of the allele (Lim et al. 2012).

The most striking sleep–wake phenotypes have emerged from relatively rare single-point mutations in the coding regions of clock genes. Toh and collaborators studied subjects with familial advance sleep-phase syndrome (FASPS); these subjects present a pathologically advanced rest–activity cycle, going to bed at very early evening hours (~7:30 PM) and waking up spontaneously at around 4:30 AM. The authors identified a mutation in the *PER2* gene that leads to a serine to glycine substitution in the PER2 binding site for CKI, leading to a hypophosphory-lated PER2 (Toh et al. 2001). A later study by the same group identified a second mutation leading to FASPS; this missense mutation in the *CKIδ* gene codes for a CKIδ protein with reduced phosphorylating activity (Xu et al. 2005). Interestingly, patients carrying the FASPS PER2 mutation present a very similar phenotype to animal models with the same mutation or with mutations in the *CKIε* gene that lead to lower phosphorylation of the PER proteins (Ralph and Menaker 1988; Lowrey et al. 2000; Xu et al. 2007).

He and collaborators characterized a mutation, DEC2-P385R, in a family in which two members are short sleepers. The mutation leads to an amino acid substitution in the DEC2 protein, a negative regulator of CLOCK.BMAL1-mediated transcription (He et al. 2009). Individuals carrying the mutation naturally sleep 1.5 h less than noncarriers in the same family. Remarkably, when the authors express the mutant protein in mice and *Drosophila*, the mutants also had shorter daily sleep times. Although the mutation affects a protein that directly interacts with the core molecular circadian clock, the phenotype of mutant mice reveals a homeostatic deficit rather than a circadian one. Compared to wild-type mice, they exhibit reduced baseline sleep characterized by reduced REM and NREM sleep, but they also show much less REM and NREM sleep rebound after sleep deprivation. The sleep phenotype of the DEC2-P385R mutant mice clearly challenges the molecular genetic boundaries between the homeostatic and circadian processes of sleep, underscoring that the two processes may not be as independent as classically thought. Indeed, other mutations in clock genes, including total deletions of clock components that severely affect the expression of circadian rhythms, also affect the homeostatic response to sleep deprivation (see review by Franken and Dijk 2009).

11.8 Neuroanatomy of Sleep Regulation

The alternation between the different states of consciousness and brain activity represented by sleep and wakefulness is the result of the interplay between sleep- and arousal-inducing centers. The main arousal systems are represented by cholinergic and monoaminergic (noradrenergic and serotonergic) ascending pathways originated in the rostral pons and caudal brainstem and orexin/hypocretin neurons in the posterior lateral hypothalamus (Saper et al. 2001, 2005). These regions are in turn under differential inhibition by sleep-promoting neurons within the preoptic area and the pons, which not only promote sleep by inhibiting wake but also determine which specific sleep stages will be manifested. Specific neural or genetic lesions of these neuronal populations can increase the amount of sleep or wake or can impair the expression of a specific stage under both baseline conditions and after sleep deprivation.

The sleep- and wake-promoting centers do not provide a neural substrate for the circadian modulation of sleep. In mammals, the suprachiasmatic nucleus (SCN) of the hypothalamus is the site of a master circadian clock that regulates circadian rhythms. The role of the SCN in the circadian regulation of sleep has been widely supported by a variety of experimental approaches including neuroanatomical lesions, tract tracing involving SCN efferent projections to sleep centers, and genetic manipulations of the circadian molecular clockwork (see review by Mistlberger 2005). While it is difficult to examine the neuroanatomical substrate for circadian regulation of sleep in humans, a case study of a SCN-lesioned patient revealed disrupted sleep–wake cycle and body temperature rhythm (Schwartz et al. 1986).

The ablation of the SCN results in arrhythmicity in the sleep–wake cycle. In rats, SCN lesions cause the loss of sleep consolidation and circadian variation under standard 12:12 light–dark cycles and constant light or dark conditions (Ibuka and Kawamura 1975; Ibuka et al. 1977; Mouret et al. 1978; Trachsel et al. 1992; Wurts and Edgar 2000). Sleep consolidation is also lost in SCN-lesioned mice (Easton et al. 2004). Among primates, squirrel monkeys who have complete SCN ablation lose circadian rhythmicity of sleep–wake (Edgar et al. 1993).

In rodents, the circadian rhythm in sleep and wake can be disrupted through abnormal light exposure. Rats exposed to long-term constant light conditions

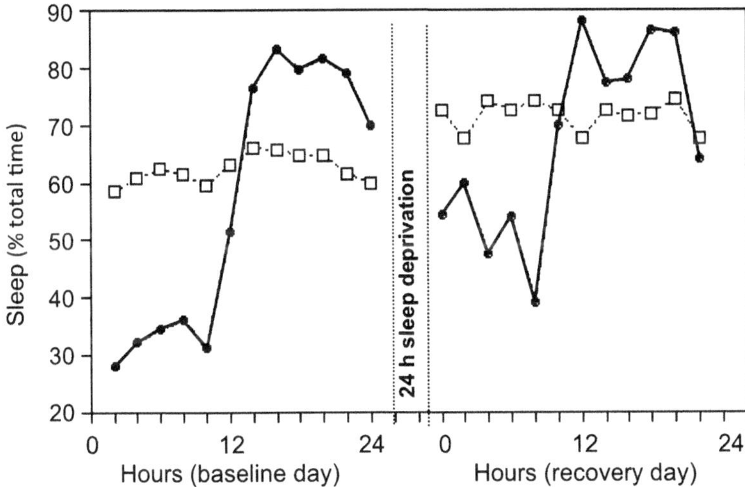

Fig. 11.6 Percent of time asleep in intact (*bold line, closed circles*) and SCN-lesioned (*dashed line, open squares*) rats over a 24-h period before and following a 24-h sleep deprivation. Lights off from hours 1 to 12, lights on from 13 to 24. See text for details. *Source*: Mistlberger et al. (1983)

express dampened circadian rhythms in locomotor activity and the sleep–wake rhythm, which often become arrhythmic (Eastman and Rechtschaffen 1983), and this effect likely results from the desynchronization among SCN neuronal oscillators (Ohta et al. 2005).

Studies in sleep-deprived SCN-lesioned rats provide further insight into the role of the SCN in sleep regulation. SCN-ablated rats show arrhythmic sleep–wake behavior, while intact rats exhibit high and low sleep propensity during the light phase (or subjective day) and dark phase (or subjective night), respectively. When exposed to a 24-h sleep deprivation, intact and SCN-lesioned rats increase sleep duration in both phases, indicating that the lesion of the central circadian pacemaker does not impair the ability of the animal to respond to sleep deprivation. However, the amount of recovery sleep in the intact rat during the dark-wake phase is less than that of a SCN-lesioned rat's baseline sleep (Mistlberger et al. 1983) (Fig. 11.6). This result shows that the SCN promotes wakefulness during the dark phase. Conversely, during the light phase, baseline sleep in intact rats is higher than in sleep-deprived SCN-lesioned rats, showing that the SCN promotes sleep during the light-phase. Thus, the evidence at least for nocturnal rodents supports the notion that the circadian timing of sleep by the SCN is the result of both subjective day induction of sleep, via stimulation of sleep centers or inhibition of wake centers, and subjective night induction of wake, via inhibition of sleep centers or stimulation of wake centers (see Mistlberger 2005 for an in-depth discussion of this topic).

The studies outlined above suggest the ablation of the SCN does not alter the brain's ability to respond to sleep deprivation with a rebound of sleep. On the other hand, the activity of SCN neurons changes in association with specific sleep stages

(Deboer et al. 2003). Interestingly, SWS during recovery sleep after SWS deprivation is associated with a decrease in SCN neuronal activity, suggesting that the master circadian pacemaker may be regulated by the homeostatic sleep response.

Studies using circadian disruptions in neurally intact animals have further characterized the role of the SCN on circadian regulation of sleep. In the rat, FD using 22-h LD cycles or abrupt phase shifts of the LD cycle that mimic jet lag desynchronize neurons within two anatomically distinct subregions of the SCN, the ventrolateral (vl) and dorsomedial (dm) SCN (de la Iglesia et al. 2004; Albus et al. 2005). These subregions differ in their pattern of afferent and efferent projections, their response to entraining stimuli, and the neurotransmitters they produce. Under a 22-h LD cycle FD protocol, the vlSCN is entrained to the LD cycle, whereas the dmSCN is dissociated from it and oscillates with a >24 h period. The timing of SWS is associated with the clock gene expression rhythm within both the vlSCN and dmSCN; however, REM sleep only oscillates in synchrony with the clock gene expression within the dmSCN (Cambras et al. 2007; Lee et al. 2009). After an abrupt 6-h phase delay of the LD cycle, the vlSCN rapidly entrains to a 6-h phase delay, while the dmSCN requires several cycles to entrain to the new LD schedule (Albus et al. 2005). This transient desynchrony between the SCN subregions is associated with abnormal sleep regulation and sleep architecture, namely, the misalignment between REM sleep and SWS depicted in Fig. 11.4. These studies support the notion that the SCN not only can act as a sleep-promoting region during the day in a nocturnal animal but also that the SCN and, more specifically, the dmSCN can time the manifestation of specific sleep stages such as REM sleep.

References

Achermann P, Borbely AA (2003) Mathematical models of sleep regulation. Front Biosci 8:s683–s693

Akerstedt T, Gillberg M (1981) The circadian variation of experimentally displaced sleep. Sleep 4:159–169

Albus H, Vansteensel MJ, Michel S et al (2005) A GABAergic mechanism is necessary for coupling dissociable ventral and dorsal regional oscillators within the circadian clock. Curr Biol 15:886–893

Andretic R, Franken P, Tafti M (2008) Genetics of sleep. Annu Rev Genet 42:361–388

Archer SN, Robilliard DL, Skene DJ et al (2003) A length polymorphism in the circadian clock gene Per3 is linked to delayed sleep phase syndrome and extreme diurnal preference. Sleep 26:413–415

Aschoff J (1965) Circadian rhythms in man. Science 148:1427–1432

Aserinsky E, Kleitman N (1953) Regularly occurring periods of eye motility, and concomitant phenomena, during sleep. Science 118:273–274

Bianchi MT, Cash SS, Mietus J et al (2010) Obstructive sleep apnea alters sleep stage transition dynamics. PLoS One 5:e11356

Bjorness TE, Greene RW (2009) Adenosine and sleep. Curr Neuropharmacol 7:238–245

Boivin DB (2000) Influence of sleep-wake and circadian rhythm disturbances in psychiatric disorders. J Psychiatry Neurosci 25:446–458

Boivin DB, Boudreau P, James FO et al (2012) Photic resetting in night-shift work: impact on nurses' sleep. Chronobiol Int 29:619–628

Borbely AA (1982) A two process model of sleep regulation. Hum Neurobiol 1:195–204

Borbely AA, Dijk DJ, Achermann P et al (2001) Processes underlying the regulation of sleep-wake cycle. In: Takahashi JS, Turek FW, Moore RY (eds) Handbook of behavioral neurobiology: circadian clocks. Kluwer/Plenum, New York, pp 457–479

Borges JL (1999) Libro de los sueños. Continental Book. Denver, CO

Cambras T, Weller JR, Angles-Pujoras M, Lee ML, Christopher A, Diez-Noguera A, Krueger JM, de la Iglesia HO (2007) Circadian desynchronization of core body temperature and sleep stages in the rat. Proc Natl Acad Sci U S A 104:7634–7639

Campuzano A, Vilaplana J, Cambras T et al (1998) Dissociation of the rat motor activity rhythm under T cycles shorter than 24 hours. Physiol Behav 63:171–176

Cirelli C (2009) The genetic and molecular regulation of sleep: from fruit flies to humans. Nat Rev Neurosci 10:549–560

Czeisler CA, Dijk DJ (2001) Human circadian physiology and sleep-wake regulation. In: Takahashi JS, Turek FW, Moore RY (eds) Handbook of behavioral neurobiology: circadian clocks. Kluwer/Plenum, New York, pp 531–569

Czeisler CA, Zimmerman JC, Ronda JM et al (1980a) Timing of REM sleep is coupled to the circadian rhythm of body temperature in man. Sleep 2:329–346

Czeisler CA, Weitzman E, Moore-Ede MC et al (1980b) Human sleep: its duration and organization depend on its circadian phase. Science 210:1264–1267

Czeisler CA, Shanahan TL, Klerman EB et al (1995) Suppression of melatonin secretion in some blind patients by exposure to bright light. N Engl J Med 332:6–11

Davis H, Davis PA, Loomis AL et al (1937) Changes in human brain potentials during the onset of sleep. Science 86:448–450

de la Iglesia HO, Cambras T, Schwartz WJ et al (2004) Forced desynchronization of dual circadian oscillators within the rat suprachiasmatic nucleus. Curr Biol 14:796–800

de Souza L, Benedito-Silva AA, Pires ML et al (2003) Further validation of actigraphy for sleep studies. Sleep 26:81–85

Deboer T, Vansteensel MJ, Detari L et al (2003) Sleep states alter activity of suprachiasmatic nucleus neurons. Nat Neurosci 6:1086–1090

Dijk DJ, Czeisler CA (1994) Paradoxical timing of the circadian rhythm of sleep propensity serves to consolidate sleep and wakefulness in humans. Neurosci Lett 166:63–68

Dijk DJ, Czeisler CA (1995) Contribution of the circadian pacemaker and the sleep homeostat to sleep propensity, sleep structure, electroencephalographic slow waves, and sleep spindle activity in humans. J Neurosci 15:3526–3538

Dijk DJ, Cajochen C, Tobler I, Borbely AA (1991) Sleep extension in humans – sleep stages, EEG power spectra and body-temperature. Sleep 14:294–306

Dijk DJ, Duffy JF, Riel E et al (1999) Ageing and the circadian and homeostatic regulation of human sleep during forced desynchrony of rest, melatonin and temperature rhythms. J Physiol 516(Pt 2):611–627

Eastman C, Rechtschaffen A (1983) Circadian temperature and wake rhythms of rats exposed to prolonged continuous illumination. Physiol Behav 31:417–427

Easton A, Meerlo P, Bergmann B et al (2004) The suprachiasmatic nucleus regulates sleep timing and amount in mice. Sleep 27:1307–1318

Edgar DM, Dement WC, Fuller CA (1993) Effect of SCN lesions on sleep in squirrel monkeys: evidence for opponent processes in sleep-wake regulation. J Neurosci 13:1065–1079

Fernandez-Duque E, de la Iglesia HO, Erkert HG (2010) Moonstruck primates: owl monkeys (Aotus) need moonlight for nocturnal activity in their natural environment. PLoS One 5:e12572

Fido A, Ghali A (2008) Detrimental effects of variable work shifts on quality of sleep, general health and work performance. Med Princ Pract 17:453–457

Fisher SP, Godinho SI, Pothecary CA et al (2012) Rapid assessment of sleep-wake behavior in mice. J Biol Rhythms 27:48–58

Franken P, Dijk DJ (2009) Circadian clock genes and sleep homeostasis. Eur J Neurosci 29:1820–1829

He Y, Jones CR, Fujiki N et al (2009) The transcriptional repressor DEC2 regulates sleep length in mammals. Science 325:866–870

Horne JA (1988) Why we sleep: the functions of sleep in humans and other mammals. Oxford University Press, Oxford

Iber C, Ancoli-Israel S, Chesson AL et al (2007) The AASM manual for the scoring of sleep and associated events: rules, terminology and technical specifications, 1st edn. American Academy of Sleep Medicine, Westchester, IL

Ibuka N, Kawamura H (1975) Loss of circadian rhythm in sleep-wakefulness cycle in the rat by suprachiasmatic nucleus lesions. Brain Res 96:76–81

Ibuka N, Inouye SI, Kawamura H (1977) Analysis of sleep-wakefulness rhythms in male rats after suprachiasmatic nucleus lesions and ocular enucleation. Brain Res 122:33–47

Jones CR, Huang AL, Ptacek LJ et al (2013) Genetic basis of human circadian rhythm disorders. Exp Neurol 243:28–33

Juda M, Vetter C, Roenneberg T (2013) Chronotype modulates sleep duration, sleep quality, and social jet lag in shift-workers. J Biol Rhythms 28:141–151

Katzenberg D, Young T, Finn L et al (1998) A CLOCK polymorphism associated with human diurnal preference. Sleep 21:569–576

Kleitman N (1939) Sleep and wakefulness as alternating phases in the cycle of existence. The University of Chicago Press, Chicago, IL

Kleitman N, Kleitman E (1953) Effect of non-twenty-four-hour routines of living on oral temperature and heart rate. J Appl Physiol 6:283–291

Klerman EB, Davis JB, Duffy JF et al (2004) Older people awaken more frequently but fall back asleep at the same rate as younger people. Sleep 27:793–798

Lee ML, Swanson BE, de la Iglesia HO (2009) Circadian timing of REM sleep is coupled to an oscillator within the dorsomedial suprachiasmatic nucleus. Curr Biol 19:848–852

Lewy AJ, Lefler BJ, Emens JS et al (2006) The circadian basis of winter depression. Proc Natl Acad Sci U S A 103:7414–7419

Lewy AJ, Rough JN, Songer JB et al (2007) The phase shift hypothesis for the circadian component of winter depression. Dialogues Clin Neurosci 9:291–300

Lim AS, Chang AM, Shulman JM et al (2012) A common polymorphism near PER1 and the timing of human behavioral rhythms. Ann Neurol 72:324–334

Lockley SW, Skene DJ, Butler LJ et al (1999) Sleep and activity rhythms are related to circadian phase in the blind. Sleep 22:616–623

Lockley SW, Arendt J, Skene DJ (2007) Visual impairment and circadian rhythm disorders. Dialogues Clin Neurosci 9:301–314

Lockley SW, Dijk DJ, Kosti O et al (2008) Alertness, mood and performance rhythm disturbances associated with circadian sleep disorders in the blind. J Sleep Res 17:207–216

Loomis AL, Harvey EN, Hobart G (1935) Potential rhythms of the cerebral cortex during sleep. Science 81:597–598

Lowrey PL, Shimomura K, Antoch MP et al (2000) Positional syntenic cloning and functional characterization of the mammalian circadian mutation tau. Science 288:483–492

McCarthy MJ, Welsh DK (2012) Cellular circadian clocks in mood disorders. J Biol Rhythms 27:339–352

McShane BB, Galante RJ, Biber M et al (2012) Assessing REM sleep in mice using video data. Sleep 35:433–442

Mistlberger RE (2005) Circadian regulation of sleep in mammals: role of the suprachiasmatic nucleus. Brain Res Rev 49:429–454

Mistlberger RE, Bergmann BM, Waldenar W et al (1983) Recovery sleep following sleep deprivation in intact and suprachiasmatic nuclei-lesioned rats. Sleep 6:217–233

Mohawk JA, Green CB, Takahashi JS (2012) Central and peripheral circadian clocks in mammals. Annu Rev Neurosci 35:445–462

Moses JM, Hord DJ, Lubin A et al (1975) Dynamics of nap sleep during a 40 hour period. Electroencephalogr Clin Neurophysiol 39:627–633

Moses J, Lubin A, Naitoh P et al (1978) Circadian variation in performance, subjective sleepiness, sleep, and oral temperature during an altered sleep-wake schedule. Biol Psychol 6:301–308

Mouret J, Coindet J, Debilly G et al (1978) Suprachiasmatic nuclei lesions in the rat: alterations in sleep circadian rhythms. Electroencephalogr Clin Neurophysiol 45:402–408

Neubauer DN (1999) Sleep problems in the elderly. Am Fam Physician 59(2551–2558):2559–2560

O'Donnell D, Silva EJ, Munch M et al (2009) Comparison of subjective and objective assessments of sleep in healthy older subjects without sleep complaints. J Sleep Res 18:254–263

Ohta H, Yamazaki S, McMahon DG (2005) Constant light desynchronizes mammalian clock neurons. Nat Neurosci 8:267–269

Rahman SA, Shapiro CM, Wang F et al (2013) Effects of filtering visual short wavelengths during nocturnal shiftwork on sleep and performance. Chronobiol Int 30:951–962

Ralph MR, Menaker M (1988) A mutation of the circadian system in golden hamsters. Science 241:1225–1227

Rechtschaffen A, Gilliland MA, Bergmann BM et al (1983) Physiological correlates of prolonged sleep deprivation in rats. Science 221:182–184

Riemann D, Voderholzer U, Berger M (2002) Sleep and sleep-wake manipulations in bipolar depression. Neuropsychobiology 45(Suppl 1):7–12

Sadeh A, Sharkey KM, Carskadon MA (1994) Activity-based sleep-wake identification: an empirical test of methodological issues. Sleep 17:201–207

Saper CB, Chou TC, Scammell TE (2001) The sleep switch: hypothalamic control of sleep and wakefulness. Trends Neurosci 24:726–731

Saper CB, Scammell TE, Lu J (2005) Hypothalamic regulation of sleep and circadian rhythms. Nature 437:1257–1263

Schmidt TM, Chen SK, Hattar S (2011) Intrinsically photosensitive retinal ganglion cells: many subtypes, diverse functions. Trends Neurosci 34:572–580

Schwartz WJ, Busis NA, Hedley-Whyte ET (1986) A discrete lesion of ventral hypothalamus and optic chiasm that disturbed the daily temperature rhythm. J Neurol 233:1–4

Sehgal A, Mignot E (2011) Genetics of sleep and sleep disorders. Cell 146:194–207

Siegel JM (2005) Clues to the functions of mammalian sleep. Nature 437:1264–1271

Summers-Brenmer E (2008) Insomnia. A cultural history. Reaktion, London

Toh KL, Jones CR, He Y et al (2001) An hPer2 phosphorylation site mutation in familial advanced sleep phase syndrome. Science 291:1040–1043

Trachsel L, Edgar DM, Seidel WF et al (1992) Sleep homeostasis in suprachiasmatic nuclei-lesioned rats: effects of sleep deprivation and triazolam administration. Brain Res 589:253–261

Wehr TA, Wirz-Justice A, Goodwin FK et al (1979) Phase advance of the circadian sleep-wake cycle as an antidepressant. Science 206:710–713

Wirz-Justice A (2006) Biological rhythm disturbances in mood disorders. Int Clin Psychopharmacol 21:S11–S15

Wurts SW, Edgar DM (2000) Circadian and homeostatic control of rapid eye movement (REM) sleep: promotion of REM tendency by the suprachiasmatic nucleus. J Neurosci 20:4300–4310

Wyatt JK, Ritz-De Cecco A, Czeisler CA et al (1999) Circadian temperature and melatonin rhythms, sleep, and neurobehavioral function in humans living on a 20-h day. Am J Physiol 277:R1152–R1163

Xu Y, Padiath QS, Shapiro RE et al (2005) Functional consequences of a CKI delta mutation causing familial advanced sleep phase syndrome. Nature 434:640–644

Xu Y, Toh KL, Jones CR et al (2007) Modeling of a human circadian mutation yields insights into clock regulation by PER2. Cell 128:59–70

Zhang EE, Kay SA (2010) Clocks not winding down: unravelling circadian networks. Nat Rev Mol Cell Biol 11:764–776

Zulley J, Wever R, Aschoff J (1981) The dependence of onset and duration of sleep on the circadian rhythm of rectal temperature. Pflugers Arch 391:314–318

Zverev YP, Misiri HE (2009) Perceived effects of rotating shift work on nurses' sleep quality and duration. Malawi Med J 21:19–21

Part III
Clinical Relevance of Circadian Rhythmicity

Chapter 12
Chronostasis: The Timing of Physiological Systems

Raúl Aguilar-Roblero

Abstract Even though the biological relevance of circadian rhythmicity has been recognized since the middle of the past century, the majority of the medical community has remained unaware of the circadian rhythms or their relevance during most of this time, perhaps because of their apparent opposition to homeostasis. It was not until the 1980s that circadian rhythms begun to be noticed among the physicians, but even now, 30 years later, circadian rhythms are still not an integral part of medical physiology. The cause of the neglect and even rejection of the relevance of circadian biology to human health, from most medical practitioners until the last quarter of the twentieth century, may have involved among other factors: (1) the heuristic power of homeostasis as a general physiological process to understand health, and its unbalance as a major cause of disease; (2) the fact that disruption of circadian rhythmicity was not ostensibly associated at the time with any pathological entity; and, last but not least, (3) the technical problems associated with collecting relevant physiological data for a long time period before the age of electronic devices and massive digital information processing.

This chapter presents a perspective for integrating circadian rhythmicity and its mechanisms with medical physiology. We start by a brief review of the concepts of homeostasis and rheostasis in physiology, then introduce the concept of chronostasis as the mechanism to timing physiological processes. Finally, we provide the main implications of the concept of chronostasis in human health and disease, in light of the recent advances in molecular genetics related to the so-called clock genes.

R. Aguilar-Roblero (✉)
División de Neurociencias, Instituto de Fisiología Celular, Universidad Nacional Autónoma de México, Circuito exterior S/N, Ciudad Universitaria, Coyoacán, México, D.F. 04510, Mexico

Departamento de Fisiología, Facultad de Medicina, Universidad Nacional Autónoma de México, Coyoacán, Mexico, D.F. Mexico
e-mail: raguilar@ifc.unam.mx

© Springer International Publishing Switzerland 2015
R. Aguilar-Roblero et al. (eds.), *Mechanisms of Circadian Systems in Animals and Their Clinical Relevance*, DOI 10.1007/978-3-319-08945-4_12

12.1   Homeostasis

Claude Bernard proposed by 1860 the concept of a "milieu intérieur" as the actual environment where biological processes occur and the constancy of this internal environment as a prerequisite to preserve physiological functions within an optimal range (Bernard 1865). By 1930, Walter Cannon moved forward Bernard's ideas and developed the concept of *homeostasis* as the processes involved in maintaining the internal environment within an optimal physiological range as a steady state reached between the organism and its environment (Cannon 1929, 1936). Wiener and Rosenblueth then proposed that the mechanism of homeostasis consists of negative feedback loops that will detect and correct deviations from key physiological parameters from the optimal range, such optimal range being encoded as a set point in a controller element of the system (Wiener 1948; Rosenblueth et al. 1943). According to the feedback model (Fig. 12.1), key physiological parameters are continuously sensed by specialized autonomic neuronal receptors, such as baroreceptors for blood pressure and chemoreceptors for O_2, CO_2, pH, or glucose that generate a feedback signal which is compared to the set point values in the controller elements. These set point values could be coded as the density of chemical receptors or ionic channels in specialized cells, which correspond to the controller elements of the model and will in turn generate control signals that would drive the effector elements, such as neuronal circuits or muscular and secretory cells, to actually generate the physiological parameters under control.

The heuristic value of the concept of homeostasis is reflected in the changes in ideas about the causes and treatment of disease, which have led to an increase in life expectancy from the last quarter of the nineteenth century to the first half of the twentieth century. So strong was the concept of homeostasis, and the misconception that during healthy conditions only minor deviations from the set points could take place, that the first reports on daily fluctuations in the value of physiological parameters, associated with biological rhythms, were received with indifference and skepticism by most members of the medical community. The concept of homeostasis remained unchanged and basically unchallenged until 1977, when Nicolaïdis provided evidence that under certain circumstances the values of the set point of homeostatic systems could be altered.

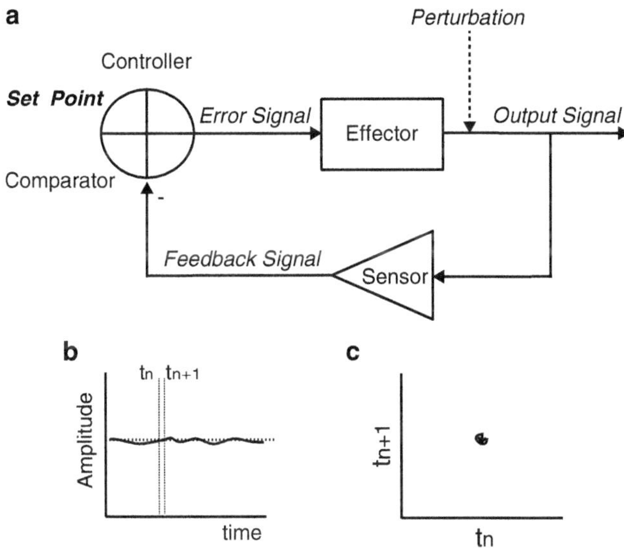

Fig. 12.1 Regulatory negative feedback system and graphs from its outcome versus time and as a phase portrait. (**a**) Diagram shows the main components of a regulatory negative feedback system: the effector, which generates the regulated parameter or output signal; the sensor, which detects the value of the regulated parameter and generates a signal that feeds back into the controller; the controller, which encompasses information regarding the optimal value or set point to which the parameter is being regulated; and a comparator, which determines the difference between the set point and the feedback signal. Such difference is translated into an error signal that modifies the output of the effector to compensate any deviation of the regulated parameter from its optimal value or range. The information flowing through the system is indicated in *italic letters*; the elements of the system are indicated with *non-italic font*. The set point is in ***bold italic letters*** to indicate that it could be either a constitutive element or information flowing through the system. (**b**) Schematic graph of the variation of a regulated parameter in time shows that any deviation from the set point (*dotted horizontal line*) is corrected within a short interval. (**c**) Schematic graph of the phase portrait from the data graph in **b**; in this graph the value of the parameter at time *n* (*tn*) is plotted against the value of the parameter at the next time *n* + 1 (*tn* + 1), as indicated by the *vertical dotted lines* in **b**; *tn* + 1 is then reset to *tn* and the process continues. In this type of graph the output of a negative feedback-regulated system is plotted within a *small circular* or *elliptical area*; the better the system is regulated the smaller is the area of the *circle* or *ellipse*

12.2   Rheostasis (Allostasis)

Even although Cannon never suggested that the set points of the homeostatic system were fixed throughout life, it was a common and unsubstantiated assumption; nevertheless, evidence that indicated that the set point could change from one value to another opened new perspectives to the understanding of physiology. Hammel (1968) proposed that the set point was adjustable to explain acute thermoregulatory corrections. Later on, Nicolaïdis (1977) proposed the term *homeorheusis* to refer to

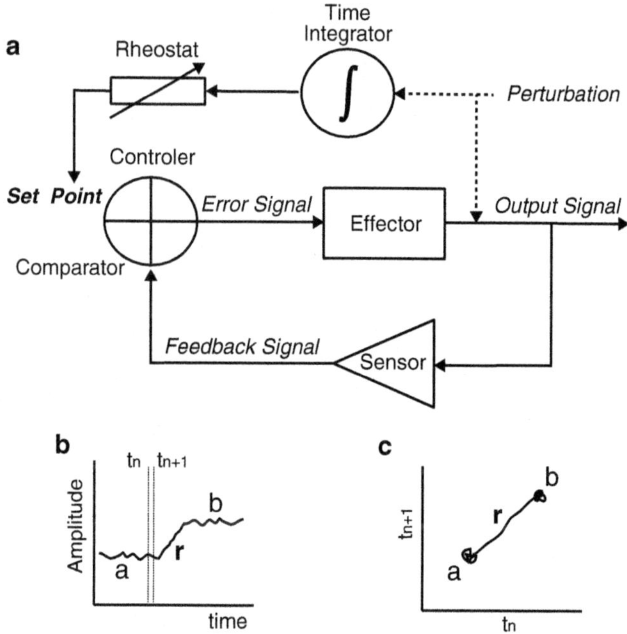

Fig. 12.2 Rheostatic regulatory system and graphs from its outcome versus time and as a phase portrait. (**a**) Diagram indicates a negative feedback system regulated by rheostatic processes. The set point of the feedback system varies according to sustained changes in environmental conditions or to the needs of the organism. Both environmental and internal changes are here indicated as perturbations. In this model we postulate that only two essential elements form the rheostatic regulatory system: a time integrator to evaluate whether there is a sustained perturbation which requires a change in set point values, and a rheostat element to generate a signal to modify the set point of the system. The rest of the elements of the system are as in Fig. 12.1. (**b**) Schematic graph of the rheostatic modification (*r*) of the set point of a regulated variable from *a* to *b*; at all other times any small deviation of the regulated parameter from the set point is corrected within a short interval. **c** Schematic graph of the phase portrait from the data graph in **b**. In this graph we can appreciate the trajectory of the regulated parameter from point *a* to *b* and the trajectory of the regulated change *r*

long-term changes in the set point of body weight occurring through life. Mrosovsky (1990) renamed this phenomenon *rheostasis*, and further developed this concept by providing an extensive review of the different conditions in which changes of the set point occur (Fig. 12.2).

During their lifespan organisms frequently face changes in their physiological or environmental conditions; in some of these events, maintaining a physiological parameter within the narrow range dictated by an immovable set point would challenge the survival of the individual by preventing adequate physiological responses. To prevent the failure of homeostatic regulation, in such situations the organism would require processes or mechanisms that allow switching set point values under regulated conditions. Rheostasis refers to such processes and mechanisms that allow adjustment of homeostatic processes when facing changes in physiological or environmental conditions. Two types of rheostasis are recognized: when the change in set point is triggered by influences external to the organism it is considered as *reactive rheostasis*; in contrast, when these changes are triggered by genetic programs, such as those occurring at specific times during development, it is considered as *programmed rheostasis*. It is worth noting that Mrosovsky considered circadian rhythmicity as programed rheostasis; however, for reasons that are presented next (see Sect. 12.3), we consider circadian and other biological rhythms as regulatory processes different from rheostasis.

Sterling and Eyer, while studying the physiological changes induced by stressful life situations, developed a concept similar to that of rheostasis but independent from Nicolaïdis and Mrosovsky. They named the stability attained through change *allostasis*, that is, "to maintain stability an organism must vary all the parameters of its internal mileu and match them appropriately to environmental demands" (Sterling and Eyer 1988; quoted in Boulos and Rosenwasser 2005, p. 228). This definition is essentially similar to stating that the set point in homeostatic systems should be modified to adjust the physiological and behavioral responses of an individual to changes in its environment or physiological needs. Thus, although for all practical purposes we may consider the two terms, rheostasis and allostasis, as equivalent concepts, nevertheless we prefer to use the term rheostasis.

12.3 Chronostasis

Periodic variations in human functions that repeat every 24 h have been known for almost 60 years (Aschoff 1955, cited in Aschoff 1965), but its impact on human physiology and medicine was not noticeable until about 30 years later (Moore-Ede et al. 1982). It was not until the clock genes were characterized in rodents and their human homologues identified that widespread attention was attracted to the role of circadian rhythmicity as a probable cause of disease (Lowrey and Takahashi 2004; Turek et al. 2005). In spite of increased awareness of the relevance of circadian rhythmicity for health and disease, integration of circadian timing within a coherent conceptual framework on physiological regulation is still lacking. Such a conceptual framework will allow us to insert circadian rhythms in the perspective of integrative physiology, as well as to integrate recent findings related to clock genes and their interaction with genes related to regulation of the cell cycle or metabolism originally studied in the context of cell biology and genetic regulation.

As previously mentioned, Mrosovsky (1990) considered circadian and other biological rhythms as programed rheostasis, and other authors have accepted this perspective without much further elaboration (Boulos and Rosenwasser 2005), whereas some others considered circadian rhythms as part of the homeostatic process (Moore-Ede et al. 1982; Perreau-Lenz et al. 2004; Woods and Ramsay 2007). The first conceptual model that attempted to join homeostatic regulation with circadian rhythmicity was developed by Alexander Borbély (1982) to account for the nonlinear sleep rebound that occurred after increasing duration of sleep deprivation. This model proposed that sleep regulation involves an interaction between circadian and homeostatic processes. The model, later formalized by Daan et al. (1984), also provided computer simulations that could reproduce a number of phenomena related to recovery from sleep deprivation and internal desynchronization induced by temporal isolation in humans. Internal desynchronization refers to the loss of the time relationship among physiological parameters such as temperature and sleep onset, which occurs when the subject is maintained isolated in an environment free of time cues and thus loses track of the time. More recently, Kalsbeek et al. (2006) have provided abundant evidence regarding glucose regulation involving both circadian and homeostatic processes. Nevertheless, Kalsbeek considers circadian rhythms as guarding homeostasis, thus missing the opportunity to recognize the wider spectrum of parameters (amplitude, period, and phase of rhythmicity; see following) regulated by circadian modulation of the homeostatic systems and the cooperative aspects among time regulation (chronostasis) and homeostasis.

The concept of an adjustable set point in a feedback system as proposed by Mrosovsky for rheostasis is a necessary step to incorporate time regulation into physiological systems. Nevertheless, the dependence of circadian rhythmicity on biological clocks or oscillators provides particular characteristics about this type of physiological regulation that justify its consideration as a process independent from rheostasis. Furthermore, in the past 15 years our knowledge of the cellular and molecular mechanisms of circadian clocks has advanced to allow us to set apart the physiological time regulation or chronostasis from the rest of the processes included in the concept of rheostasis and homeostasis.

The term chronostasis was coined to refer the cyclic timing (circadian or otherwise) of physiological systems (Fig. 12.3). We followed the reasoning of Mrosovsky for the term rheostasis, but here the prefix *homeo or rheo* was substituted by *chrono* to indicate that the steady state guarded by homeostatic and rheostatic processes is time modulated. Chronostasis is used to refer to the cyclic time regulation of physiological systems, which includes the period, amplitude, and phase of the oscillations in any physiological variable, as well as the phase relationship among different physiological systems and also with environmental cycles. The period of a rhythmic phenomenon refers to the elapsed time for a complete cycle to occur; in natural (entrained) conditions, the period of the circadian rhythmicity is 24 h. Amplitude refers to the change in the value of the variable from its highest value (peak) to its lowest (trough). In the context of the cosinor analysis (which is an statistical procedure used to fit cyclic phenomena to a cosine function), amplitude refers to the change in intensity of the variable from the mean value of the variable throughout the cycle (mesor) to the highest value of the best fitting cycle (acrophase). Finally, the phase of the rhythm refers to the time of the cycle at which any particular value of the cyclic variable occurs; for example, the time of waking or sleep initiation corresponds to a particular phase of the sleep–wake cycle. Also, the time at which the peak or acrophase occurs is also a clear-cut phase reference for cyclic or rhythmic phenomena. Finally, because each rhythmic variable has specific phase descriptors, the time relationship between two variables is best referred to as a phase relationship. Thus, it is clear that time regulation of physiological variables broadens the

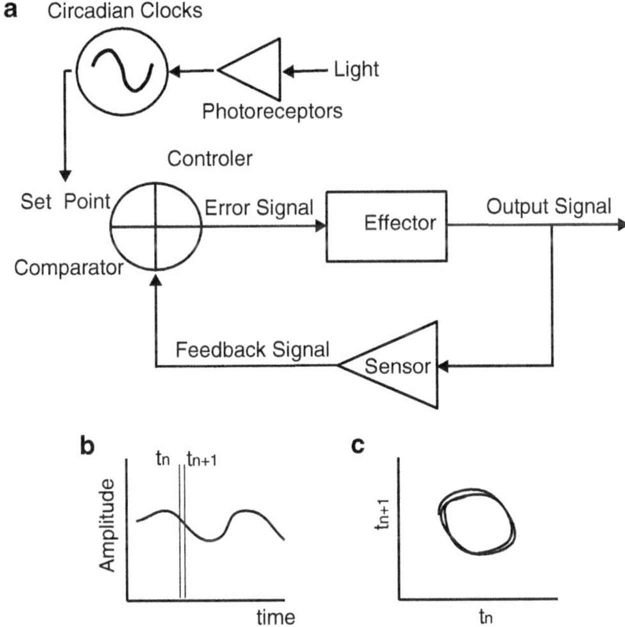

Fig. 12.3 Chronostatic regulatory system and graphs from its outcome versus time and as a phase portrait. (**a**) Diagram indicates a negative feedback system regulated by chronostatic processes. The set point of the feedback system varies according to the time signal generated by the circadian clock. In this scheme the clock is entrained to the light–dark cycle by the effects of light on specialized photoreceptors, which in mammals generate a glutamatergic signal that is received by the suprachiasmatic nuclei (SCN) to adjust the phase and period of the circadian molecular clocks. (**b**) Schematic graph of the regulated signal shows circadian rhythm in the regulated parameter. (**c**) Schematic phase portrait from the data graph in **b** shows the limit cycle of the regulated parameter; the dispersion from a perfect circular or elliptical trajectory reflects mainly the perturbations occurring throughout the cycle. At any given time the phase of the limit cycle is regulated within a narrow range by the negative feedback regulatory system

scope of parameters under study and analysis, beyond the changes in their mean values and their reset to new values, as seen from homeostatic and rheostatic perspectives, respectively. These considerations provide further support to the view of chronostasis as an independent regulatory process in physiology.

12.3.1 A Review of the Set Point Concept

Because the adjustable set point from a regulatory feedback loop is a key concept in rheostasis and chronostasis, a brief review of this concept is required before any further elaboration on chronostasis. According to Cabanac (2006) a set point is "an information input that may be determined by an external signal to which the regulated variable is compared, or it may be determined by the structural characteristics of the system itself." Thus, the set point refers to the information used by a regulatory system to compare with the actual value of the regulated variable, to generate a correction signal to actively counteract any deviation presented in the regulated variable from a desired value, which is somehow encoded by (or in) the set point.

The notion of the set point as an "external signal" most not be misunderstood as external to the organism, but rather as external to the regulated system (or subsystem) although originating from the organism itself. On the other hand, nowadays the notion of the set point as an "structural characteristic" of the system extends beyond the architecture of the regulatory feedback loops, either from neural or endocrine networks (Russek and Cabanac 1983), but also includes the concentration of hormones or neuroactive substances, the density of channels and chemical receptors, or even the constitutive level of expression of some genes (Boulant 2006; Roth et al. 2006; Jéquier and Tappy 1999; Jéquiere 2002; Speakman et al. 2011; Viola et al. 2007; Wright et al. 2008; Goel et al. 2009).

The concept of set point has been questioned by some physiologists, who suggest it should be discarded because it is confusing and misleading to the point of oversimplification, whereas regulatory systems are quite complex and involve several subsystems and regulatory loops. A deeper review of such criticisms is beyond the scope of this chapter, but for those interested in this controversy we suggest referring to Harris (1990), Werner (2010), Romanovsky (2004, 2006), Cabanac (2006), Jéquier and Tappy (1999), and Speakman et al. (2011). Although we agree with Werner that in the end we must examine the actual experimental data to draw any conclusion on how a physiological system is regulated, the idea of set point allows us to examine the hypothesis of a possible integrative mechanism underlying time regulation of physiological processes, at different levels of biological complexity. Thus, we shall not discard the concept but embrace it in its complexity.

Here the concept of chronostasis is proposed as the set of mechanisms involved in the rhythmic timing of physiological regulation. This model incorporates circadian clock mechanisms, to the feedback regulatory loops used in homeostatic and rheostatic regulatory systems, which feedforward a time signal to modulate the set points of the system (Fig. 12.3). Thus, the feedback-regulated variables are compared to different optimal values that depend on the time of the day (Aguilar-Roblero and Díaz-Muñoz 2010).

12.3.2 Time Regulation of the Set Point

In mammals, the major circadian clock has been located in the suprachiasmatic nuclei (SCN), in the anterior hypothalamus, and interconnected to autonomic and behavioral regulatory systems located mainly in the hypothalamus (Klein et al. 1991). At the molecular level the circadian clock is formed by a set of genes, known as clock genes, functionally arranged in a transcription–translation feedback loop with a delay that oscillates with a period close to 24 h. The clock genes have been found in brain regions other than the SCN (Abe et al. 2002; Yamazaki et al. 2000) and also outside the brain in most organs of the body (Yamazaki et al. 2000, 2002), which indicates the presence of multiple circadian oscillators collectively known as peripheral circadian oscillators. It has been suggested that the circadian system is a hierarchical multioscillatory system, coordinated by neural and endocrine systems driven by the SCN. This hierarchical multioscillatory circadian system would enable different physiological systems to resonate at a circadian frequency (Aguilar-Roblero and Díaz-Muñoz 2010). Thus, the clock mechanisms involved in chronostatic time modulation can be recognized at a systemic level in the SCN and its neuronal network that modulates regulatory neural and neuroendocrine networks. At a cellular level chronostatic time modulation occurs throughout the body by clock genes from peripheral oscillators, coordinated in time by neural and endocrine signaling, interacting with regulatory systems of metabolic intracellular signaling, and gene expression.

Time modulation of homeostatic and rheostatic feedback loops by biological clocks may occur at any level of the loop, but we hypothesize that it occurs mainly at the set point value of each system because at this level the time signal permits an update of the complete regulatory system, without modifying (1) the responsiveness of the system to perturbations, produced by either external or internal factors; (2) the output of the physiological systems; or (3) the hierarchical organization of the regulatory subsystems in the complete organism. Furthermore, by its resonance to the central clock(s), clock genes at the cellular level may regulate directly cell metabolism and gene expression of the entire organism.

Thus, we propose that chronostatic modulation of the set point of physiological systems, generated by the circadian clocks and peripheral oscillators, allows the feedback signals of the regulated systems to be compared to different optimal values at each time of the day. These optimal set point values are the result of environmental adaptive pressures acting upon the organisms throughout several thousand years of human evolution. Nevertheless, humans have drastically changed their environment, at least in the past 1,000 years. Because adaptation through natural selection to such changes in the environment seems to operate at a much longer time scale than the rate of human-induced environmental changes, it seems that the optimal set point values are outdated for our present lifestyle and environmental conditions, and this is also true for homeostatic and rheostatic regulatory systems. In relationship to chronostatic regulation, some recognizable examples of diseases related to nonoptimal set points are some sleep and physiological disorders resulting from transmeridian flights, nocturnal work, or shift-work schedules. A review of recent literature shows that metabolic alterations leading to obesity and diabetes (Turek et al. 2005; Sahar and Sassone-Corsi 2012), some types of cancer (Borgs et al. 2009; Stevens et al. 2011), mental disorders such as depression (Lewy et al. 2009), schizophrenia (Mansour et al. 2006), Alzheimer's (Chen et al. 2013), and some cardiovascular diseases (Tenkanen et al. 1997; Morris et al. 2012; Martino et al. 2008), to mention those most cited, are now linked to circadian clock genes, which is consistent with alterations in chronostatic regulation.

12.4 Integration of Regulatory Systems in Physiology

From the present perspective, the concept of chronostasis integrates circadian rhythmicity into physiology, not as a phenomenon but as the expression of a general regulatory system. Physiological regulation reveals itself as a much more complex process than was envisioned by Bernard or Cannon. Thus, the concept of chronostatic regulation builds up with the concepts of homeostasis and rheostasis to attempt to unravel yet another aspect of this complexity. The integrated model of physiological regulation proposed here involves at a primary level the homeostatic regulation of all physiological variables in a continuous manner, and at subsequent levels chronostatic and rheostatic adjustment of the set points from feedback loops regulating specific physiological variables (Fig. 12.4).

At any given time physiological parameters are under strict homeostatic regulation by means of negative feedback systems, as already described (see Sect. 12.1). Throughout the day, the values of the set points of physiological parameters are modified under the influence of circadian clocks. Thus, chronostatic regulation allows the circadian expression of physiological parameters and yet homeostatic processes continue operating at all times (Fig. 12.5). Rheostatic regulation adjusts the set point, independently from chronostasis, to behavioral state changes such as wakefulness and sleep, physiological status such as sexual maturation or preg-

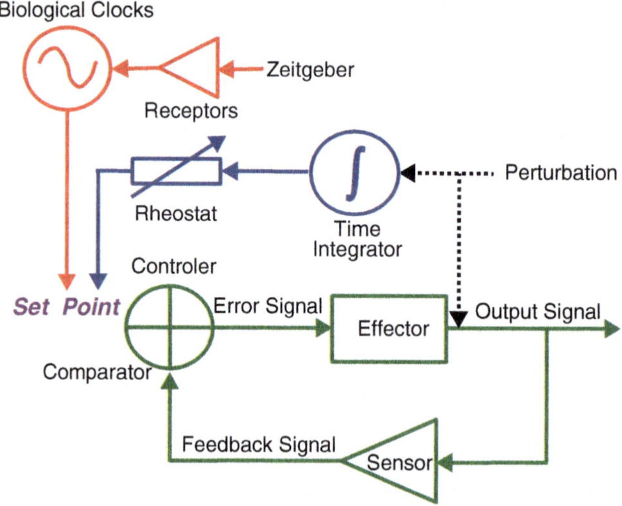

Fig. 12.4 Illustration of the integrative view of regulatory systems in physiology. At a primary level, the homeostatic regulation (shown in *green*) of all physiological variables operates in a continuous manner; at subsequent levels rheostatic regulation (*blue lines*) and chronostatic regulation (*red lines*) adjust the set point (*purple*) from feedback loops regulating specific physiological variables. In this scheme we include all biological clocks or oscillators, not only circadian elements, because the principles of time regulation could operate also at frequencies other than once a day (see also Fig. 12.6). Other elements of the scheme are as in Figs. 12.1, 12.2, and 12.3

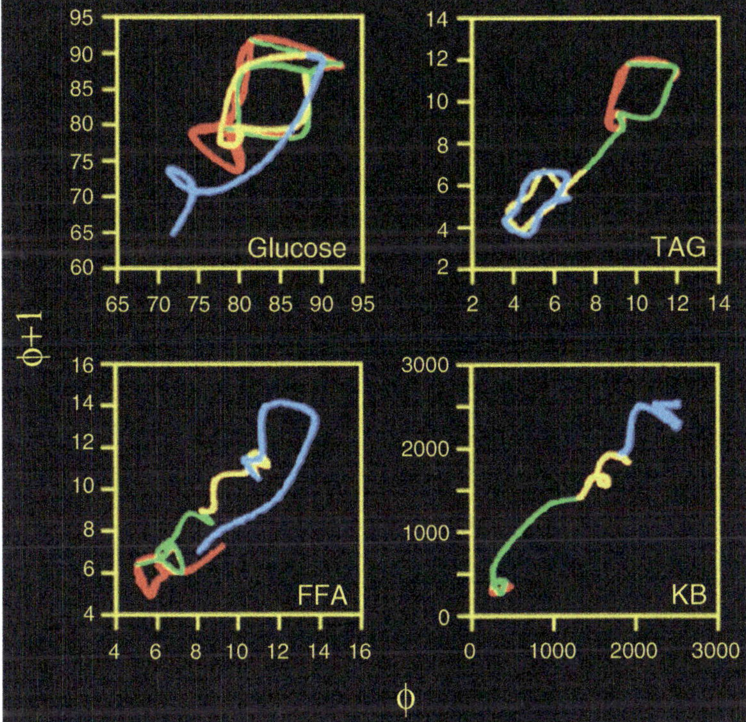

Fig. 12.5 Phase portraits of changes in plasma metabolites. Mean values (mg/dl at 3-h intervals) from fed rats during the night (*red*) and after 72-h fasting; each color represents successive 24-h fasting (*green, yellow, blue*). Glucose is clearly under chronostatic regulation, which sustains its plasma concentration within optimal values even after 60 h of fasting. Triacyl glycerides (TAG) are also under chronostatic regulation, but after only 12 h of fasting there is a change in the level of regulation upon which the chronostasis operates. Such change indicates the operation of a rheo-static regulation upon the TAG set point triggered by the fasting conditions. Free fatty acids (FFA) also show a chronostatic regulation under basal conditions but after 12-h fasting switch the limit cycle to 60 h oscillation. Finally, ketone bodies (KB) are under homeostatic regulation, kept at very low concentrations in basal conditions; shortly after fasting FAA increase their levels in direct relationship to the time of fast. Altogether, these results demonstrate a combination of the results predicted from the integrated model of physiological regulation shown in **c** from Figs. 12.1, 12.2, and 12.3

nancy, or chronic environmental changes leading to malnourishment or obesity; so that the physiological parameters, although expressing different values in each of these states, are continuously under strict homeostatic and chronostatic regulation (Fig. 12.6).

Because time regulation in human physiology as a result of the operation of circadian clocks is now widely known, the present model provides a congruent and integrated explanation of circadian phenomena with previous physiological concepts. The main unsolved issues regarding the concept of chronostasis remain related to the interaction of circadian clocks with the homeostatic and rheostatic

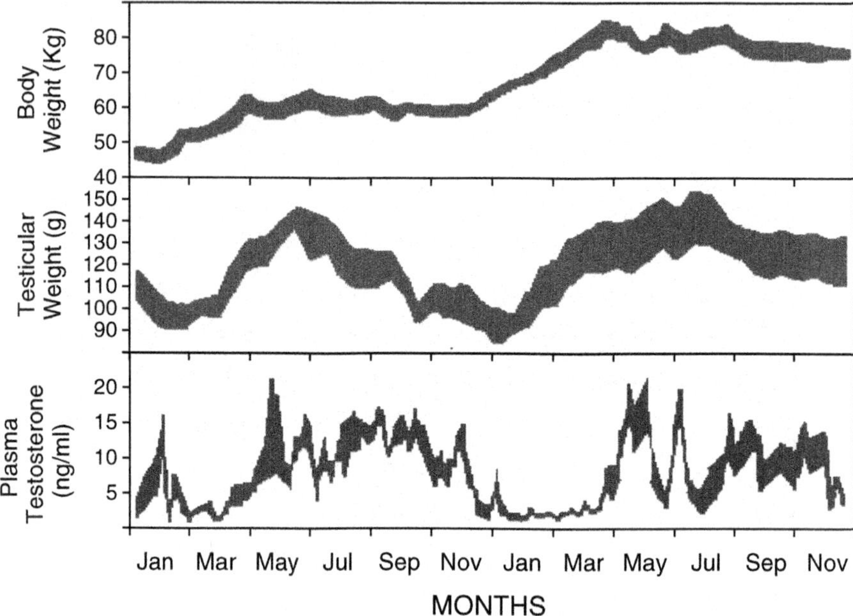

Fig. 12.6 Seasonal variations in body weight, testicular weight, and plasma testosterone from goats in natural photoperiods in northern México (26° N). Body weight illustrates rheostatic increases from January to April whereas homeostatic regulation maintains weight within a narrow range from May to December. Testicular weight and plasma testosterone show circannual rhythmicity that illustrates chronostatic regulation of reproduction with a period different from 24 h. Compare this graphs with **b** of Figs. 12.1, 12.2, and 12.3. These data suggest that the principle of chronostatic regulation could be valid for biological clocks other than circadian. The dependence of seasonal regulation of gonadal regression on photoperiod may also indicate interactions between rheostatic and chronostatic regulation processes that deserve further analysis. (Redrawn from Fig. 2 from Delgadillo et al. 1999)

regulation of physiological parameters. As previously mentioned, the set point is a useful concept to explain the level at which circadian clock(s) interact with the mechanisms of homeostatic regulation. SCN efferents to hypothalamic regions involved in endocrine and autonomic regulation are consistent with the model of chronostatic regulation acting upon the set point level of homeostatic regulation (Watts 1991). At a first level of complexity, the set point refers to the neuronal groups integrating autonomic and endocrine feedback information with information regarding an optimal value for the physiological parameter (Nakamura 2011; Bi et al. 2012; Tonsfeldt and Chappell 2011). At another level of complexity, the questions are whether (and how) the information of optimal values is stored by the cells regulating endocrine or autonomic functions, or whether the optimal value information is transmitted to these neurons from an alternative input (Fredholm et al. 2011; Nakamura 2011; Bi et al. 2012; Kotz et al. 2012; Tonsfeldt and Chappell 2011). Yet at another level of complexity, the question would be to determine the role of intracellular signaling pathways in gene expression as part of coding an optimal value for a physiological parameter (Firsov et al. 2011; Baeza-Raja et al. 2013; Goel et al. 2009; Lowrey and Takahashi 2004; Turek et al. 2005).

12.5 Concluding Remarks

As most conceptual models, the one proposed here is aimed (1) to provide new explanations of previous observations of general physiological regulation, collected in the past decades, (2) to induce hypothesis to verify the postulates and consequences of the model itself, and (3) to guide us in the collection and interpretation of new observations. Regarding the first issue, this model makes explicit the current view on how circadian rhythms influence physiological processes throughout the organism, and further emphasizes the operation of clocks as part of the regulatory system, thus differentiating it from programmed rheostasis. This model also refers to the implications of oscillation-related parameters, such as phase, amplitude, and phase relationships, to fully describe the physiological status of an individual, which parameters are not considered now in most textbooks of physiology, with the exception of hormonal patterns. With respect to the second and third issues, the assumption of three independent yet cooperative regulatory processes, homeostasis, chronostasis and rheostasis, which operate simultaneously on physiological systems, could provide a number of specific hypothesis susceptible to experimental verification. Examples of the latter are the studies of the interaction between homeostatic and circadian processes on sleep (Yasenkov and Deboer 2012) or the regulation of glucose metabolism (Kalsbeek et al. 2010). Current advances on neurobiology and gene regulation have shown a complex scenario for physiological regulation, which is characterized by intermingling of neuronal circuits and signaling cascades involved in apparently independent physiological processes such as sleep, glucose metabolism, temperature, and hydroelectrolytic balance. It is expected that the present model would contribute to unraveling the complexity of physiological regulation.

Acknowledgments The author thanks José Luis Chavez-Juarez and Ana Maria Escalante for critical reading of the manuscript and assistance obtaining bibliographic material. Supported by Grants IN-204811 from PAPIIT/DGAPA/UNAM and CB-2009-01-128528 from CONACyT.

References

Abe M, Herzog E, Yamazaki T et al (2002) Circadian rhythms in isolated brain regions. J Neurosci 22:350–356

Aguilar-Roblero R, Díaz-Muñoz M (2010) Chronostatic adaptations in the liver to restricted feeding: the FEO as an emergent oscillator. Sleep Biol Rhythms 8:9–17

Aschoff J (1955) Der Tagesgang der Korpertemperatur beim Menschen. Klin Wochenschr 33:545–551

Aschoff J (1965) Circadian rhythms in man. Science 148:1427–1432

Baeza-Raja B, Eckel-Mahan K, Zhang L et al (2013) p75 Neurotrophin receptor is a clock gene that regulates oscillatory components of circadian and metabolic networks. J Neurosci 33:10221–10234

Bernard C (1865) Introduction â l'etude de la Médicine Experimentale. J. B. Baillièère et Fils, Paris

Bi S, Kim J, Zheng F (2012) Dorsomedial hypothalamic NPY and energy balance control. Neuropeptides 46:309–314

Borbély A (1982) A two-process model of sleep regulation. I. Physiological basis and outline. Hum Neurobiol 1:195–204

Borgs L, Beukelaers P, Vandenbosch R et al (2009) Cell "circadian" cycle. New role for mammalian core clock genes. Cell Cycle 8(6):832–837

Boulant JA (2006) Neuronal basis of Hammel's model for set-point thermoregulation. J Appl Physiol 100:1347–1354

Boulos Z, Rosenwasser AM (2005) A chronobiological perspective on allostasis and its application to shift work. In: Shulkin J (ed) Allostasis, homeostasis and the costs of physiological adaptation. Cambridge University Press, Cambridge, UK, pp 228–301

Cabanac M (2006) Adjustable set point: to honor Harold T. Hammel. J Appl Physiol 100: 1338–1346

Cannon WB (1929) Organization for physiological homeostasis. Physiol Rev 9:399–431

Cannon WB (1936) The wisdom of the body. W.W. Norton, New York

Chen HF, Huang CQ, You C et al (2013) Polymorphism of CLOCK gene rs 4580704 COG is associated with susceptibility of Alzheimer's disease in a Chinese population. Arch Med Res 44:203–207

Daan S, Beersma DM, Borbély AA (1984) Timing of human sleep: recovery process gated by a circadian pacemaker. Am J Physiol 246(2 Pt 2):R161–R178

Delgadillo JA, Canedo GA, Chemineau P et al (1999) Evidence for an annual reproductive rhythm independent of food availability in male Creole goats in subtropical northern Mexico. Theriogenology 52:727–737

Firsov D, Tokonami N, Bonny O (2011) Role of the renal circadian timing system in maintaining water and electrolytes homeostasis. Mol Cell Endocrinol 349:51–55

Fredholm B, Johansson S, Wang Y-Q (2011) Adenosine and the regulation of metabolism and body temperature. Adv Pharmacol 61:77–94

Goel N, Banks S, Mignot E, Dinges DF (2009) PER3 polymorphism predicts cumulative sleep homeostatic but not neurobehavioral changes to chronic partial sleep deprivation. PLoS One 4:e5874

Hammel HT (1968) Regulation of internal body temperature. Annu Rev Physiol 30:641–710

Harris RBS (1990) Role of set-point theory in regulation of body weight. FASEB J 4:3310–3318

Jéquier E, Tappy L (1999) Regulation of body weight in humans. Physiol Rev 79:451–480

Jéquiere E (2002) Leptin signaling, adiposity, and energy balance. Ann N Y Acad Sci 967: 379–388

Kalsbeek A, Ruiter M, La Fleur SE et al (2006) The hypothalamic clock and its control of glucose homeostasis. Prog Brain Res 153:283–307

Kalsbeek A, Yi C-X, La Fleur SE, Fliers E (2010) The hypothalamic clock and its control of glucose homeostasis. Trends Endocrinol Metab 21:402–410

Klein DC, Moore RY, Reppert SM (1991) Suprachiasmatic nucleus: the mind's clock. Oxford University Press, New York, pp 467–472

Kotz C, Nixon J, Butterick T et al (2012) Brain orexin promotes obesity resistance. Ann N Y Acad Sci 1264:72–86

Lewy AJ, Emens JS, Songer JB, Sims N, Laurie AL, Fiala SC, Buti AL (2009) Winter depression: integrating mood, circadian rhythms, and the sleep/wake and light/dark cycles into a bio-psycho-social-environmental model. Sleep Med Clin 4:285–299

Lowrey PL, Takahashi JS (2004) Mammalian circadian biology: elucidating genome-wide levels of temporal organization. Annu Rev Genomics Hum Genet 5:407–441

Mansour HA, Wood J, Logue T, Chowdari KV, Dayal M, Kupfer DJ, Monk TH, Devlin B, Nimgaonkar VL (2006) Association study of eight circadian genes with bipolar I disorder, schizoaffective disorder and schizophrenia. Genes Brain Behav 5:150–157

Martino TA, Oudit GY, Herzenberg AM, Tata N, Koletar MM, Kabir GM, Belsham DD, Backx PH, Ralph MR, Sole MJ (2008) Circadian rhythm disorganization produces profound cardiovascular and renal disease in hamsters. Am J Physiol 294:R1675–R1683

Moore-Ede MC, Sulzman FM, Fuller CA (1982) The clocks that time us: physiology of the circadian timing system. Harvard University Press, Cambridge, MA, pp 448

Morris CJ, Yang JN, Scheer FA (2012) The impact of the circadian timing system on cardiovascular and metabolic function. Prog Brain Res 199:337–358

Mrosovsky N (1990) Rheostasis: the physiology of change. Oxford University Press, New York

Nakamura K (2011) Central circuitries for body temperature regulation and fever. Am J Physiol Regul Integr Comp Physiol 301:R1207–R1228

Nicolaïdis S (1977) Physiologie du comportement alimentaire. In: Meyer P (ed) Physiologie humaine. Flammarion, Paris, pp 908–922

Perreau-Lenz S, Pevet P, Buijs RM, Kalsbeek A (2004) The biological clock: the bodyguard of temporal homeostasis. Chronobiol Int 21:1–25

Romanovsky A (2004) Do fever and anapyrexia exist? Analysis of set point-based definitions. Am J Physiol Regul Integr Comp Physiol 287:R992–R995

Romanovsky A (2006) Thermoregulation: some concepts have changed. Functional architecture of the thermoregulatory system. Am J Physiol Regul Integr Comp Physiol 292:R37–R46

Rosenblueth A, Wiener N, Bigelow J (1943) Behavior, purpose and teleology. Philos Sci 10: 18–24

Roth J, Rummel C, Barth S, Gerstberger R, Hübschle T (2006) Molecular aspects of fever and hyperthermia. Neurol Clin 24:421–439

Russek M, Cabanac M (1983) Regulacion y Control en Biología. Instituto Politécnico Nacional, México, 148

Sahar S, Sassone-Corsi P (2012) Regulation of metabolism: the circadian clock dictates the time. Trends Endocrinol Metab 23:1–8

Speakman J, Levitsky D, Allison D (2011) Set points, settling points and some alternative models: theoretical options to understand how genes and environments combine to regulate body adiposity. Dis Model Mech 4:733–745

Sterling P, Eyer J (1988) Allostasis: a new paradigm to explain arousal pathology. In: Fisher S, Reason J (eds) Handbook of life stress cognition and health. Wiley, New York, pp 629–650

Stevens RG, Hansen J, Costa G et al (2011) Considerations of circadian impact for defining 'shift work' in cancer studies: IARC Working Group Report. Occup Environ Med 68:154–162

Tenkanen L, Sjoblom T, Kalimo R, Alikoski T, Harma M (1997) Shift work, occupation and coronary heart disease over 6 years of follow-up in the Helsinki Heart Study. Scand J Work Environ Health 23:257–265

Tonsfeldt K, Chappell P (2011) Clocks on top: the role of the circadian clock in the hypothalamic and pituitary regulation of endocrine physiology. Mol Cell Endocrinol 349:3–12

Turek FW, Joshu C, Kohsaka A et al (2005) Obesity and metabolic syndrome in circadian Clock mutant mice. Science 308:1043–1045

Viola A, Archer S, James L (2007) PER3 polymorphism predicts sleep structure and waking performance. Curr Biol 17:613–618

Watts AG (1991) The efferent projections of the suprachiasmatic nucleus: anatomical insights into the control of circadian rhythms. In: Klein DC, Moore RY, Reppert SM (eds) Suprachiasmatic nucleus: the mind's clock. Oxford University Press, New York, pp 77–106

Werner J (2010) System properties, feedback control and effector coordination of human temperature regulation. Eur J Appl Physiol 109:13–25

Wiener N (1948) Cybernetics or control and communication in the animal and the machine. MIT Press, Cambridge, MA, p 212
Woods SC, Ramsay DS (2007) Homeostasis: beyond Curt Richter. Appetite 49:388–398
Wright CL, Burgoon P, Bishop P, Boulant JA (2008) Cyclic GMP alters the firing rate and thermosensitivity of hypothalamic neurons. Am J Physiol Regul Integr Comp Physiol 294: R1704–R1715
Yamazaki S, Numano R, Abe M et al (2000) Resetting central and peripheral circadian oscillators in transgenic rats. Science 288:682–685
Yamazaki S, Straume M, Tei H, Sakaki Y, Menaker M, Block GD (2002) Effects of aging on central and peripheral mammalian clocks. Proc Natl Acad Sci U S A 99:10801–10806
Yasenkov R, Deboer T (2012) Circadian modulation of sleep in rodents. Prog Brain Res 199: 203–218

Chapter 13
Circadian Rhythm and Food/Nutrition

Yu Tahara and Shigenobu Shibata

Abstract Scheduled food access during the daytime for nocturnal mice or rats entrains the food-entrainable oscillator (FEO) in the brain and the food-entrainable peripheral oscillator (FEPO) in the peripheral tissues. FEO and FEPO are not regulated by the central clock, which is in the suprachiasmatic nucleus. FEO produces food anticipatory activity (FAA) 2–3 h before the scheduled feeding time initiates. FEPO produces entrainment in the rhythm of peripheral clock gene expression and in the rhythm of food-metabolic functions in peripheral organs. At present, the mechanisms of the FEO and FEPO are not completely understood, despite many studies that have been performed in this field. In addition, circadian clocks affect metabolism of nutrition (absorption, distribution, metabolism, excretion). In this review, we describe and review the characteristics and biological implications of FEO and FEPO and the mechanism of metabolism of nutrition with day–night differences. We call this relationship between nutrition and chronobiology "Chrono-nutrition," which is an important study field to understand how our body clocks contribute to our health throughout the day.

13.1   Introduction

The rotation of the earth around its axis causes 24-h changes in environmental conditions, including temperature, food availability, light, and darkness. In addition, seasonal changes in the length of day occur because of the earth orbiting the sun. To cope with and anticipate these changes, most organisms across the plant and animal kingdoms possess a circadian timing system. The circadian timing system in mammals is divided into two systems: suprachiasmatic nucleus (SCN)-dependent rhythm and SCN-independent rhythm (Mistlberger 2009, for review). Environmental light–dark signals can entrain the SCN circadian rhythm, and the SCN governs the

Y. Tahara • S. Shibata (✉)
Laboratory of Physiology and Pharmacology, School of Advanced Science and Engineering, Waseda University, Twins, Wakamatsu-cho 2-2, Shinjuku-ku, Tokyo 162-8480, Japan
e-mail: shibatas@waseda.jp

© Springer International Publishing Switzerland 2015
R. Aguilar-Roblero et al. (eds.), *Mechanisms of Circadian Systems in Animals and Their Clinical Relevance*, DOI 10.1007/978-3-319-08945-4_13

circadian rhythm in the entire body through the light entrainable oscillator (LEO) (Meijer et al. 2007, for review). Therefore, ablation of the SCN disrupts almost all circadian phenomena, such as locomotor activity and body temperature rhythms. Clock genes in mammals were discovered in 1997, and the expression patterns of these genes have since been elucidated (King et al. 1997). In mice and rats, circadian clock gene expression was observed not only in the SCN but also in other brain areas, such as the cerebral cortex and hippocampus, and in peripheral organs such as the liver, heart, lung, and kidney (Dardente and Cermakian 2007). Genes involved in rhythm generation, such as *Clock, brain and muscle Arnt-like protein-1 (Bmal1), cryptochrome (Cry)*, and *period (Per)*, form negative feedback loops that result in circadian rhythm expression in the SCN as well as peripheral tissues (Gachon et al. 2004).

The synchronization, or entrainment, of the circadian rhythm with the environmental cycle under natural light–dark conditions is usually achieved by making daily adjustments in the phase and period of the circadian oscillator (Daan 2000). The effect of light on the circadian rhythm in locomotor activity has been intensively studied in a large number of species (Daan and Pittendrigh 1976; Honma et al. 1985). Short light pulses of a few minutes, when applied during the night, can adjust the phase of the circadian rhythm. The generation of phase shifts is dependent on the time when the light pulses are given within the circadian cycle. Although there is no phase response to light pulses during the day, phase delays in the rhythm occur after light exposure during early night, and phase advances occur after light pulses during the late night (Pittendrigh 1988).

Scheduled food (SF) during the day increases locomotor activity and induces the food anticipatory activity (FAA) rhythm, which is defined as locomotor activity that increases 2–3 h before the restricted feeding time begins (Mistlberger 2009). Because FAA persists in SCN-ablated animals, the food-entrainable oscillator (FEO) that regulates FAA is believed to be an SCN-independent circadian timing system. In addition to FAA, daily SF can also entrain peripheral circadian clocks, the food-entrainable peripheral oscillator (FEPO) (Damiola et al. 2000; Hara et al. 2001; Stokkan et al. 2001) (Fig. 13.1).

Drugs also have the power to entrain circadian systems. Some drugs such as benzodiazepines and melatonin receptor agonist can entrain circadian rhythms. Many chemical compounds such as CK1 inhibitors and GSK3B inhibitors have been reported to affect the period of circadian rhythms. CK1 inhibitors prolong the circadian period, and GSK3b inhibitors such as lithium shorten it. On the other hand, the circadian clock influences the main effects and side effects of some drugs. Because many proteins or receptors have transcriptional or translational circadian rhythms, drug administration times should be considered in order to maximize the desired effects of the drug and reduce side effects. For example, antiasthmatic drugs are preferably taken early in the morning because airway obstruction is high during this time. As another example, antihyperlipidemia drugs such as statins are ideally taken early in the evening because these drugs inhibit HMG CoA reductase, the

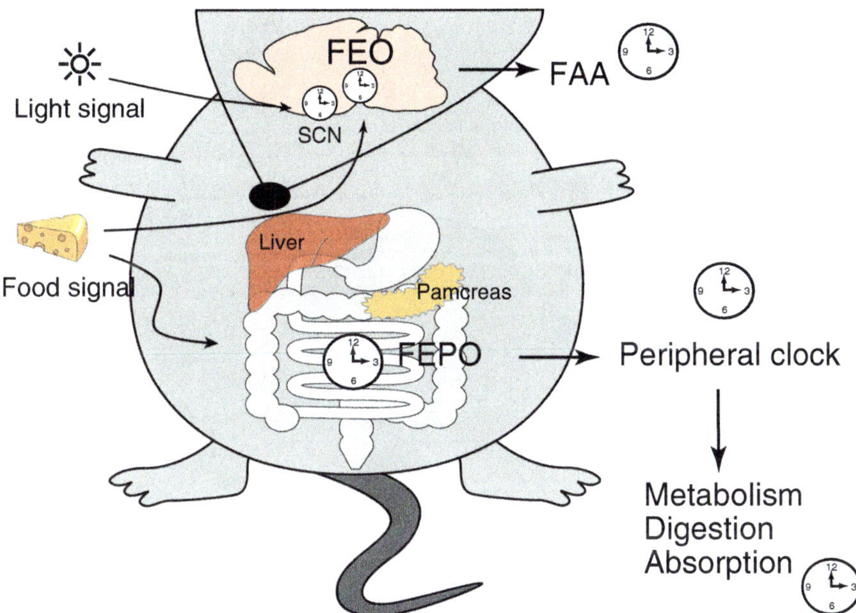

Fig. 13.1 Schematic diagram explaining the food-entrainable oscillator (FEO) and food-entrainable peripheral oscillator (FEPO). Light signals can entrain the light entrainable oscillator (LEO) by activating the suprachiasmatic nucleus (SCN). On the other hand, scheduled feeding can entrain the FEO or FEPO. The FEO produces food anticipatory activity (FAA), which appears 2–3 h before food timing. The FEPO produces phase entrainment of clock gene expression rhythms and clock-regulated metabolic gene expression rhythms in peripheral tissues

rate-limiting cholesterol synthase enzyme, whose activity is high during this time. Thus, in the past decade, scientific study of the connection between the circadian clock and pharmacology, chrono-pharmacology, has become a common medical research field.

Similarly, there are interactions between nutrients and the circadian system, which we call chrono-nutrition (Fig. 13.2). As an example of chrono-nutrition, SF can act as a signal to entrain the circadian system in rodents. The effect of food entrainment in humans is still unknown because of the lack of reports to date. However, eating times may function as an entrainment factor in humans. Because there are circadian variations in metabolic gene expression throughout the day, the food content and meal timing may affect health. Late night dinners increase fat synthesis and phase-shifted clock gene expression rhythms in peripheral clocks. Just as bright light exposure therapy in the morning has become a common treatment for circadian sleep disorders or seasonal affective depression, food-timing therapy may become a common therapy for many circadian disorders.

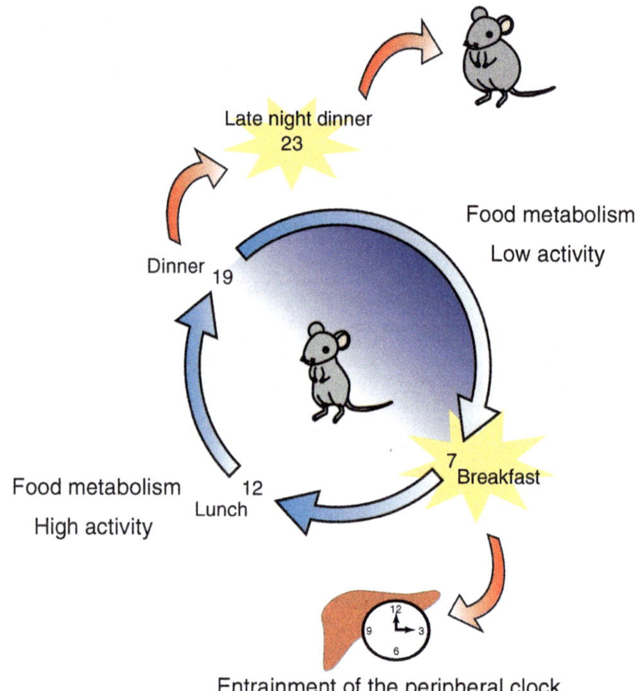

Fig. 13.2 Schematic diagram explaining chrono-nutrition. Chrono-nutrition is a new field of study, like chrono-pharmacology. There are two aspects. (1) Time-restricted food access can be an entrainable factor for the circadian system. (2) Peripheral clocks regulate day–night differences of clock-regulated metabolic genes. Therefore, food timing is an important factor for maintaining health

13.2 Mechanism of the Food-Entrainable Oscillator in the Brain

13.2.1 Role of the Dorsomedial Hypothalamus in the FEO

Neural centers in the hypothalamus are putative sites of the FEO in the brain. These centers are central to the regulation of energy homeostasis and arousal; they include the arcuate nucleus, paraventricular nucleus, dorsomedial hypothalamus (DMH), and ventromedial hypothalamus (Cone et al. 2001; Rodgers et al. 2002; Saper et al. 2005a, b; Ribeiro et al. 2007; Gooley et al. 2006). Daily SF can entrain the peripheral circadian and cerebral cortex clocks. In contrast, the SCN clock uses the light–dark (LD) cycle as the entraining stimulus (Castillo et al. 2004); it cannot be entrained by SF conditions. SF-induced entrainment of the peripheral clock was observed under constant lighting with ablation of the SCN (Damiola et al. 2000; Hara et al. 2001; Stokkan et al. 2001; Wakamatsu et al. 2001). After starvation, re-feeding increased *Per1* and *Per2* gene expression in the DMH of mice

(Moriya et al. 2009; Mieda et al. 2006). These studies suggest that the DMH may be the brain region responsible for FAA formation. In other experiments, behavioral FAA rhythm persisted in rats (Landry et al. 2006, 2007) and mice (Moriya et al. 2009; Tahara et al. 2010) with lesions in the DMH. However, Gooley et al. (2006) reported that a cell-specific lesion in the DMH blocked SF-induced FAA and body temperature rhythms in rats. Therefore, the role of the DMH in FAA maintenance remains controversial. In a recent study, we demonstrated that a relatively large lesion in the medial basal hypothalamus (MBH) attenuated FAA, suggesting that MBH regions, including the DMH, arcuate nucleus, and ventromedial hypothalamus, may be involved in the formation of FAA (Tahara et al. 2010).

13.2.2 *Role of Limbic Brain Areas in the FEO*

SF shifts the phase of clock gene mRNA expression in the cerebral cortex and hippocampus as well as PER2 protein levels in the bed nucleus, stria terminalis, amygdale, and dentate gyrus of mice (Wakamatsu et al. 2001; Verwey et al. 2007; Shibata et al. 2009). Limited daily access to sucrose or saccharine in freely fed rats or scheduled access to saline in sodium-deprived rats had no effect on PER2 rhythms (Waddington Lamont et al. 2007). These data suggest that PER2 rhythms in the limbic forebrain are sensitive to signals that arise from the alleviation of a negative metabolic state associated with SF. Therefore, access to rewarding substances in the absence of food deprivation or metabolic challenge is not sufficient to alter the rhythm of PER2 protein in the limbic forebrain (Amir and Stewart 2009). Gestation appears to play some role in the protein level of PER2 because rats and hamsters entrain to the presence of palatable foods and demonstrate FAA even when regular rodent chow is available ad libitum (Mendoza et al. 2005).

13.2.3 *Role of Other Brain Areas in the FEO*

Ablation of sensory inputs that carry gustatory and satiety sensation appears to influence circadian FAA (Mistlberger 1994) because FAA is attenuated in the ghrelin receptor-deficient mouse (Lesauter et al. 2009). Destruction of the hindbrain parabrachial region, which processes visceral and gustatory information, also attenuates FAA (Davidson et al. 2000). These data suggest that the biological clock within the SCN that regulates circadian patterns of appetite and feeding is itself sensitive to the presence or absence of food. To investigate the brain regions that are involved in FAA, Nakahara et al. (2004) examined c-Fos expression before and after feeding in rats subjected to restricted feeding (RF) for 2 h. The thalamic paraventricular nucleus (PVT) was the only region in which c-Fos expression was higher before feeding than after feeding. After FAA was established, lesions created in the PVT attenuated this rhythm but did not affect the LD-entrained rhythm.

The anticipatory increase of blood corticosterone levels was not established in long-term PVT-lesioned rats. These results suggest that the PVT is involved in the expression of FAA under a SF regimen.

Recently, Mendoza et al. (2010) reported that FAA was markedly reduced in mice injected intracerebroventricularly with an immunotoxin that depletes Purkinje cells in the cerebellum (OX7-saporin). In addition, Grid2 (ho/ho) mice did not show any FAA or SF-induced changes in cerebellar clock gene expression. However, in hypothalamic arcuate and DMH, shifts in PER1 expression in response to SF were similar in the cerebellar mutant and wild-type mice. The study concluded that the cerebellar oscillator is required for FAA formation.

13.2.4   Role of Clock Genes in the FEO

FAA rhythm formation was attenuated by mutations in the $Per2^{Brdm1}$ gene but not the $Per1^{Brdm1}$ gene (Feillet et al. 2006). In contrast to the phenotype in $Per2^{Brdm1}$ mutant mice, no obvious alteration of behavioral FAA has been reported in mice carrying $Per2^{ldc}$, which is considered a null-mutant allele (Storch and Weitz 2009). This difference may be due to the redundant function of paralog genes and/or compensatory developmental mechanisms present in the mice bearing the loss-of-function mutations, which are absent in the mice bearing point mutations or a short deletion in the gene (Takahashi et al. 2008, for review). $Per2^{ldc}$ mutation, but not $Per2^{Brdm1}$, may compensate for $Per1$ and/or $Per3$, although the mechanism for this is still unknown. Fuller et al. (2008) also recently reported that cell-specific downregulation of $Bmal1$ gene expression in the DMH reduced FAA formation. Our data, showing that mice with a $Cry1/Cry2$ double mutation exhibit FAA with several unstable changes, support these findings (Iijima et al. 2005). However, $Bmal1$ knockout mice exhibit normal FAA rhythm formation (Mistlberger et al. 2008). Multiple food-entrainable circadian clocks have been discovered in the brain and periphery, raising strong expectations that one or more of them underlie FAA. Mutant mice lacking a known circadian clock function ($Bmal1$, $Per1$, $Per2$, $Clock$, and $Per1/Per2$ double mutants) in all tissues exhibit normal FAA in an LD cycle and in constant darkness (Storch and Weitz 2009). Interestingly, a T-cycle experiment, exposing mice to non-24-h cycles, showed that FAA was entrained by $T = 20$ h to $T = 28$ h in $Bmal1$ mutant mice but not in wild-type control mice (Takasu et al. 2012). Therefore, normal clock gene expression is necessary to entrain to SF stimulation with a circadian cycle range, suggesting that the FEO is dependent on the known circadian clock. These studies indicate that FAA is the output of an oscillator or the output of a known circadian clock oscillator. As mentioned above, only a few clock genes have been considered within the framework of FEO. Specifically, the involvement of important components of the current model for molecular clockwork, such as $Dec1$, $Dec2$, Rev-$erb\alpha/\beta$, and $Ror\alpha/\beta$, remains to be investigated. The latter two genes, members of the orphan nuclear receptor superfamily, encode negative and positive regulators of transcription, respectively. These transcription factors are intracellular sensors for circulating lipids; they participate in diverse pathways of lipid and carbohydrate metabolism in the liver, skeletal muscle, and adipose tissues (Duez and Staels 2008). *Rev-erbs* and *Rors* genes are, therefore, involved in the circadian control of cellular metabolism (Yang et al. 2006; Teboul et al. 2008). It would be interesting to investigate whether these genes are necessary for the formation of FAA.

13.2.5   Role of NMDA Receptors in the FEO

Glutamate receptors are divided into three types depending on the secondary signal transduction pathway used: AMPA, NMDA, and mGluR. NMDA receptor antagonists impair learning/memory in passive avoidance, water maze, and radial maze learning performance. In circadian rhythm experiments, the NMDA receptor antagonist MK-801 attenuated the light-induced phase shift of behavioral rhythm and light-induced *Per1* gene expression in the SCN (Moriya et al. 2000). Several days are required to establish the FAA, suggesting that the process of FAA formation is involved in the learning/memory process. In order to examine the role of NMDA receptor activation in the acquisition of FAA, we administered MK-801 to rats every day after 4 h of restricted feeding. MK-801 strongly inhibited FAA formation (Ono et al. 1996). Injection of MK-801 immediately before re-feeding attenuated feeding-induced *Per1* gene expression in the mouse DMH (Moriya et al. 2009). Taken together, these data suggest that feeding-induced NMDA receptor activation is involved in FAA formation and increased *Per1* gene expression.

13.2.6   Role of the Arousal Process in the FEO

Orexin knockout or destruction of orexin-positive cells impairs the maintenance of arousal and food-seeking behavior in mice (Akiyama et al. 2004; Mieda et al. 2004). These data suggest that activation of orexin neurons may be involved in FAA rhythm formation because it was attenuated by the destruction of orexin-positive cells (Akiyama et al. 2004) (Fig. 13.3). When orexin knockout mice were placed on an SF schedule, core temperature and activity entrained to the feeding schedule. Therefore, orexin is not required for the entrainment of activity and temperature to an SF schedule, but is required for the robust expression of gross locomotor activity seen in anticipation of the scheduled feeding (Kaur et al. 2008).

13.2.7   Role of Aging in the FEO

Aging affects learning/memory formation, and, in general, aging animals exhibit impaired learning/memory. Aging rats (≥ 1.5 years old) showed impaired FAA formation (Shibata et al. 1994). In other experiments, the effects of aging on the

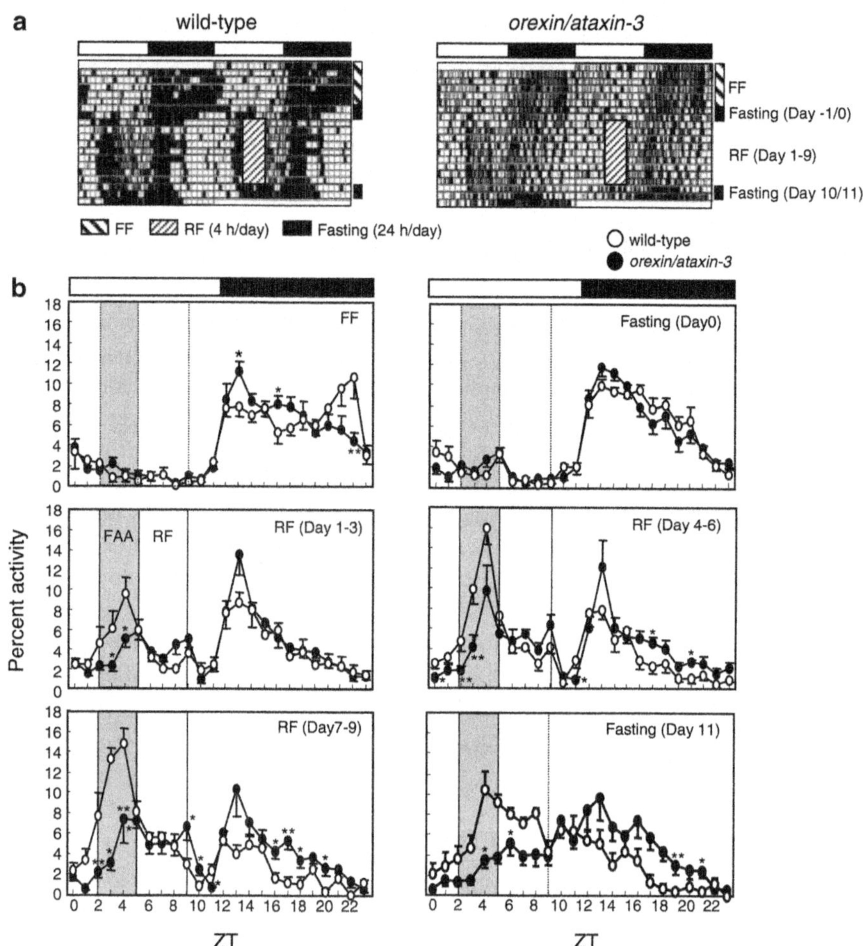

Fig. 13.3 Reduced FAA formation in orexin neuron-ablated mice (data from Akiyama et al. 2004). (**a**) Representative double-plotted actograms of the locomotor activity of restricted feeding (RF)-treated, wild-type (*left*), and orexin/ataxin-3 (*right*) mice (C57BL/6 background and 12 weeks old). The *hatched box* indicates RF treatment for 9 days (Zeitgeber time (ZT) 5–9 of days 1–9), during which food was available from ZT 5–9. Mice were fasted for 2 days before (days −1 and 0) and after (days 10 and 11) RF (fasting). (**b**) Averaged activity counts per hour as a fraction of the activity count in one whole day were plotted for wild-type (s, $n=5$) and orexin/ataxin-3 (days, $n=5$) mice. FF, averaged count under free-feeding (FF) conditions; fasting (day 0) and fasting (day 11), averaged count of fasted mice on day 0 and day 11, respectively; RF (days 1–3, days 4–9, and days 7–9), averaged activity count for each 3 consecutive days during RF treatment; $*P<0.05$; $**P<0.01$ vs. wild-type mice (Student's t-test)

development and persistence of feeding-associated circadian rhythms were examined with respect to the pre-feeding plasma corticosterone peak levels under an RF schedule. Aging (12- to 20-month-old rats) impaired the development and persistence of feeding-associated circadian rhythm in rats (Walcott and Tate 1996).

We examined the attenuation effect of bifemelane hydrochloride (a drug used in the treatment of senile dementia) on the impairment of FAA using aged and MK-801-treated rats (Shibata et al. 1995). FAA was impaired in 24-month-old rats and MK-801-treated rats. Daily injections of bifemelane or aniracetam (an experimental drug with potential cognition-enhancing effects) for six successive days significantly attenuated the impairment of FAA in a dose-dependent manner (Shibata et al. 1995; Tanaka et al. 2000). These results suggest that bifemelane and aniracetam enhance learning and memory performance, such as spatial and temporal perception.

13.2.8   Role of Nutrients in the FEO

Several circulating metabolic and hormonal factors driven by the SCN are also driven by food intake, including glucose, free fatty acids, glucocorticoid, thyroid hormones, and other factors (Escobar et al. 2009). Le Minh et al. (2001) demonstrated that adrenalectomized rats entrained faster to SF, suggesting that an induced corticosterone rhythm alone is not sufficient to drive FAA. At the behavioral level, glucose was the strongest entraining factor related to feeding schedules. In addition, a caloric-rich diet had a stronger influence on FAA induction than a fiber-rich or fat-rich diet (Stephan 1997). Because of the direct link to glucose, insulin secretion has also been examined for a potential role in FAA. However, because insulin did not induce or influence FAA, it has been disregarded as a main entraining signal for peripheral oscillators (Davidson and Stephan 1999; Kudo et al. 2004). It is worth mentioning here that mice fed a ketogenic diet show a phase-advanced rhythm of locomotor activity (Shirai et al. 2007; Oishi et al. 2009). Other food components, including sodium chloride, cholesterol, and proteins, do not have the capacity to phase-shift clock gene oscillation (Mohri et al. 2003; Iwanaga et al. 2005). Under ketogenic diet conditions, shortage of energy from carbon hydrate may happen in the SCN, and AMPK activation may affect CRY1 degradation (Lamia et al. 2009), resulting in free-running period changes. This may be one of the mechanisms by which ketogenic diet causes phase advance.

13.2.9   Role of Ghrelin in the FEO

Ghrelin is an orexigenic peptide predominantly secreted from the stomach. It acts on hypothalamic targets such as the ventromedial hypothalamus or arcuate nuclei to trigger food intake (Zigman et al. 2006). Plasma ghrelin is increased in calorie-restricted laboratory rats (Johansson et al. 2008) and gray mouse lemurs (Giroud et al. 2009). Stimulation of the ghrelinergic system can affect SCN activity in rodents (Yannielli et al. 2007). Interestingly, studies in mice with a targeted mutation in the ghrelin receptor gene show a reduction in FAA formation (Blum et al. 2009; LeSauter et al. 2009). Furthermore, ghrelin has phase-advancing effects in fasted but not fed mice (Yannielli et al. 2007). Therefore, ghrelin is considered a chronomodulating hormone possibly involved in the changes of activity timing during calorie restriction.

13.2.10 Role of Multiple Meal Times in the FEO

SF once per day has been a useful tool for investigating the FEO, but SF may not accurately represent the food availability patterns of nocturnal rodents in the wild. An alternative approach to assess the effects of multiple feeding times like in the wild is to schedule smaller meals throughout the day. Previously, meals were delivered 2–6 times per day to understand whether multiple FEO pacemakers are used to track each mealtime. Rats exhibit FAA when fed two meals per day separated by 5 h or more; both activity bouts exhibited properties of circadian timing rather than interval timing (Mistlberger et al. 2009; Stephan 1989a, b; Davidson et al. 2003). Some rats displayed partial FAA when they received two or three meals per day, but none anticipated all three meals simultaneously (Stephan 1989a). In addition, delivery of six meals per day did not lead to the same development of coordinated FAA (Mendoza et al. 2008). Interestingly, rats exhibited FAA twice with two meals per day (daytime and nighttime) and showed entrainment in the liver and gastrointestinal organs that was consistent with only the nighttime meal (Davidson et al. 2003). However, this study did not explain why only the nighttime meal entrained the peripheral clock or why the peripheral organ clock displayed a unimodal peak, but not a bimodal peak, rhythm when the rats were fed two meals per day. Because wild animals and humans have more complex feeding styles, it is necessary to examine the effect of multiple feeding times and volumes on SF-induced entrainment. In the future, we will examine the effect of two or three meals per day on FAA formation by using different food volumes at each meal as well as different starvation intervals between each mealtime.

13.3 Mechanism of the Food-Entrainable Peripheral Oscillator

13.3.1 Role of the Hypothalamus in the FEPO

SCN lesions do not affect FAA formation. However, it is unclear whether SCN lesions affect the SF-induced entrainment of the liver circadian clock. We discovered that SF causes a phase advance in the expression rhythm of the liver clock genes in mice with or without an SCN (Hara et al. 2001). Our recent experiments demonstrated that DMH lesions do not affect the SF-induced phase shift of *Per2* gene expression in the cerebral cortex and arcuate nucleus (Moriya et al. 2009). Therefore, the DMH may not be involved in restricted feeding-induced entrainment of the brain clock. When we monitored *Per2* gene expression rhythm using PER2::LUCIFERASE knock-in mice, restricted feeding-induced phase advance of *Per2* gene expression in the liver was unaffected by MBH lesions (Tahara et al. 2010).

13.3.2 Effect of Food Volume, Starvation Period, and Food Frequency on the FEPO

Restricted feeding for 4 h during daytime caused phase advance of the mouse liver clock (Feillet et al. 2006); hence, we measured food intake for 4 h after a 24-h fasting regimen in PER2::LUCIFERASE mice (Hirao et al. 2009). Consumption of food caused phase advancement of the liver clock in a food volume-dependent manner. Under our experimental conditions, fasting itself caused a marked increase in phase advancement of the liver clock. The liver clock phase was advanced in a feeding day-dependent manner. Previous studies showed that the volume of food produced large phase shifts in the FAA rhythm in rats; these shifts occurred in a food volume-dependent manner, whereas substitution of the diet with non-nutritive bulk food failed to produce phase shifts (Stephan and Davidson 1998). When rats were entrained to 20 g of food per day, approximately 6 g was necessary to produce phase shifts (Stephan and Davidson 1998). In our present study, we found that phase advance of the liver clock was dependent on the volume of food (0.7, 1.4, or 2.0 g/day). Mice consume 4–6 g of food per day under free-feeding conditions, and 1.4 g of food per day is required to cause phase shifts. Therefore, we estimated that at least 30 % of the food intake under free-feeding conditions is required to cause phase shifts in the liver clock in mice and rats. After the mice were fed 1.4 g of food per day, a marked increase in the phase advance of the bioluminescence rhythm was observed on the second day but not the first day. A previous study also reported clear phase shifts induced by restricted feeding schedules on the second day of daytime feeding (Le Minh et al. 2001). Therefore, entrainable stimulation for at least 2 days may be necessary to produce a phase advance, and entrainable stimulation for 4 days is needed to produce a maximum phase advance of the liver clock.

When mice are fed two or three meals per day, food administered after longer intervals of fasting has more power to entrain peripheral clock phase than food administered after other food timings. In fact, mice fed at Zeitgeber time (ZT) 12, 18, and 1 were entrained by food at ZT 12, because food at ZT12 came 11 h after the last meal at ZT1, but the other meals came 6 or 7 h after the last meal (Kuroda et al. 2012). However, mice fed at ZT 12, 18, and 4 were entrained by food at ZT 4, because food at ZT 4 came 10 h after the last meal at ZT 18, but food at ZT 12 came 8 h after the last meal. In this paper, food timing mimicked human eating patterns: food at ZT 12 as breakfast, food at ZT 18 as lunch, and food at ZT 1 or ZT 4 as dinner, with ZT 4 as late dinner, in particular. This suggests that dinner has the power to the change clock phase if dinner occurs at late nighttime. Such a phase change may cause late dinner-induced obesity.

13.3.3   Effect of Nutrients on the FEPO

Glucose induces phase shifts in peripheral tissue clocks. In diabetic rats lacking insulin, the phase of cardiac circadian gene expression is advanced by approximately 3 h (Young et al. 2002), suggesting that high blood glucose levels can cause phase shifts in the peripheral tissue clocks. Clock resetting by downregulation of *Per1* and *Per2* has been observed in Rat-1 fibroblasts treated with glucose (Hirota et al. 2002). Glucose caused phase shifts in locomotor activity, such as FAA, when the rats were treated with SF (Stephan and Davidson 1998). The phase shift induced by glucose is probably not due to the higher energy content because the consumption of vegetable oil did not produce phase shift in rats on SF schedules (Stephan and Davidson 1998). Taken together, these data suggest that glucose may be at least one factor for peripheral tissue clock gene entrainment, as well as FAA, during SF conditions. We recently examined the effect of carbohydrates, including sugar and starch, on the entrainment of the circadian liver clock by using PRE2::LUCIFERASE knock-in mice (Hirao et al. 2009). Humans generally consume an appropriately balanced diet containing starch, protein, and oil, but not simple nutrients, including glucose. Therefore, to apply our study results to humans, appropriate dietary guidelines were developed. In order to elucidate the role of nutrition in inducing phase shifts, components of the AIN-93M diet were partially or completely substituted. In particular, we focused on cornstarch and sugar because glucose is a good candidate for inducing entrainment signals (Hirao et al. 2009).

It has been reported that fasting upregulates *Per1* and downregulates *Per2*, *Dec1*, and *Bmal1* in the mouse liver and that re-feeding prevents this fasting-induced change in gene expression (Kawamoto et al. 2006; Kobayashi et al. 2004). In our experiments, liver *Per2* gene expression markedly increased when mice were re-fed after 24 h of starvation (Tahara et al. 2011). The increase of *Per2* by re-feeding was dependent on the release of insulin from the pancreas. In addition to *Per2*, we also detected downregulation of *Rev-erbα* in the liver (Fig. 13.4). Therefore, the acute changes in clock gene expression that occur after re-feeding may cause a phase resetting of the peripheral clock.

13.3.4   Effect of Nutrient Substitution on the FEPO

Each component of the control AIN-93M diet was completely substituted as a single nutrient (Hirao et al. 2009). An insignificant, weak phase advance of the liver clock was observed with 100 % cornstarch. Similarly, an insignificant, weak phase advance of the liver clock was observed in mice administered 0.3 ml of soybean oil on the first day and 0.43 ml of soybean oil on the second day, compared to the mice from the 2-day fasting group. Because neither a 100 % glucose diet nor a 100 % sucrose diet caused phase advances, other components in AIN-93M may cooperatively play a role in the action of sugar on the phase advance of the liver clock (Fig. 13.5). The combination of 86 % glucose with 14 % casein significantly ($P < 0.05$ vs. 100 % glucose) increased the phase advance of the liver bioluminescence rhythm.

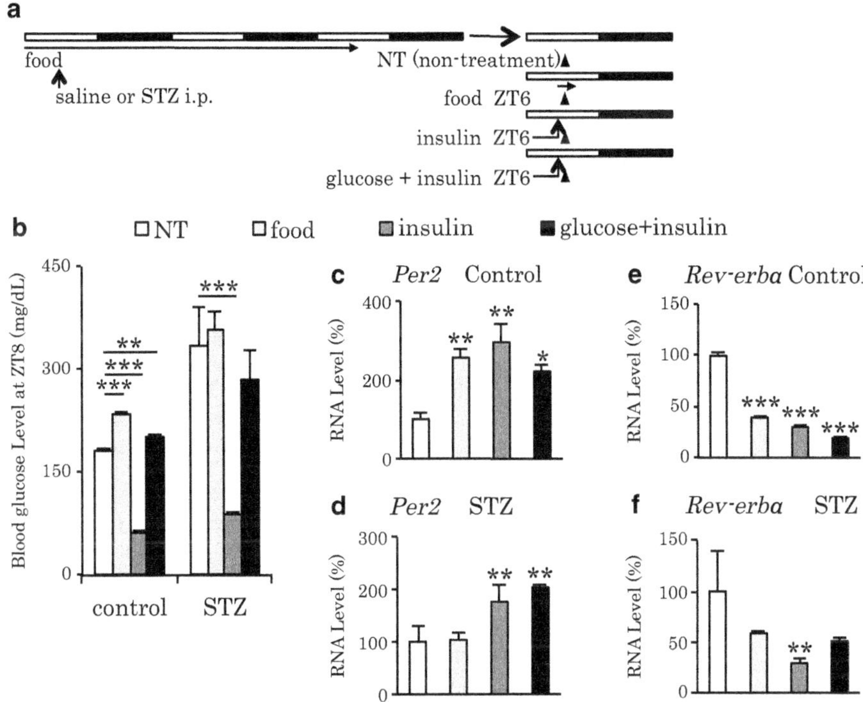

Fig. 13.4 Insulin-dependent, re-feeding-induced upregulation of *Per2* and downregulation of *Rev-erba* mRNA in the liver (data from Tahara et al. 2011). (**a**) Experimental schedule. Mice were treated with saline as control or streptozotocin (STZ; 200 mg/kg, i.p.) on day 0. On day 4, we prepared four groups. After 24-h fasting, mice were not treated (NT), fed (1 g), injected with insulin (5 U/kg, i.p.), or injected with insulin and glucose (1 g/ml/mouse, p.o.) at ZT 6 on day 4. Each *arrowhead* indicates the sampling point at ZT 8. Next, liver samples were analyzed for *Per2* and *Rev-erba* gene expression by real-time polymerase chain reaction. Three mice were assessed in the NT group, and four were assessed in the other groups. (**b**) Blood glucose levels were measured at the time of tissue sampling at ZT 8 on day 4. **$P < 0.01$; ***$P < 0.001$ vs. the NT group or the STZ condition, as determined by Fisher's protected least significant difference (PLSD) test. (**c**, **e**) *Per2* and *Rev-erba* mRNA levels in the liver of control mice, respectively. (**d**, **f**) *Per2* and *Rev-erba* mRNA levels in the liver of STZ-treated mice, respectively. (**c**, **f**) The values of the NT group at ZT 8 were set to 100 %. *$P < 0.05$; **$P < 0.01$; ***$P < 0.001$ vs. NT group by Fisher's PLSD test. Data are presented as the mean ± SEM

This study suggests that a balanced diet such as the control AIN-93M diet is important for the entrainment of the liver circadian clock. The present data demonstrated that simple diets such as 100 % sugar, 100 % protein, and 100 % oil are inadequate for inducing entrainment signals. Therefore, we suggest that a balanced diet such as the control AIN-93M diet not only maintains the health and metabolism of mice but also induces entrainment signals in the peripheral circadian clock. Although 100 % glucose, 100 % casein, 100 % starch, or 100 % soybean oil failed to cause a significant phase advance in the present experiment, 86 % glucose + 14 % casein caused a significant phase advance, compared to 100 % sucrose alone. Ingestion of meals rich

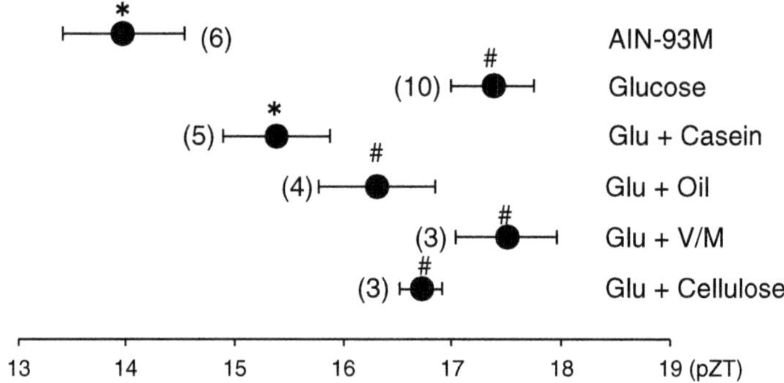

Fig. 13.5 The effects of complete substitution of a single nutrient component in the AIN-93M diet on the phase of the liver circadian clock in PER2::LUCIFERASE knock-in mice (data from Hirao et al. 2009). Mice were administered a diet tablet containing 0.6 g/10 g body weight (BW) of each nutrient in AIN-93M on the first day and 0.85 g/10 g BW of each nutrient on the second day after 24-h food deprivation. Some mice were administered 0.3 ml of soybean oil on the first day and 0.43 ml of soybean oil on the second day after 24-h food deprivation. The values indicate the mean ± SEM. The *horizontal axis* indicates the ZT at the peak of the bioluminescence rhythm. Fisher's PLSD test. *$P < 0.05$ (vs. fasting); fasting, 2 days fast #$P < 0.05$, (vs. AIN-93M). The *numbers in parentheses* indicate the number of mice tested

in protein and fat leads to the secretion of gut hormones such as cholecystokinin, secretin, and peptide YY from the upper small intestine (Dockray 2009; Arora 2006; Green et al. 1989). At present, we do not know the role of such gut hormones on the phase shift of the liver clock; however, some papers have suggested a role of cholecystokinin in circadian systems (Lundberg et al. 2007; Shimazoe et al. 2008).

In one behavioral study, glucose shifted the phase of FAA when the rats were subjected to a restricted feeding schedule (Stephan and Davidson 1998). We have found that in addition to the AIN-93M diet, 100 % glucose and 86 % glucose with 14 % casein caused FAA under a 2-day SF schedule, suggesting that 100 % glucose has the ability to produce FAA without affecting the liver clock. Dissociation of behavioral FAA rhythms and peripheral clock gene expression has been reported (Feillet et al. 2006; Hirao et al. 2009). Although the mechanism underlying the dissociation by glucose was not elucidated in the study, this is the first successful report of dissociation by diet. A combination of glucose and amino acids increased *Per1* and *Per2* gene expression and changed the phase of clock gene expression rhythms (Oike et al. 2011). However, we still do not know whether a specific amino acid or several amino acids in combination facilitate the glucose-induced phase shift.

13.3.5 *Interaction of Entrainment Signals with Light, Food, and Corticosterone*

Although light, food, and corticosterone can entrain peripheral clocks, there are some interactions among these entraining signals. SF-induced FEPO is observed under light–dark conditions and SCN-lesioned conditions, suggesting food-entraining signals are more powerful than light signals in the peripheral clock. Re-entrainment by a shift of the light–dark cycle was potentiated by adrenal organ depletion (Le Minh et al. 2001), suggesting that corticosterone released by the adrenal organ may negatively affect light–dark cycle entrainment. Chronic administration of prednisolone (a synthetic steroid) abolished the circadian rhythm of the peripheral clock. Recently, peripheral organ-specific entrainment was observed. Daily injection of corticosterone entrained the circadian clock of the kidney and lung, but not that of the liver. On the other hand, the liver clock was entrained by daily SF, even when daily corticosterone injections were given at antiphase. Thus, organ- and cellular-specific entrainments may exist.

13.3.6 Cellular Redox and the FEPO

Energy metabolism, including glucose availability, modifies the redox potential (defined as the NAD^+ to NADH ratio) in cells (Sato et al. 2004). Molecular oscillators in the liver and other peripheral organs are directly influenced by elements of the redox potential, which influence the DNA binding of the CLOCK-BMAL1 or NPAS2-BMAL1 heterodimers. This process is highly sensitive to the proportion of NAD^+ cofactors (Rutter et al. 2001, 2002). In rats exhibiting FAA, the cytoplasm and mitochondria from liver cells have a reduced redox potential that immediately reverts to an oxidized state after feeding (Díaz-Muñoz et al. 2000). Importantly, the functional connections between cellular metabolism and the circadian clock involve the redox-sensing histone deacetylase sirtuin-1 (SIRT1) (Asher et al. 2008; Nakahata et al. 2009; Ramsey et al. 2009). SIRT1 inhibits *Per2* gene expression by interfering with the DNA binding of CLOCK-BMAL1. Stimulation of adenosine monophosphate-activated protein kinase (AMPK) destabilizes cryptochromes and alters circadian rhythms. In addition, genetic disruption of the AMPK pathway in mice alters the peripheral clock (Lamia et al. 2009). Therefore, phosphorylation by AMPK enables the cryptochrome to transduce nutrient signals to circadian clocks in mammalian peripheral organs. Fasting and low glucose levels may affect SIRT1 and peroxisome proliferator-activated receptor-γ coactivator 1 (PGC1) gene and/or protein expression through AMPK phosphorylation. Recently, it was reported that fasting upregulated *Per1* and downregulated *Per2*, *Dec1*, and *Bmal1* in the mouse liver; re-feeding prevented the fasting-induced changes in gene expression (Kobayashi et al. 2004; Kawamoto et al. 2006). Therefore, acute changes in clock gene expression by re-feeding may be one factor in RF-induced entrainment of the peripheral clock.

13.3.7 Dissociation of the FEO and the FEPO

Dissociation of behavioral FAA rhythms regulated by the FEO and peripheral clock gene expression regulated by the FEPO has been reported. Feillet et al. (2006) demonstrated that *Per2* mutant mice exhibit very weak FAA rhythms but a strong phase shift in peripheral clock gene expression. Davidson et al. (2003) also reported that fasting reestablished FAA in mice fed ad libitum for 1 week after establishment of an FAA rhythm. However, these mice did not show a phase change in peripheral clock gene expression rhythm under a reestablished FAA. Therefore, we do not know whether the DMH has different roles in restricted feeding-induced FAA formation and restricted feeding-induced entrainment of the peripheral circadian clock. As mentioned previously, MBH lesions attenuate the formation of FAA, but RF-induced entrainment of the liver clock was unaffected by the lesions. Therefore, the medial basal hypothalamus may participate in FAA formation but not restricted feeding-mediated peripheral entrainment.

13.4 Chrono-nutrition

In a study of adult humans in a laboratory setting, insufficient sleep for five consecutive days increased the total daily energy expenditure, but energy intake after dinner was increased, and it exceeded the energy needed to maintain energy balance (Markwald et al. 2013). In addition, humans and rats were more likely to select food in fat at dinnertime compared with breakfast time (Lax et al. 1998; Westerterp-Plantenga et al. 1996), suggesting that the late night nutritional preference accelerates obesity. Taken together, these findings suggest that a late dinner increases the risk of obesity in humans.

The same findings were reported in rodents. Scheduled food intake during only the light period caused higher weight gain than food intake during night period (Arble et al. 2009). Food intake during nighttime increased in mice maintained in dim light conditions at nighttime (light at night); the mice had increased body weight compared with mice maintained under normal light–dark conditions (Fonken et al. 2010). Consumption of a high-fat diet at the end of the dark period increased weight gain, adiposity, glucose intolerance, hyperinsulinemia, hypertriglyceridemia, and hyperleptinemia (Bray et al. 2012; Wu et al. 2011). Similarly, we reported that mice fed only breakfast showed body weight gain, hyperinsulinemia, hyperleptinemia, and decreased expression of β-oxidation-related genes in adipose tissue and liver compared with mice fed a bigger breakfast and a small dinner (Fuse et al. 2012). Therefore, the timing of food intake is an important factor for maintaining appropriate body weight. The reason, though, is still unknown.

13.4.1 Circadian Control of Food/Nutrient Digestion and Absorption

Digestion and absorption in the stomach and intestine follow circadian rhythms in mammals, and these functions are regulated by rhythmically expressed clock genes in the gut and by daily food intake (Bron and Furness 2009; Scheving 2000). For chrono-nutrition, we can consider digestion and absorption activities.

The circadian expression rhythm of clock genes in digestive organs has been investigated carefully. The phases of clock gene expression rhythm were different along the craniocaudal axis of the gut (Polidarová et al. 2009). The phase in the upper part of the gut was phase-advanced compared with that in the lower part of the gut, suggesting that the upper part of the gut is entrained faster than the lower part of the gut by the different speed at which they receive food or nutrition. In the distal colon, microarray analysis revealed that 3.7 % of genes had a circadian pattern of gene expression; these genes were related to cell signaling, differentiation, proliferation, and cell death (Hoogerwerf et al. 2008). Scheduled feeding in the daytime induced a phase shift of clock gene expression rhythm in the gastrointestinal tract of nocturnal mice (Hoogerwerf et al. 2007). Therefore, nutrient signals can affect the gut circadian systems.

Colonic motility in humans has a circadian rhythm, with movement during the day and seldom during the night. Mice also showed a day–night rhythm of colonic motility; this rhythm was regulated by clock genes and by neuronal nitric oxide synthase (nNOS) activity (Hoogerwerf et al. 2010). Stool weight, the colonic contractile response of acetylcholine (measured by colonic organ culture), and intracolonic pressure (measured by a telemetry system) followed clear circadian rhythms in wild-type mice, but the rhythmicity was disrupted in *Per1* and *Per2* double-knockout mice or in nNOS knockout mice. The intestinal digestive enzyme sucrase also has circadian changes in activity; the activity peaks before feeding time (Stevenson and Fierstein 1976). Therefore, the digestive system undergoes circadian changes in rodents and humans.

Circadian variation in the intestinal absorption of glucose, peptides, lipids, and drugs through several transporters has been studied. Isolated rat small intestine showed glucose and water absorption, with higher activity in the nighttime and lower activity in the daytime (Fisher and Gardner 1976). The expression of sodium/glucose cotransporter 1 (*Sglt1*), glucose transporter 2 (*Glut2*), and *Glut5* undergoes clear circadian oscillation (Tavakkolizadeh et al. 2005; Houghton et al. 2008), regulated by clock genes through E-box activity (Iwashina et al. 2011). *Sglt1* is also regulated by PER1 activity independent of the E-box (Balakrishnan et al. 2010). Additionally, *Sglt1* and H(+)/peptide cotransporter 1 (*Pept1*) were phase-entrained by scheduled feeding. Therefore, feeding conditions directly regulate these transporters (Pan et al. 2004). In clock-mutant mice, peptide transport was reduced, but lipid absorption was high (Pan and Hussain 2009). In contrast, knockout of *Nocturnin*, a clock-regulated deadenylase, in mice lowered lipid absorption by reducing chylomicron transit (Douris et al. 2011). The expression of a sodium pump (*Atpa1a*), a channel (*γENac*), transporters (*Dra*, *Ae1*, and *Nhe3*), and Na(+)/H(+) exchanger regulatory factor (*Nherf1*) in rat colonic mucosa showed circadian variations, suggesting NaCl absorption in colon was under circadian regulation (Soták et al. 2011). Among the drug transporters, multidrug resistance 1 (*Mdr1*), monocarboxylate transporter 1 (*Mct1*), multidrug resistance-associated protein 2 (*Mrp2*), peptide transporter 1 (*Pept1*), and breast cancer resistance protein (*Bcrp*) showed circadian expression in rat jejunal mucosa (Stearns et al. 2008). Taken together, many important transporters are under circadian regulation, and circadian disruption leads to abnormal absorption.

13.4.2 Circadian Control of Metabolism and Energy Expenditure

The circadian system in mammals tightly regulates energy metabolism. Several studies of clock gene mutation or deletion reported dysfunction of energy metabolism (Albrecht 2012). Clock-mutant mice showed attenuated feeding rhythm and obesity with a regular diet or high-fat diet (Turek et al. 2005). *Per2*$^{-/-}$ mice showed disrupted feeding rhythm and obesity with a high-fat diet, which was caused by disrupted glucocorticoid rhythm (Yang et al. 2009). *Per2*$^{-/-}$ mice also have disrupted circadian rhythm of alpha-melanocyte-stimulating hormone (α-MSH), which regulates feeding behavior in the hypothalamus. *Rev-erbα* and *Rev-erbβ*, which function as nuclear receptors and core clock genes, were reported to regulate lipid metabolism (Delezie et al. 2012; Cho et al. 2012). Antagonists of the *Rev-erbs* can improve or prevent high-fat diet-induced obesity and circadian disruption in mice (Solt et al. 2012). *Bmal1* KO mice were obese, and their insulin secretion was lower than that of wild-type mice (Lamia et al. 2008). Adipocyte-specific deletion of *Bmal1* resulted in obesity and, interestingly, a shift in the timing of food intake from nocturnal to diurnal, which was caused by changes in circulating polyunsaturated fatty acids and nonesterified polyunsaturated fatty acids in hypothalamic neurons (Paschos et al. 2012). Thus, circadian clock disruption causes energy metabolism dysfunction, suggesting the circadian system tightly regulates metabolic function.

Metabolic factors such as AMPK, SIRT1, PPAR, and PGC-1 have circadian rhythms in activity. They act as important regulators of core circadian mechanisms. AMPK, a nutrient sensor in peripheral tissues, destabilizes CRY in the core circadian system (Lamia et al. 2009). SIRT1, an antiaging protein regulated by nicotinamide adenine dinucleotide (NAD$^+$), regulates the histone acetyltransferase activity of CLOCK (Nakahata et al. 2009; Ramsey et al. 2009) and promotes the deacetylation and degradation of PER2 (Asher et al. 2008). PPARα, a nuclear receptor for lipid metabolism in the liver, binds with PER2 (Schmutz et al. 2010) and promotes *Bmal1* expression through the peroxisome proliferator response element in the promoter of *Bmal1* (Inoue et al. 2005). PGC-1α, a transcriptional coactivator for the regulation of energy metabolism, also included in circadian regulation, promotes the expression of *Bmal1* and *Rev-erbα* through a RORE site (Liu et al. 2007). Thus, core metabolic genes are tightly related to the clock systems and their activity changes in a circadian manner.

13.5 Conclusion

SF induces the FEO and FEPO, but the mechanisms in the brain and peripheral tissues may be different. With FAA as the index of FEO activity, research investigating the location and mechanism of the FEO remains complicated because many compensation pathways appear when some functions cease working. It is easier to investigate the mechanism of the FEPO because the index of FEPO activity is clock gene expression rhythm in peripheral tissues. The FEPO establishment is regulated by nutrient, feeding time, and feeding frequency. Circadian clocks control the food/nutrition metabolism through absorption and excretion.

References

Akiyama M, Yuasa T, Hayasaka N et al (2004) Reduced food anticipatory activity in genetically orexin (hypocretin) neuron-ablated mice. Eur J Neurosci 20:3054–3062

Albrecht U (2012) Timing to perfection: the biology of central and peripheral circadian clocks. Neuron 74:246–260

Amir S, Stewart J (2009) Motivational modulation of rhythms of the expression of the clock protein PER2 in the limbic forebrain. Biol Psychiatry 65:829–834

Arble DM, Bass J, Laposky AD et al (2009) Circadian timing of food intake contributes to weight gain. Obesity (Silver Spring) 17:2100–2102

Arora S (2006) Role of neuropeptides in appetite regulation and obesity. Neuropeptides 40:375–401

Asher G, Gatfield D, Stratmann M et al (2008) SIRT1 regulates circadian clock gene expression through PER2 deacetylation. Cell 134:317–328

Balakrishnan A, Stearns AT, Ashley SW et al (2010) Restricted feeding phase shifts clock gene and sodium glucose cotransporter 1 (SGLT1) expression in rats. J Nutr 140:908–914

Blum ID, Patterson Z, Khazall R et al (2009) Reduced anticipatory locomotor responses to scheduled meals in ghrelin receptor deficient mice. Neuroscience 164:351–359

Bray MS, Ratcliffe WF, Grenett MH et al (2012) Quantitative analysis of light-phase restricted feeding reveals metabolic dys-synchrony in mice. Int J Obes (Lond) 37:843–852

Bron R, Furness JB (2009) Rhythm of digestion: keeping time in the gastrointestinal tract. Clin Exp Pharmacol Physiol 36:1041–1048

Castillo MR, Hochstetler KJ, Tavernier RJ Jr et al (2004) Entrainment of the master circadian clock by scheduled feeding. Am J Physiol Regul Integr Comp Physiol 287:R551–R555

Cho H, Zhao X, Hatori M et al (2012) Regulation of circadian behaviour and metabolism by REV-ERB-α and REV-ERB-β. Nature 485:123–127

Cone RD, Cowley MA, Butler AA et al (2001) The arcuate nucleus as a conduit for diverse signals relevant to energy homeostasis. Int J Obes Relat Metab Disord 25 Suppl 5:S63–S67

Daan S (2000) The Colin S. Pittendrigh Lecture. Colin Pittendrigh, Jürgen Aschoff, and the natural entrainment of circadian systems. J Biol Rhythms 15:195–207

Daan S, Pittendrigh CS (1976) A functional analysis of circadian pacemakers in nocturnal rodents II. The variability of phase response curves. J Comp Physiol 106:253–266

Damiola F, Le Minh N, Preitner N et al (2000) Restricted feeding uncouples circadian oscillators in peripheral tissues from the central pacemaker in the suprachiasmatic nucleus. Genes Dev 14:2950–2961

Dardente H, Cermakian N (2007) Molecular circadian rhythms in central and peripheral clocks in mammals. Chronobiol Int 24:195–213

Davidson AJ, Stephan FK (1999) Plasma glucagon, glucose, insulin, and motilin in rats anticipating daily meals. Physiol Behav 66:309–315

Davidson AJ, Cappendijk SL, Stephan FK (2000) Feeding-entrained circadian rhythms are attenuated by lesions of the parabrachial region in rats. Am J Physiol Regul Integr Comp Physiol 278:R1296–R1304

Davidson AJ, Poole AS, Yamazaki S et al (2003) Is the food-entrainable circadian oscillator in the digestive system? Genes Brain Behav 2:32–39

Delezie J, Dumont S, Dardente H et al (2012) The nuclear receptor REV-ERBα is required for the daily balance of carbohydrate and lipid metabolism. FASEB J 26:3321–3335

Díaz-Muñoz M, Vázquez-Martínez O, Aguilar-Roblero R et al (2000) Anticipatory changes in liver metabolism and entrainment of insulin, glucagon, and corticosterone in food-restricted rats. Am J Physiol Regul Integr Comp Physiol 279:R2048–R2056

Dockray GJ (2009) Cholecystokinin and gut-brain signalling. Regul Pept 155:6–10

Douris N, Kojima S, Pan X et al (2011) Nocturnin regulates circadian trafficking of dietary lipid in intestinal enterocytes. Curr Biol 21:1347–1355

Duez H, Staels B (2008) Rev-erb alpha gives a time cue to metabolism. FEBS Lett 582:19–25

Escobar C, Cailotto C, Angeles-Castellanos M et al (2009) Peripheral oscillators: the driving force for food-anticipatory activity. Eur J Neurosci 30:1665–1675

Feillet CA, Ripperger JA, Magnone MC et al (2006) Lack of food anticipation in Per2 mutant mice. Curr Biol 16:2016–2022

Fisher RB, Gardner ML (1976) A diurnal rhythm in the absorption of glucose and water by isolated rat small intestine. J Physiol 254:821–825

Fonken LK, Workman JL, Walton JC et al (2010) Light at night increases body mass by shifting the time of food intake. Proc Natl Acad Sci U S A 107:18664–18669

Fuller PM, Lu J, Saper CB (2008) Differential rescue of light- and food-entrainable circadian rhythms. Science 320:1074–1077

Fuse Y, Hirao A, Kuroda H et al (2012) Differential roles of breakfast only (one meal per day) and a bigger breakfast with a small dinner (two meals per day) in mice fed a high-fat diet with regard to induced obesity and lipid metabolism. J Circadian Rhythms 10:4. doi:10.1186/1740-3391-10-4

Gachon F, Nagoshi E, Brown SA et al (2004) The mammalian circadian timing system: from gene expression to physiology. Chromosoma 113:103–112

Giroud S, Perret M, Le Maho Y et al (2009) Gut hormones in relation to body mass and torpor pattern changes during food restriction and re-feeding in the gray mouse lemur. J Comp Physiol B 179:99–111

Gooley JJ, Schomer A, Saper CB (2006) The dorsomedial hypothalamic nucleus is critical for the expression of food-entrainable circadian rhythms. Nat Neurosci 9:398–407

Green GM, Taguchi S, Friestman J et al (1989) Plasma secretin, CCK, and pancreatic secretion in response to dietary fat in the rat. Am J Physiol 256:G1016–G1021

Hara R, Wan K, Wakamatsu H et al (2001) Restricted feeding entrains liver clock without participation of the suprachiasmatic nucleus. Genes Cells 6:269–278

Hirao A, Tahara Y, Kimura I et al (2009) A balanced diet is necessary for proper entrainment signals of the mouse liver clock. PLoS One 4:e6909

Hirota T, Okano T, Kokame K et al (2002) Glucose down-regulates Per1 and Per2 mRNA levels and induces circadian gene expression in cultured Rat-1 fibroblasts. J Biol Chem 277(46): 44244–44251

Honma K, Honma S, Hiroshige T (1985) Response curve, free-running period, and activity time in circadian locomotor rhythm of rats. Jpn J Physiol 35:643–658

Hoogerwerf WA, Hellmich HL, Cornélissen G et al (2007) Clock gene expression in the murine gastrointestinal tract: endogenous rhythmicity and effects of a feeding regimen. Gastroenterology 133:1250–1260

Hoogerwerf WA, Sinha M, Conesa A et al (2008) Transcriptional profiling of mRNA expression in the mouse distal colon. Gastroenterology 135(6):2019–2029

Hoogerwerf WA, Shahinian VB, Cornélissen G et al (2010) Rhythmic changes in colonic motility are regulated by period genes. Am J Physiol Gastrointest Liver Physiol 298:G143–G150

Houghton SG, Iqbal CW, Duenes JA et al (2008) Coordinated, diurnal hexose transporter expression in rat small bowel: implications for small bowel resection. Surgery 143:79–93

Iijima M, Yamaguchi S, van der Horst GT et al (2005) Altered food-anticipatory activity rhythm in Cryptochrome-deficient mice. Neurosci Res 52:166–173

Inoue I, Shinoda Y, Ikeda M et al (2005) CLOCK/BMAL1 is involved in lipid metabolism via transactivation of the peroxisome proliferator-activated receptor (PPAR) response element. J Atheroscler Thromb 12:169–174

Iwanaga H, Yano M, Miki H et al (2005) Per2 gene expressions in the suprachiasmatic nucleus and liver differentially respond to nutrition factors in rats. JPEN J Parenter Enteral Nutr 29:157–161

Iwashina I, Mochizuki K, Inamochi Y et al (2011) Clock genes regulate the feeding schedule-dependent diurnal rhythm changes in hexose transporter gene expressions through the binding of BMAL1 to the promoter/enhancer and transcribed regions. J Nutr Biochem 22:334–343

Johansson A, Fredriksson R, Winnergren S et al (2008) The relative impact of chronic food restriction and acute food deprivation on plasma hormone levels and hypothalamic neuropeptide expression. Peptides 29:1588–1595

Kaur S, Thankachan S, Begum S et al (2008) Entrainment of temperature and activity rhythms to restricted feeding in orexin knock out mice. Brain Res 1205:47–54

Kawamoto T, Noshiro M, Furukawa M et al (2006) Effects of fasting and re-feeding on the expression of Dec1, Per1, and other clock-related genes. J Biochem 140:401–408

King DP, Zhao Y, Sangoram AM et al (1997) Positional cloning of the mouse circadian clock gene. Cell 89:641–653

Kobayashi H, Oishi K, Hanai S et al (2004) Effect of feeding on peripheral circadian rhythms and behaviour in mammals. Genes Cells 9:857–864

Kudo T, Akiyama M, Kuriyama K et al (2004) Night-time restricted feeding normalizes clock genes and Pai-1 gene expression in the db/db mouse liver. Diabetologia 47:1425–1436

Kuroda H, Tahara Y, Saito K et al (2012) Meal frequency patterns determine the phase of mouse peripheral circadian clocks. Sci Rep 2:711

Lamia KA, Storch KF, Weitz CJ (2008) Physiological significance of a peripheral tissue circadian clock. Proc Natl Acad Sci U S A 105:15172–15177

Lamia KA, Sachdeva UM, DiTacchio L et al (2009) AMPK regulates the circadian clock by cryptochrome phosphorylation and degradation. Science 326:437–440

Landry GJ, Simon MM, Webb IC et al (2006) Persistence of a behavioral food-anticipatory circadian rhythm following dorsomedial hypothalamic ablation in rats. Am J Physiol Regul Integr Comp Physiol 290:R1527–R1534

Landry GJ, Yamakawa GR, Webb IC et al (2007) The dorsomedial hypothalamic nucleus is not necessary for the expression of circadian food-anticipatory activity in rats. J Biol Rhythms 22:467–478

Lax P, Larue-Achagiotis C, Martel P et al (1998) Repeated short-fasting modifies the macronutrient self-selection pattern in rats. Physiol Behav 65:69–76

Le Minh N, Damiola F, Tronche F et al (2001) Glucocorticoid hormones inhibit food-induced phase-shifting of peripheral circadian oscillators. EMBO J 20:7128–7136

LeSauter J, Hoque N, Weintraub M et al (2009) Stomach ghrelin-secreting cells as food-entrainable circadian clocks. Proc Natl Acad Sci U S A 106:13582–13587

Liu C, Li S, Liu T et al (2007) Transcriptional coactivator PGC-1alpha integrates the mammalian clock and energy metabolism. Nature 447:477–481

Lundberg K, Josefsson S, Nordin C (2007) Diurnal and seasonal variation of cholecystokinin peptides in humans. Neuropeptides 41:59–63

Markwald RR, Melanson EL, Smith MR et al (2013) Impact of insufficient sleep on total daily energy expenditure, food intake, and weight gain. Proc Natl Acad Sci U S A 110:5695–5700

Meijer JH, Michel S, Vansteensel MJ (2007) Processing of daily and seasonal light information in the mammalian circadian clock. Gen Comp Endocrinol 152:159–164

Mendoza J, Ángeles-Castellanos M, Escobar C (2005) A daily palatable meal without food deprivation entrains the suprachiasmatic nucleus of rats. Eur J Neurosci 22:2855–2862

Mendoza J, Drevet K, Pévet P et al (2008) Daily meal timing is not necessary for resetting the main circadian clock by calorie restriction. J Neuroendocrinol 20:251–260

Mendoza J, Pévet P, Felder-Schmittbuhl MP et al (2010) The cerebellum harbors a circadian oscillator involved in food anticipation. J Neurosci 30:1894–1904

Mieda M, Williams SC, Sinton CM et al (2004) Orexin neurons function in an efferent pathway of a food-entrainable circadian oscillator in eliciting food-anticipatory activity and wakefulness. J Neurosci 24:10493–10501

Mieda M, Williams SC, Richardson JA et al (2006) The dorsomedial hypothalamic nucleus as a putative food-entrainable circadian pacemaker. Proc Natl Acad Sci U S A 103:12150–12155

Mistlberger RE (1994) Circadian food-anticipatory activity: formal models and physiological mechanisms. Neurosci Biobehav Rev 18:171–195

Mistlberger RE (2009) Food-anticipatory circadian rhythms: concepts and methods. Eur J Neurosci 30:1718–1729

Mistlberger RE, Yamazaki S, Pendergast JS et al (2008) Comment on "Differential rescue of light- and food-entrainable circadian rhythms". Science 322:675

Mistlberger RE, Kent BA, Weinberg A et al (2009) Anticipation of two daily meals: circadian properties and neural correlates. Proc Eur Soc Biol Rhythms 10:139

Mohri T, Emoto N, Nonaka H et al (2003) Alterations of circadian expressions of clock genes in Dahl salt-sensitive rats fed a high-salt diet. Hypertension 42:189–194

Moriya T, Horikawa K, Akiyama M et al (2000) Correlative association between N-methyl-D-aspartate receptor-mediated expression of period genes in the suprachiasmatic nucleus and phase shifts in behavior with photic entrainment of clock in hamsters. Mol Pharmacol 58:1554–1562

Moriya T, Aida R, Kudo T et al (2009) The dorsomedial hypothalamic nucleus is not necessary for food-anticipatory circadian rhythms of behavior, temperature or clock gene expression in mice. Eur J Neurosci 29:1447–1460

Nakahara K, Fukui K, Murakami N (2004) Involvement of thalamic paraventricular nucleus in the anticipatory reaction under food restriction in the rat. J Vet Med Sci 66:1297–1300

Nakahata Y, Sahar S, Astarita G et al (2009) Circadian control of the NAD+ salvage pathway by CLOCK-SIRT1. Science 324:654–657

Oike H, Nagai K, Fukushima T et al (2011) Feeding cues and injected nutrients induce acute expression of multiple clock genes in the mouse liver. PLoS One 6:e23709

Oishi K, Uchida D, Ohkura N et al (2009) Ketogenic diet disrupts the circadian clock and increases hypofibrinolytic risk by inducing expression of plasminogen activator inhibitor-1. Arterioscler Thromb Vasc Biol 29:1571–1577

Ono M, Shibata S, Minamoto Y et al (1996) Effect of the noncompetitive N-methyl-D-aspartate (NMDA) receptor antagonist MK-801 on food-anticipatory activity rhythm in the rat. Physiol Behav 59:585–589

Pan X, Hussain MM (2009) Clock is important for food and circadian regulation of macronutrient absorption in mice. J Lipid Res 50:1800–1813

Pan X, Terada T, Okuda M et al (2004) The diurnal rhythm of the intestinal transporters SGLT1 and PEPT1 is regulated by the feeding conditions in rats. J Nutr 134:2211–2215

Paschos GK, Ibrahim S, Song WL et al (2012) Obesity in mice with adipocyte-specific deletion of clock component Arntl. Nat Med 18:1768–1777

Pittendrigh CS (1988) The photoperiodic phenomena: seasonal modulation of the "day within". J Biol Rhythms 3:173–188

Polidarová L, Soták M, Sládek M et al (2009) Temporal gradient in the clock gene and cell-cycle checkpoint kinase Wee1 expression along the gut. Chronobiol Int 26:607–620

Ramsey KM, Yoshino J, Brace CS et al (2009) Circadian clock feedback cycle through NAMPT-mediated NAD+ biosynthesis. Science 324:651–654

Ribeiro AC, Sawa E, Carren-LeSauter I et al (2007) Two forces for arousal: pitting hunger versus circadian influences and identifying neurons responsible for changes in behavioral arousal. Proc Natl Acad Sci U S A 104:20078–20083

Rodgers RJ, Ishii Y, Halford JC et al (2002) Orexins and appetite regulation. Neuropeptides 36:303–325

Rutter J, Reick M, Wu LC et al (2001) Regulation of clock and NPAS2 DNA binding by the redox state of NAD cofactors. Science 293:510–514

Rutter J, Reick M, McKnight SL (2002) Metabolism and the control of circadian rhythms. Annu Rev Biochem 71:307–331

Saper CB, Lu J, Chou TC et al (2005a) The hypothalamic integrator for circadian rhythms. Trends Neurosci 28:152–157

Saper CB, Scammell TE, Lu J (2005b) Hypothalamic regulation of sleep and circadian rhythms. Nature 437:1257–1263

Sato TK, Panda S, Miraglia LJ et al (2004) A functional genomics strategy reveals Rora as a component of the mammalian circadian clock. Neuron 43:527–537

Scheving LA (2000) Biological clocks and the digestive system. Gastroenterology 119:536–549

Schmutz I, Ripperger JA, Baeriswyl-Aebischer S et al (2010) The mammalian clock component PERIOD2 coordinates circadian output by interaction with nuclear receptors. Genes Dev 24:345–357

Shibata S, Minamoto Y, Ono M et al (1994) Age-related impairment of food anticipatory locomotor activity in rats. Physiol Behav 55:875–878

Shibata S, Ono M, Minamoto Y et al (1995) Attenuating effect of bifemelane on an impairment of mealtime-associated activity rhythm in aged and MK-801-treated rats. Pharmacol Biochem Behav 50:207–210

Shibata S, Hirao A, Tahara Y (2009) Restricted feeding-induced entrainment of activity rhythm and peripheral clock rhythm. Sleep Biol Rhythms 8:18–27

Shimazoe T, Morita M, Ogiwara S et al (2008) Cholecystokinin-A receptors regulate photic input pathways to the circadian clock. FASEB J 22:1479–1490

Shirai H, Oishi K, Kudo T et al (2007) PPARalpha is a potential therapeutic target of drugs to treat circadian rhythm sleep disorders. Biochem Biophys Res Commun 357:679–682

Solt LA, Wang Y, Banerjee S et al (2012) Regulation of circadian behaviour and metabolism by synthetic REV-ERB agonists. Nature 485:62–68

Soták M, Polidarová L, Musílková J et al (2011) Circadian regulation of electrolyte absorption in the rat colon. Am J Physiol Gastrointest Liver Physiol 301:G1066–G1074

Stearns AT, Balakrishnan A, Rhoads DB et al (2008) Diurnal rhythmicity in the transcription of jejunal drug transporters. J Pharmacol Sci 108:144–148

Stephan FK (1989a) Entrainment of activity to multiple feeding times in rats with suprachiasmatic lesions. Physiol Behav 46:489–497

Stephan FK (1989b) Forced dissociation of activity entrained to T cycles of food access in rats with suprachiasmatic lesions. J Biol Rhythms 4:467–479

Stephan FK (1997) Calories affect zeitgeber properties of the feeding entrained circadian oscillator. Physiol Behav 62:995–1002

Stephan FK, Davidson AJ (1998) Glucose, but not fat, phase shifts the feeding-entrained circadian clock. Physiol Behav 65:277–288

Stevenson NR, Fierstein JS (1976) Circadian rhythms of intestinal sucrase and glucose transport: cued by time of feeding. Am J Physiol 230:731–735

Stokkan KA, Yamazaki S, Tei H et al (2001) Entrainment of the circadian clock in the liver by feeding. Science 291:490–493

Storch KF, Weitz CJ (2009) Daily rhythms of food-anticipatory behavioral activity do not require the known circadian clock. Proc Natl Acad Sci U S A 106:6808–6813

Tahara Y, Hirao A, Moriya T et al (2010) Effects of medial hypothalamic lesions on feeding-induced entrainment of locomotor activity and liver Per2 expression in Per2::luc mice. J Biol Rhythms 25:9–18

Tahara Y, Otsuka M, Fuse Y et al (2011) Refeeding after fasting elicits insulin-dependent regulation of Per2 and Rev-erbα with shifts in the liver clock. J Biol Rhythms 26:230–240

Takahashi JS, Shimomura K, Kumar V (2008) Searching for genes underlying behavior: lessons from circadian rhythms. Science 322:909–912

Takasu NN, Kurosawa G, Tokuda IT et al (2012) Circadian regulation of food-anticipatory activity in molecular clock-deficient mice. PLoS One 7:e48892

Tanaka Y, Kurasawa M, Nakamura K (2000) Recovery of diminished mealtime-associated anticipatory behavior by aniracetam in aged rats. Pharmacol Biochem Behav 66:827–833

Tavakkolizadeh A, Ramsanahie A, Levitsky LL et al (2005) Differential role of vagus nerve in maintaining diurnal gene expression rhythms in the proximal small intestine. J Surg Res 129:73–78

Teboul M, Guillaumond F, Gréchez-Cassiau A et al (2008) The nuclear hormone receptor family round the clock. Mol Endocrinol 22:2573–2582

Turek FW, Joshu C, Kohsaka A et al (2005) Obesity and metabolic syndrome in circadian Clock mutant mice. Science 308:1043–1045

Verwey M, Khoja Z, Stewart J et al (2007) Differential regulation of the expression of Period2 protein in the limbic forebrain and dorsomedial hypothalamus by daily limited access to highly palatable food in food-deprived and free-fed rats. Neuroscience 147:277–285

Waddington Lamont E, Harbour VL et al (2007) Restricted access to food, but not sucrose, saccharine, or salt, synchronizes the expression of Period2 protein in the limbic forebrain. Neuroscience 144:402–411

Wakamatsu H, Yoshinobu Y, Aida R et al (2001) Restricted-feeding-induced anticipatory activity rhythm is associated with a phase-shift of the expression of mPer1 and mPer2 mRNA in the cerebral cortex and hippocampus but not in the suprachiasmatic nucleus of mice. Eur J Neurosci 13:1190–1196

Walcott EC, Tate BA (1996) Entrainment of aged, dysrhythmic rats to a restricted feeding schedule. Physiol Behav 60:1205–1208

Westerterp-Plantenga MS, IJedema MJ, Wijckmans-Duijsens NE (1996) The role of macronutrient selection in determining patterns of food intake in obese and non-obese women. Eur J Clin Nutr 50:580–591

Wu T, Sun L, Zhuge F et al (2011) Differential roles of breakfast and supper in rats of a daily three-meal schedule upon circadian regulation and physiology. Chronobiol Int 28:890–903

Yang X, Downes M, Yu RT et al (2006) Nuclear receptor expression links the circadian clock to metabolism. Cell 126:801–810

Yang S, Liu A, Weidenhammer A et al (2009) The role of mPer2 clock gene in glucocorticoid and feeding rhythms. Endocrinology 150:2153–2160

Yannielli PC, Molyneux PC, Harrington ME et al (2007) Ghrelin effects on the circadian system of mice. J Neurosci 27:2890–2895

Young ME, Wilson CR, Razeghi P et al (2002) Alterations of the circadian clock in the heart by streptozotocin-induced diabetes. J Mol Cell Cardiol 34:223–231

Zigman JM, Jones JE, Lee CE et al (2006) Expression of ghrelin receptor mRNA in the rat and the mouse brain. J Comp Neurol 494:528–548

Chapter 14
Physiopathology of Circadian Rhythms: Understanding the Biochemical Mechanisms of Obesity and Cancer

Manuel Miranda-Anaya, Christian Molina-Aguilar, Olivia Vázquez-Martínez, and Mauricio Díaz-Muñoz

Abstract Obesity and cancer have become urgent and pressing health problems in industrialized societies. Both diseases are interrelated: overweight and obesity favor tumorogenesis in a variety of tissues and organs, mainly supported in the biochemical alterations caused by a diabetogenic state and the pro-inflammatory condition associated. An additional perspective involves disturbances in circadian rhythms which are associated with a metabolic dysfunction; circadian disruption has also been considered as a carcinogenic condition for humans based in evidence that link dysfunctional control of cell division and differentiation to an altered circadian molecular clock. In this chapter, we review recent findings that demonstrate the interplay between metabolism and circadian rhythmicity, as well as the repercussions involved in the physiopathology of obesity and cancer.

14.1 Introduction

Among the most prevalent pathologies in recent years, obesity and cancer have become urgent and pressing health problems in industrialized societies. Each condition presents its own complexity, but it is now evident that these two diseases are

Manuel Miranda-Anaya, Christian Molina-Aguilar, and Olivia Vázquez-Martínez contributed equally to this work.

M. Miranda-Anaya
Facultad de Ciencias, Unidad Multidisciplinaria de Docencia e Investigación,
Universidad Nacional Autónoma de México, Querétaro 76230, QRO, México

Department of Cellular and Molecular Neurobiology, Neurobiology Institute,
Campus UNAM-Juriquilla, Querétaro 76230, QRO, México

C. Molina-Aguilar • O. Vázquez-Martínez • M. Díaz-Muñoz (⌧)
Department of Cellular and Molecular Neurobiology, Neurobiology Institute,
Campus UNAM-Juriquilla, Querétaro 76230, QRO, México
e-mail: mdiaz@comunidad.unam.mx

© Springer International Publishing Switzerland 2015
R. Aguilar-Roblero et al. (eds.), *Mechanisms of Circadian Systems in Animals and Their Clinical Relevance*, DOI 10.1007/978-3-319-08945-4_14

interrelated: overweight and obesity favor tumorogenesis in a variety of tissues and organs (Calle and Kaaks 2004). The connection is mainly supported by the biochemical alterations caused by a diabetogenic state and the pro-inflammatory condition associated with chronically overactive adipose tissue. In this context, at least 20 % of cancer-related deaths (colon, breast, liver, pancreas, and stomach) are directly linked to increased adiposity (Calle et al. 2003).

A newly identified factor to be considered in the onset and development of obesity and cancer is altered circadian rhythmicity. Disturbances in the 24-h timing system have been linked to metabolic dysfunction, which leads to an altered handling of energetic reserves and an excess of lipid deposits (Delezie and Challet 2011). Similarly, in 2007, the International Agency for Research in Cancer (IARC) categorized the circadian disruption as probably carcinogenic in humans, making evident the connection between dysfunctional control of cell division and differentiation, and a modified 24-h circadian molecular clock (Savvidis and Koutsilieris 2012). In this chapter, we review recent findings that demonstrate the interplay between metabolism and circadian rhythmicity, as well as its repercussions for the physiopathology of obesity and cancer.

14.2 Foundations

The following concepts and definitions provide a more complete perspective of the connections among circadian rhythmicity, metabolic control, obesity, and cancer.

14.2.1 Timing System

All living beings must adapt to the changing environmental conditions, including the cyclic succession of day and night, and the seasons of the year. In general, this adaptation allows the organisms to optimize feeding, mating, and growth. However, the process not only reacts to the environmental fluctuations, but also involves an important anticipatory component. Along the phylogenetic scale, the endogenous capacity to measure time has arisen in several groups, with different strategies to contend with the daily ecological challenges (Dunlap 1999).

At the cellular level, the circadian molecular clock is constituted by a network of genes and proteins forming positive and negative regulatory loops (Fig. 14.1). The pacing of the clock is governed by transcriptional and translational processes that complete a cycle in approximately 24 h (Ukai and Ueda 2010). However, some reports have shown the existence of a timing system independent of genetic material, and formed exclusively by biochemical reactions within metabolic pathways (Reddy and O'Neill 2011).

In mammals, the timing system also involves the coordination of a set of oscillators; the suprachiasmatic nucleus (SCN) serves as pacemaker, whereas the peripheral

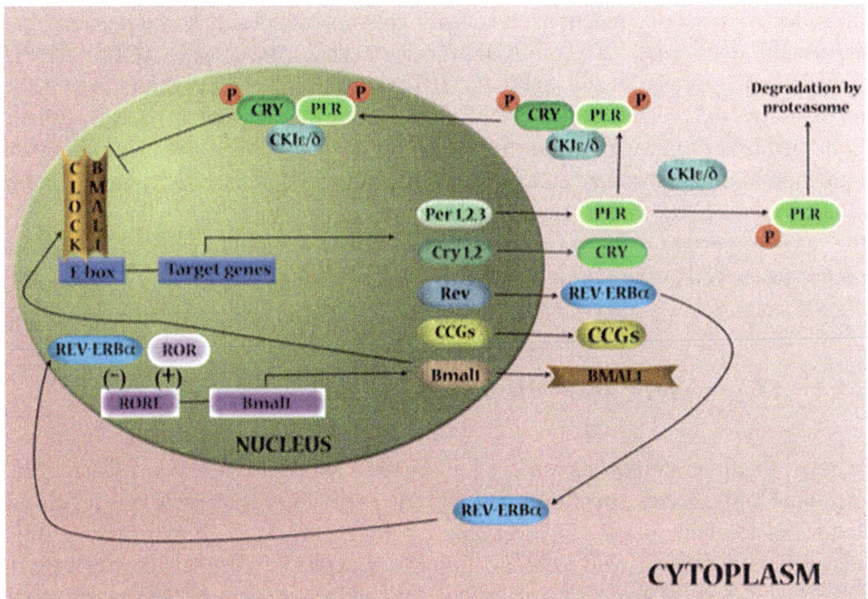

Fig. 14.1 The mammalian molecular clock. This circadian molecular clock has a negative feedback loop involving the genes per1-2, cry1-2, bmal1, and clock. Clock and Bmal1 are transcription factors that activate per and cry expression by binding to their E-box promoter region. Per and Cry inhibit Clock/Bmal-induced gene expression. The expression of Rev-Erba represses Bmal 1 transcription through the RAR-related orphan receptor (ROR). Picture taken from Golombek and Rosenstein (2010)

oscillators (in other brain regions as well as in thoracic and abdominal organs) play a subordinated role in basal conditions. A healthy circadian physiology depends on the correct reciprocal communication and harmonization between the SCN and the peripheral oscillators (Schibler et al. 2003).

14.3 Metabolic Regulation

The generic term "metabolism" is defined as the set of biochemical reactions that take place within the cell. It involves the synthesis and degradation of hundreds of intermediary metabolites by means of catalytic proteins known as enzymes. Metabolic networks are controlled at several levels, but three are very important: (1) energy charge, (2) redox state, and (3) cellular compartments (Liu and Zhang 1985). Energy charge refers to the proportion of adenine nucleotides within the internal milieu. The ratio ATP/AMP drives the regulation of key allosteric proteins to enhance or reduce metabolic fluxes in response to the cellular requirements to grow, differentiate, or adapt to energetic challenges (Atkinson 1968). Redox regulation concerns the regulated transit of electrons that enable the control of the oxidative–catabolic or the reductive–anabolic reactions. It is based on pairs of molecules (redox pairs) that are usually coenzymes: $NAD^+/NADH$ for degradative processes, $NADP^+/NADPH$ for synthetic pathways, and GSSG/GSH (oxidized and reduced glutathione) for modulating the redox state of protein-bound thiol groups (Veech 2006). Endomembranes make possible the formation of well-defined organelles such as mitochondria, peroxisomes, endoplasmic reticulum, and others. These intracellular compartments allow the creation of gradients of ions or metabolites that are one of the most important driving forces to control the directionality of the metabolic pathways as well as the state of membrane polarization (Elias 2010).

14.4 Overweight and Obesity

Most of the time, organisms reach equilibrium between the calories they ingest at mealtime and calories spent during their daily activities. The result is a regulated control of the body mass. However, under certain pathological circumstances this equilibrium is broken, with the concomitant consequence of under- or overweight. Obesity develops when energy intake exceeds energy expenditure. The health problem associated with obesity is not only the increase in fatty deposits in the subcutaneous white adipose tissue, but more importantly, the increases in visceral fat and the fat deposited "ectopically" in organs like the liver, pancreas, heart, or skeletal muscles that tend to cause a pro-inflammatory condition with negative health consequences (Item and Konrad 2012).

Obesity is an organic condition caused by multiple factors including genetic predisposition, altered endocrine signaling, and metabolic abnormalities; however, in order to be expressed, it requires an environmental influence, usually a change in the daily access to calories or performance of physical activity (Faith and Kral 2006).

To gain more understanding regarding the metabolic, physiological, and genetic mechanisms that underlie obesity, several experimental models and approaches have been implemented (reviewed in Kanasaki and Koya 2011). Experimental models of obesity in mice are divided into monogenic and polygenic. Among monogenic mice, the most popular are those with defects in leptin signaling, the *ob/ob* and *db/db* mice that lack the gene for leptin or its receptor, respectively. The altered aguti gene/protein (lethal yellow mutant) was the first obesity gene characterized at the molecular level more than 20 years ago. Although monogenic models provide important information about the biology of obesity, human obesity is most likely mediated by multiple genes. Therefore, polygenic models could be much more relevant to human obesity. The New Zealand obese mouse exhibits hyperphagia, reduced energy expenditure, and type 2 diabetes (only in males). The Tsumura Suzuki obese diabetic mouse shows marked hyperglycemia and hyperinsulinemia, and eventually displays lesions similar to diabetic nephropathy and neuropathy. M16 mice exhibit hyperphagia, hyperinsulinemia, and hyperleptinemia. Kuo Kond mice also show type 2 diabetes with a marked insulin resistance preceding the onset of obesity. *Neotomodon alstoni* is a Mexican mouse that spontaneously develops overweight and obesity when fed with a regular diet for laboratory rodents, especially the females. These mice show modified circadian rhythmicity and altered parameters related to the metabolic syndrome (Carmona-Alcocer et al. 2012).

Three types of rat have been used as experimental models for obesity: Zucker, Wistar fatty, and Otsuka Long Evans Tokushima fatty. All of them show abnormalities in glucose handling and are prone to the diabetogenic condition. Usually, the males are more likely than females to develop more severe anomalies.

Manipulating the dietary input of calories is also a very popular way to produce obese animals in biomedical laboratories. Depending on the rodent strain, a high-fat diet induces overweight and abnormalities similar to human metabolic syndrome. C57BL/6J mice show more severe irregularities than C57BL/KsJ, whereas C57BL/J6 mice, sand mice, and spiny mice exhibit a clear obese phenotype with type 2 diabetes characteristics (Collins et al. 2004). A diet supplemented with sucrose in the drinking water has also been used to produce overweight in rats and is accompanied by metabolic alterations similar to human pathologies (Larqué et al. 2011).

14.5 Cancer

Cancer is defined as an undifferentiated cell population with abnormal metabolism and uncontrolled cell division. The role of onco-proteins and tumor suppressor proteins in proliferative signaling, cell cycle regulation, and altered adhesion is well established. Chemicals, viruses, and radiation are generally accepted as agents that commonly induce mutations in genes encoding these cancer-inducing proteins, thereby giving rise to cancer. More recent evidence indicates the importance of two additional key factors imposed on proliferating cells—hypoxia and/or lack of glucose. These two additional triggers can initiate and promote the process of malignant transformation, when a low percentage of cells escape cellular senescence. Unregulated cell proliferation leads to the formation of cellular masses that extend beyond the resting vasculature, causing oxygen and nutrient deprivation. The resulting hypoxia triggers a number of critical adaptations that enable cancer cell survival. The process of apoptosis is suppressed, and glucose metabolism is altered. Aerobic glycolysis or the Warburg effect links the high rate of glucose fermentation to cancer. Together with glutamine, glycolysis of glucose provides the carbon skeletons, NADPH, and ATP to build new cancer cells that persist in hypoxia, which in turn rewires metabolic pathways for cell growth and survival (Dang 2012). Recent investigations suggest that oxygen depletion stimulates mitochondria to compensate increased reactive oxygen species (ROS). Oxygen depletion activates signaling pathways, such as hypoxia-inducible factor 1, that promote cancer cell survival and tumor growth (Hahn et al. 1999).

There are several approaches to study the biology of cancer in experimental models (reviewed in Zhao et al. 2004): (1) use of established cell lines derived from cancerous organisms; (2) immortalization and transformation using chemical, physical, and viral agents; (3) immortalization and transformation using defined combinations of genetic elements such as oncogenes; (4) immortalization by upregulation of telomere reverse transcriptase, and (5) introduction of SV40 large T antigen with an oncogenic allele of *RAS*.

14.6 Metabolic Interactions

There are several examples of how metabolic networks interact with transcriptional programs to regulate circadian functions that have a direct impact on the handling of cellular energy and are affected during states of obesity and cancer (Fig. 14.2).

14.6.1 Nuclear Receptors

Nuclear receptors belong to a family of transcriptional factors that recognize specific ligands, usually of lipophilic nature. These ligands can be hormones (steroid and thyroid hormones), vitamins (β-carotene and cholecalciferol and ergocalciferol),

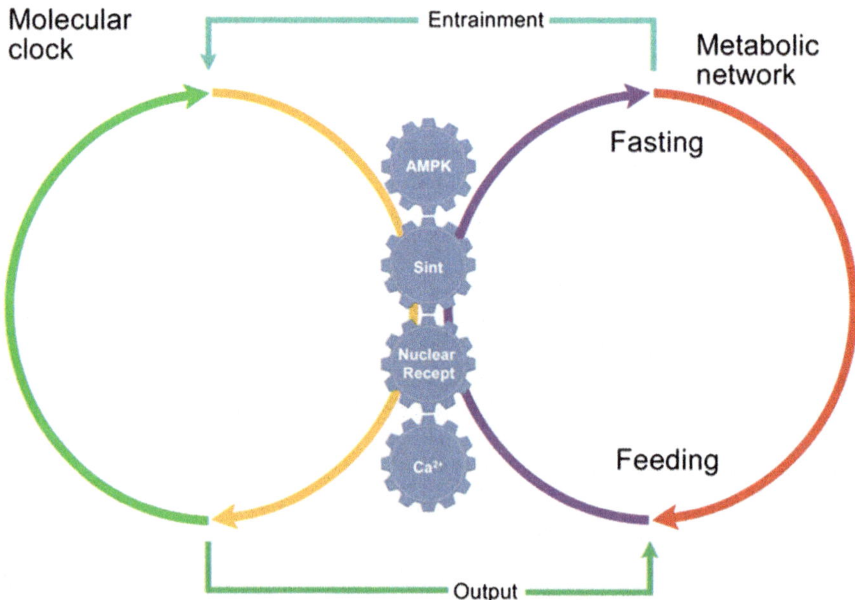

Fig. 14.2 Communication by several metabolic outputs between the molecular circadian clock and the metabolic network. The genes in the molecular clock are connected with the metabolic network by different metabolic signaling pathways; their regulation depends on whether the organism is in a fasting or feeding condition

metabolic intermediates (fatty acids and derivates, bile salts, and sterols), and xeno-biotics. The protein clocks ROR (α, β, and γ) and REV-ERB (α and β) are part of this family. Recent studies reveal some of the nuclear receptor genes as being direct targets of the circadian clock. Peroxisome proliferator-activated receptor-α (PPARα) and PPARγ regulate lipid metabolism and energy homeostasis by coordinated actions in a variety of tissues. Both are under the direct control of the timing system (Yang 2010). Deregulation of PPARs is involved in the onset and development of obesity (Evans et al. 2004). Nuclear receptor dysfunction has been implicated in diverse pathological states, genetic syndromes, and cancer. Co-regulators have broad physiological and pathological functions that make them promising new drug targets for diseases such as hormone-dependent cancers (Lonard and O'Malley 2012).

14.6.2 AMP-Dependent Kinase

AMPK acts as a sensor for the energy status of the cell. It is a trimeric protein that is directly regulated by the ratio AMP/ATP as well as by the cellular metabolic stress response (LKB signaling). When AMPK is active, catabolic processes are upregulated, and anabolic pathways are arrested. This enzyme induces a phase advance of circadian expression of clock genes by degrading one of the isoforms of PER (PER2) through phosphorylation of Casein kinase Iε at Ser389. AMPK also contributes to metabolic entrainment of peripheral clocks by phosphorylating and destabilizing CRY1 (Um et al. 2011). KO models of AMPK result in overweight and development of hyperinsulinemia and glucose intolerance (Steinberg et al. 2010). It has been observed that cancerous tissues show low AMPK activity, suggesting a connection among nutrient sensing, metabolic modulation, and control of cellular division in neoplasic diseases (Korsse et al. 2013).

14.6.3 Sirtuins

Sirtuins are a family of enzymes (from SIRT1 to SIRT9) that show deacetylase or ADP ribosyl-transferase activity. They are present in several compartments of the cell, including the nucleus, mitochondria, and cytosol. Their activity is dependent on the availability of NAD^+, and hence, sirtuins are sensitive to the cytoplasmic and mitochondrial redox state (Denu and Gottesfeld 2012). Experimental evidence has linked the activity of sirtuins to increased longevity and the arrest of cellular decline and metabolic dysfunction associated with aging. SIRT1, through its regulation of PPAR-γ and PGC-1α activity, has a significant regulatory role in fat mobilization and fatty acid oxidation. The observation that SIRT1 can directly deacetylate core components of the circadian clock machinery is interesting, as the ultimate goal of circadian physiology is to coordinate and match intracellular metabolism to external food availability. SIRT1 is required for transcription of several core clock genes, including *Bmal1*, *Rorγ*, *Per2*, and *Cry1* (Jung-Hynes and Ahmad 2009). Interestingly, the clock protein CLOCK has intrinsic histone acetyltransferase activity and operates within a large nuclear complex with other chromatin remodelers. Some substrates acetylated by CLOCK are the histone H3 and the other clock protein BMAL1. SIRT1 counterbalances the enzymatic activity of CLOCK, and hence, it is an important metabolic regulator of the molecular clock (Bellet et al. 2011). Recent studies have shown SIRT1 to regulate fatty acid oxidation in the liver, sense nutrient availability in the hypothalamus, and influence obesity-induced inflammation in macrophages (Schug and Li 2011). It has been reported that SIRT3 can function as a tumor promoter (it is overexpressed in oral cancer cells and tissues) or suppressor (it is downregulated in breast cancer), depending on the cell and tumor type and the presence of different stress or cell death stimuli (Alhazzazi et al. 2011).

14.6.4 Calcium Dynamics

Intracellular calcium handling involves the coordinated action of a variety of proteins responsible for calcium mobilization, responsible for both outward and inward calcium mobilization. There are 2 principal intracellular calcium-release channels: the inositol 1,4,5-trisphosphate receptor (IP_3R) (types 1–3) and the ryanodine receptor (RyR) (types 1–3). The metabolic pumps that extrude cytosolic calcium are the sarco/endoplasmic reticulum calcium ATPase (SERCA) and the plasma membrane calcium ATPase (PMCA) (Gaspers and Thomas 2005). Intracellular calcium fluctuates with different frequencies and to various extents among the cellular organelles. It has been suggested that calcium transients could be a kind of signaling code to coordinate metabolic and transcriptional activities (Berridge et al. 2000). Intracellular calcium is thought to play diverse roles as a component of the timing system (for review Báez-Ruiz and Díaz-Muñoz 2011) by regulating the entrainment process, clock gene expression, and output signaling. Diurnal fluctuations of cytoplasmic calcium as well as of RyR and IP_3R have been observed in the SCN. Supporting the importance of intracellular calcium dynamics in the rhythmicity of the SCN, calcium buffered with >40 µM BAPTA-AM or with low calcium levels (Ca^{2+}-free media) promoted damping of *Per1-luc* expression in SCN explants. In NIH3T3 and rat1-fibroblast cell cultures, thapsigargin and calimycin (a calcium ionophore) increased *Per1* expression during the first hours of treatment. Enhanced adhesion and calcium dynamics are reported to be favorable for the invasion and extravasation required for malignant progression in prostate cancer (Bastatas et al. 2012). Munaron et al. in 2008 postulated that intracellular calcium plays a key role in tumor angiogenesis, an important step in the physiopathology of many cancerous growths.

14.7 Understanding the Mechanisms of Obesity

Obesity is currently a major health problem and is increasing worldwide; its prevalence is mainly a consequence of interactions among multiple genes, physiology, and lifestyle factors. Defined in humans as a body mass index (BMI, weight in kilograms divided by the square of the height in meters) above 30, obesity is currently associated with premature death through increased risk of many chronic diseases, including type 2 diabetes, cardiovascular disease, and cancer (Cheung and Mao 2012). For the past several years, various approaches have been used to identify the physiological and genetic bases of obesity; one of the strongest pieces of evidence involves mechanisms regulating food intake and mediated by hormones and neurotransmitters controlling the sensations of satiety in specific centers within the brain, resulting in chronic energy imbalance such that calories consumed exceed calories expended (Shin et al. 2009). This condition has also been linked to many environments that offer an abundance of calorie-rich foods and few opportunities for physical activities. Recent clinical and laboratory studies suggest that circadian biology also plays an important role in the pathogenesis of obesity, type 2 diabetes, and the metabolic syndrome (Maury et al. 2010).

Circadian rhythms are evolutionarily conserved and represent an adaptation of the organism to its environment; diurnal or nocturnal activity involves behaviors that include feeding–fasting cycles, which are also inputs to the circadian system. The circadian system in mammals is hierarchically organized the hypothalamic suprachiasmatic nucleus (SCN), identified as the dominant circadian pacemaker, imposes period and phase on peripheral oscillators, which are all cells in the body that express "clock genes." However, internal entrainment requires a tight coordination between neural and humoral inputs, driven directly or indirectly by the SCN. These "peripheral clocks" differ in the manner in which they are reset as well as in the outputs under their control, suggesting that the system must be accurately synchronized (Mohawk et al. 2012; Menaker et al. 2013).

Studies on these peripheral clocks, especially as they affect metabolism, may give us a vision of circadian alignment for internal synchrony in normal physiology as well as in disease. The mechanisms, including this internal circadian synchrony, may include signaling at the transcriptional level, posttranscriptional, and posttranslational regulation of the proteins encoded by clock genes, and epigenetic influences of clock genes on chromatin modifications that affect gene regulation (Green et al. 2008; Bass and Takahashi 2010; Froy 2010; Bass 2012).

In normal conditions the SCN maintains temporal organization of activity, body temperature, and feeding, in such a way that local and systemic circadian signals present a particular alignment. However, when food is provided on a regular basis at a different time, out of the animal's activity phase, anticipatory locomotor activity prior to the arrival of the meal indicates that the internal timing also depends on the coincidence of different zeitgebers, such as light and food cycles, and that a different alignment of peripheral oscillators occurs (Pezuk et al. 2012; Mistlberger 2011).

Unorganized activity–rest and feeding–fasting cycles in humans may cause circadian misalignment, and also may facilitate the occurrence of metabolic disorders (Scheer et al. 2009). In addition, disorders in the sleep–wake cycles are associated with irregular feeding schedules, and may be one of the multiple causes of metabolic disorders (Laposky et al. 2008).

Misalignment of the endogenous circadian phase can occur with forced or artificial non-circadian cycles that do not coincide with retinal light exposure, meal times, and the times of wakefulness/activity, work/school, exercise, and sleep; the main examples are shift work and jet lag (Bass 2012). Circadian disregulation and obesity have an initial behavioral reference; artificial bright-light exposure during the evening and hypercaloric diets consumed at night are signals that may work as "false" zeitgebers in conflict with the natural cycles. Together with reduced physical activity and sleep disorders, such behaviors favor an increase in body weight that, over a number of years, results in obesity and metabolic syndrome.

The link between obesity and clock mechanisms is just beginning to be understood. Studies performed in animal models offer unique and powerful experimental approaches and allow us to understand the underlying mechanisms of metabolic disorders and circadian disruption, as well as to develop therapeutic strategies (Arble et al. 2010). Circadian misalignment in rodents may be illustrated by experiments simulating jet lag, where some tissues are able to re-entrain to jet lag more quickly than others (Yamazaki et al. 2000). Mice with mutations in the circadian *Clock* gene develop obesity and metabolic syndrome (Turek et al. 2005), and genetic polymorphisms within the core clock genes *Clock* and *Bmal1* are also associated with obesity, hypertension, and type 2 diabetes (Marcheva et al. 2010). On the other hand, just changing the feeding schedule facilitates overweight and obesity in mice. When food access occurs during the rest/fasting phase of the cycle (photophase in nocturnal rodents), a gradual increase of body weight is clearly noted in comparison with the animals that have access only during the night or ad libitum (Arble et al. 2009). In a rat model susceptible to the development of type 2 diabetes (HIP rats), disruption of circadian rhythms accelerates the loss of beta-cell function and the mass characteristic of this condition (Gale et al. 2011). Feeding time determines the targets and phases of rhythmic transcripts in liver. Thus, there is a synergistic interaction between clock and temporarily restricted feeding patterns in many important metabolic genes, including those involved in mitochondrial function. By using a Restricted Feeding protocol, it is possible to cause a completely new phase relationship between the light-entrained (SCN) and a food-entrainable oscillator (FEO). Mice and rats are naturally nocturnal feeders, and limiting food access to the light phase causes some behaviors, such as feeding-dependent hormonal rhythms and the expression of circadian clock genes and clock-controlled genes within metabolic tissues (e.g., the liver), to become entrained to the novel feeding time.

Clock gene expression in the liver is highly correlated with circadian feeding rhythms, while the SCN remains synchronized to light cues (Stokkan et al. 2001). This dissociation offers an exciting window through which new insights about the link between circadian disruption and metabolism can be explored. On the other hand, misalignment produced in mice by protocols resembling shift-work protocols may result in metabolic disturbances, such as the loss of glucose rhythmicity, inverted triglyceride rhythm, and an increase in body weight (Yoon et al. 2012).

Besides the timing of feeding, hyperlipidic diets also have an impact on the clock outputs, such as the free running period, amplitude of circadian clock protein oscillations, and phase-shift responses to light (Kohsaka et al. 2007; Mendoza et al. 2011). High-fat diets consumed ad libitum may affect the clock genes, reducing protein transcript levels, hormone receptors that regulate the clock, and clock-controlled genes involved in fuel utilization in the hypothalamus and peripheral tissues (Mendoza et al. 2008, 2011), also affecting the phase relationship between central and peripheral circadian oscillators (Pendergast et al. 2013).

Obese mice often show a less organized circadian activity, with bursts of locomotion more often observed during the rest phase of the cycle, more fragmented sleep, and a reduced amplitude of activity during the night (Laposky et al. 2009; Carmona-Alcocer et al. 2012; Fuentes-Granados et al. 2012). The way in which obesity affects circadian rhythms may also differ between sexes; obese female mice present more acute changes of circadian parameters than those observed in males, in an animal model where obesity occurs in an environment with free access to food but with limited space (Carmona-Alcocer et al. 2012).

The understanding of the mechanisms by which obesity is linked to the circadian clock is in its early stages. Different levels of organization in circadian regulation and metabolism can be observed. At the level of hypothalamic center that regulates feeding, the SCN projects axons to the arcuate nucleus, where the main integration of feeding–fasting cycles takes place (Green et al. 2008). Two main hormones, Ghrelin and Leptin, participate in feeding–fasting behavior through hypothalamic mediators. Ghrelin is an orexigenic hormone from the periphery (released mainly by the oxyntic cells within the stomach) that stimulates food intake and is elevated under conditions of physiological demand such as fasting (Schellekens et al. 2012). Leptin is mainly secreted by adipose tissue, and it is recognized as an endocrine signal related to the satiety process. It signals pro-opiomelanocortin (POMC) neurons in hypothalamus (De Jonghe et al. 2012). Both leptin and ghrelin also exert feedback on the SCN and are endocrine signals that are out of balance in conditions of obesity (Froy 2010).

At the cell level, chemical signals that couple internal clock function to nutrient state involve AMP kinase (AMPK). Stimulation of AMPK leads to degradation of CRY (Lamia et al. 2009), suggesting that AMPK is a key sensor and integrator of hormonal and nutritional signals with neurochemical and neurophysiological responses (Schneeberger and Claret 2012) that affect circadian transcriptional cycles. Glucocorticoids have also been shown to modulate the expression of lipogenic genes (Cho et al. 2012) and REV-ERB-α (Torra et al. 2000), a core clock gene involved together with PPARs in lipolytic metabolic pathways. NAD+ oscillation, redox flux, ATP availability, and mitochondrial function can also participate in tuning the clock by posttranslational modification of transcription factors (Bass 2012). NAD+/NADH+H oscillation regulates a protein deacetylase, Sirtuin 1, recently recognized as a key metabolic sensor in several tissues; it modulates a diversity of cellular processes, including glucose and lipid metabolism, fat mobilization, insulin secretion, sensing of nutrient availability in the hypothalamus, and the activity of the circadian clock in metabolic tissues (Li 2013).

Increasing research leads us to conclude that the relationships between circadian clock genes, central and peripheral oscillators as well as inputs and outputs to the SCN, represent a multifactorial system that must eventually be taken into consideration for treatments of obesity.

In humans, the meal timing as well as the food quality and quantity may be considered key factors contributing to metabolic impairment, seen as circadian disruption from consuming calories during the "wrong" circadian time. Circadian misalignment, such as occurs with jet lag or chronically with shift work, increases postprandial glucose, insulin, and mean arterial pressure, while it decreases leptin and sleep efficiency and completely inverts the cortisol profile across the behavioral cycle (Scheer et al. 2009).

Better planning and control of our environment (sleep cycles, time of eating and exercise, light exposure, etc.) will allow us to reduce the chances of circadian misalignment and may also help reduce the severity of various other diseases (Mikhail 2009).

14.8 Chronobiology of Cancer

Cancer is a group of diseases characterized by chronic proliferation of cells having many competetive advantages compared with normal cells. Two models have been proposed to explain the heterogeneous potential of tumor cells and the process of metastasis: (1) the stochastic model, in which a distinct population of cells acquires some somatic mutations and develops metastatic capability, and (2) the hierarchical model, in which primary tumors and metastatic cancer are initiated by occasional stem cells with cancerous properties (Reya et al. 2001; Li et al. 2007).

Hanahan and Weinberg in 2000 proposed to characterize the cancerous process by six hallmarks that constitute a logical framework for understanding the remarkable diversity of neoplasic diseases (Fig. 14.3).

14.8.1 Proliferative Signaling

The most fundamental characteristic of cancer cells is their ability to maintain chronic proliferation. Normal tissues control the release of growth-promoting signals that direct the progression through the cell cycle, therefore ensuring the correct tissue architecture and function. Cancer cells send signals to stimulate normal cells to support the tumor-associated stroma (Cheng et al. 2008; Bhowmick et al. 2004). Receptor signaling can also be deregulated by elevating the levels of receptor proteins displayed at the cancer cell surface, rendering such cells hyperresponsive to otherwise limiting amounts of growth factor ligands.

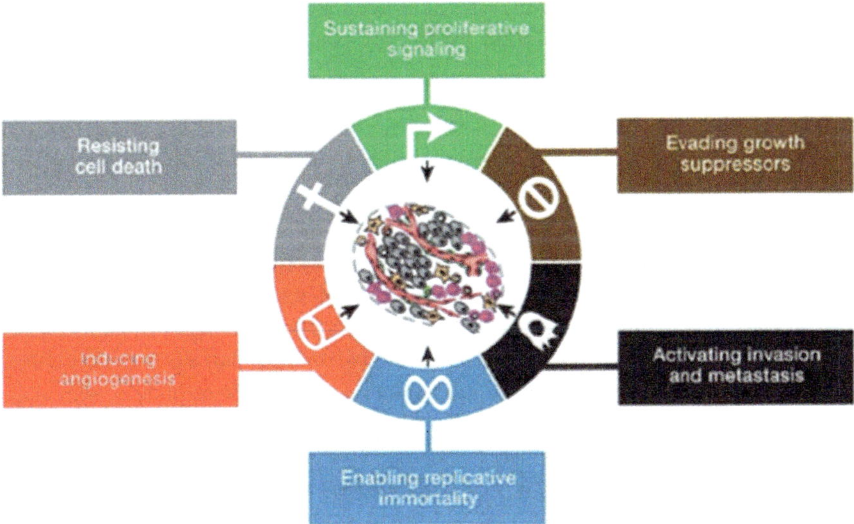

Fig. 14.3 Six hallmark capabilities proposed by Hanahan and Weinberg in 2000 to understand the multistep process of transformation from a healthy cell to neoplasic cell

14.8.2 *Evading Growth Suppressors*

Cancer cells must also circumvent powerful programs that downregulate cell proliferation, mainly by the actions of tumor suppressor genes. Dozens of these genes have been discovered through their effects in one or another type of animal or human cancer; many of these genes have been validated by gain- or loss-of-function experiments in mice. Two prototypical tumor suppressors, RB (retinoblastoma-associated) and TP53 (tumoral p53) proteins, operate as central controls that govern the decisions of cells to proliferate or, alternatively, activate senescence and apoptotic programs (Hanahan and Weinberg 2011).

14.8.3 *Resisting Cell Death*

Experimental and clinical evidence has revealed that apoptosis is attenuated in tumors that progress to states of high-grade malignancy and resistance to therapy. The loss of TP53 tumor suppressor function is one of the principal factors associated with cellular immortalization. Alternatively, tumors may also increase the expression of anti-apoptotic regulators (Bcl-2, Bcl-xL) or of survival signals (IGF-1/2), or downregulate pro-apoptotic factors (Bax, Bim, Puma). The multiplicity of apoptosis-avoiding mechanisms presumably reflects the diversity of apoptosis-inducing signals that cancer cell populations adapt during their evolution to malignancy.

The concept of "programmed cell death" has been broadened to include other forms of cellular death that are inhibited during the cancerous process (Hanahan and Weinberg 2011).

14.8.4   Enabling Replicative Immortality

Cancer cells require unlimited replicative potential in order to generate macroscopic tumors. This capability contrasts with the behavior of normal cell lineages, which are able to pass through only a limited number of successive cell growth-and-division cycles. This limitation has been associated with two distinct barriers to proliferation: senescence, a typically irreversible entrance into a non-proliferative but viable state, and crisis, which involves cell death. Accordingly, when cells are propagated in culture, repeated cycles of cell division lead first to the induction of senescence and then, for those cells that succeed in circumventing this barrier, to a crisis phase, in which the great majority of cells in the population die. On rare occasions, cells emerge from a population in crisis with unlimited replicative potential. This transition has been known as immortalization (Hanahan and Weinberg 2011). Multiple lines of evidence indicate that telomeres protecting the ends of chromosomes are centrally involved in the capacity for unlimited proliferation (Blasco 2005; Shay and Wright 2000).

14.8.5   Inducing Angiogenesis

Like normal tissues, tumors require nutrients and oxygen as well as the ability to evacuate metabolic waste and carbon dioxide. The tumor-associated new vasculature, generated by the process of angiogenesis, fulfills these needs. During tumor progression, an "angiogenic switch" is activated, causing quiescent vasculature to continually sprout new vessels that sustain expanding neoplastic growths (Hanahan and Folkman 1996).

14.8.6   Activating Invasion and Metastasis

As cancer develops and becomes metastasic, there are alterations in its shape as well as in its attachment to other cells and to the extracellular matrix (ECM). The best-characterized alteration involves the loss of E-cadherin, a key cell-to-cell adhesion molecule. By forming adherent junctions between epithelial cells, E-cadherin helps to maintain the quiescent state of the cells. Increased expression of E-cadherin is protective against cancer invasion and metastasis, whereas its

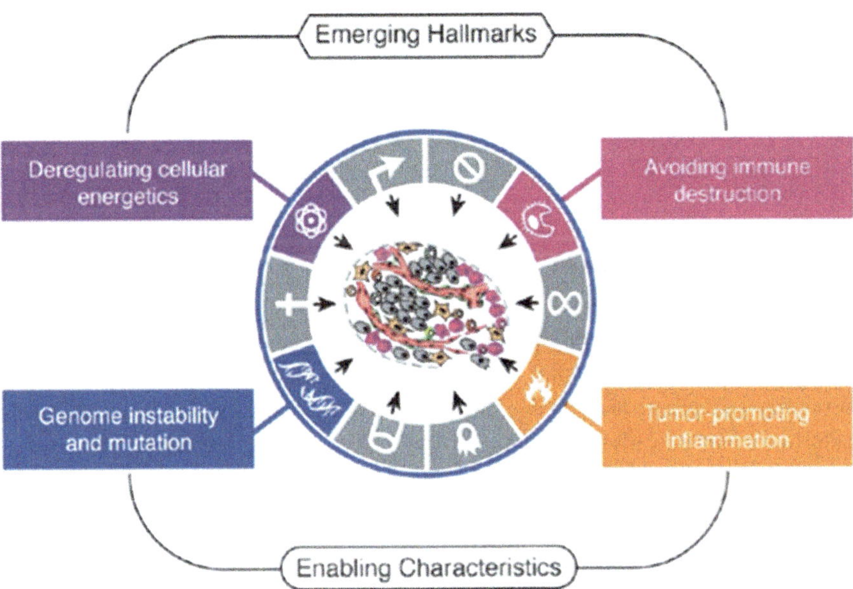

Fig 14.4 This diagram incorporates two additional hallmarks for cancer: (1) deregulation of cellular energetics (metabolism) and (2) avoiding immune destruction. It also illustrates two enabling characteristics involved in transforming healthy cells into neoplasic cells

reduction potentiates these phenotypes. The frequently observed downregulation and occasional mutational inactivation of E-cadherin in human carcinomas support its role as a key suppressor of metastasis (Berx and van Roy 2009; Cavallaro and Christofori 2004). Additionally, expression of genes encoding other cell-to-cell and cell-to-ECM adhesion molecules has been shown to be altered in some highly aggressive carcinomas.

In 2011, Hanahan and Weinberg added another two hallmarks to the cancerous process (Fig. 14.4).

14.8.7 *Reprogramming of Energy Metabolism*

A neoplasic cell adjusts its energy metabolism to fuel cell growth and division. Normal cells process glucose by transforming it into pyruvate via the glycolytic pathway; next, pyruvate goes into mitochondria as one of the main substrates for oxidative phosphorylation. In the neoplasic condition, even in oxidative conditions, glycolysis is favored, and pyruvate is converted to lactate in a process called lactic fermentation (Warburg 1930, 1956a, b). In 1931, the biochemist Otto H. Warburg (1883–1970) received the Nobel Prize for showing how tumor cells changed their metabolic pathways. In his book "The Metabolism of Tumors," Warburg demonstrated that all forms of cancer are characterized by two basic conditions: acidosis and the ability to survive in a hypoxic environment. He reported that cancer cells metabolize approximately tenfold more glucose to lactate than normal tissues (Koppenol et al. 2011).

14.8.8 Evading Immune Destruction

This point is still surrounded by polemics, but increasing evidence suggests that cancer involves an impaired ability of the immune system to prevent or eradicate the formation and progression of incipient neoplasias, late-stage tumors, and micrometastases. The long-standing theory of immune surveillance says that every cell is constantly monitored by the immune system to recognize and eliminate the incipient cancer cells. According to this idea, solid tumors have avoided immune detection or have been able to limit the extent of immunological response (Hanahan and Weinberg 2011).

There are also two enabling characteristics:

14.8.8.1 Tumor-Promoting Inflammation

It is a chronic state of inflammation induced by many processes, such as very active metabolism, formation of reactive species of oxygen (ROS), synthesis of interleukins, and frequent necrotic events. This pro-oxidant condition constitutes a tumor-promoting milieu.

14.8.8.2 Genome Instability and Mutation

Acquisition of the multiple hallmarks already mentioned depends in large part on the alterations in the genomes of the neoplasic cells, reported to include point mutations, enhanced chromosomal translocation, and loss of genetic material. These alterations are globally known as chromosomal instability.

14.9 Circadian Rhythms and Cancer

Recent epidemiological studies suggest that circadian system disorders constitute a risk factor for the development of cancer. A study of pilots and flight attendants who frequently suffer situations of jet lag found that they had a higher incidence of cancer in breast, skin, prostate, and colon (Rafnsson et al. 2001). Other studies reported that women working at night had a higher incidence of breast cancer (Davis et al. 2001).

A 10-year study involving more than 78,000 women identified 2,400 cases with breast cancer that showed a significant correlation with the number of years of nocturnal working. These results suggest that alterations in circadian rhythms may be as important as family history in determining the risk of developing breast cancer (Schernhammer et al. 2001).

14.9.1 *Circadian Molecular Clock and Cellular Proliferation*

Besides measuring time cues, the timing system also modulates many genes of the stress response, metabolism, and cell cycle. Cell division is frequently associated with specific times of day, but just how the circadian clock controls this timing is unclear. In the eukaryotic cell, cellular duplication oscillates between DNA synthesis and mitosis. A few decades ago, the cyclin-dependent kinases (CDKs), their associated proteins, the cyclins, and cyclin-dependent kinases inhibitors (CKIs) (e.g., p21 and p27) emerged as the main regulatory factors in the control of cell cycle events (Fig. 14.5) (Sherr and Roberts 1999; Obaya and Sedivy 2002). The expression of genes that code for proteins that regulate cell proliferation, such as c-myc, cyclin D1, and Wee-1, is influenced by the clock genes (Canaple et al. 2003).

It is noteworthy that a number of cell cycle genes involved in either G2-M or G1-S transition contain the circadian promoters known as E boxes. For example, Wee-1 has 3 E boxes in its promoter and is under the direct control of CLOCK: BMAL1 (Matsuo et al. 2003; Hirayama et al. 2005). The Wee-1 gene encodes a protein kinase that phosphorylates and inactivates the CDC2/Cyclin B1 complex, an event that delays or prevents entry into mitosis. Wee-1 levels robustly oscillate in the mammalian liver, being coordinately regulated with clock genes, even though there is normally little cell proliferation in this organ (Fig. 14.6) (Matsuo et al. 2003).

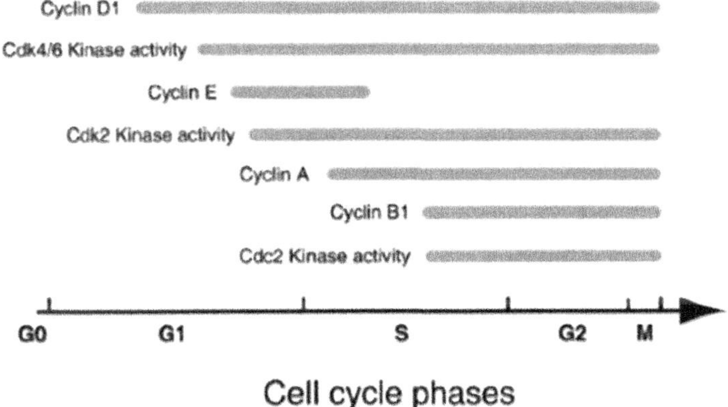

Fig. 14.5 Expression patterns of cyclins and Cdk activities during the cell cycle. Note that Cdk2 can be activated by cyclin E or cyclin A, and Cdc2 can be activated by cyclin A or cyclin B (Obaya and Sedivy 2002)

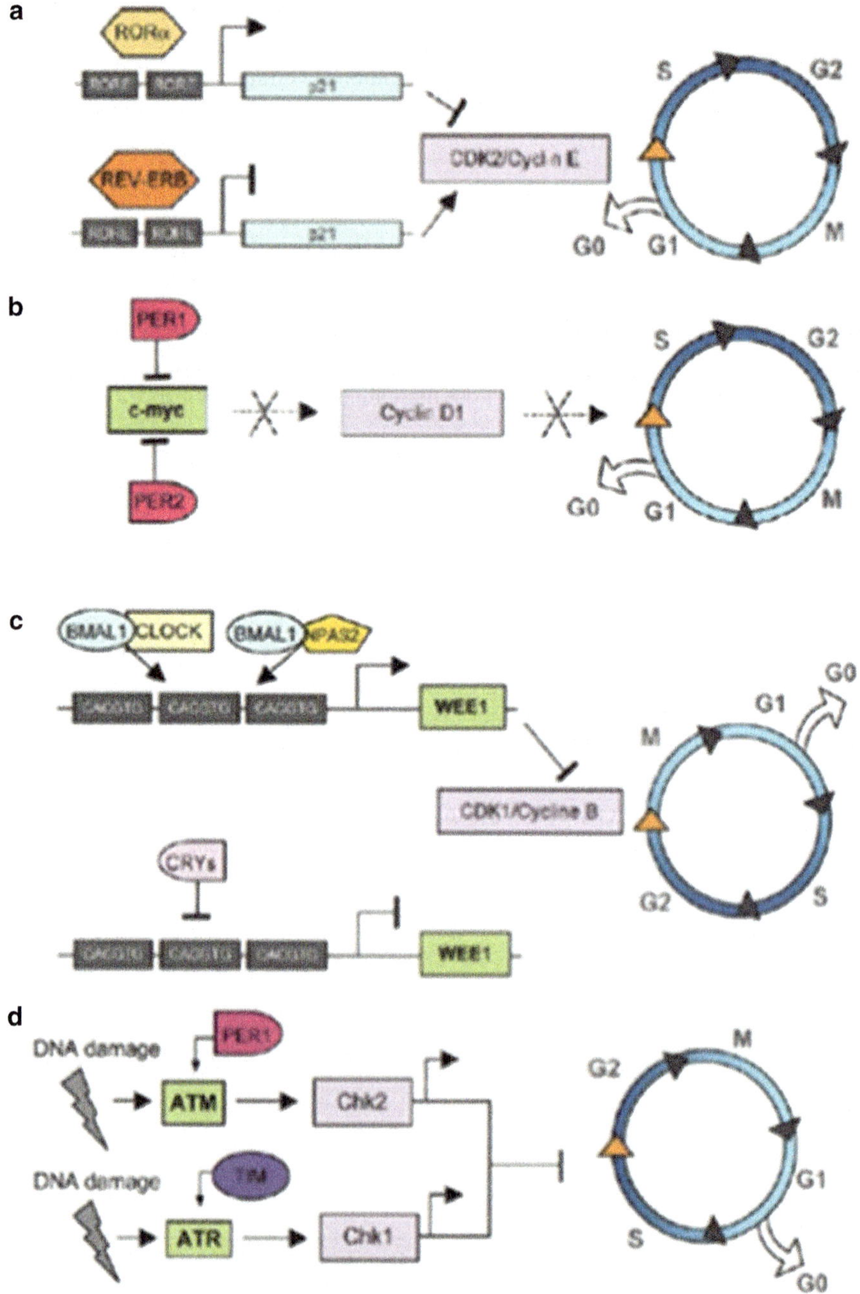

Fig. 14.6 Schematic view of mechanisms implicated in the regulation of cell cycle progression by core clock genes. (**a**) RORα and REV-ERBα bind the same RORE element present in the p21 promoter. RORα activates transcription of p21, which in turn inhibits CDK2/cyclin E and represses G1/S

The p21WAF1/CIP1 cell cycle inhibitor complex is regulated by the clock protein Rev-Erv-α/β, which enables circadian regulation of the progression between the G1 and the S phase of the cell cycle (Gréchez-Cassiau et al. 2008). By reducing c-Myc, the clock protein Bmal provides an additional mechanism to decrease cell proliferation, even in the absence of DNA damage (Filipski et al. 2009). In both liver and skeletal muscle, a significant number of CLOCK-regulated genes are associated with the cell cycle and cell proliferation (Miller et al. 2007).

14.9.2 Circadian Disruption in Carcinogenesis

Recent studies have suggested that components of the circadian pacemaker, such as Clock and Per2, regulate a wide variety of processes, including obesity, sensitization to cocaine, cancer susceptibility, and morbidity to chemotherapeutic agents. Some facts that link circadian physiology with the onset and progression of cancer are: (1) mutations of circadian clock genes are associated with enhanced rates of cancer; (2) clock gene expression is altered in tumors; and (3) mutant mice with disrupted circadian clock genes show increased cancer incidence under basal aging conditions, following gamma-irradiation, and in cancer-prone models (Yu and Weaver 2011).

The influence of Clock gene function on the cytotoxicity of the carcinogenic agent diethyl-nitrosamine (DEN) was shown using wild-type and Clk^-/Clk^- mouse primary hepatocytes. After DEN exposure, wild-type cells underwent apoptosis but not necrosis. This cellular response seems to eliminate damaged cells and prevent carcinogenesis induced by DEN. In contrast, Clk^-/Clk^- cells were resistant to DEN-induced apoptosis. Because apoptosis is important for suppressing carcinogenesis, the mechanisms that underlie the resistance to DEN-induced apoptosis in Clk^-/Clk^- mouse hepatocytes are of great interest (Matsunaga et al. 2011).

Temporal expression of genes involved in cell cycle regulation and tumor suppression, such as c-Myc, Cyclin D1, Cyclin A, Mdm-2, and Gadd45a, are deregulated in Per2 mutant mice (Lee 2006). These animals are more susceptible to damage upon radiation with γ rays, showing salivary gland hyperplasia and aggressive lymphoma (Fu et al. 2002).

Downregulation of Per1 expression increases cancer cell growth in vitro and tumor growth in vivo by enhancing the amplitude of the daily tumor growth peaks.

Fig. 14.6 (continued) progression. REV-ERBα inhibits p21 transcription, leading to the activation of the CDK2/cyclin E complex and to the progression of G1/S. (b) PERs inhibit c-myc transcription, blocking its ability to activate cyclin D1. Thus, PERs indirectly repress the G1/S progression. (c) BMAL1/CLOCK or BMAL1/NPAS2 heterodimers and CRYs recognize E-boxes present in the Wee1 promoter, thereby activating or repressing Wee1 transcription, respectively. WEE1 inhibits the CDK1/cyclin B complex and thus represses G2/M progression. (d) PER1 and TIM act as cofactors that activate ATM or ATR, which in turn phosphorylate Chk2 or Chk1, respectively, leading to cell cycle arrest (Sancar et al. 2010)

The data suggest that Per1 has a tumor-suppressor function that diminishes cancer proliferation and tumor growth, but only at specific times of day.

It has been reported that a functional polymorphism of the PER3 gene is associated with longer survival in patients with hepatocarcinoma (Zhao et al. 2012).

Several investigations have shown increased incidence of some types of cancer in patients with disruption in the circadian system, giving evidence that circadian genes are involved in tumorigenesis by modulating many aspects of the cell cycle, metabolism, and the response to DNA damage.

A study of 104 patients with metastatic breast cancer showed that the risk of death at 4 years was twice as high in the patients with an abnormal diurnal rhythm of salivary cortisol as compared to those with a normal pattern (Sephton et al. 2000). Another study of 200 patients with metastatic colorectal cancer found that severe alterations of the rest–activity circadian rhythm were predictive of a fivefold increase in the risk of death after 2 years (Mormont et al. 2000).

An animal model (mice B6D2F1) was developed to study the consequence of SCN destruction on tumor growth. SCN destruction abolished the circadian rhythmicity and promoted faster growth in tumors, showing that disruption of circadian physiology favors tumor progression (Filipski et al. 2002).

Disruption of circadian rhythm in SCN-injured mice promoted accelerated growth of malignant tumors implanted in experimental models as well as in mice under a chronic jet lag (CJL) protocol. CJL also suppresses or significantly alters the expression rhythms of clock genes in liver and tumors by downregulating p53 and upregulating c-Myc, thus favoring cellular proliferation (Filipski et al. 2009). Regular timing of food access over 24 h counterbalances this effect of CJL on tumor growth; it prevented the circadian disruption produced by CJL and slowed tumor growth (Filipski and Lévi 2009).

14.9.3 Dysregulated Metabolism in Cancer

Cancer cells exhibit enhanced glycolysis, reflected in excess lactate production (Warburg Effect) with respect to normal cells, even in the presence of O_2. Cancer cells show a great demand for energy and precursors to synthesize macromolecules in order to duplicate their genome and biomass. It has been suggested that these intermediate metabolites are supplied by glycolysis. Part of the same hypothesis says that the Krebs cycle is reduced by diminution of anaplerotic reactions, since this type of reactions act in the replenishment of Krebs cycle intermediates (Dell' Antone 2012).

It has been also postulated that glutamine, an important fuel for cancer cells, could play a critical role supporting anaplerosis, via its conversion to glutamate by glutaminase and to α-ketoglutarate via glutamate dehydrogenase. In this way, part of the glutamine structure can be incorporated directly to the Krebs cycle to secure the production of NADH and the functioning of the oxidative phosphorylation process. Malic enzyme could be supplied NADPH for anabolic redox reactions (Dell' Antone 2012).

Activation of glycolysis and overexpression of glucose transporters are also necessary to meet the energy demands of proliferating cells. Increased glucose transport in malignant cells has been associated with enhanced expression of glucose transporter proteins such as GLUT1 and GLUT3, and high levels of GLUT1 expression in human tumors have been correlated with poor survival. Studies indicate that GLUT1 and GLUT3 do not fully explain glucose transport in breast cancer, suggesting involvement of another glucose transporter. Recently, a novel glucose transporter protein, GLUT12, has been found in breast and prostate cancers. GLUT12 is located in the plasma membrane but also in endomembranes. Trafficking of GLUT12 to the plasma membrane could therefore contribute to glucose uptake. Several factors have been implicated in the regulation of glucose transporter expression in breast cancer. Hypoxia can increase GLUT1 levels and glucose uptake. Estradiol and epidermal growth factor also increase GLUT12 protein levels in cultured breast cancer cells (Mancheda et al. 2005).

Fifty years ago, it was understood that aerobic glycolysis and uptake of glutamine and glycine allow cancer cells to produce ATP and the nucleotides, amino acids, and lipids required for cell proliferation. There are reports of specific alterations in several enzymes related to energy metabolism, such as isocitrate dehydrogenase 1 and 2 (IDH1 and IDH2), succinate dehydrogenase (SDH), fumarate hydratase (FH), and pyruvate kinase M2 (PKM2). These enzymes are now considered to be proto-oncogenes. PKM2 expression in cancer cells stimulates aerobic glycolysis. Among intermediary metabolism enzymes, mutations in SDH occur in gastrointestinal stromal tumors and result in a pseudo-hypoxic metabolic milieu. FH mutations are characteristic of renal cell carcinoma. Isocitrate dehydrogenase (IDH1/2) mutations have been found in leukemias, gliomas, prostate cancer, colon cancer, thyroid cancer, and sarcomas (Teicher et al. 2012).

Altered lipid metabolism has been increasingly recognized as another hallmark of cancer cells. The changes of expression and activity of lipid metabolizing enzymes are directly regulated by oncogenic signals. Like glucose metabolism, lipid metabolism in cancer cells is regulated by the common oncogenic signaling pathways, and is believed to be important for the initiation and progression of tumors. The exacerbated lipogenesis in cancer cells is also directly coupled to other common metabolic pathways and their associated cell signaling pathways (Tennant et al. 2010; Cairns et al. 2011).

Interestingly, similar to embryonic cells, many cancer cells synthesize 95 % of their own fatty acids (Vazquez-Martin et al. 2008). The expression and activity of many enzymes involved in fatty acid synthesis, i.e., ATP citrate lyase (ACL), acetyl-CoA carboxylase (ACC), and fatty acid synthase (FASN) are upregulated in many types of cancer.

Many connections between the circadian clock and cellular metabolism have been identified, including chromatin remodeling. This suggests that abnormal metabolism in cancer could also be a consequence of a disrupted circadian clock.

Therefore, a comprehensive understanding of the molecular links that connect the circadian clock to the cell cycle and metabolism could provide therapeutic benefits against certain human neoplasias (Sahar and Sassone-Corsi 2009).

The acetyl-transferase activity of CLOCK, and the counterbalancing deacetylase activity of SIRT1 seem to regulate and integrate the circadian clock and metabolism, with many implications for neoplasic processes (Sahar and Sassone-Corsi 2009).

It has long been known that the toxicity, efficacy, metabolism, and elimination of many drugs change over a 24-h period (Gachon and Firsov 2011). Recent knowledge about clock genes has given rise to the field of chronopharmacology, the study of the interaction of biological rhythms and drugs, and chronopharmaceutics, the design and evaluation of delivering drugs in a time-controlled manner and rhythm that matches biological requirements (Youan 2004). Diseases that are related to known circadian rhythms, including cancer and other diseases such as asthma, arthritis, diabetes, hypertension, and hypercholesterolemia, are currently studied from the point of view of chronotherapy.

Research in rodents with restricted feeding schedules showed ~40 % inhibition of cancer growth as compared with the control ad libitum group, irrespective of calorie intake. Clock gene transcription remained arrhythmic in tumors and was independent of feeding schedule or diet. However, restricted feeding upregulated or downregulated the expression of 423 tumor genes. The tumor growth curves of experimental animals indicated that restricted feeding significantly slowed the tumor growth. Survival of rodents increased with MT (meal timing) during the light phase versus the Ad Libitum condition or MT during the dark period (Li et al. 2010).

Restriction of dietary calories reduces cancer in experimental animals and probably also in humans. This effect is generally attributed to the inhibitory effect of fasting on cell proliferation.

14.10 Concluding Remarks

The recent studies and concepts summarized in this text have advanced our understanding of the physiopathology of obesity and cancer. The interaction among metabolic networks, cell cycle, and the timing system has supported a new perspective regarding the importance of circadian physiology to explain the molecular mechanisms that underlie obesity and cancer. Further studies using multidisciplinary approaches applied to complex systems would eventually provide a better insight into the intricate control of cellular bioenergetics and differentiation exerted by clock protein/genes.

Acknowledgments We are in debt with Nutriologist Fernando López Barrera for design of Fig. 14.2 and Dr. Dorothy Pless for revision and enrich the English redaction. Research in M. D.-M.'s laboratory is supported by projects 129-511 (CONACyT) and 202412-23 (PAPIIT, UNAM).

References

Alhazzazi TY, Kamarajan P, Verdin E et al (2011) SIRT3 and cancer: Tumor promoter or suppressor? Biochem Biophys Acta 1816:80–88

Arble DM, Bass J, Laposky AD et al (2009) Circadian timing of food intake contributes to weight gain. Obesity 17(11):2100–2102

Arble DM, Ramsey KM, Bass J et al (2010) Circadian disruption and metabolic disease: findings from animal models. Best Pract Res Clin Endocrinol Metab 24(5):785–800

Atkinson DE (1968) The energy charge of the adenylate pool as a regulatory parameter. Interaction with feedback modifiers. Biochemistry 7:4030–4034

Báez-Ruiz A, Díaz-Muñoz M (2011) Chronic inhibition of endoplasmic reticulum calcium-release channels and calcium-ATPase lengthens the period of hepatic clock gene Per1. J Circadian Rhythms 9:6

Bass J (2012) Circadian topology of metabolism. Nature 491:348–356

Bass J, Takahashi JS (2010) Circadian integration of metabolism and energetics. Science 330(6009):1349–1354

Bastatas L, Martínez-Marín D, Matthews J et al (2012) AFM nano-mechanics and calcium dynamics of prostate cancer cells with distinct metastatic potential. Biochem Biophys Acta 1820:1111–1120

Bellet MM, Orozco-Solis R, Sahar S et al (2011) The time of metabolism: NAD+, SIRT1, and the circadian clock. Cold Spring Harb Symp Quant Biol 76:31–38

Berridge MJ, Lipp P, Bootman MD (2000) The versatility and universality of calcium signaling. Nature 1:11–21

Berx G, van Roy F (2009) Involvement of members of the cadherin superfamily in cancer. Cold Spring Harb Perspect Biol 1(6):a003129. doi:10.1101/cshperspect.a003129

Bhowmick NA, Neilson EG, Moses HL (2004) Stromal fibroblasts in cancer initiation and progression. Nature 432:332–337

Blasco MA (2005) Telomeres and human disease: ageing, cancer and beyond. Nat Rev Genet 6:611–622

Cairns RA, Harris IS, Mak TW (2011) Regulation of cancer cell metabolism. Nat Rev Cancer 11:85–95

Calle EE, Kaaks R (2004) Overweight, obesity and cancer: epidemiological evidence and proposed mechanisms. Nat Rev Cancer 4:579–591

Calle EE, Rodríguez C, Walker-Thurmond K et al (2003) Overweight, obesity, and mortality from cancer in a prospectively studied cohort of U. S. adults. N Engl J Med 348:1625–1638

Canaple L, Kakizawa T, Laudet V (2003) The days and nights of cancer cells. Cancer Res 63(22):7545–7552

Carmona-Alcocer V, Fuentes-Granados C, Carmona-Castro A, Aguilar-González I, Cárdenas-Vázquez R, Miranda-Anaya M (2012) Obesity alters circadian behavior and metabolism in sex dependent manner in the volcano mouse Neotomodon alstoni. Physiol Behav 105:727–733

Cavallaro U, Christofori G (2004) Cell adhesion and signalling by cadherins and Ig-CAMs in cancer. Nat Rev Cancer 4(2):118–132

Cheng N, Chytil A, Shyr Y et al (2008) Transforming growth factor-beta signaling-deficient fibroblasts enhance hepatocyte growth factor signaling in mammary carcinoma cells to promote scattering and invasion. Mol Cancer Res 6:1521–1533

Cheung WW, Mao P (2012) Recent advances in obesity: genetics and beyond. ISRN Endocrinol 2012:536905. doi:10.5402/2012/536905, Epub 2012 Mar 5

Cho H, Zhao X, Hatori M et al (2012) Regulation of circadian behaviour and metabolism by REV-ERB-α and REV-ERB-β. Nature 485:123–127

Collins S, Martin TL, Surwit RS et al (2004) Genetic vulnerability to diet-induced obesity in the C57BL/6J mouse: physiological and molecular characteristics. Physiol Behav 81:243–248

Dang CV (2012) Links between metabolism and cancer. Genes Dev 23:877–890

Davis S, Mirick DK, Stevens RG (2001) Night shift work, light at night, and risk of breast cancer. J Natl Cancer Inst 93:1557–1562

De Jonghe BC, Hayes MR, Kanoski SE et al (2012) Food intake reductions and increases in energetic responses by hindbrain leptin and melanotan II are enhanced in mice with POMC-specific PTP1B deficiency. Am J Physiol Endocrinol Metab 303:E644–E651

Delezie J, Challet E (2011) Interactions between metabolism and circadian clocks: reciprocal disturbances. Ann N Y Acad Sci 1243:30–46

Dell' Antone P (2012) Energy metabolism in cancer cells: How to explain the Warburg and Crabtree effects? Med Hypotheses 79:388–392

Denu JM, Gottesfeld JM (2012) Minireview series on sirtuins: from biochemistry to health and disease. J Biol Chem 287:42417–42418

Dunlap JC (1999) Molecular bases for circadian clocks. Cell 96:271–290

Elias M (2010) Patterns and processes in the evolution of the eukaryotic endomembrane system. Mol Membr Biol 27:479–489

Evans RM, Barish GD, Wang YX (2004) PPARs and the complex journey to obesity. Nat Med 10:355–361

Faith MS, Kral TVE (2006) Social environmental and genetic influences on obesity and obesity-promoting behaviors: fostering research integration. In: Institute of Medicine (US) Committee on Assessing Interactions among Social, Behavioral, and Genetic Factors in Health; Nández LM, Blazer DG (eds). Genes, behavior, and the social environment: moving beyond the nature/nurture debate. National Academies Press, Washington, DC

Filipski L, Lévi F (2009) Circadian disruption in experimental cancer processes. Integr Cancer Ther 8(4):298–302

Filipski E, King VM, Li X-M et al (2002) Disruption of circadian coordination accelerates malignant growth in mice. Pathol Biol 51:216–219

Filipski E, Subramanian P, Carrière J et al (2009) Circadian disruption accelerates liver carcinogenesis in mice. Mutat Res 680:95–105

Froy O (2010) Metabolism and circadian rhythms-implications for obesity. Endocr Rev 31(1):1–24

Fu L, Pelicano H, Liu J et al (2002) The circadian gene Period2 plays an important role in tumor suppression and DNA damage response in vivo. Cell 111:41–50

Fuentes-Granados C, Duran P, Carmona-Castro A et al (2012) Obesity alters the daily sleep homeostasis and metabolism of the volcano mouse Neotomodon alstoni. Biol Rhythm Res 43(1):39–47

Gachon F, Firsov D (2011) The role of circadian timing system on drug metabolism and detoxification. Expert Opin Drug Metab Toxicol 7:147–158

Gale JE, Cox HI, Qian J et al (2011) Disruption of circadian rhythms accelerates development of diabetes through pancreatic beta-cell loss and dysfunction. J Biol Rhythms 26(5):423–433

Gaspers LD, Thomas AP (2005) Calcium signaling in the liver. Cell Calcium 38:329–342

Golombek DA, Rosenstein RE (2010) Physiology of the circadian entrainment. Physiol Rev 90(3):1063–1102

Gréchez-Cassiau A, Rayet B, Guillaumond F, Teboul M, Delaunay F (2008) The circadian clock component BMAL1 is a critical regulator of p21WAF1/CIP1 expression and hepatocyte proliferation. J Biol Chem 283(8):4535–4542

Green CB, Takahashi J, Bas J (2008) The meter of metabolism. Cell 134:728–742

Hahn WC, Counter CM, Lundberg AS et al (1999) Creation of human tumor cells with well-defined genetic elements. Nature 400:464–468

Hanahan D, Folkman J (1996) Patterns and emerging mechanisms of the angiogenic switch during tumorigenesis. Cell 86:353–364

Hanahan D, Weinberg RA (2000) The hallmarks of cancer. Cell 100:57–70

Hanahan D, Weinberg RA (2011) Hallmarks of cancer: the next generation. Cell 144:646–674

Hirayama J, Cardone L, Doi M, Sassone-Corsi P (2005) Common pathways in circadian and cell cycle clocks: light-dependent activation of Fos/AP-1 in zebrafish controls CRY-1a and WEE-1. Proc Natl Acad Sci USA 102(29):10194–10199

Item F, Konrad D (2012) Visceral fat and metabolic inflammation: the portal theory revisited. Obes Rev 13(Suppl 2):30–39

Jung-Hynes B, Ahmad N (2009) SIRT1 controls circadian clock circuitry and promotes cell survival: a connection with age-related neoplasms. FASEB J 23:2803–2809

Kanasaki K, Koya D (2011) Biology of obesity: lessons from animal models of obesity. Biomed Res Inter 2011:Article ID 197636, 11 pages, doi:10.1155/2011/197636

Kohsaka A, Laposky AD, Ramsey KM et al (2007) High- fat diet disrupts behavioral and molecular circadian rhythms in mice. Cell Metab 6:414–421

Koppenol WH, Bounds PL, Dang CV (2011) Otto Warburg's contributions to current concepts of cancer metabolism. Nat Rev Cancer 11:325–337

Korsse SE, Peppelenbosch MP, van Veelen W (2013) Targeting LKB1 signaling in cancer. Biochim Biophys Acta 1835(2):194–210

Lamia KA, Sachdeva UM, DiTacchio L et al (2009) AMPK regulates the circadian clock by cryptochrome phosphorylation and degradation. Science 326(5951):437–440

Laposky AD, Bradley MA, Williams DL et al (2008) Sleep-wake regulation is altered in leptin-resistant (db/db) genetically obese and diabetic mice. Am J Physiol Regul Integr Comp Physiol 295(6):R2059–R2066

Larqué C, Velasco M, Navarro-Tableros V et al (2011) Early endocrine and molecular changes in metabolic syndrome models. IUBMB Life 63(10):831–839

Lee CC (2006) Tumor suppression by the mammalian Period genes. Cancer Causes Control 17:525–530

Li X (2013) SIRT1 and energy metabolism. Acta Biochim Biophys Sin (Shanghai) 45(1):51–60. doi:10.1093/abbs/gms108

Li F, Tiede B, Massague J et al (2007) Beyond tumorigenesis: cancer stem cells in metastasis. Cell Res 17:3–14

Li X-M, Delaunay F, Dulong S et al (2010) Cancer inhibition through circadian reprogramming of tumor transcriptome with meal timing. Cancer Res 70(8):3351–3360

Liu MS, Zhang JN (1985) Glycolytic and tricarboxylic acid cycle intermediates in dog livers during endotoxic shock. Biochem Med 34:335–343

Lonard DM, O'Malley BW (2012) Nuclear receptors coregulators: Modulators of pathologies and therapeutic targets. Nat Rev Endocrinol 8(10):598–604

Mancheda LM, Rogers S, Best DJ (2005) Molecular and cellular regulation of glucose transporter (GLUT) proteins in cancer. J Cell Physiol 202:654–662

Marcheva B, Ramsey KM, Buhr ED et al (2010) Disruption of the clock components CLOCK and BMAL1 leads to hypoinsulinaemia and diabetes. Nature 466(7306):627–631

Matsunaga N, Kohno Y, Kakimoto K et al (2011) Influence of CLOCK on cytotoxicity induced by diethylnitrosamine in mouse primary hepatocytes. Toxicology 280:144–151

Matsuo T, Yamaguchi S, Mitsui S et al (2003) Control mechanism of the circadian clock for timing of cell division in vivo. Science 302:255–259

Maury E, Ramsey KM, Bass J (2010) Circadian rhythms and metabolic syndrome: from experimental genetics to human disease. Circ Res 106(3):447–462

Menaker M, Murphy ZC, Sellix MT (2013) Central control of peripheral circadian oscillators. Curr Opin Neurobiol 23:1–6

Mendoza J, Pevet P, Challet E (2008) High-fat feeding alters the clock synchronization to light. J Physiol 586(24):5901–5910

Mendoza J, Lopez-Lopez C, Revel FG et al (2011) Dimorphic effects of leptin on the circadian and hypocretinergic systems of mice. J Neuroendocrinol 23(1):28–38

Mikhail N (2009) The metabolic syndrome: insulin resistance. Curr Hypertens Rep 11(2):156–158

Miller BH, McDearmon EL, Panda S et al (2007) Circadian and CLOCK-controlled regulation of the mouse transcriptome and cell proliferation. Proc Natl Acad Sci U S A 104(9): 3342–3347

Mistlberger RE (2011) Neurobiology of food anticipatory circadian rhythms. Physiol Behav 104(4):535–545

Mohawk JA, Green CB, Takahashi JS (2012) Central and peripheral circadian clocks in mammals. Annu Rev Neurosci 35:445–462

Mormont MC, Waterhouse J, Bleuzen P et al (2000) Marked 24-h rest/activity rhythms are associated with better quality of life, better response, and longer survival in patients with metastatic colorectal cancer and good performance status. Clin Cancer Res 6:3038–3045

Munaron L, Tomatis C, Fiorio PA (2008) The secret marriage between calcium and tumor angiogenesis. Technol Cancer Res Treat 7:335–339

Obaya AJ, Sedivy JM (2002) Regulation of cyclin-Cdk activity in mammalian cells. Cell Mol Life Sci 59:126–142

Pendergast JS, Branecky KL, Yang W et al (2013) High-fat diet acutely affects circadian organization and eating behavior. Eur J Neurosci 37:1350–1356

Pezuk P, Mahawk JA, Yoshikawa T et al (2012) Circadian organization is governed by extra SCN pacemakers. J Biol Rhythms 25:432–441

Rafnsson V, Tulinius H, Jónasson JG et al (2001) Risk of breast cancer in female flight attendants: a population-based study (Iceland). Cancer Causes Control 12(2):95–101

Reddy AB, O'Neill JS (2011) Metaclocks. EMBO Rep 12:612

Reya T, Morrison SJ, Clarke MF et al (2001) Stem cells, cancer, and cancer stem cells. Nature 414:105–111

Sahar S, Sassone-Corsi P (2009) Metabolism and cancer: the circadian clock connection. Nat Rev Cancer 9:886–896

Sancar A, Lindsey-Boltz LA, Kang TH, Reardon JT, Lee JH, Ozturk N (2010) Circadian clock control of the cellular response to DNA damage. FEBS Lett 584(12):2618–2625

Savvidis C, Koutsilieris M (2012) Circadian rhythm disruption in cancer biology. Mol Med 18:1249–1260

Scheer FA, Hilton MF, Mantzoros CS et al (2009) Adverse metabolic and cardiovascular consequences of circadian misalignment. Proc Natl Acad Sci U S A 106(11):4453–4458

Schellekens H, Finger BC, Dinan TG, Cryan JF (2012) Ghrelin signalling and obesity: at the interface of stress, mood and food reward. Pharmacol Ther 135(3):316–326

Schernhammer ES, Laden F, Speizer FE et al (2001) Rotating night shifts and risk of breast cancer in women participating in the nurses' health study. J Natl Cancer Inst 93:1563–1568

Schibler U, Ripperger J, Brown SA (2003) Peripheral circadian oscillators in mammals: time and food. J Biol Rhythms 18:250–260

Schneeberger M, Claret M (2012) Recent insights into the role of hypothalamic AMPK signaling cascade upon metabolic control. Front Neurosci 6:185. doi:10.3389/fnins.2012.00185

Schug XX, Li T (2011) Sirtuin 1 in lipid metabolism and obesity. Ann Med 43:198–211

Sephton SE, Sapolsky RM, Kraemer HC et al (2000) Diurnal cortisol rhythm as a predictor of breast cancer survival. J Natl Cancer Inst 92:994–1000

Shay JW, Wright WE (2000) Hayflick, his limit, and cellular ageing. Nat Rev Mol Cell Biol 1:72–76

Sherr CJ, Roberts JM (1999) CDK inhibitors: positive and negative regulators of G1-phase progression. Genes Dev 13:1501–1512

Shin AC, Zheng H, Berthoud HR (2009) An expanded view of energy homeostasis: neural integration of metabolic, cognitive, and emotional drives to eat. Physiol Behav 97(5):572–580

Steinberg GR, O'Neill HM, Dzamko NL et al (2010) Whole body deletion of AMP-activated protein kinase $\beta 2$ reduces muscle AMPK activity and exercise capacity. J Biol Chem 285(48):37198–37209

Stokkan KA, Yamazaki S, Tei H et al (2001) Entrainment of the circadian clock in the liver by feeding. Science 291(5503):490–493

Teicher BA, Marston Linehan W et al (2012) Targeting cancer metabolism. Clin Cancer Res 18(20):5537–5545

Tennant DA, Durán RV, Gottlieb E (2010) Targeting metabolic transformation for cancer therapy. Nat Rev Cancer 10:267–277

Torra IP, Tsibulsky V, Delaunay F et al (2000) Circadian and glucocorticoid regulation of Rev-erbα expression in liver. Endocrinology 141:3799–3806

Turek FW, Joshu C, Kohsaka A et al (2005) Obesity and metabolic syndrome in circadian Clock mutant mice. Science 308(5724):1043–1045

Ukai H, Ueda HR (2010) Systems biology of mammalian circadian clocks. Annu Rev Physiol 72:579–603

Um JH, Pendergast JS, Springer DA et al (2011) AMPK regulates circadian rhythms in a tissue- and isoform-specific manner. PLoS One 6(3):e18450. doi:10.1371/journal.pone.0018450

Vazquez-Martin A, Colomer R, Brunet J et al (2008) Overexpression of fatty acid synthase gene activates HER1/HER2 tyrosine kinase receptors in human breast epithelial cells. Cell Prolif 41:59–85

Veech RL (2006) The determination of the redox states and phosphorylation potential in living tissues and their relationship to metabolic control of disease phenotypes. Biochem Mol Biol Educ 34:168–179

Warburg OH (1930) The metabolism of tumours: investigations from the Kaiser Wilhelm Institute for Biology, Berlin-Dahlem. Arnold Constable, London

Warburg O (1956a) On the origin of cancer cells. Science 123:309–314

Warburg O (1956b) On respiratory impairment in cancer cells. Science 124:269–270

Yamazaki S, Numano R, Abe M et al (2000) Resetting central and peripheral circadian oscillators in transgenic rats. Science 288(5466):682–685

Yang X (2010) A wheel of time: the circadian clock, nuclear receptors, and physiology. Genes Dev 24:741–747

Yoon JA, Han DH, Noh JY et al (2012) Meal time shift disturbs circadian rhythmicity along with metabolic and behavioral alterations in mice. PLoS One 7(8):e44053

Youan B-B (2004) Chronopharmaceutics: gimmick or clinically relevant approach to drug delivery? J Control Release 98:337–353

Yu E, Weaver D (2011) Disrupting the circadian clock: Gene-specific effects on aging, cancer and other phenotypes. Aging 3:479–493

Zhao B, Lu J, Yin J et al (2012) A functional polymorphism in PER3 gene is associated with prognosis in hepatocellular carcinoma. Liver Int 32(9):1451–1459

Zhao JJ, Roberts TM, Hahn WC (2004) Functional genetic and experimental models of human cancer. Trends Mol Med 10(7):344–350

Chapter 15
Effects of Circadian Disruption on Physiology and Pathology: From Bench to Clinic (and Back)

Juan J. Chiesa, José M. Duhart, Leandro P. Casiraghi, Natalia Paladino, Ivana L. Bussi, and Diego A. Golombek

Abstract Nested within the hypothalamus, the suprachiasmatic nuclei (SCN) represent a central biological clock that regulates daily and circadian (i.e., close to 24 h) rhythms in mammals. Besides the SCN, a number of peripheral oscillators throughout the body control local rhythms and are usually kept in pace by the central clock. In order to represent an adaptive value, circadian rhythms must be entrained by environmental signals or *zeitgebers*, the main one being the daily light–dark (LD) cycle. The SCN adopt a stable phase relationship with the LD cycle that, when challenged, results in abrupt or chronic changes in overt rhythms and, in turn, in physiological, behavioral, and metabolic variables. Changes in entrainment, both acute and chronic, may have severe consequences in human performance and pathological outcome. Indeed, animal models of desynchronization have become a useful tool to understand such changes and to evaluate potential treatments in human subjects. Here we review a number of alterations in circadian entrainment, including jet lag, social jet lag (i.e., desynchronization between body rhythms and normal time schedules), shift work, and exposure to nocturnal light, both in human subjects and in laboratory animals. Finally, we focus on the health consequences related to circadian/entrainment disorders and propose a number of approaches for the management of circadian desynchronization.

15.1 Circadian Entrainment

The circadian clock located in the hypothalamic suprachiasmatic nucleus (SCN) is the central oscillator that regulates daily rhythms in mammals. Peripheral oscillators coupled to the SCN are synchronized by neural and humoral pathways and are able

J.J. Chiesa • J.M. Duhart • L.P. Casiraghi • N. Paladino • I.L. Bussi • D.A. Golombek (✉)
Chronobiology Lab, National University of Quilmes (UNQ)/CONICET,
Roque Saénz Peña 352, Bernal B1876BXD, Buenos Aires, Argentina
e-mail: dgolombek@unq.edu.ar

© Springer International Publishing Switzerland 2015
R. Aguilar-Roblero et al. (eds.), *Mechanisms of Circadian Systems in Animals and Their Clinical Relevance*, DOI 10.1007/978-3-319-08945-4_15

to achieve stable temporal relationships with the environment. Moreover, internal temporal order ensures synchronization among the diverse oscillators in the body (Mohawk et al. 2012). This means that most, if not all, physiological, behavioral, and cognitive variables will be fine-tuned with cyclic environmental resources, in order to ultimately optimize metabolic processes and energy use (Menaker 2006). Since the period of the circadian clock (τ) spontaneously deviates from 24 h, daily synchronization of its oscillation is required for the temporal adjustment of the organism with the 24 h light:dark (LD) cycle (i.e., $\tau = 24$ h). Photic information from the LD cycle induces daily changes in circadian period and/or phase, either by responding to the complete photoperiod and/or to acute light stimulation during the light–dark transitions. Indeed, short light exposition (i.e., a light-pulse) during the night can either delay or advance the clock during the early or late night, respectively. This "phase-dependent" changes of clock activity configure a phase response curve (PRC), which has a similar pattern both for nocturnal and diurnal animals (reviewed in Johnson et al. 2003). Under normal conditions, this entrainment process (reviewed in Roenneberg et al. 2003) is limited by the amplitude of the PRC. Changes in this phase-relationship are environmentally imposed both by abrupt LD shifts (e.g., as in jet lag), or by shifting the activity–rest cycle in relation to the LD cycle (i.e., as in shift work). Such conflicts lead to a transient desynchronization of the circadian clock, which gradually tends to recover its normal phase-relationship with the environment. Changes in circadian entrainment may have severe consequences in human performance and pathological outcome. Indeed, animal models of desynchronization have become a useful tool to understand such changes and to evaluate potential treatments in human subjects. In this chapter we discuss human and experimental models of entrainment and desynchronization, focusing on their effects on health and, ultimately, quality of life for human and animal subjects.

15.1.1 Physiology of Entrainment

Most of the individual SCN cells contain autoregulatory transcription–translation loops that drive circadian oscillations in metabolic and electrical activity (Reppert and Weaver 2001). The positive arm protein components, Circadian Locomotor Output Cycles Kaput (CLOCK), and brain-muscle aryl hydrocarbon receptor nuclear translocator-like one (BMAL1) heterodimerize and bind to E-boxes in the promoter region of *period* (*per*) genes, Cryptochrome (Cry) genes and Rev-Erbα, activating their transcription. As PER and CRY proteins accumulate in the cytoplasm, they dimerize and translocate to the nucleus where they inhibit the activity of CLOCK–BMAL, thus inhibiting their own transcription. At the tissue level, each bilateral SCN is generally composed of two separate regions with neurons expressing different neurotransmitter phenotypes (Antle and Silver 2005; Welsh et al. 2010). A ventral "core" region receives retinal efferents and is composed of cells expressing vasoactive intestinal peptide (VIP), calretinin, and gastrin-releasing peptide (GRP), which project densely to a dorsal-medial "shell"-shaped region. In this shell region most neurons exhibit oscillatory properties and many express arginine–vasopresin peptide (AVP). All SCN neurons also express gamma-aminobutyric acid (GABA). Although these cellular oscillators exhibit a distribution of periods when dispersed in culture, they are synchronized in the whole SCN to maintain a coherent circadian output, by both chemical (VIP, AVP, and GABA-ergic) and electrical (gap-junctional) synapses, as well as paracrine calcium and nitric oxide signaling (Aton and Herzog 2005; Honma et al. 2012; Plano et al. 2010).

Photic stimuli from the environment are transmitted to the ventral region of the SCN from intrinsically photoreceptive retinal ganglion cells, through efferents in the retinohypothalamic tract, releasing glutamate, substance P, and pituitary adenylyl cyclase activating peptide (PACAP) (Golombek and Rosenstein 2010). Phase-advancing or delaying light-pulses are differentially transduced in these retinorecipient neurons. Downstream from the glutamate–NMDA interaction, an intracellular cascade of second messengers is activated, involving extracellular calcium inflow and the sequential increase in the catalytic activity of calmodulin kinase and nitric oxide (NO) synthase (increasing NO levels). The pathway then bifurcates and can lead to guanilyl-cyclase (increasing cyclic-guanosyl monophosphate, cGMP levels) and cGMP-dependent protein-kinase (PKG) activation, in the case of light-induced phase advances (Golombek et al. 2004). NO can also activate both types 1 and 2 ryanodine channels in the endoplasmic reticulum membrane, leading to an intracellular calcium increase and delaying the phase of the circadian clock (Mercado et al. 2009). In addition, the mitogen-activated protein kinase (MAPKs) pathway is also activated in response to light, involving cyclic-adenosin monophosphate (cAMP), cAMP-responding element (CRE), and CRE-binding protein (CREB) (Butcher et al. 2005; Dziema et al. 2003; Obrietan et al. 1999; Pizzio et al. 2003). Transcription factors (e.g., CREB) are ultimate effectors changing the expression of *period 1–2* and other clock genes, regulators of circadian phase. Photic resetting is integrated by the communication of this molecular change to the whole SCN by neurotransmitter release (VIP, GABA, GRP) and paracrine signals (NO) from the ventral to the dorsal region (Golombek and Rosenstein 2010). Furthermore, the SCN controls overt rhythms such as activity–rest, body temperature, feeding, and hormone release by both humoral and autonomic neuronal outputs (Li et al. 2012), which can in turn synchronize peripheral body clocks. Coordination at both the molecular and systems level is important for circadian function, as is entrainment to the 24 h environmental and local time (Takahashi et al. 2008). One of the main outputs of the circadian system, which is an increasing challenge for life in contemporary society, is the sleep–wake cycle.

The circadian system interacts closely with the sleep–wake cycle in order to control the levels of alertness and fatigue and sleep architecture, which are determined by a dual circadian-homeostatic regulation (Borbely and Achermann 1999). These two processes interact in such a way that nocturnal sleep onset is strongly determined by the circadian clock, while sleep duration depends more on the homeostatic demand. The homeostatic sleep drive increases with the duration of prior wakefulness, in relation to acute sleep deprivation or chronic short sleep schedules. Higher homeostatic sleep drive results in impaired cognition, increased sleepiness, and increased propensity for sleep (Reid et al. 2011). In general, an altered sleep–wake rhythm affects the activation of the hypothalamic–pituitary–adrenal (HPA) axis, a major neuroendocrine component regulating cardiometabolic responses and the immune system (Morris et al. 2012a, b). Importantly, these circadian and homeostatic processes interact to influence the quality of both wake and sleep.

15.1.2 *Entrainment of Peripheral Oscillators*

Upon the discovery of the molecular components of the central biological clock, it became evident that several tissues and cells outside the SCN could function as autonomous circadian oscillators. Indeed, the molecular oscillator of peripheral clocks shares most of its components with the one found in SCN cells. Robust circadian oscillations in clock gene expression has been described in a variety of tissues explants, including liver, lung, kidney, spleen, pancreas, heart, stomach, skeletal muscle, cornea, thyroid gland, adrenal gland, esophagus, and thymus (Davidson et al. 2009; Dibner et al. 2010), as well as in circulating peripheral blood mononuclear cells and peritoneal macrophages (Boivin et al. 2003; Hayashi et al. 2007). Noteworthy, circadian rhythms in redox status has been described in erythrocytes, which lack transcriptional activity, setting the basis for noncanonical circadian clocks in mammalian tissues (O'Neill and Reddy 2011). Besides oscillating clock gene expression, several physiological functions in peripheral tissues are under circadian control, such as carbohydrate and lipid metabolism in the liver, muscle, and adipose tissue (Lamia et al. 2008; Shostak et al. 2013), detoxification metabolism in the liver and kidney (Gachon et al. 2006), immune response (Scheiermann et al. 2013), and cardiovascular functions (Morris et al. 2012b).

It is generally accepted that in order to maintain phase coherence among the different oscillators throughout the body, and of all these clocks with the light–dark cycle, the SCN must orchestrate synchronizing signals to the peripheral "slave" oscillators. This can be achieved through indirect pathways such as rhythmic feeding and temperature, which are controlled by the SCN and serve as synchronizing cues to peripheral oscillators. Daily feeding seems to be the strongest zeitgeber in the liver, kidney, pancreas, and heart, modulating clock gene expression in this tissue. Interestingly, restricted daytime feeding in mice (which feed mostly during the dark phase) inverts the rhythm of clock gene expression in peripheral organs (Damiola et al. 2000). Similarly, inverted temperature cycles (warm during the night, cold during the day) resulted in inverted clock gene expression in the liver, while SCN oscillations remained unaltered (Brown et al. 2002).

In addition, the SCN also utilizes direct control pathways, such as neural and endocrine signaling, to coordinate circadian timing in peripheral oscillators. Hypothalamic–pituitary axis control of adrenal glucocorticoid release is tightly controlled by the SCN, resulting in robust plasma corticosterone oscillations. Glucocorticoid signaling has been shown to reset circadian timing in peripheral tissues, thus transmitting the time cues originating from the SCN (Balsalobre et al. 2000). In addition, although the SCN does not express glucocorticoid receptors, indirect feedback signals might reach the central oscillators, since glucocorticoid rhythms have been described to influence photic entrainment, as well as rhythms in the expression of the glial-specific protein GFAP in the SCN (Becquet et al. 2008; Kiessling et al. 2010; Sage et al. 2004). The SCN can also communicate timing signals to peripheral tissues through the autonomic nervous system. The hormone melatonin, synthesized and released mainly by the pineal gland, presents clear diurnal and circadian rhythms in its synthesis and circulating levels. These rhythms are regulated by the SCN through autonomic PVN neurons that project to pre-ganglionic neurons in the spinal cord, which in turn project to noradrenergic neurons in the superior cervical ganglion (Golombek 2012). Moreover, besides the neuroendocrine control of the adrenal cortex by the SCN described above, an important neuronal SCN–PVN–sympathetic nervous system–adrenal cortex axis also controls corticosterone release by modulating the responsiveness to ACTH (Buijs et al. 1999). Finally recent work indicates that the circadian control on leukocyte migration is regulated locally by sympathetic nerves, and, although direct participation of the SCN in this process has not been proved, rhythms in leukocyte trafficking are altered by photic cues, suggesting a central orchestration of these rhythms (Scheiermann et al. 2012).

15.2   Alterations in Circadian Entrainment

An entrained sleep–wake pattern will be phase-locked to a diurnal activity–rest cycle, with predictive circadian functions: a morning increase of corticotrophin and cortisol, together with an increase in core body temperature, heart rate, blood pressure, and behavioral activity. Conversely, a decrease in these variables during the night is followed by a rise in melatonin, contemporary with minimal core temperature values. Alterations in normal circadian entrainment and its consequences in human health underscore the importance of maintaining a strict internal temporal order, i.e., the normal 24 h phase-relationships between physiological rhythms (Moore-Ede and Sulzman 1981). In general, these alterations are due to (1) the failure to process the LD cycle information, as in blindness (Skene and Arendt 2007) or senescence (Gibson et al. 2009) or (2) the conflict between the circadian phase and the LD cycle. This conflict can be originated by:

(a) Exogenous alterations, which might be achieved by shifting LD cycles with respect to the activity–rest cycle (i.e., in jet lag), by shifting the spontaneous phase-relationship between the social pressure and endogenous circadian rhythmicity (i.e., in "social" jet lag), or by continuously shifting the activity–rest cycle with respect to the normal LD (i.e., in shift work)
(b) Endogenous alterations of the normal phase-relationship, as in genetic disorders of the circadian clock.

We will focus on both exogenous and endogenous alterations, in real-world situations as well as in laboratory experiments. Figure 15.1 depicts several experimental models of circadian desynchronization, while Table 15.1 summarizes the effects on health of circadian disruption protocols and clock gene mutations in animal models.

15.2.1 Exogenous Disruption of Circadian Rhythms

15.2.1.1 Alterations Related to Human Activities

Jet lag Transmeridian flights produce a mismatch between the circadian timing at destination respect to that of the departure location or, in other words, a transient desynchronization between internal and external time. Depending upon the time zones crossed, travelers might experience an inversion of the LD cycle with respect to their normal sleep schedule. This will result in conflictive light information appearing at the subjective night, forcing the circadian clock to be gradually re-entrained to the new local LD schedule. During the immediate days after arrival, a disruption of the sleep–wake pattern is generated, with decreasing alertness and increasing fatigue during the day, and even gastrointestinal symptoms (Sack 2009). In addition, internal desynchronization between physiological rhythms might persist for several days (Harrington 2010; Sellix et al. 2012). As a general rule, it typically requires one day to adjust for each time zone crossed, but more days to re-entrain are needed when traveling in the Eastward direction. The extent and severity of jet lag can be also worsened by age and affected by a host of personal and experiential factors (Arendt 2009; Waterhouse et al. 2007). Although jet lag is a temporary inconvenient for occasional travelers, it can be a health concern for frequent travelers as airline workers if flight schedules are not planned considering circadian factors, as well as enough time to allow for sleep recovery.

Social Jet lag Humans exhibit large differences in their temporal preferences (i.e., the "chronotype"), with extreme morning larks and evening owls and most people falling in between these extremes (Roenneberg et al. 2012). However, social schedules impose temporal demands on most individuals without considering their circadian preferred timing. This alteration in the phase-relationship between the activity–rest rhythm (synchronized by social zeitgebers) and the sleep–wake rhythm (synchronized by the circadian clock) is called "social jet lag" (Foster 2012; Wittmann et al. 2006). Population studies indicate that sleep timing in late chronotypes is subjected to the largest differences between work and free days (Wittmann et al. 2006). In addition, social jet lag can be a risk factor for several health troubles, including depression (Levandovski et al. 2011) and obesity (Roenneberg et al. 2012).

Shift work The International Labour Organization defines working in shifts as "a method of organization of working time in which workers succeed one another at the workplace so that the establishment can operate longer than the hours of work of individual workers" (IARC 2010). Several types of shift work exist and can be

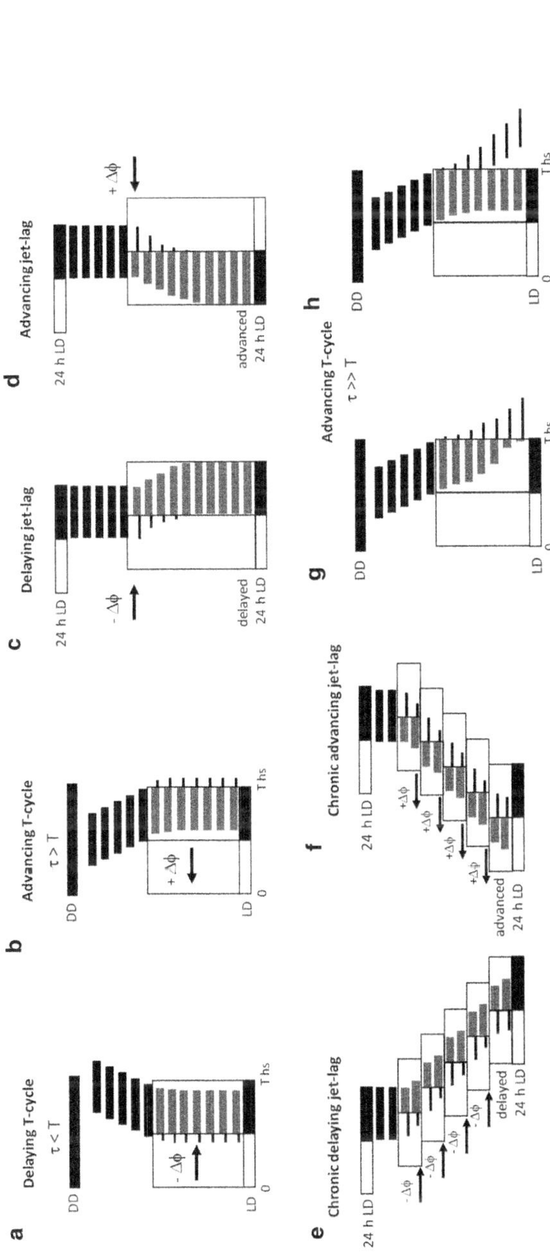

Fig. 15.1 Experimental models of circadian desynchronization. Schematic representation of behavioral rhythms under different environmental conditions as observed in the laboratory. *Light-dark cycles* different from 24 h (i.e., T-cycles) whose periods (T) are set within the entrainment range are normally compensated by the circadian clock. (**a**) When T is longer than the circadian period (τ), entrainment occurs by phase-delays ($-\Delta\phi$). As predicted by the phase response curve (PRC), light pulses tend to be aligned at the beginning of the subjective night (i.e., locomotor activity onset). (**b**) An advancing T-cycle is imposed with T-cycles shorter than 24 h, generating phase-advances ($+\Delta\phi$) of the clock. Stable entrainment can be challenged by an abrupt shift in the phase of the 24 h LD cycle, i.e., a jet lag simulation. Indeed, either a delay (**c**) or an advance (**d**) of the LD cycle will impose a delaying or advancing light-input to the clock, respectively, which will gradually entrain to the new conditions. Finally, chronic jet lag schedules (**e, f**) or T-cycles outside the entrainment range (**g, h**) (see Casiraghi et al. 2012), are used in the laboratory to study the circadian system under extreme entraining conditions. While the clock can entrain to gradual changes in the LD cycle, when animals are subjected to frequent phase advances of the LD cycle, relative coordination and/or forced desynchronization of behavioral rhythms can be observed

Table 15.1 Observed effects on health of protocols of circadian disruption and circadian genes mutations in animal models

Model	Observed alterations	References
Rats under CJL 8/2	• Accelerated development of osteosarcoma and hepatocarcinoma tumors • Altered locomotor activity, temperature, and corticosterone rhythms	Filipski et al. (2004, 2009)
Mice under CJL 8/2	• Fast development of Lewis lung and lung metastasis • Altered lung mechanics	Wu et al. (2012) Hadden et al. (2012)
Mice under CJL 4/1	• Increased body weight • Altered blood levels of leptin and insulin • Decreased prefrontal cortex neurons complexity and impaired cognitive abilities	Karatsoreos et al. (2011)
Rats under CJL 6/2	• Accelerated growth of lung tumors • Disrupted rhythms of NK cells function	Logan et al. (2012a)
Mice under CJL 6/2	• Increased body weight and diverse metabolic alterations	Casiraghi et al. (unpublished)
wt, cry1/2–/–, per1/2–/– mice under CJL ± 8/3	• High rates of spontaneous and γ-radiation-induced tumors • Hyperplasia of the salivary and preputial gland. • Spontaneous lymphoma development and liver and ovarian cancer	Lee et al. (2010)
p53–/– mice under CJL ± 8/3	• Liver and salivary gland hyperplasia • Accelerated development of lymphoma and osteosarcoma • Kidney failure	Lee et al. (2010)
Hamsters under CJL 6/3	• Decreased hippocampal neurogenesis • Impaired learning and memory formation	Gibson et al. (2010)
HIP rats under CJL 6/3	• Increased beta cell apoptosis and diabetes development	Gale et al. (2011)
Mice under CJL 6/7	• Increased lipopolysaccharide induced mortality and cytokines expression	Castañon-Cervantes et al. (2010)
Aged mice under CJL 6/7 or 6/4	• Lifespan shortened	Davidson et al. (2006)
Rats under CJL 6/7	• Reduced hippocampal neurogenesis	Kott et al. (2012)
Mice under Chronic LD inversions	• Accelerated growth of carcinoma and sarcoma tumors • Dramatic worsening of dextran sulfate-induced colitis	Li and Xu (1997) Preuss et al. (2008)
Rats under LL	• Higher rates of hepatocarcinoma development	van den Heiligenberg et al. (1999)
Rats chronically exposed to light at night	• Accelerated aging process • Increased spontaneous tumor development	Vinogradova et al. (2010)

(continued)

Table 15.1 (continued)

Model	Observed alterations	References
	• Faster xenograft tumor growth	Blask et al. (2009)
Mice chronically exposed to light at night	• Increased body weight • Impaired glucose tolerance	Fonken et al. (2010)
Rats forced to be active during light phase	• Disrupted rhythms of plasma glucose, TAG, and corticosterone • Increased weight gain	Salgado-Delgado et al. (2008, 2010a, b)
SCN lesion in mice	• Accelerated development of osteosarcoma and adenocarcinoma tumors	Filipski et al. (2003)
tau mutant hamster	• Cardiac and renal disease	Martino et al. (2008)
per1 (Brd) mutant mice	• Altered glucocorticoid rhythms	Dallmann et al. (2006)
bmal−/−, clock mutant mice	• Altered glucose regulation	Rudic et al. (2004)
mper2 mutant mice	• Accelerated tumorigenesis • Increased cancer development after γ-irradiation	Wood et al. (2008) Fu et al. (2002)
Clock mutant mice and mper1/2/3 triple deficient mice	• Symptoms of metabolic syndrome	Turek et al. (2005), Dallmann and Weaver (2010)
cry1/2−/−, per1/2−/− mice	• Increased spontaneous and radiation-induced tumorigenesis	Lee et al. (2010)

described as follows: (a) permanent: people work regularly on one shift only, i.e. morning or afternoon or night; or rotating: people alternate more or less periodically on different shifts; (b) continuous: all days of the week are covered; or discontinuous: interruption on weekends or on Sundays; (c) with or without night work: the working time can be extended to all or part of the night, and the number of nights worked per week/month/year can vary considerably. Moreover, the definition of "period of night work" varies from country to country, i.e., in some countries it ranges from 8, 9, or 10 pm to 5, 6, or 7 am, and in many others from 11 or 12 pm to 5 or 6 am (IARC 2010). Types of shift work that affect normal rest hours will interfere with the circadian and homeostatic regulation of sleep, leading to cardiovascular disease, accidents, and even cancer (Knutsson 2003). It is generally observed that although most shift workers can behaviorally adapt in order to be awake at night and asleep during the day, they manifest reduced alertness and sleep efficiency (Akerstedt and Wright 2009). Furthermore, the majority of shift workers do not display a corresponding shift in core temperature and melatonin rhythms (Folkard 2008).

Nocturnal Lighting Light pollution during the night is a feature of the industrialized urban environment where the major of human population resides. Light at night might induce acute effects in the human, including alterations in sleep, alertness, and melatonin levels. Recent work indicates that even low levels of illumination at night can reset circadian phase and suppress melatonin release in humans (Gooley et al. 2011; Zeitzer et al. 2000), suggesting that the relatively common exposure to artificial light may in fact affect circadian function. Even the use of low levels of artificial lighting at home and work may alter circadian function. These considerations are of importance since exposure to chronic nocturnal lighting will occur in shift work with nighttime work. However, the use of artificial light at circadian night in humans might compensate photoperiod shortening in higher latitudes, reducing depression symptoms in people suffering seasonal affective disorder (SAD) (Howland 2009).

15.2.1.2 Exogenous Alterations in the Laboratory

Acute and chronic jet lag schedules, constant light conditions and counter-phase forced activity, which may be considered model paradigms for shift- and night-work conditions in humans, as well as rodent circadian mutants, have been employed to study circadian disruption and its consequences on physiology, health, and disease development (see Fig. 15.1).

Simulated Jet Lag Experimental jet lag is a standard protocol designed for studying circadian re-entrainment dynamics in the laboratory, by advancing or delaying the LD cycle (simulating Eastward or Westward flights, respectively). LD-advancing schedules usually require more days for the re-entrainment of circadian rhythms both in humans (Aschoff et al. 1975) and rodent models (Reddy et al. 2002; Yamazaki et al. 2000). In murine models, simulated jet lag induces transient internal desynchronization at all physiological levels. At the SCN clock, shifts in the LD cycle produce transient desynchrony between the dorso-medial and the ventro-lateral regions, revealing differential capabilities of each region to resynchronize to the new LD cycle (Davidson et al. 2009; Nagano et al. 2003; Nakamura et al. 2005; Sellix et al. 2012). LD-shifts also lead to uncoupling between electrical activity and clock gene expression in the SCN (Reddy et al. 2002; Vansteensel et al. 2003). Outputs downstream from the SCN show a differential re-entrainment rate according to its relationship with the activity–rest cycle; for example, the pineal melatonin peak takes 5 days to re-entrain after an 8 h LD-advance of the LD cycle (Drijfhout et al. 1997). Moreover, shifts in the LD cycle also produce a desynchronization between peripheral oscillators and the SCN, with differential clock gene expression pattern and speed of re-entrainment between organs (Davidson et al. 2009; Kiessling et al. 2010; Yamazaki et al. 2000). Of particular attention is the role of the adrenal circadian clock which regulates the speed of resynchronization after advances of the LD cycle through glucocorticoid secretion. Indeed, chronopharmacological manipulation of glucocorticoid secretion accelerated re-entrainment to the new LD cycle in mice (Kiessling et al. 2010), suggesting a possible therapy for ameliorating the effects of jet lag.

In addition, we have developed a chronic jet lag schedule consisting of a 6 h LD-advance each two days, which generated a stable desynchronization of two activity components in mice (Casiraghi et al. 2012) and exerted severe metabolic

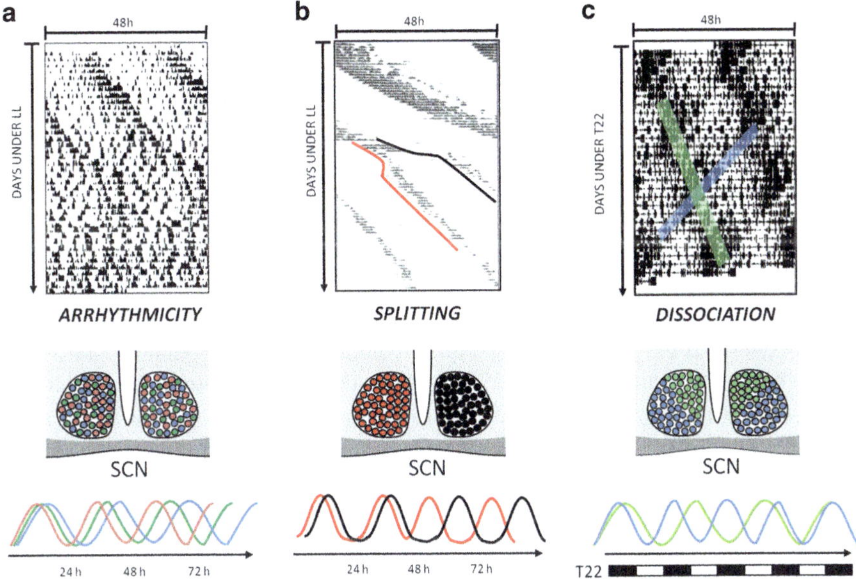

Fig. 15.2 Behavioral models of desynchronization: effects in the SCN. (**a**) Constant-light (LL) induced behavioral arrhythmicity in the mouse. Individual SCN cells (depicted in different colors in the *middle*-panel cartoon and in the representation of their activity in the *lower panel*) express their own rhythmic pattern, with different periods and phases, which results in an arrhtyhmic output such as locomotor activity (after Ohta et al. 2005). (**b**) Under chronic LL schedules, hamster might exhibit "split" locomotor activity rhythms, with two independent components of different circadian periods. These components represent the activity of each SCN which tend to be active in opposite phase to each other (*middle* and *lower* panels) (after De la Iglesia et al. 2000). (**c**) Rats under a *T* = 22 cycle might develop a spontaneous dissociation into two locomotor activity components, one following the T cycle and the other one related to the endogenous clock. This reflects the activity of the dorsomedial (depicted in *green*) or ventrolateral (*blue*) areas of the SCN (*middle* and *lower* panels) (after De la Iglesia et al. 2004)

alterations (unpublished results). Moreover, animals under similar chronic jet lag protocols showed biphasic corticosterone rhythms (Filipski et al. 2004), attenuated AVP and Pk2 expression in the SCN (Yan 2011), and alterations in PER2 expression rhythms in the SCN, thymus, and liver (Castañon-Cervantes et al. 2010). These persistent alterations due to chronic desynchronization have important consequences in health (see below) and might reflect the risks to which individuals working in frequent rotating shifts are subjected.

Experimental Shift Work The physiological effects of working schedules involving night shifts have been assessed in an animal model in which rats were forced to stay awake and active for 8 h during their resting phase (i.e., lights-on period) for five consecutive days, alternating with two "weekend" days with no disturbance. Rats under this protocol exhibited a progressive decrease of their nocturnal activity, and a phase shift in food intake and in circulating triacylglycerols rhythm toward the working hours

(Salgado-Delgado et al. 2008). In addition, this schedule induced uncoupling between rhythms in hypothalamic nuclei involved in metabolic homeostasis, wakefulness, and arousal (which are shifted by the shift-work protocol), and the SCN, which remained locked with the LD cycle (Salgado-Delgado et al. 2010b). This internal desynchrony might contribute to the metabolic disorders that arises in both animals subjected to this type of protocols and workers under shifting schedules, which are discussed later.

LD Cycles Outside the Range of Entrainment LD cycles with a T period (i.e., T cycles) can be experimentally set outside the entrainment limits, where the $\tau - T$ difference is not normally compensated with a given phase-relationship (Plano et al. 2010; Schwartz et al. 2011). Classical experiments in humans used these T-cycle protocols in order to evidence endogenous circadian regulation (Aschoff and Wever 1981). In rats, imposing a T cycle of 22 h outside the limits of entrainment led to the appearance of two rhythmic components of locomotor activity with different periods, one coincident with the LD period, and the other free running with a period >24 h (Campuzano et al. 1998). Each component was linked with clock gene expression in different regions of the SCN: the light-entrained component was related to the ventro-lateral region and the free running component to the dorso-medial region (de la Iglesia et al. 2004; see Fig. 15.2c). Moreover, under this desynchronization paradigm several physiological variables, such as temperature rhythms, rapid-eye movement sleep, and melatonin rhythms could be uncoupled from the LD scheme (Cambras et al. 2007; Lee et al. 2009; Schwartz et al. 2009).

Abnormal Lighting Chronic light exposure has been shown to induce internal desynchrony of the activity-rest and core body temperature rhythms (Aschoff 1965). In nocturnal rodents, constant lighting lengthens circadian period, and when prolonged it can cause arrhythmia of behavioral (Chiesa et al. 2010) hormonal (i.e., corticosterone, Waite et al. 2012) and physiological rhythms with desynchronized SCN neurons (Ohta et al. 2005; see Fig. 15.2a). Moreover, in some animals subjected to constant light, activity rhythms become "splitted" into two components separated 12 h from each other, and each of them correlated with left- or right-sided SCN clock gene oscillations (de la Iglesia et al. 2000; Ohta et al. 2005; see Fig. 15.2b). Recent studies have emphasized the effects of nocturnal light pollution by providing rodents with low levels of light at night (typically ~5 lux) under an otherwise normal LD cycle (Bedrosian et al. 2011; Fonken et al. 2010).

In humans, light-pulses at night induced acute melatonin suppression and subjective sleepiness in a dose-dependent response to light exposure duration (Chang et al. 2012).

15.2.2 Endogenous Alterations in Human and Animal Models

Correct entrainment of the circadian clock can be disturbed due to genetic alterations, producing sleep-wake disorders. For instance, in advanced sleep-phase syndrome, specific mutations in the human clock gene *Per2* and in *Casein kinase Iδ* (CKIδ) leads to daytime somnolence and difficulties to sleep and night. These problems arise due to the misalignment between the internal clock and environmental conditions, with persistent 3–4 h advanced sleep onsets and awakening times relative to conventional and desired times (Dauvilliers and Tafti 2008; Reid and Zee 2009; Toh et al. 2001; Xu et al. 2005). The alteration of the activity of CKIε and CKIδ produce changes in the period of the circadian rhythms in mammals (Lowrey et al. 2000; Meng et al. 2008; Ralph and Menaker 1988; Xu et al. 2005) due to the accumulation of PER-CRY dimers (Akashi et al. 2002; Eide et al. 2002; Lee et al. 2001).

Current treatment for this syndrome includes phototherapy and the use of chronobiotics such as melatonin (or its analogs ramelteon and tasimelton) or armodafinil (Hirai et al. 2005; Revell and Eastman 2005; Zisapel 2001). The hamster is an animal model for this syndrome and presents a single autosomal mutation in the enzyme CKIε, which leads to a shorter period in activity (Ralph and Menaker 1988) and endocrine rhythms (Lucas et al. 1999) from 24 to about 20 h. This mutation accelerates the degradation of both nuclear and cytoplasmatic PER proteins (Dey et al. 2005).

Another disorder associated with endogenous factors is the delayed sleep phase syndrome, which has been associated with a polymorphism in PER3 (Archer et al. 2003; Ebisawa et al. 2001; Pereira et al. 2005). In this syndrome the timing of the sleep episode occurs later than desired and is associated with problems to fall asleep and awakening in the mornings (e.g., to meet work or school obligations), and daytime sleepiness, with a particularly higher prevalence among adolescents and young adults (Okawa and Uchiyama 2007). Current treatments for this disorder include melatonin administration and bright light exposure during the morning. In addition, it should be noted that mouse mutants of other clock genes, such as Clock, Cry1, Cry2, Npas2, Dbp, and Bmal1, also have alterations in sleep duration and/or homeostasis (Reviewed in Takahashi et al. 2008), adding examples in which circadian clock genes can clearly affect the circadian and homeostatic control of sleep.

15.3 Health Consequences of Circadian Disruption

Misalignment between circadian oscillators and the 24 h LD cycle promote a wide set of health alterations, in general dominated by disruptions in the sleep–wake pattern. A recent study in humans shows that insufficient sleep at night generates circadian alterations in multiple genes in the blood transcriptome, including those associated with circadian rhythms (e.g., Per1–3, cry, clock, rora), sleep homeostasis (e.g., interleukin-6), oxidative stress (peroxyredoxin 2 and 5), and metabolism (e.g., ghrelin, glucose transporter 2, glucose/fructose transporter 5, cholesterol efflux regulatory protein), thus affecting several processes including chromatin modification, gene expression, macromolecular metabolism, and inflammatory, immune and stress responses (Moller-Levet et al. 2013). Indeed, several of such alterations depend on the deynchrony between the central (SCN) and peripheral clocks, as suggested in Fig. 15.3. Moreover, chronic desynchrony, such as the one observed in shift work schedules, might result in profound health disorders, as represented in Fig. 15.4.

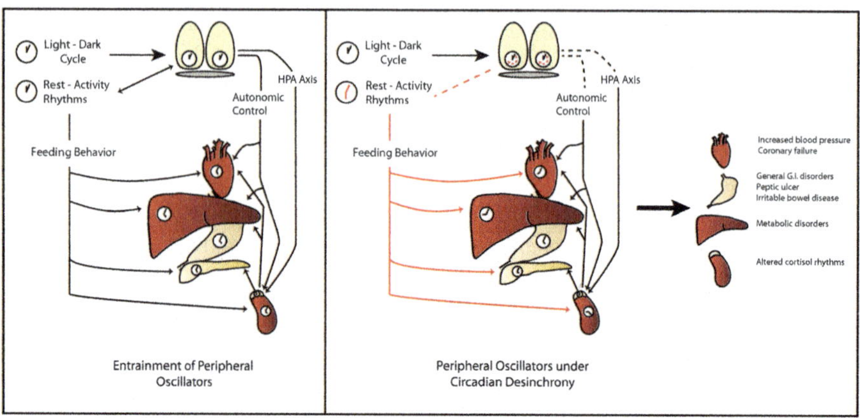

Fig. 15.3 Circadian desynchronization in peripheral organs. *Left panel*: Schematic representation of circadian entrainment in peripheral tissues. The suprachiasmatic nuclei (SCN) are entrained by photic cues, while peripheral organs receive time cues from the SCN through the autonomous nervous system and the hypothalamo–pituitary–adrenal (HPA) axis. In addition, food availability acts as a strong zeitgeber in several organs. *Right panel*: Schematic representation of the effects of circadian desynchrony in peripheral organs and its consequence in health status. Under circadian desynchronization, such as shift-work, the activity-rest cycle adopts a changing phase relation with the *light–dark cycle*. The SCN might remain synchronized by the *light–dark cycle*; however, exposure to light at night and feedback from downstream outputs might alter circadian timing. On the other hand, feeding behavior sets time cues for peripheral organs that are in conflict with the SCN circadian control, leading to desynchronization within peripheral clocks, and between these and the SCN. This alterations might lead to metabolic, cardiovascular, endocrine, and gastrointestinal malfunctions

15.3.1 Metabolic Alterations

Although it has been known for several decades that many metabolic processes, such as glucose and cholesterol metabolism, are regulated by the circadian clock, it is only in the past few years that the severe metabolic consequences of circadian disruption have emerged. The SCN exerts a hierarchical daily control on metabolic rhythms by neurohumoral efferents to other brain regions (mainly several hypothalamic nuclei) and to autonomic centers, which drive hormones (e.g., glucocorticoids) and behavior (e.g., feeding) producing a set of physiological signals that synchronize peripheral tissues (reviewed in Delezie and Challet 2011; Huang et al. 2011). In addition, food-entrainable oscillators exert a secondary daily control on peripheral clocks (reviewed in Delezie and Challet 2011) and fine-tune the balance between energy intake and expenditure. Pancreatic (Marcheva et al. 2010) and thyroid hormones regulating carbohydrate and lipid metabolism (and thus basal metabolic rate) are under circadian control, as well as rate-limiting enzymes of metabolic pathways (Panda et al. 2002). Several molecular clock components are found to be regulators of cell metabolism through the so-called nuclear receptor transcription factors (as Retinoic acid-related orphan receptor α, RORα, reverse viral

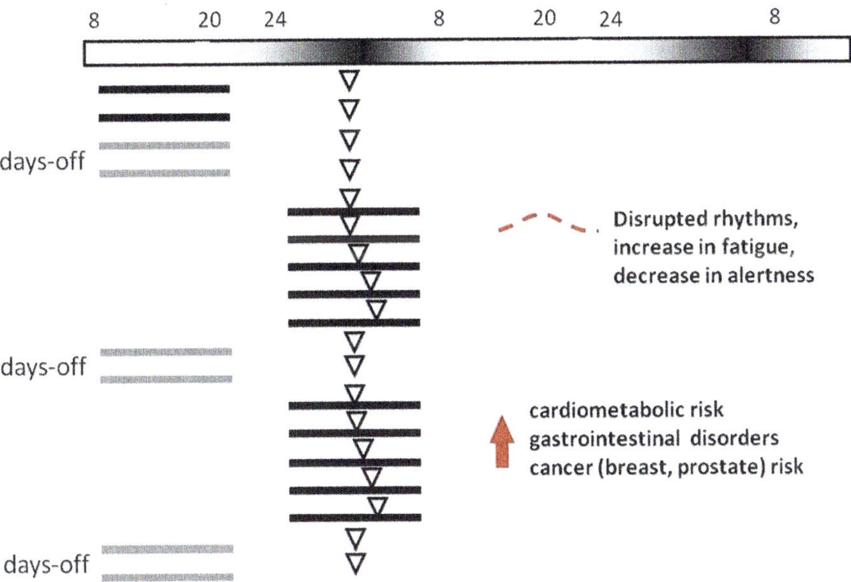

Fig. 15.4 Schematic representation of a typical shift work situation, with nocturnal working hours, and its expected disruptions on circadian rhythms and health. The *horizontal top* bar represents the relative light intensity perceived by the subjects, exposed to artificial light during the night. *Black lines* represents activity (*grey* on "days-off"), while *white triangle* represents a circadian marker (e.g., melatonin or core body temperature). Acute exposure to nightwork, with subsequent sleep deprivation, incoherent feeding/fasting timing, nocturnal light exposure, and increased sleep drive during the day, will lead to disrupted circadian rhythms in melatonin, cortisol, temperature, activity/rest, sleep/wake, and metabolism. Chronic exposure to abnormal entraining conditions might increase the risk of several health disorders

erythroblastis oncogene products, REV-ERBα, and peroxisome proliferator-activated receptors, PPAR) which sense metabolites such as fat-soluble hormones, vitamins, lipids, oxysterols, bile acids, and xenobiotics. For instance, CLOCK and BMAL1 act as transcription factors of PPAR (α, γ, δ) and estrogen-related receptor (ERR) family members (α, β, γ). Conversely, REV-ERBα and PPARα directly regulate BMAL1 expression, and relevant cellular (nutrient-redox) metabolic sensors are able to interact with clock genes (Sahar and Sassone-Corsi 2012). For instance NADH/NAD+balance regulates CLOCK/BMAL1 and NPAS2/BMAL1 heterodimer activity (Rutter et al. 2001) and, in addition, the NAD+-deacetylase dependent SIRT1 induced by food restriction can directly bind to CLOCK/BMAL1 (Nemoto et al. 2004). This physiological and molecular cross-talk of both clockwork and metabolic networks provides a mechanistic framework to understand how systematic circadian disruption will lead to metabolic alterations (Eckel-Mahan and Sassone-Corsi 2009). For example, circadian Clock mutant mice develop obesity (Turek et al. 2005), and diabetes was found in Clock-Bmal1 double-mutants due to impaired insulin secretion (Marcheva et al. 2010) and gluconeogenesis suppression

(Rudic et al. 2004). Moreover, liver-specific deletion of Bmal1 in mice confirmed its strong involvement in hepatic glucose metabolism (Lamia et al. 2008).

Impaired entrainment in the laboratory also generates metabolic alterations, as when exposing mice to T cycles of 20 h, leading to larger body mass gain and increased insulin/glucose ratio (Karatsoreos et al. 2011). In human subjects exposed to controlled 28 h sleep–wake cycles under dim light, misalignment between the free-running central circadian pacemaker and the behavioral sleep–wake and fasting/feeding rhythms, resulted in increased plasma glucose and insulin concentrations, increased blood pressure, and decreased plasma leptin concentrations (Scheer et al. 2009). In addition, constant exposure of mice to bright or dim light led to increased body mass and reduced glucose tolerance as compared to mice housed under a normal LD cycle, despite equivalent levels of caloric intake and total daily activity output (Fonken et al. 2010). There is a well-documented increased risk for metabolic syndrome in humans subjected to shift work (reviewed in Zimberg et al. 2012) perhaps as the result of pathological adaptation to chronically sleeping and eating at abnormal circadian times. Disrupted sleep must be taken as an important factor in glucose metabolism alterations increasing the risk for obesity (Gangwisch 2009; Nedeltcheva et al. 2009; Spiegel et al. 2009) as show in a recent study with humans provided with 5 days with insufficient sleep, showing weight gain due to increased food intake exceeding energy needs (Markwald et al. 2013). On the other hand, a "night-eating" syndrome was detected in women, with a delayed circadian pattern of melatonin and food intake, presenting evening hyperphagia and frequent awakenings accompanied by food intake, while retaining a normal sleep–wake cycle. This alteration in feeding behavior is accompanied by delayed and low-amplitude rhythms in cortisol and several nutrients intake, and altered rhythms of metabolic hormones as insulin, ghrelin, and leptin (Goel et al. 2009). In addition, a failure to manage cholesterol in shift workers could lead to cardiovascular alterations (Ha and Park 2005) such as the ones discussed in the following section.

15.3.2 Cardiovascular Alterations

Autonomic nuclei that regulate the cardiovascular system are directly controlled by neural efferents from SCN and/or by circadian hypothalamic–pituitary factors acting on corticoid stress hormones, as well as by neurohumoral effectors on vascular tone. Cardiovascular risk markers as cortisol, nonlinear dynamic heart rate control, and hemostatic factors are higher during the morning, when adverse cardiovascular events are most likely to occur (reviewed in Morris et al. 2012b; Ruger and Scheer 2009). These circadian acute cardiovascular events are also suggested to be dependent on specific phase-relationship between peripheral oscillators at cardiovascular tissues (Davidson et al. 2005). Epidemiological studies show a consistent association for shift work as a factor for cardiovascular disease (reviewed in Knutsson and Boggild 2000), such as coronary failure (Nakamura et al. 1997; Tenkanen et al. 1997) or increased blood pressure in night workers (Lo et al. 2010; Su et al. 2008).

A meta-analysis study of almost 2,000,000 shift workers detected higher risk for coronary events (especially myocardial infarct) in the nocturnal shift (Vyas et al. 2012). It is suggested that cardiovascular risk markers might respond unfavorably to exogenous stressors such as increased nocturnal activity (Morris et al. 2012b), promoting alterations in shift workers with misalignment between the circadian and the activity–rest cycles.

Disruption of circadian regulation in the laboratory increases several risk factors of cardiovascular disease. Chronic LD-shifts increased plasmatic plasminogen activator inhibitor-1 (PAI-1) levels (Oishi and Ohkura 2013), which promotes hypofibrinolysis. The aforementioned study of Scheer et al. (2009) in humans forced to follow a 28 h sleep–wake cycle indicated increased blood pressure during wakefulness. Altered cardiovascular functions are found in the 22 h short-period hamster τ, with a point mutation in Ck1ε circadian regulatory gene, which exhibits abnormal 24 h LD behavioral entrainment and disrupted rhythms. These animals had cardiomyopathy, extensive fibrosis, and severely impaired contractility with shortened life-expectancy (Martino et al. 2008), while under T22 cycles (i.e., their normal endogenous period) behavioral rhythms and cardiac phenotype are normalized. Several vascular dysfunctions were found in clock gene-deficient mice. In Bmal1 knock-out mice, abnormal thrombosis and vascular remodeling responses were found, while Clock mutants showed impeded responses to vascular injury; moreover, in Bmal1, Clock, and Per2 mutant mice, vascular endothelial dysfunction was found (Anea et al. 2009; Viswambharan et al. 2007). Also, blood pressure was altered in Cry 1–2 null mice, showing hypertension due to deregulation of salt homeostasis caused by abnormal high synthesis of adrenal mineralocorticoid aldosterone (Doi et al. 2010).

15.3.3 Gastrointestinal Alterations

The gastrointestinal tract (GT) displays circadian rhythms in several functions and variables, such as motility, maintenance and replacement of the protective barrier, absorption rates, and enzyme secretion rhythms (reviewed in Konturek et al. 2011). Furthermore, rhythmic clock gene expression has been found in peripheral oscillators throughout the GT, which can be shifted, or entrained, by timed meals (Hoogerwerf et al. 2007; Mazzoccoli et al. 2012; Polidarova et al. 2011). The correct phase-relationship between feeding time and the activity–rest cycle led to normal entrainment of the GT, by a myriad of neural and humoral signals related with the interplay between SCN activity, food-entrainable oscillators, and metabolism (Stenvers et al. 2012). There are evidences showing that circadian disruption leads to alterations in gastrointestinal physiology. Increased general symptoms for gastrointestinal disorders were more prevalent among shift workers as compared with day workers (Koller 1983), and aircrew members, especially those on long-distance flights, had more upper-GT symptoms than ground staff (Enck et al. 1995). The consequences on gastrointestinal functions of shifting working schedules might be related to an increase in food consumption during the rest cycle, since time of feed has been shown to be the strongest Zeitgeber in gastrointestinal oscillators (Malloy et al. 2012; Polidarova et al. 2011), resulting in a misalignment between circadian clocks among the body. There is a higher prevalence on peptic ulcer among shift workers (Higashi et al. 1988; Segawa et al. 1987), which might be related to higher incidences in *Helicobacter pylori* infections in shift workers (Pietroiusti et al. 2006; Zober et al. 1998). Irritable bowel disease prevalence was also higher in nurses working rotating shifts (Nojkov et al. 2010).

Circadian disruption by chronic lighting in rats fed *ad libitum* abolished rhythmic clock gene expression in the liver and colon, while a restricted feeding schedule restored these rhythms (Polidarova et al. 2011). Similar results were found in SCN-ablated mice, where restricted feeding re-established locomotor, feeding, and stool output rhythms (Malloy et al. 2012). Per1–2 double knockout mice exhibited abolished rhythms in colonic functions, such as motility, intracolonic pressure changes, stool output, and circular muscle contraction (Hoogerwerf 2010).

15.3.4 Immune Alterations

The circadian system exerts a tight control over the immune system and its function, which in turn feedbacks over the circadian clock (reviewed in Cermakian et al. 2013; Scheiermann et al. 2013). Robust rhythms in the number of immune cells, plasmatic levels of proinflammatory cytokines, and immune cell-produced peptide hormones, as well as in major humoral responses, have been described vastly, in humans and animal models (Fortier et al. 2011; Haus and Smolensky 1999; Silver et al. 2012). In addition, circadian disruption has profound effects on the immune system, which has been evidenced both in animal models and in human studies. Female nurses working in a three-shift rotating system showed a progressive decrease in natural killer (NK) cell activity and CD16[+] CD56[+] cells number (Nagai et al. 2011). Shift working has also been linked to a higher risk for common infections, with the highest risk in workers going through three different shifts (Mohren et al. 2002). Impeding normal LD-entrainment by chronic jet lag under weekly 6 h phase-advances altered the inflammatory response of mice to an experimental challenge with lipopolysaccharide (LPS). After four consecutive shifts, jet lagged mice displayed a fivefold increase in mortality and severe hypothermia due to LPS, as compared to LD-entrained controls. In this chronic jet lag the expression of the cytokines IL-1β, IL-10, IL-12, IL-13, GM-CSF, and TNF-α was significantly increased 24 h after LPS injection, and in vitro macrophage stimulation with LPS produced higher levels of IL-6 (Castañon-Cervantes et al. 2010). Chronic jet lag also impaired NK cells cytolytic activity, as well as altered rhythms in clock gene and cytokine expression (Logan et al. 2012b). Deleterious effects of abnormal entrainment were also reported in a mice model of experimental colitis, where chronic inversions of the LD cycle every 5 days led to increased symptoms as reduced body mass, abnormal intestinal histopathology, and potentiated inflammatory response (Preuss et al. 2008).

Changes in the sleep–wake rhythm, which are frequently concomitant with circadian disruption also have profound effects on the immune system. An altered sleep–wake pattern changed the HPA axis activity increasing cortisol, and consequently affecting the expression of proinflammatory cytokines, such as IL-6 or TNF-α (Vgontzas and Chrousos 2002). Additionally, NK cell activity, antigen uptake, circulating immune complexes, and secondary antibody responses were altered by sleep loss (Krueger et al. 2003), while IL-1β and TNF-α levels increased during sleep deprivation (Majde and Krueger 2005). In addition, sleep deprivation leads to the suppression of immune surveillance, which may permit the establishment and/or growth of malignant clones.

Alterations of clock genes, especially those of the *period* family, disrupt immune functions. Per3 clock gene polymorphisms have also been associated with circadian disruption and with increased cancer risk together with elevated IL-6 concentrations (Guess et al. 2009). Homozygous Per1 mutant mice exhibited altered cytokine rhythms (e.g., interferon-γ) and cytolytic factors (e.g., perforin and granzyme B) in splenic NK cells (Logan et al. 2013). On the other hand, Per2-deficient mice were more resistant to LPS-induced endotoxic shock than control wild-type mice. Moreover, serum levels of IFN-α and IL-1β were dramatically decreased in Per2−/− mice following an LPS challenge, attributable to defective NK and NKT cell function (Liu et al. 2006). In addition mice carrying hematopoietic Cry1−/−Cry2−/− mutant cells presented a potentiated immune response to LPS administration, with higher levels of circulating proinflammatory cytokines (Narasimamurthy et al. 2012).

Chronodisruption may result in altered immune responses, such as aberrant immune cell trafficking and abnormal cell proliferation cycles, which can lead to cancer (Mormont and Levi 1997). Per2 mutant mice have been shown to present enhanced cancer susceptibility (Fu et al. 2002; Wood et al. 2008), along with disrupted circadian rhythms of NK cell gene expression and function.

15.3.5 *Endocrine Alterations*

There is an intimate connection between the circadian clock and multiple endocrine axes, which involve several aspects of the daily regulation of hormonal release. A highly specialized organization configures the SCN neural output to both hypothalamic neuroendocrine neurons (i.e., cells containing or releasing corticotrophin-releasing hormone, CRH, tyrotrophin-releasing hormone, TRH, or gonadotrophin-releasing hormone GnRH) and pre-autonomic neurons, as well as to other brain structures (Kalsbeek and Fliers 2013). Such output ensures the coordination between behavioral, endocrine, and metabolic rhythms needed for the circadian adjustment of the organism. For instance, glucose homeostasis is tightly regulated in order to predict the energy needs for the circadian onset of behavioral activity, by increasing hypothalamic CRH and adrenal cortisol release, and by activating hypothalamic orexins at the right time of day (Kalsbeek et al. 2011).

One of the best known examples of endocrine circadian regulation is represented by the synthesis and release of pineal melatonin, which is under the control of sympathetic autonomic efferents commanded by the SCN (Moore 1996). Blood melatonin increases during the night, it is suppressed by light (Zawilska et al. 2009), and its amplitude and duration are proportional to photoperiod duration (Hazlerigg 2012). Indeed, melatonin can be considered a daily and calendar signal which communicates circadian and circannual cues to the body. In recent years, additional evidence suggested an antioxidant role for this hormone, which falls outside the scope of this chapter (Reiter et al. 2010).

It is well known that sleep disruption affects autonomic and endocrine responses. A representative example is the fact that cortisol rhythms are altered in night-working emergency physicians (Machi et al. 2012), shift-working nurses (Korompeli et al. 2009) and in police officers working night or afternoon shifts than day shifts, with maximal differences occurring after 5 days of shift work (Wirth et al. 2011). Moreover, flights involving 7 h of jet lag produced a decrease in cortisol levels that lasted for 9 days after the flight (Pierard et al. 2001). Furthermore, in laboratory conditions, in which subjects were kept in a 28 h forced desynchronization protocol, cortisol was more strongly coupled to the circadian endogenous component than to the environmental "day." Indeed, an inverted pattern of cortisol secretion was observed when sleep–wake and circadian cycles were misaligned. On the other hand, leptin rhythm followed only the sleep–wake cycle component, with conflict phases of misalignment between the circadian and behavioral component resulting in suppression of circulating leptin levels (Scheer et al. 2009). More information is needed (and on the way) regarding the role of disturbed hormonal cycles on circadian synchronization, both at central and peripheral level.

15.3.6 Cognitive Effects of Circadian Disruption

Cognitive and mood-related behaviors vary with time of day. Brain and behavioral mechanisms underlying this variation include prior amount of wakefulness, prior sleep history, and internal circadian timing (Wright et al. 2012). Indeed, circadian rhythms can be affected by several factors such as alteration on ambient conditions and genetic factors which lead to impaired cognitive and affective functions.

Animal models with mutations or disruptions in the circadian machinery show changes in mood and cognition (see Table 15.1). For example, *clock* mutant mice showed a behavioral profile similar to human mania including hyperactivity, low anxiety levels, and an increased reward response to cocaine and sucrose (Roybal et al. 2007). Deficiency of the transcription factor BMAL1 in mice produced hyperactivity in novel environments and defects in short- and long-term memory formation (Kondratova et al. 2010). Npas2-deficient mice exhibited deficits in the acquisition of long-term memory and contextual fear task (García et al. 2000). In addition, CRY1,2 knockout mice were unable to learn time-place associations (Van der Zee et al. 2008).

In humans, mood spectrum disorders are generally accompanied by circadian rhythm dysfunction (Aydin et al. 2013). In addition, single nucleotide polymorphisms found in clock genes were associated with these mood-related diseases. In particular, polymorphisms found in CRY1 were correlated with major depression (Soria et al. 2010). Seasonal affective disorder was associated with polymorphisms in Npas2, Bmal1, and Per2 genes (Partonen et al. 2007). Moreover, Bmal1 and Per3 are related with bipolar disorder (Nievergelt et al. 2006). Even though there are multiple evidences indicating cognitive performance and learning are influenced by circadian processes (Wright et al. 2012), there are still no clear correlations between these processes and clock genes polymorphisms in humans.

In the laboratory, animals can be exposed to experimental conditions which affect circadian rhythms in order to study their effect on cognitive and affective functions. For example, animals under altered entrainment conditions such as a T7 cycle (3.5 h/3.5 h LD) showed a disability in establish long term potentiation and an increased depression-like behavior (LeGates et al. 2012). Accordingly, mice under a T20 cycle revealed a decrease in cognitive flexibility and increased anxiety-like behavior (Karatsoreos et al. 2011). Moreover, chronic exposure to dim light at night had deleterious effects on learning abilities and memory consolidation and produced depression-like behaviors in nocturnal and diurnal rodents (Fonken et al. 2012). A 6 h LD-advance produced a decrease in hippocampal cell proliferation and neurogenesis, which was associated with deficits in memory and learning in hamsters and mice (Gibson et al. 2010; Kott et al. 2012) and altered memory consolidation in mice (Loh et al. 2010). A chronic jet lag schedule with 3 h phase advances per day during six days induced alteration in context memory formation in rats (McDonald et al. 2013). Additionally, constant light conditions, which lead to arrhythmicity in the SCN, temperature, and behavioral outputs, produced impaired spatial memory in rats (Ma et al. 2007).

Humans are subjected periodically to changes in environmental conditions or social factors that alter circadian rhythms, including subjective-day lengthening in social jet lag, rotating shift-work schedules, and transmeridian travels, with frequent fluctuations in the timing of the sleep–wake schedule, inducing sleep deprivation and exposure to nighttime lighting. These alterations will produce circadian misalignment between biological rhythms and the environment, thus resulting in mood and cognitive disorders (LeGates et al. 2012). For example, it was shown that chronic jet lag produced by consecutive transmeridian flights produced temporal lobe atrophy and impairment in short-term memory and cognitive processing in flight attendants (Cho et al. 2000). In addition, shift workers showed lower cognitive scores in memory and attention tests, as compared to nonrotating controls (Rouch et al. 2005). Accordingly, a decrease in cognitive performance (especially in short-term memory) was found in emergency physicians during the course of both day and night shifts (Machi et al. 2012). Meanwhile, sleep deprivation affected learning, attention and decision-making, increasing impulsivity toward negative stimuli (Anderson and Platten 2011; Wright et al. 2006).

15.4 Treatment of Circadian Desynchronization

Our current productive system depends on maximization of temporal resources. Indeed, circadian pressure as exemplified by social jet lag and shift work clearly reduce quality of life and might be accompanied by health hazards and pathologies. In general, strategies to achieve a better adaptation to such conflicts depend on minimizing the number of days needed to re-entrain the clock, along with the concurrent internal desynchronization in the organism (reviewed in Arendt 2009, 2010). Short-term night shift can be managed by maintaining daytime circadian phase, while using strategies such as alerting stimulants (caffeine and possibly modafinil), as well quiet and dark sleeping rooms (and possibly hypnotics) to preserve sleep and performance. In terms of transmeridian flights, for short stopovers (1–2 days) there is no real benefit of a complete adjustment of the circadian clock, while this is obviously necessary if an extended permanence at the destination is planned. Importantly, the human clock phase can be modified by light stimuli of adequate intensity (in the order of 10^2 lux) and wavelength (480-500 nm), with largest phase delays and advances at the beginning and end of the subjective night, respectively (Czeisler and Dijk 1995; Paul et al. 2009; Revell and Eastman 2005). Consequently, light exposure should be managed in an adequate and timely fashion in order to allow (or avoid) the desired phase-changes. Also melatonin, a chronobiotic drug modulating SCN clock activity, can be used to advance the circadian phase (Paul et al. 2010; Revell and Eastman 2005), induce daytime sleep (Rajaratnam et al. 2004) or even synchronize circadian activity in blind subjects (reviewed in Skene and Arendt 2007). Chronic nocturnal work is the worse situation, since it involves uncontrolled nocturnal lighting exposure, forced arousal and nocturnal feeding, with a myriad of psychophysiological variables out of normal phase. Schedules should be designed and administered based on chronotypes and circadian typology, including light-therapy and efficient sleep hygiene, which tend to stabilize sleep–wake rhythm to nocturnal activity/diurnal rest (reviewed in Adan et al. 2012). By controlled manipulation of the sleep–rest schedule and light exposure both in laboratory and field studies, an adequate circadian phase management (i.e., a complete shift of melatonin onset) was found in shift workers, improving fatigue, nocturnal performance, mood, and late nighttime sleep on days off (Boivin and James 2002; Lee et al. 2006; Smith et al. 2009a, b). In general, avoidance of early morning light and exposure to late-morning and afternoon light alone or in conjunction with bedtime melatonin, can accelerate re-entrainment by phase advances following Eastward travel. For Westward travel, a circadian phase delay can be achieved after arrival with afternoon and early-evening light with bedtime melatonin (Kolla and Auger 2011; Zee and Goldstein 2010). A combination of afternoon melatonin, morning intermittent bright light, and a gradually advancing sleep schedule advanced circadian rhythms by almost 1 h/day (Revell et al. 2006; Revell and Eastman 2005).

Modulation of light input can be a promising strategy. This can be done either indirectly, for instance by enhancing serotonergic transmission from the raphe nuclei to the SCN (Lall et al. 2012), which accelerates the re-entrainment to a 6 h

LD-advance (Kessler et al. 2008) or directly by pharmacological targeting of photic second messengers, as increasing the cGMP-GC-PKG signal transduction by inhibiting cGMP enzymatic degradation, which accelerates re-entrainment after an abrupt 6 h LD-advance (Agostino et al. 2007; Plano et al. 2012).

In summary, circadian entrainment is an important feature of environmental adaptation. Disruption of such synchronization leads not only to circadian disruption but also to diverse pathological situations, especially when entrainment is compromised chronically. The use of animal models, as well as controlled laboratory studies of human studies, is fundamental for a deep understanding of the profound consequences of messing up with biological timing, which at this point of civilization can be considered with justice a sign of the times.

References

Adan A, Archer SN, Hidalgo MP et al (2012) Circadian typology: a comprehensive review. Chronobiol Int 29:1153–1175

Agostino PV, Plano SA, Golombek DA (2007) Sildenafil accelerates reentrainment of circadian rhythms after advancing light schedules. Proc Nat Acad Sci USA 104:9834–9839

Akashi M, Tsuchiya Y, Yoshino T et al (2002) Control of intracellular dynamics of mammalian period proteins by casein kinase I epsilon (CKIepsilon) and CKIdelta in cultured cells. Mol Cell Biol 22:1693–1703

Akerstedt T, Wright KP Jr (2009) Sleep loss and fatigue in shift work and shift work disorder. Sleep Med Clin 4:257–271

Anderson C, Platten CR (2011) Sleep deprivation lowers inhibition and enhances impulsivity to negative stimuli. Behav Brain Res 217:463–466

Anea CB, Zhang M, Stepp DW et al (2009) Vascular disease in mice with a dysfunctional circadian clock. Circulation 119:1510–1517

Antle MC, Silver R (2005) Orchestrating time: arrangements of the brain circadian clock. Trend Neurosci 28:145–151

Archer SN, Robilliard DL, Skene DJ et al (2003) A length polymorphism in the circadian clock gene Per3 is linked to delayed sleep phase syndrome and extreme diurnal preference. Sleep 26:413–415

Arendt J (2009) Managing jet-lag: Some of the problems and possible new solutions. Sleep Med Rev 13:249–256

Arendt J (2010) Shift work: coping with the biological clock. Occup Med (Lond) 60:10–20

Aschoff J (1965) Circadian rhythms in man: a self-sustained oscillator with an inherent frequency underlies human 24-hour periodicity. Science 148:1427–1432

Aschoff J, Wever R (1981) The circadian system of man. In: Aschoff J (ed) Handbook of Behavioral Neurobiology. Plenun Press, New York, pp 311–350

Aschoff J, Hoffmann K, Pohl H et al (1975) Re-entrainment of circadian rhythms after phase-shifts of the Zeitgeber. Chronobiol 2:23–78

Aton SJ, Herzog ED (2005) Come together, right...now: synchronization of rhythms in a mammalian circadian clock. Neuron 48:531–534

Aydin A, Selvi Y, Besiroglu L et al (2013) Mood and metabolic consequences of sleep deprivation as a potential endophenotype' in bipolar disorder. J Affect Dis 150:284–294

Balsalobre A, Brown SA, Marcacci L et al (2000) Resetting of circadian time in peripheral tissues by glucocorticoid signaling. Science 289:2344–2347

Becquet D, Girardet C, Guillaumond F et al (2008) Ultrastructural plasticity in the rat suprachiasmatic nucleus. Possible involvement in clock entrainment Glia 56:294–305

Bedrosian TA, Fonken LK, Walton JC et al (2011) Chronic exposure to dim light at night suppresses immune responses in Siberian hamsters. Biol Lett 7:468–471

Blask DE, Dauchy RT, Brainard GC et al (2009) Circadian stage-dependent inhibition of human breast cancer metabolism and growth by the nocturnal melatonin signal: consequences of its disruption by light at night in rats and women. Int Cancer Ther 8:347–353

Boivin DB, James FO (2002) Circadian adaptation to night-shift work by judicious light and darkness exposure. J Biol Rhyth 17:556–567

Boivin DB, James FO, Wu A et al (2003) Circadian clock genes oscillate in human peripheral blood mononuclear cells. Blood 102:4143–4145

Borbely AA, Achermann P (1999) Sleep homeostasis and models of sleep regulation. J Biol Rhyth 14:557–568

Brown SA, Zumbrunn G, Fleury-Olela F et al (2002) Rhythms of mammalian body temperature can sustain peripheral circadian clocks. Curr Biol 12:1574–1583

Buijs RM, Wortel J, Van Heerikhuize JJ et al (1999) Anatomical and functional demonstration of a multisynaptic suprachiasmatic nucleus adrenal (cortex) pathway. Eur J Neurosci 11:1535–1544

Butcher GQ, Lee B, Cheng HY et al (2005) Light stimulates MSK1 activation in the suprachiasmatic nucleus via a PACAP-ERK/MAP kinase-dependent mechanism. J Neurosci 25:5305–5313

Cambras T, Weller JR, Angles-Pujoras M et al (2007) Circadian desynchronization of core body temperature and sleep stages in the rat. Proc Nat Acad Sci USA 104:7634–7639

Campuzano A, Vilaplana J, Cambras T et al (1998) Dissociation of the rat motor activity rhythm under T cycles shorter than 24 hours. Physiol Behav 63:171–176

Casiraghi LP, Oda GA, Chiesa JJ et al (2012) Forced desynchronization of activity rhythms in a model of chronic jet-lag in mice. J Biol Rhythm 27:59–69

Castañon-Cervantes O, Wu M, Ehlen JC et al (2010) Dysregulation of inflammatory responses by chronic circadian disruption. J Immunol 185:5796–5805

Cermakian N, Lange T, Golombek D et al (2013) Crosstalk between the circadian clock circuitry and the immune system. Chronobiol Int 30:870–888

Chang AM, Santhi N, St Hilaire M et al (2012) Human responses to bright light of different durations. J Physiol 590:3103–3112

Chiesa JJ, Cambras T, Carpentieri AR et al (2010) Arrhythmic rats after SCN lesions and constant light differ in short time scale regulation of locomotor activity. J Biol Rhythm 25:37–46

Cho K, Ennaceur A, Cole JC et al (2000) Chronic jet-lag produces cognitive deficits. J Neurosci 20:RC66–RC78

Czeisler CA, Dijk DJ (1995) Use of bright light to treat maladaptation to night shift work and circadian rhythm sleep disorders. J Sleep Res 4:70–73

Dallmann R, Weaver DR (2010) Altered body mass regulation in male mPeriod mutant mice on high-fat diet. Chronobiol Int 27:1317–1328

Dallmann R, Touma C, Palme R et al (2006) Impaired daily glucocorticoid rhythm in Per1 (Brd) mice. J Comp Physiol A 192:769–775

Damiola F, Le Minh N, Preitner N et al (2000) Restricted feeding uncouples circadian oscillators in peripheral tissues from the central pacemaker in the suprachiasmatic nucleus. Genes Develop 14:2950–2961

Dauvilliers Y, Tafti M (2008) The genetic basis of sleep disorders. Curr Pharmaceut Des 14:3386–3395

Davidson AJ, London B, Block GD et al (2005) Cardiovascular tissues contain independent circadian clocks. Clin Exp Hypertens 27:307–311

Davidson AJ, Sellix MT, Daniel J et al (2006) Chronic jet-lag increases mortality in aged mice. Curr Biol 16:R914–916

Davidson AJ, Castañon-Cervantes O, Leise TL et al (2009) Visualizing jet-lag in the mouse suprachiasmatic nucleus and peripheral circadian timing system. Eur J Neurosci 29:171–180

de la Iglesia HO, Meyer J, Carpino A Jr et al (2000) Antiphase oscillation of the left and right suprachiasmatic nuclei. Science 290:799–801

de la Iglesia HO, Cambras T, Schwartz WJ et al (2004) Forced desynchronization of dual circadian oscillators within the rat suprachiasmatic nucleus. Curr Biol 14:796–800

Delezie J, Challet E (2011) Interactions between metabolism and circadian clocks: reciprocal disturbances. Ann New York Acad Sci 1243:30–46

Dey J, Carr AJ, Cagampang FR et al (2005) The tau mutation in the Syrian hamster differentially reprograms the circadian clock in the SCN and peripheral tissues. J Biol Rhythm 20:99–110

Dibner C, Schibler U, Albrecht U (2010) The mammalian circadian timing system: organization and coordination of central and peripheral clocks. Ann Rev Physiol 72:517–549

Doi M, Takahashi Y, Komatsu R et al (2010) Salt-sensitive hypertension in circadian clock-deficient Cry-null mice involves dysregulated adrenal Hsd3b6. Nat Med 16:67–74

Drijfhout WJ, Brons HF, Oakley N et al (1997) A microdialysis study on pineal melatonin rhythms in rats after an 8-h phase advance: new characteristics of the underlying pacemaker. Neuroscience 80:233–239

Dziema H, Oatis B, Butcher GQ et al (2003) The ERK/MAP kinase pathway couples light to immediate-early gene expression in the suprachiasmatic nucleus. Eur J Neurosci 17:1617–1627

Ebisawa T, Uchiyama M, Kajimura N et al (2001) Association of structural polymorphisms in the human period3 gene with delayed sleep phase syndrome. EMBO Rep 2:342–346

Eckel-Mahan K, Sassone-Corsi P (2009) Metabolism control by the circadian clock and vice versa. Nat Struc Mol Biol 16:462–467

Eide EJ, Vielhaber EL, Hinz WA et al (2002) The circadian regulatory proteins BMAL1 and cryptochromes are substrates of casein kinase Iepsilon. J Biol Chem 277:17248–17254

Enck P, Muller-Sacks E, Holtmann G et al (1995) Gastrointestinal problems in airline crew members. Zeit Gastroenterol 33:513–516

Filipski E, King VM, Li X et al (2003) Disruption of circadian coordination accelerates malignant growth in mice. Pathol Biol 51:216–219

Filipski E, Delaunay F, King VM et al (2004) Effects of chronic jet-lag on tumor progression in mice. Cancer Res 64:7879–7885

Filipski E, Subramanian P, Carriere J et al (2009) Circadian disruption accelerates liver carcinogenesis in mice. Mut Res 680:95–105

Folkard S (2008) Do permanent night workers show circadian adjustment? A review based on the endogenous melatonin rhythm. Chronobiol Int 25:215–224

Fonken LK, Workman JL, Walton JC et al (2010) Light at night increases body mass by shifting the time of food intake. Proc Natl Acad Sci USA 107:18664–18669

Fonken LK, Kitsmiller E, Smale L et al (2012) Dim nighttime light impairs cognition and provokes depressive-like responses in a diurnal rodent. J Biol Rhythm 27:319–327

Fortier EE, Rooney J, Dardente H et al (2011) Circadian variation of the response of T cells to antigen. J Immunol 187:6291–6300

Foster RG (2012) Biological clocks: who in this place set up a sundial? Curr Biol 22:R405–407

Fu L, Pelicano H, Liu J et al (2002) The circadian gene Period2 plays an important role in tumor suppression and DNA damage response in vivo. Cell 111:41–50

Gachon F, Olela FF, Schaad O et al (2006) The circadian PAR-domain basic leucine zipper transcription factors DBP, TEF, and HLF modulate basal and inducible xenobiotic detoxification. Cell Met 4:25–36

Gale JE, Cox HI, Qian J et al (2011) Disruption of circadian rhythms accelerates development of diabetes through pancreatic beta-cell loss and dysfunction. J Biol Rhythm 26:423–433

Gangwisch JE (2009) Epidemiological evidence for the links between sleep, circadian rhythms and metabolism. Obes Rev 10:37–45

García JA, Zhang D, Estill SJ et al (2000) Impaired cued and contextual memory in NPAS2-deficient mice. Science 288:2226–2230

Gibson EM, Williams WP 3rd, Kriegsfeld LJ (2009) Aging in the circadian system: considerations for health, disease prevention and longevity. Exp Gerontol 44:51–56

Gibson EM, Wang C, Tjho S et al (2010) Experimental 'jet-lag' inhibits adult neurogenesis and produces long-term cognitive deficits in female hamsters. PloS One 5:e15267

Goel N, Stunkard AJ, Rogers NL et al (2009) Circadian rhythm profiles in women with night eating syndrome. J Biol Rhythm 24:85–94

Golombek DA (2012) Chapter 32 – Circadian rhythms and autonomic function. In: David R, Italo B, Geoffrey B, Phillip AL, Julian FR, Paton A, David R, Julian FRP (eds) Primer on the autonomic nervous system, 3rd edn. Academic, San Diego, pp 157–159

Golombek DA, Rosenstein RE (2010) Physiology of circadian entrainment. Physiol Rev 90:1063–1102

Golombek DA, Agostino PV, Plano SA et al (2004) Signaling in the mammalian circadian clock: the NO/cGMP pathway. Neuroch Int 45:929–936

Golombek DA, Casiraghi LP, Agostino PV et al (2013) The times they're a-changing: effects of circadian desynchronization on physiology and disease. J Physiol Paris 107:310–322

Gooley JJ, Chamberlain K, Smith KA et al (2011) Exposure to room light before bedtime suppresses melatonin onset and shortens melatonin duration in humans. J Clin Endocrinol Met 96:463–472

Guess J, Burch JB, Ogoussan K et al (2009) Circadian disruption, Per3, and human cytokine secretion. Int Cancer Ther 8:329–336

Ha M, Park J (2005) Shiftwork and metabolic risk factors of cardiovascular disease. J Occup Health 47:89–95

Hadden H, Soldin SJ, Massaro D (2012) Circadian disruption alters mouse lung clock gene expression and lung mechanics. J Appl Physiol 113:385–392

Harrington M (2010) Location, location, location: important for jet-lagged circadian loops. J Clin Invest 120:2265–2267

Haus E, Smolensky MH (1999) Biologic rhythms in the immune system. Chronobiol Int 16:581–622

Hayashi M, Shimba S, Tezuka M (2007) Characterization of the molecular clock in mouse peritoneal macrophages. Biol Pharm Bull 30:621–626

Hazlerigg D (2012) The evolutionary physiology of photoperiodism in vertebrates. Progr Brain Res 199:413–422

Higashi T, Sakurai H, Satoh T et al (1988) Absenteeism of shift and day workers with special reference to peptic ulcer. Asia-Pacific J Public Health/Asia-Pacific Academic Consortium for Public Health 2:112–119

Hirai K, Kita M, Ohta H et al (2005) Ramelteon (TAK-375) accelerates reentrainment of circadian rhythm after a phase advance of the light-dark cycle in rats. J Biol Rhythm 20:27–37

Honma S, Ono D, Suzuki Y et al (2012) Suprachiasmatic nucleus: cellular clocks and networks. Progr Brain Res 199:129–141

Hoogerwerf WA (2010) Role of clock genes in gastrointestinal motility. Am J Physio (Gast Liver Physiol) 299:G549–G555

Hoogerwerf WA, Hellmich HL, Cornelissen G et al (2007) Clock gene expression in the murine gastrointestinal tract: endogenous rhythmicity and effects of a feeding regimen. Gastroenterology 133:1250–1260

Howland RH (2009) An overview of seasonal affective disorder and its treatment options. Phys Sport Med 37:104–115

Huang W, Ramsey KM, Marcheva B et al (2011) Circadian rhythms, sleep, and metabolism. J Clin Invest 121:2133–2141

IARC WHO (2010) Monographs on the evaluation of carcinogenic risks to humans Painting, firefighting, and shiftwork. Lyon, France

Johnson CH, Elliott JA, Foster R (2003) Entrainment of circadian programs. Chronobiol Int 20:741–774

Kalsbeek A, Fliers E (2013) Daily regulation of hormone profiles. Handb Exp Pharmacol 217:185–226

Kalsbeek A, Scheer FA, Perreau-Lenz S et al (2011) Circadian disruption and SCN control of energy metabolism. FEBS Lett 585:1412–1426

Karatsoreos IN, Bhagat S, Bloss EB et al (2011) Disruption of circadian clocks has ramifications for metabolism, brain, and behavior. Proc Natl Acad Sci USA 108:1657–1662

Kessler EJ, Sprouse J, Harrington ME (2008) NAN-190 potentiates the circadian response to light and speeds re-entrainment to advanced light cycles. Neuroscience 154:1187–1194

Kiessling S, Eichele G, Oster H (2010) Adrenal glucocorticoids have a key role in circadian resynchronization in a mouse model of jet-lag. J Clin Invest 120:2600–2609

Knutsson A (2003) Health disorders of shift workers. Occup Med 53:103–108

Knutsson A, Boggild H (2000) Shiftwork and cardiovascular disease: review of disease mechanisms. Rev Environ Health 15:359–372

Kolla BP, Auger RR (2011) Jet-lag and shift work sleep disorders: how to help reset the internal clock. Cleveland Clin J Med 78:675–684

Koller M (1983) Health risks related to shift work. An exam of long time-contingent effects of long-term stress. Int Arch Occup Environ Health 53(1):59–75

Kondratova AA, Dubrovsky YV, Antoch MP et al (2010) Circadian clock proteins control adaptation to novel environment and memory formation. Aging 2:285–297

Konturek PC, Brzozowski T, Konturek SJ (2011) Gut clock: implication of circadian rhythms in the gastrointestinal tract. J Physiol Pharmacol 62:139–150

Korompeli A, Sourtzi P, Tzavara C et al (2009) Rotating shift-related changes in hormone levels in intensive care unit nurses. J Adv Nurs 65:1274–1282

Kott J, Leach G, Yan L (2012) Direction-dependent effects of chronic "jet-lag" on hippocampal neurogenesis. Neurosci Lett 515:177–180

Krueger JM, Majde JA, Obal F (2003) Sleep in host defense. Brain Behav Immun 17(Suppl 1):S41–S47

Lall GS, Atkinson LA, Corlett SA et al (2012) Circadian entrainment and its role in depression: a mechanistic review. J Neural Transm 119:1085–1096

Lamia KA, Storch KF, Weitz CJ (2008) Physiological significance of a peripheral tissue circadian clock. Proc Natl Acad Sci USA 105:15172–15177

Lee C, Etchegaray JP, Cagampang FR et al (2001) Posttranslational mechanisms regulate the mammalian circadian clock. Cell 107:855–867

Lee C, Smith MR, Eastman CI (2006) A compromise phase position for permanent night shift workers: circadian phase after two night shifts with scheduled sleep and light/dark exposure. Chronobiol Int 23:859–875

Lee ML, Swanson BE, de la Iglesia HO (2009) Circadian timing of REM sleep is coupled to an oscillator within the dorsomedial suprachiasmatic nucleus. Curr Biol 19:848–852

Lee S, Donehower LA, Herron AJ et al (2010) Disrupting circadian homeostasis of sympathetic signaling promotes tumor development in mice. PloS One 5:e10995

LeGates TA, Altimus CM, Wang H et al (2012) Aberrant light directly impairs mood and learning through melanopsin-expressing neurons. Nature 491:594–598

Levandovski R, Dantas G, Fernandes L et al (2011) Depression scores associate with chronotype and social jetlag in a rural population. Chronobiol Int 28:771–778

Li JC, Xu F (1997) Influences of light-dark shifting on the immune system, tumor growth and life span of rats, mice and fruit flies as well as on the counteraction of melatonin. Biol Signal 6:77–89

Li JD, Hu WP, Zhou QY (2012) The circadian output signals from the suprachiasmatic nuclei. Progr Brain Res 199:119–127

Liu J, Malkani G, Shi X et al (2006) The circadian clock Period 2 gene regulates gamma interferon production of NK cells in host response to lipopolysaccharide-induced endotoxic shock. Infect Immun 74:4750–4756

Lo SH, Lin LY, Hwang JS et al (2010) Working the night shift causes increased vascular stress and delayed recovery in young women. Chronobiol Int 27:1454–1468

Logan RW, Zhang C, Murugan S et al (2012a) Chronic shift-lag alters the Circadian Clock of NK cells and promotes lung cancer growth in rats. J Immunol 188:2572–2582

Logan RW, Zhang C, Murugan S et al (2012b) Chronic shift-lag alters the circadian clock of NK cells and promotes lung cancer growth in rats. J Immunol 188:2583–2591

Logan RW, Wynne O, Levitt D et al (2013) Altered circadian expression of cytokines and cytolytic factors in splenic natural killer cells of Per1(−/−) mutant mice. J Interf Cytok Res 33:108–114

Loh DH, Navarro J, Hagopian A et al (2010) Rapid changes in the light/dark cycle disrupt memory of conditioned fear in mice. PloS One 5:e18540

Lowrey PL, Shimomura K, Antoch MP et al (2000) Positional syntenic cloning and functional characterization of the mammalian circadian mutation tau. Science 288:483–492

Lucas RJ, Stirland JA, Darrow JM et al (1999) Free running circadian rhythms of melatonin, luteinizing hormone, and cortisol in Syrian hamsters bearing the circadian tau mutation. Endocrinology 140:758–764

Ma WP, Cao J, Tian M et al (2007) Exposure to chronic constant light impairs spatial memory and influences long-term depression in rats. Neurosci Res 59:224–230

Machi MS, Staum M, Callaway CW et al (2012) The relationship between shift work, sleep, and cognition in career emergency physicians. Acad Emerg Med 19:85–91

Majde JA, Krueger JM (2005) Links between the innate immune system and sleep. J Allerg Clin Immunol 116:1188–1198

Malloy JN, Paulose JK, Li Y et al (2012) Circadian rhythms of gastrointestinal function are regulated by both central and peripheral oscillators. Am J Physiol Gastrointest Liver Physiol 303:G461–473

Marcheva B, Ramsey KM, Buhr ED et al (2010) Disruption of the clock components CLOCK and BMAL1 leads to hypoinsulinaemia and diabetes. Nature 466:627–631

Markwald RR, Melanson EL, Smith MR et al (2013) Impact of insufficient sleep on total daily energy expenditure, food intake, and weight gain. Proc Natl Acad Sci USA 110:5695–5700

Martino TA, Oudit GY, Herzenberg AM et al (2008) Circadian rhythm disorganization produces profound cardiovascular and renal disease in hamsters. Am J Physiol Reg Integ Compar Physiol 294:R1675–1683

Mazzoccoli G, Francavilla M, Pazienza V et al (2012) Differential patterns in the periodicity and dynamics of clock gene expression in mouse liver and stomach. Chronobiol Int 29: 1300–1311

McDonald RJ, Zelinski EL, Keeley RJ et al (2013) Multiple effects of circadian dysfunction induced by photoperiod shifts: Alterations in context memory and food metabolism in the same subjects. Physiol Behav 118:14–24

Menaker M (2006) Circadian organization in the real world. Proc Natl Acad Sci USA 103: 3015–3016

Meng QJ, Logunova L, Maywood ES et al (2008) Setting clock speed in mammals: the CK1 epsilon tau mutation in mice accelerates circadian pacemakers by selectively destabilizing PERIOD proteins. Neuron 58:78–88

Mercado C, Díaz-Muñoz M, Alamilla J et al (2009) Ryanodine-sensitive intracellular Ca2+ channels in rat suprachiasmatic nuclei are required for circadian clock control of behavior. J Biol Rhythm 24:203–210

Mohawk JA, Green CB, Takahashi JS (2012) Central and peripheral circadian clocks in mammals. Ann Rev Neurosci 35:445–462

Mohren DC, Jansen NW, Kant IJ et al (2002) Prevalence of common infections among employees in different work schedules. J Occup Environ Med/Am Coll Occup Environ Med 44:1003–1011

Moller-Levet CS, Archer SN, Bucca G et al (2013) Effects of insufficient sleep on circadian rhythmicity and expression amplitude of the human blood transcriptome. Proc Natl Acad Sci USA 110:1132–1141

Moore RY (1996) Neural control of the pineal gland. Behav Brain Res 73:125–130

Moore-Ede M, Sulzman F (1981) Internal temporal order. In: Aschoff J (ed) Biological rhythms. Springer, New York, NY, pp 215–241

Mormont MC, Levi F (1997) Circadian-system alterations during cancer processes: a review. Int J Cancer 70:241–247

Morris CJ, Aeschbach D, Scheer FA (2012a) Circadian system, sleep and endocrinology. Mol Cell Endocrinol 349:91–104

Morris CJ, Yang JN, Scheer FA (2012b) The impact of the circadian timing system on cardiovascular and metabolic function. Progr Brain Res 199:337–358

Nagai M, Morikawa Y, Kitaoka K et al (2011) Effects of fatigue on immune function in nurses performing shift work. J Occup Health 53:312–319

Nagano M, Adachi A, Nakahama K et al (2003) An abrupt shift in the day/night cycle causes desynchrony in the mammalian circadian center. J Neurosci 23:6141–6151

Nakamura K, Shimai S, Kikuchi S et al (1997) Shift work and risk factors for coronary heart disease in Japanese blue-collar workers: serum lipids and anthropometric characteristics. Occup Med (Lond) 47:142–146

Nakamura W, Yamazaki S, Takasu NN et al (2005) Differential response of Period 1 expression within the suprachiasmatic nucleus. J Neurosci 25:5481–5487

Narasimamurthy R, Hatori M, Nayak SK et al (2012) Circadian clock protein cryptochrome regulates the expression of proinflammatory cytokines. Proc Natl Acad Sci USA 109: 12662–12667

Nedeltcheva AV, Kessler L, Imperial J et al (2009) Exposure to recurrent sleep restriction in the setting of high caloric intake and physical inactivity results in increased insulin resistance and reduced glucose tolerance. J Clin Endocrinol Metabol 94:3242–3250

Nemoto S, Fergusson MM, Finkel T (2004) Nutrient availability regulates SIRT1 through a forkhead-dependent pathway. Science 306:2105–2108

Nievergelt CM, Kripke DF, Barrett TB et al (2006) Suggestive evidence for association of the circadian genes PERIOD3 and ARNTL with bipolar disorder. Am J Med Gen Part B 141B:234–241

Nojkov B, Rubenstein JH, Chey WD et al (2010) The impact of rotating shift work on the prevalence of irritable bowel syndrome in nurses. Am J Gastroenterol 105:842–847

Obrietan K, Impey S, Smith D et al (1999) Circadian regulation of cAMP response element-mediated gene expression in the suprachiasmatic nuclei. J Biol Chem 274:17748–17756

Ohta H, Yamazaki S, McMahon DG (2005) Constant light desynchronizes mammalian clock neurons. Nat Neurosci 8:267–269

Oishi K, Ohkura N (2013) Chronic circadian clock disruption induces expression of the cardiovascular risk factor plasminogen activator inhibitor-1 in mice. Blood Coag Fibrinol 24:106–108

Okawa M, Uchiyama M (2007) Circadian rhythm sleep disorders: characteristics and entrainment pathology in delayed sleep phase and non-24 h sleep-wake syndrome. Sleep Med Rev 11:485–496

O'Neill JS, Reddy AB (2011) Circadian clocks in human red blood cells. Nature 469:498–503

Panda S, Antoch MP, Miller BH et al (2002) Coordinated transcription of key pathways in the mouse by the circadian clock. Cell 109:307–320

Partonen T, Treutlein J, Alpman A et al (2007) Three circadian clock genes Per2, Arntl, and Npas2 contribute to winter depression. Ann Med 39:229–238

Paul MA, Miller JC, Love RJ et al (2009) Timing light treatment for eastward and westward travel preparation. Chronobiol Int 26:867–890

Paul MA, Miller JC, Gray GW, Love RJ, Lieberman HR, Arendt J (2010) Melatonin treatment for eastward and westward travel preparation. Psychopharmacol 208:377–386

Pereira DS, Tufik S, Louzada FM et al (2005) Association of the length polymorphism in the human Per3 gene with the delayed sleep-phase syndrome: does latitude have an influence upon it? Sleep 28:29–32

Pierard C, Beaumont M, Enslen M et al (2001) Resynchronization of hormonal rhythms after an eastbound flight in humans: effects of slow-release caffeine and melatonin. Eur J Appl Physiol 85:144–150

Pietroiusti A, Forlini A, Magrini A et al (2006) Shift work increases the frequency of duodenal ulcer in H pylori infected workers. Occup Environ Med 63:773–775

Pizzio GA, Hainich EC, Ferreyra GA et al (2003) Circadian and photic regulation of ERK, JNK and p38 in the hamster SCN. Neuroreport 14:1417–1419

Plano SA, Golombek DA, Chiesa JJ (2010) Circadian entrainment to light-dark cycles involves extracellular nitric oxide communication within the suprachiasmatic nuclei. Eur J Neurosci 31:876–882

Plano SA, Agostino PV, de la Iglesia HO et al (2012) cGMP-phosphodiesterase inhibition enhances photic responses and synchronization of the biological circadian clock in rodents. PloS One 7:e37121

Polidarova L, Sladek M, Sotak M et al (2011) Hepatic, duodenal, and colonic circadian clocks differ in their persistence under conditions of constant light and in their entrainment by restricted feeding. Chronobiol Int 28:204–215

Preuss F, Tang Y, Laposky AD et al (2008) Adverse effects of chronic circadian desynchronization in animals in a "challenging" environment. Am J Physiol (Reg Int Comp Physiol) 295: R2034–2040

Rajaratnam SM, Middleton B, Stone BM et al (2004) Melatonin advances the circadian timing of EEG sleep and directly facilitates sleep without altering its duration in extended sleep opportunities in humans. J Physiol 561:339–351

Ralph MR, Menaker M (1988) A mutation of the circadian system in golden hamsters. Science 241:1225–1227

Reddy AB, Field MD, Maywood ES et al (2002) Differential resynchronisation of circadian clock gene expression within the suprachiasmatic nuclei of mice subjected to experimental jet-lag. J Neurosci 22:7326–7330

Reid KJ, Zee PC (2009) Circadian rhythm disorders. Sem Neurol 29:393–405

Reid KJ, McGee-Koch LL, Zee PC (2011) Cognition in circadian rhythm sleep disorders. Progr Brain Res 190:3–20

Reiter RJ, Tan DX, Fuentes-Broto L (2010) Melatonin: a multitasking molecule. Progr Brain Res 181:127–151

Reppert SM, Weaver DR (2001) Molecular analysis of mammalian circadian rhythms. Ann Rev Physiol 63:647–676

Revell VL, Eastman CI (2005) How to trick mother nature into letting you fly around or stay up all night. J Biol Rhythm 20:353–365

Revell VL, Burgess HJ, Gazda CJ et al (2006) Advancing human circadian rhythms with afternoon melatonin and morning intermittent bright light. J Clin Endocrinol Metabol 91:54–59

Roenneberg T, Daan S, Merrow M (2003) The art of entrainment. J Biol Rhythm 18:183–194

Roenneberg T, Allebrandt KV, Merrow M et al (2012) Social jetlag and obesity. Curr Biol 22:939–943

Rouch I, Wild P, Ansiau D et al (2005) Shiftwork experience, age and cognitive performance. Ergonomics 48:1282–1293

Roybal K, Theobold D, Graham A et al (2007) Mania-like behavior induced by disruption of CLOCK. Proc Natl Acad Sci USA 104:6406–6411

Rudic RD, McNamara P, Curtis AM et al (2004) BMAL1 and CLOCK, two essential components of the circadian clock, are involved in glucose homeostasis. PLoS Biol 2:e377

Ruger M, Scheer FA (2009) Effects of circadian disruption on the cardiometabolic system. Rev Endocrine Met Dis 10:245–260

Rutter J, Reick M, Wu LC et al (2001) Regulation of clock and NPAS2 DNA binding by the redox state of NAD cofactors. Science 293:510–514

Sack RL (2009) The pathophysiology of jet-lag. Travel Med Infect Dis 7:102–110

Sage D, Ganem J, Guillaumond F et al (2004) Influence of the corticosterone rhythm on photic entrainment of locomotor activity in rats. J Biol Rhythm 19:144–156

Sahar S, Sassone-Corsi P (2012) Regulation of metabolism: the circadian clock dictates the time. Trends Endocrinol Metabol 23:1–8

Salgado-Delgado R, Angeles-Castellanos M, Buijs MR et al (2008) Internal desynchronization in a model of night-work by forced activity in rats. Neuroscience 154:922–931

Salgado-Delgado R, Angeles-Castellanos M, Saderi N et al (2010a) Food intake during the normal activity phase prevents obesity and circadian desynchrony in a rat model of night work. Endocrinology 151:1019–1029

Salgado-Delgado R, Nadia S, Ángeles-Castellanos M et al (2010b) In a rat model of night work, activity during the normal resting phase produces desynchrony in the hypothalamus. J Biol Rhythm 25:421–431

Scheer FA, Hilton MF, Mantzoros CS et al (2009) Adverse metabolic and cardiovascular consequences of circadian misalignment. Proc Natl Acad Sci USA 106:4453–4458

Scheiermann C, Kunisaki Y, Lucas D et al (2012) Adrenergic nerves govern circadian leukocyte recruitment to tissues. Immunity 37:290–301

Scheiermann C, Kunisaki Y, Frenette PS (2013) Circadian control of the immune system. Nature Rev Immunol 13:190–198

Schwartz MD, Wotus C, Liu T et al (2009) Dissociation of circadian and light inhibition of melatonin release through forced desynchronization in the rat. Proc Natl Acad Sci USA 106:17540–17545

Schwartz WJ, Tavakoli-Nezhad M, Lambert CM et al (2011) Distinct patterns of Period gene expression in the suprachiasmatic nucleus underlie circadian clock photoentrainment by advances or delays. Proc Natl Acad Sci USA 108:17219–17224

Segawa K, Nakazawa S, Tsukamoto Y et al (1987) Peptic ulcer is prevalent among shift workers. Dig Dis Sci 32:449–453

Sellix MT, Evans JA, Leise TL et al (2012) Aging differentially affects the re-entrainment response of central and peripheral circadian oscillators. J Neurosci 32:16193–16202

Shostak A, Meyer-Kovac J, Oster H (2013) Circadian regulation of lipid mobilization in white adipose tissues. Diabetes 62:2195–2203

Silver AC, Arjona A, Walker WE et al (2012) The circadian clock controls Toll-like receptor 9-mediated innate and adaptive immunity. Immunity 36:251–261

Skene DJ, Arendt J (2007) Circadian rhythm sleep disorders in the blind and their treatment with melatonin. Sleep Med 8:651–655

Smith MR, Fogg LF, Eastman CI (2009a) A compromise circadian phase position for permanent night work improves mood, fatigue, and performance. Sleep 32:1481–1489

Smith MR, Fogg LF, Eastman CI (2009b) Practical interventions to promote circadian adaptation to permanent night shift work: study 4. J Biol Rhythm 24:161–172

Soria V, Martínez-Amoros E, Escaramis G et al (2010) Differential association of circadian genes with mood disorders: CRY1 and NPAS2 are associated with unipolar major depression and CLOCK and VIP with bipolar disorder. Neuropsychopharmacol 35:1279–1289

Spiegel K, Tasali E, Leproult R et al (2009) Effects of poor and short sleep on glucose metabolism and obesity risk. Nat Rev Endocrinol 5:253–261

Stenvers DJ, Jonkers CF, Fliers E et al (2012) Nutrition and the circadian timing system. Progr Brain Res 199:359–376

Su TC, Lin LY, Baker D et al (2008) Elevated blood pressure, decreased heart rate variability and incomplete blood pressure recovery after a 12-hour night shift work. J Occup Health 50:380–386

Takahashi JS, Hong HK, Ko CH et al (2008) The genetics of mammalian circadian order and disorder: implications for physiology and disease. Nature Rev Gen 9:764–775

Tenkanen L, Sjoblom T, Kalimo R et al (1997) Shift work, occupation and coronary heart disease over 6 years of follow-up in the Helsinki Heart Study. Scand J Work Environ Health 23:257–265

Toh KL, Jones CR, He Y et al (2001) An hPer2 phosphorylation site mutation in familial advanced sleep phase syndrome. Science 291:1040–1043

Turek FW, Joshu C, Kohsaka A et al (2005) Obesity and metabolic syndrome in circadian Clock mutant mice. Science 308:1043–1045

van den Heiligenberg S, Depres-Brummer P, Barbason H et al (1999) The tumor promoting effect of constant light exposure on diethylnitrosamine-induced hepatocarcinogenesis in rats. Life Sci 64:2523–2534

Van der Zee EA, Havekes R, Barf RP et al (2008) Circadian time-place learning in mice depends on Cry genes. Curr Biol 18:844–848

Vansteensel MJ, Yamazaki S, Albus H et al (2003) Dissociation between circadian Per1 and neuronal and behavioral rhythms following a shifted environmental cycle. Curr Biol 13:1538–1542

Vgontzas AN, Chrousos GP (2002) Sleep, the hypothalamic-pituitary-adrenal axis, and cytokines: multiple interactions and disturbances in sleep disorders. Endocrinol Metabol Clin North Am 31:15–36

Vinogradova IA, Anisimov VN, Bukalev AV et al (2010) Circadian disruption induced by light-at-night accelerates aging and promotes tumorigenesis in young but not in old rats. Aging 2:82–92

Viswambharan H, Carvas JM, Antic V et al (2007) Mutation of the circadian clock gene Per2 alters vascular endothelial function. Circulation 115:2188–2195

Vyas MV, Garg AX, Iansavichus AV et al (2012) Shift work and vascular events: systematic review and meta-analysis. BMJ 345:e4800

Waite EJ, McKenna M, Kershaw Y et al (2012) Ultradian corticosterone secretion is maintained in the absence of circadian cues. Eur J Neurosci 36:3142–3150

Waterhouse J, Reilly T, Atkinson G et al (2007) Jet-lag: trends and coping strategies. Lancet 369:1117–1129

Welsh DK, Takahashi JS, Kay SA (2010) Suprachiasmatic nucleus: cell autonomy and network properties. Ann Rev Physiol 72:551–577

Wirth M, Burch J, Violanti J et al (2011) Shiftwork duration and the awakening cortisol response among police officers. Chronobiol Int 28:446–457

Wittmann M, Dinich J, Merrow M et al (2006) Social jetlag: misalignment of biological and social time. Chronobiol Int 23:497–509

Wood PA, Yang X, Taber A et al (2008) Period 2 mutation accelerates ApcMin/+tumorigenesis. Mol Cancer Res 6:1786–1793

Wright KP Jr, Hull JT, Hughes RJ et al (2006) Sleep and wakefulness out of phase with internal biological time impairs learning in humans. J Cognit Neurosci 18:508–521

Wright KP, Lowry CA, Lebourgeois MK (2012) Circadian and wakefulness-sleep modulation of cognition in humans. Front Mol Neurosci 5:50–62

Wu M, Zeng J, Chen Y et al (2012) Experimental chronic jet-lag promotes growth and lung metastasis of Lewis lung carcinoma in C57BL/6 mice. Oncol Rep 64:256–262

Xu Y, Padiath QS, Shapiro RE et al (2005) Functional consequences of a CKIdelta mutation causing familial advanced sleep phase syndrome. Nature 434:640–644

Yamazaki S, Numano R, Abe M et al (2000) Resetting central and peripheral circadian oscillators in transgenic rats. Science 288:682–685

Yan L (2011) Structural and functional changes in the suprachiasmatic nucleus following chronic circadian rhythm perturbation. Neuroscience 183:99–107

Zawilska JB, Skene DJ, Arendt J (2009) Physiology and pharmacology of melatonin in relation to biological rhythms. Pharmacol Rep 61:383–410

Zee PC, Goldstein CA (2010) Treatment of shift work disorder and jet-lag. Curr Treat Opt Neurol 12:396–411

Zeitzer JM, Dijk DJ, Kronauer R et al (2000) Sensitivity of the human circadian pacemaker to nocturnal light: melatonin phase resetting and suppression. J Physiol 526:695–702

Zimberg IZ, Fernandes-Junior SA, Crispim CA et al (2012) Metabolic impact of shift work. Work 41:4376–4383

Zisapel N (2001) Circadian rhythm sleep disorders: pathophysiology and potential approaches to management. CNS Drugs 15:311–328

Zober A, Schilling D, Ott MG et al (1998) Helicobacter pylori infection: prevalence and clinical relevance in a large company. J Occup Environ Med/Am College Occup Environ Med 40:586–594

Chapter 16
Circadian Dysfunction in Huntington's Disease

Dika Kuljis, Analyne M. Schroeder, Takashi Kudo,
Dawn H. Loh, and Christopher S. Colwell

Abstract Sleep disorders are common in Huntington's disease (HD) patients and develop early in the disease process. Among a number of possible mechanisms that underlie sleep disruption, there is evidence that circadian system dysfunction is a contributing factor. Using the BACHD mouse models of HD, we have determined that at the onset of symptoms, rhythmic spontaneous electrical activity of neurons within the suprachiasmatic nucleus (SCN) is disrupted even though the molecular clockwork is still functional. These findings suggest that reduced SCN output may underlie disrupted timing of sleep and have wide ranging impact throughout the body. The mechanism underlying this deficit is not yet known, but mitochondrial dysfunction and oxidative stress are likely involved. Our animal model findings raise the possibility that disordered sleep and circadian function experienced by HD patients may be an integral part of the disease, and we speculate that circadian dysfunction may accelerate the pathology underlying HD. If these hypotheses are correct, we should focus on treating circadian misalignment and sleep disruptions early in disease progression.

16.1 Introduction

For many patients with chronic diseases of the nervous system, a good night's sleep is hard to find (Foley et al. 2004). It is increasingly evident that in neurodegenerative disorders such as Alzheimer's, Huntington's, and Parkinson's diseases, sleep

D. Kuljis
Laboratory of Circadian and Sleep Medicine, University of California Los Angeles,
Los Angeles, CA 90025, USA

Department of Neurobiology, University of California Los Angeles,
Los Angeles, CA 90025, USA

A.M. Schroeder • T. Kudo • D.H. Loh • C.S. Colwell (✉)
Laboratory of Circadian and Sleep Medicine, University of California Los Angeles,
Los Angeles, CA 90025, USA

Department of Psychiatry and Biobehavioral Sciences, Semel Institute for Neuroscience,
University of California Los Angeles, 760 Westwood Plaza, Los Angeles, CA 90025, USA
e-mail: ccolwell@mednet.ucla.edu

© Springer International Publishing Switzerland 2015
R. Aguilar-Roblero et al. (eds.), *Mechanisms of Circadian Systems
in Animals and Their Clinical Relevance*, DOI 10.1007/978-3-319-08945-4_16

disruptions are common and occur early in the disease progression (McCurry et al. 2007; Morton 2013; Willison et al. 2013). In some cases, prodromal sleep disturbance will occur years before the onset of dementia or motor symptoms characteristic of the specific disease in question (Iranzo et al. 2006; Julein et al. 2007). A number of possible mechanisms may underlie these sleep disturbances, but based on animal model findings we propose that dysfunction of the circadian system is an integral contributing factor.

16.2 Circadian System

16.2.1 *Molecular Clockwork in Tissues Throughout the Body*

Output from the master circadian clock in the central nervous system regulates physiological rhythms throughout the body by modifying the phase and amplitude of "clock gene" rhythms. Most cells express robust rhythms in the transcription and translation of these key clock genes through an autoregulatory negative feedback loop. Over the last 20 years, circadian rhythm researchers have identified the mechanism by which the cell-autonomous molecular clockwork generates and regulates rhythms in gene expression in a circadian manner (Dibner et al. 2010; Mohawk and Takahashi 2011). Briefly, CLOCK and BMAL1 are basic helix-loop-helix PAS-containing transcription factors that heterodimerize and enhance transcription of Period genes (Per1-3) (Albrecht et al. 1997; Shearman et al. 1997; Sun et al. 1997) and Cryptochrome genes (Cry1-2) (van der Horst et al. 1999) through E-box elements (Gekakis et al. 1998; Hogenesch et al. 1998; Takahata et al. 1998). PER and CRY proteins form complexes in the cytoplasm that translocate to the nucleus where they recruit HDAC complex to repress BMAL/CLOCK-mediated transcription (van der Horst et al. 1999; Thresher et al. 1998; Siepka et al. 2007; Duong et al. 2011). This transcription inhibition ends when ubiquitination of CRYs targets them for proteasomal degradation allowing a new cycle to begin (Siepka et al. 2007; Busino et al. 2007; Godinho et al. 2007). Additionally, clock protein phosphorylation state regulates the timing and stability of this core negative feedback loop by directing the protein for degradation or nuclear translocation (Gallego and Virshup 2007; Lee et al. 2001; Harms et al. 2004), and auxiliary stabilizing transcriptional loops regulate *Bmal1* (Preitner et al. 2002; Sato et al. 2004) and other clock genes (Crumbley et al. 2010; Panda et al. 2002) through nuclear receptors from the REV-ERB and ROR families. Taken together, the network that drives circadian oscillations in gene expression output is a highly regulated, robust, and resilient timekeeper.

16.2.2   Circadian Regulation of Gene Expression Varies with Tissue

Each of the major organ systems of the body (i.e., brain, heart, liver, pancreas) is made up of a network of circadian oscillators with its own clockwork that regulates the transcription of genes important to the specific target organ (Dibner et al. 2010). DNA microarray expression profiling indicates approximately 8–12 % of genes display circadian oscillations (Panda et al. 2002; Storch et al. 2002; Hogenesch et al. 2003), and many of these rhythmically expressed genes are involved in key rate-limiting steps of biochemical pathways (Panda et al. 2002; Baggs and Hogenesch 2010). For instance, in the nervous system, a number of genes that are essential components of neuronal signaling such as peptide synthesis, secretion, and oxidative phosphorylation are transcribed with a circadian oscillation. The role of the core molecular clockwork on tissue transcription is evidenced in a mouse overexpressing REV-ERB alpha specifically in the liver. This genetic manipulation results in the arrest of the molecular clockwork, and this in turn results in a 90 % reduction of hepatocyte transcriptional rhythms (Kornmann et al. 2007). A notable caveat is that in some cases, rhythms in mRNA expression do not lead to rhythms in protein levels (Reddy et al. 2006; Deery et al. 2009), so it is important to confirm that loss of gene expression rhythms also lead to the loss of protein expression rhythms. Nevertheless, these findings indicate that even though some liver rhythms are imposed by external signals such as cortisol, the predominant driver of gene expression rhythms in the liver appears to be rhythmic clock gene expression.

16.2.3   SCN Output Coordinates an Array of Rhythmic Tissues

The SCN of the hypothalamus contains the "master" oscillatory network necessary for coordinating circadian rhythms throughout the body (Welsh et al. 2010; Mohawk and Takahashi 2011). This bilaterally paired nucleus is made up of tightly compacted, small-diameter neurons just above the optic chiasm and lateral to the third ventricle (van den Pol 1980). Anatomical studies indicate there are at least two distinct subregions of the SCN including a ventral (core) and dorsal (shell) region (Antle et al. 2009; Golombek and Rosenstei 2010). Ventral neurons are thought to integrate ambient light information from the environment from three major input pathways: the retinohypothalamic tract (RHT), the geniculohypothalamic tract from the intergeniculate leaflet of the thalamus, and the raphe nuclei (Morin and Allen 2006). Ventral input processing neurons exhibit relatively low amplitude rhythms in clock gene expression and the neuropeptides vasoactive intestinal peptide (VIP) or gastrin-releasing peptide (GRP) in addition to the neurotransmitter GABA—signaling molecules that communicate environmental information to dorsal SCN neurons. In contrast to the ventral population, neurons of the dorsal shell appear to generate robust circadian oscillations in gene expression (Yan and Okamura 2002; Hamada et al. 2004; Nakamura et al. 2005) and express vasopressin (AVP) or prokineticin 2 (PK2) in addition to GABA.

The interplay between ventral core neurons and dorsal shell neurons coordinates the output of circadian information from the SCN, as evidenced by the fact that core neurons densely project to shell neurons (Antle and Silver 2005), and projections from both the core and shell subpopulations form synapse on other medial hypothalamic structures, such as the subparaventricular zone surrounding the SCN (Abrahamson and Moore 2001; Kriegsfeld et al. 2004; Kalsbeek et al. 2006). These hypothalamic relay nuclei in turn send projections throughout the nervous and endocrine systems, providing multiple pathways by which SCN output conveys environmental time to the rest of the brain and body (Kalsbeek et al. 2006; Dibner et al. 2010). Additionally, output from physiological systems that are circadianly regulated can "feed back" to modulate the master clock in the SCN. Sleep states (Deboer et al. 2003) and locomotor activity (Meijer et al. 1997; Yamazaki et al. 1998; Schaap and Meijer 2001), among others, "feed back" to regulate SCN neural activity recorded in vivo. These in vivo studies demonstrated the difficulty in disentangling the role of the circadian system on sleep, behavior, and physiology, since altering sleep without impacting the circadian system may not be possible.

16.3 Focus on Huntington's Disease

HD is caused by an expanded CAG trinucleotide repeat region in the gene encoding huntingtin (HTT) protein that translates into polyglutamine tract (Huntington's Disease Collaborative Research Group 1993). HD disease onset is typically in middle age, but this depends on the extent of the CAG repeat expansion so that symptoms can start quite young. In fact, human genetic studies indicate that patients with a larger CAG repeat expansion exhibit earlier onset and more severe symptoms. The classical clinical features of HD are abnormal movements including both involuntary movements (chorea) and difficulty with motor coordination (Bonelli and Beal 2012). These motor symptoms appear to be due to pathological and progressive neuronal cell loss in basal ganglia and cortex (Margolis and Ross 2003; Schulte and Littleton 2011). In addition to motor symptoms, HD patients also exhibit early-onset nonmotor symptoms such as sleep disturbances, depressed mood, metabolic disorder, and cognitive dysfunction (Lavin et al. 1981; Kremer et al. 1990; Petersén et al. 2005, 2009; Hinton et al. 2007; Paulsen et al. 2008; Wood et al. 2008; Videnovic et al. 2009; Aziz et al. 2010b). In humans, many of these nonmotor symptoms begin *before* the onset of motor symptoms and are likely due to alterations in hypothalamic and endocrine system function (Duff et al. 2007; Soneson et al. 2010). The cluster of resulting neuropsychiatric symptoms includes cognitive impairment that progresses to dementia, anxiety, apathy, personality changes, and sleep disorders (Duff et al. 2007; Peavy et al. 2010; Vaccarino et al. 2011). Sleep disorders are extremely common in HD and have major detrimental effects on daily functioning and the quality of life of patients and their caregivers (Cuturic et al. 2009; Goodman and Barker 2010; Aziz et al. 2010a; Goodman et al. 2011; Morton 2013). Due to the high prevalence of these nonmotor symptoms and because they often become apparent years before the onset of motor symptoms, an improved understanding of the mechanism underlying these disruptions has important implications for early diagnosis and treatment of HD.

The possibility that deficits in SCN coordination of physiological and behavioral rhythms occur in HD is raised when one considers temporal aspects of HD nonmotor symptoms. Patient reports of sleep disruptions include increased sleep latency, decreased sleep maintenance, fragmented sleep, and excessive daytime sleepiness. These symptoms may reflect alterations in the temporal patterning of sleep which can be the result of circadian dysfunction. Similarly, the autonomic dysfunction related to cardiovascular function reported in HD is also consistent with possible circadian dysregulation. To explore the possibility that circadian dysfunction contributes to HD disease progression, we examined behavior and physiology of various mouse models of HD.

16.3.1 BACHD Mouse Model of HD

Previous work has shown that the R6/2 CAG150+ mouse model of HD exhibits a progressive and rapid breakdown of the circadian rest/activity cycle that mimics the condition observed in human patients. This condition is typified by loss of consolidated sleep, increased wakeful activity during the sleep phase, and more sleep during the active/waking phase (Morton et al. 2005; Pallier et al. 2007). No single mouse model can be expected to recapitulate all aspects of the human disease, so we sought to extend this work by examination of bacterial artificial chromosome (BAC) HD transgenic mice. This sophisticated transgenic HD mouse model expresses the entire human *HTT* gene with stable expression of 97 mixed CAA-CAG repeats in somatic and germline tissues under the control of human HD promoter (Gray et al. 2008).

16.3.2 Circadian Behavior and Temporal Distribution of Sleep Is Disrupted in the BACHD Mouse Model

The method of choice for screening mutations that influence the circadian system of small mammals is the monitoring of daily wheel-running activity. We examined the BACHD line using this method and found low amplitude, fragmented rhythms in wheel-running behavior under both a light/dark photic environment and constant darkness (Kudo et al. 2011) (Fig. 16.1). By young adulthood (3–6 months), BACHD mice

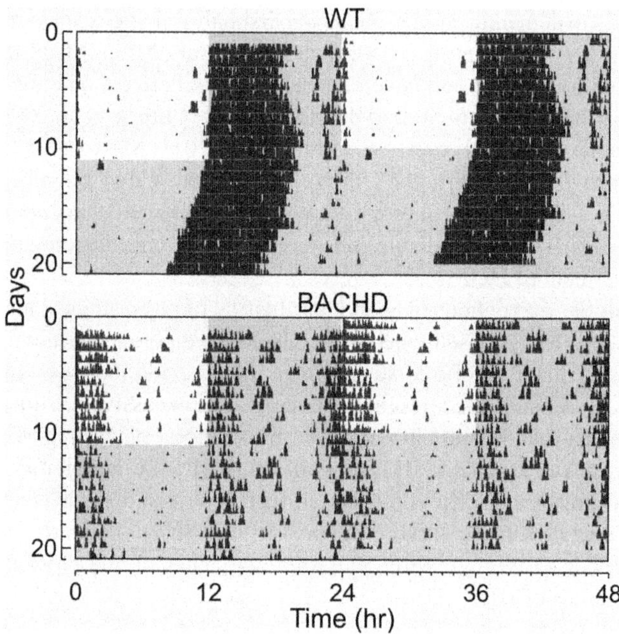

Fig. 16.1 Circadian dysfunction in locomotor activity. Mice were placed individually in cages with running wheels, and locomotor activity was recorded under different lighting conditions. Each *horizontal row* represents an activity record for a 24-h day. Successive days are plotted from *top* to *bottom*. The *grey shading* represents darkness. Mice were initially held in LD (12:12) and then released into DD. *Panels* show examples of the wheel-running activity recorded from WT (*top*) and the BACHD model of HD (*bottom*). The mice were 2–3 months of age

exhibit deficits in circadian rhythm fragmentation, rhythm power, and activity and with increasing age activity amount further deteriorates. Even though the amplitude and coherence of behavioral rhythms progressively declines over the lifespan of WT and BACHD mice, this age-related decline occurs earlier in BACHD mice, with young BACHD mice exhibiting the circadian phenotype of an older WT mouse. Additionally, video recordings were used to measure sleep distribution in BACHD mice. Although the effects of the human transgene were subtle, BACHD mice show significantly reduced sleep during the beginning of their rest phase. In other words, just like the human patient phenotype, BACHD mice show deficits in the onset of sleep.

16.3.3 Disrupted Circadian Regulation of the Cardiovascular System in the BACHD Mouse Model

Poor cardiovascular health in additional to sleep-related nonmotor symptoms has been observed in HD patients. Alterations in autonomic tone that result in cardiac dysfunction is best evidenced by reductions in heart rate variability (HRV).

Fig. 16.2 Circadian dysfunction in physiological outputs. Examples of heart rate and body temperature measured by telemetry from WT and littermate BACHD are shown. The average waveform for each genotype as measured over 10 days in LD. The mice were 5–6 months of age

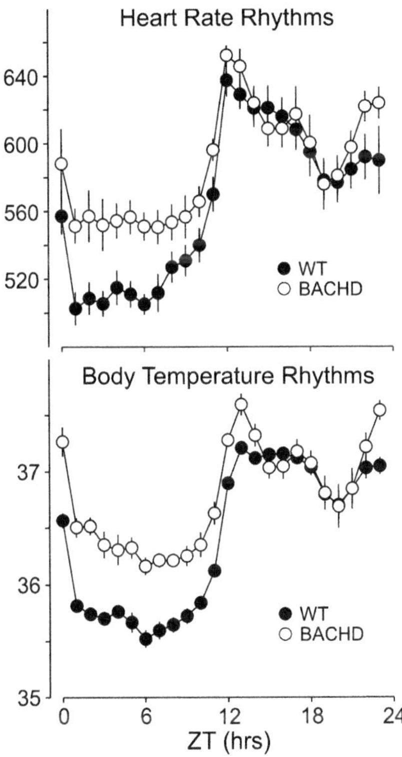

HRV is a measure of variation in the beat-to-beat interval and reflects the dynamic balance of sympathetic and parasympathetic control of heart function and normally displays a robust diurnal and circadian rhythm. Day/night differences in HRV are lost in BACHD suggestive of a loss of circadian control over the autonomic nervous system. Additionally, an overall decrease in HRV over 24 h in BACHD mice suggests this may be due to autonomic dysfunction. Consistent with this hypothesis is the fact that heart rate and body temperature measurements during the rest phase (daytime) in BACHD mice were elevated (Fig. 16.2). Notably, a similar decrease in HRV has also been reported in HD patients beginning during the presymptomatic and early stages of HD progression (Andrich et al. 2002; Kobal et al. 2004, 2010). Reduced HRV is generally considered an indication of poor cardiovascular health and a predictor for cardiovascular disease and mortality (Bigger et al. 1992; Buccelletti et al. 2009; Thayer et al. 2010). In this regard, it is worth considering that loss of temporal control and autonomic imbalance may be promoting cardiac failure—a leading cause of death among HD patients (Chiu and Alexander 1982; Lanska et al. 1988). Electrocardiograph (ECG) parameters were also examined in WT and BACHD mice. While most parameters were not different between the two genotypes, we detected a loss of day/night differences in the PR interval, or the time it takes for the cardiac electrical impulse to travel from the sinus node of the atria to the atrioventricular node. In BACHD mice, the PR interval was significantly elongated during the active period when compared to WT mice. These findings also implicate aberrant autonomic system input to the cardiovascular system, due to its important role in regulating of the PR interval (Wallick et al. 1982; Carruthers et al. 1987). Alternatively, deficits in the sensitivity of cardiomyocyte potassium handling could affect the duration of the PR interval (Akita et al. 1998). We tested whether BACHD mice have deficits in autonomic function by examining their baroreceptor reflex, which is dependent on proper autonomic function (Schroeder et al. 2011). A blunted baroreceptor reflex response in BACHD mice suggests dysfunction of both branches of the autonomic nervous system and is consistent with significantly higher daytime blood pressure and increased HR during the rest phase in BACHD mice compared to WTs. Overall, dramatic decreases in HRV, alterations in the PR interval, and an inability to appropriately decrease heart rate and blood pressure during sleep all support the hypothesis that autonomic function is compromised in BACHD mice.

16.3.4 BACHD Molecular Clockwork May Not Be Altered

To screen whether deficits in the molecular clockwork responsible for generating circadian oscillation underlie circadian deficits observed in BACHD mice, we examined PER2 expression in the SCN at peak and trough time-points, but found no differences in BACHD mice (Kudo et al. 2011) (Fig. 16.3). These findings are in contrast to prior work using the R6/2 line that suggests behavioral impairment *are* accompanied by disordered circadian clock gene expression in vivo. These findings include disrupted expression of circadian clock genes centrally in the SCN and striatum, as well as peripherally in liver where core clock gene rhythms of *Bmal1* and *Per2* were intact but *Cry1* and output gene *Dbp* rhythms were lost (Pallier et al. 2007; Maywood et al. 2010). Due to the conflicting findings between animal models, further work is required to definitively determine whether deficits in clock gene expression rhythms are a common phenotype in HD mouse models and, specifically, whether the BACHD line shows evidence of disruptions in its molecular clockwork beyond the SCN.

16.3.5 Reduced Daytime Firing Rate of SCN Neurons in BACHD Mice During the Early Stages of HD Progression

SCN neurons are spontaneously active and generate action potentials with peak firing rates during the day (Colwell 2011). SCN regulates physiological rhythms throughout the brain and body using a combination of neuronal and hormonal signaling pathways. Rhythms in SCN neuronal output are believed to be necessary for the regulation of the autonomic nervous system and downstream physiological rhythms (Kalsbeek et al. 2006; Dibner et al. 2010). In the daytime, BACHD mouse SCN

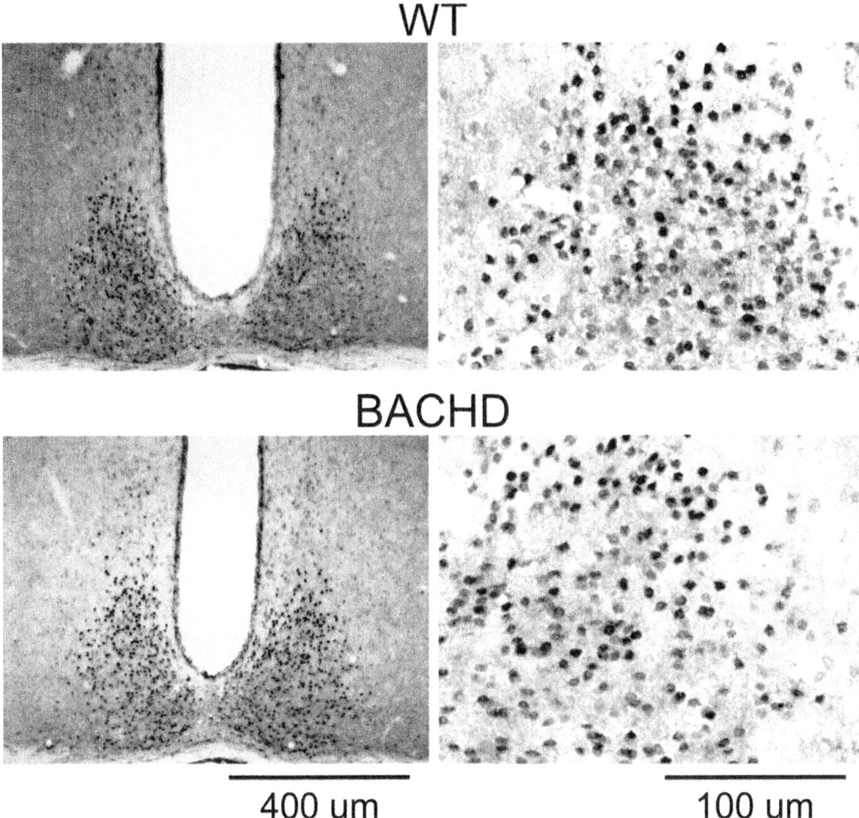

Fig. 16.3 Molecular clockwork in the SCN does not appear to be disrupted. Examples of PER2 expression in the SCN sampled from WT (*top*) and BACHD (*bottom*) mice. Mice were held in DD and wheel-running activity measured to determine circadian phase. Tissue was collected in subjective day (CT 2). IHC was used to measure PER2 immunoreactivity in the SCN. Photomicrographs of SCN tissue of each genotype in low (×10) and higher (×40) magnification. The mice were 2–3 months of age

neurons showed significantly reduced spontaneous firing rates and an overall loss in firing rate rhythms (Kudo et al. 2011). Decreased daytime electrical activity in the SCN would therefore weaken rhythms of both neural and hormonal outputs (Fig. 16.4). This observation is consistent with the hypothesis that the weakening of electrical output from the SCN is part of the pathology responsible for the circadian behavioral phenotypes observed in the BACHD mice.

Reduced daytime firing rates in the SCN of the BACHD line were measured from acute slice preparation of neurons embedded within a circuit. One potential mechanism by which mutant HTT may alter the function of the circadian system is by reducing the strength of intercellular coupling within the SCN circuit. Indeed, recent work in a variety of neurodevelopmental and psychiatric disorder mouse

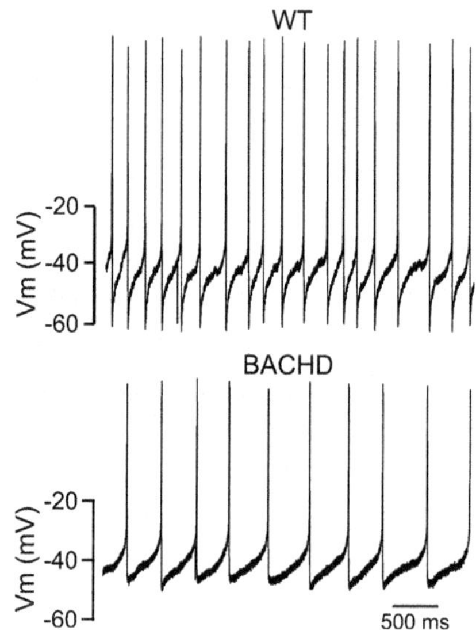

Fig. 16.4 Circadian dysfunction in neural output from SCN. Examples of spontaneous neural activity recorded from the SCN of WT (*top*) and BACHD (*bottom*) mice. The spontaneous firing rate in dorsal SCN neurons was measured during the day using the current-clamp recording technique in the cell-attached configuration. The mice were 2–3 months of age

models suggests the balance between synaptic excitation and inhibition is altered and likely underlies disease-specific pathophysiology (Dani et al. 2005; Nelson and Turrigiano 2008; Gogolla et al. 2009; Milnerwood and Raymond 2010; Shepherd and Katz 2011). Inhibitory postsynaptic currents (IPSC) have been observed in cortex and striatum of R6/2 and YAC HD mouse models early in their disease progression, and these alterations are followed by late-stage decreases in spontaneous IPSC frequency in cortical pyramidal neurons (Cummings et al. 2009). Furthermore, the amplitude of spontaneous excitatory postsynaptic currents (EPSC) increases at early stages in striatal medium spiny neurons, and then in late stages the frequencies of spontaneous EPSC are decreased (Cepeda et al. 2003). These synaptic changes may underlie increased firing rates and decreased synchrony/coupling or correlated firing in both brain regions (Rebec et al. 2006; Miller et al. 2008; Walker et al. 2008). While we have no specific evidence that HD alters synaptic transmission or coupling within the SCN circuit, similar alterations in synaptic transmission is a very plausible mechanism to explain the observed disruption in circadian function in the hypothalamus.

SCN spontaneous electrical activity is controlled by a set of currents that drives daily rhythms in action potentials. During the day, SCN neurons are relatively depolarized with a resting membrane potential (−50 to −55 mV) close to the threshold for generating an action potential (−45 mV). This relatively depolarized resting potential is the result of multiple cation currents that provide excitatory drive (Pennartz et al. 1997; Jackson et al. 2004; Kononenko et al. 2004). In response to this excitatory drive, SCN neurons exhibit sustained discharge for 4–6 h without spike adaptation during their subjective day. Prior work suggests that 3 potassium (K^+) currents including fast delayed rectifier (FDR), A-type-K^+ current (I_A), and large-conductance Ca^{2+} activated K^+ (BK) currents are critical in the regulation of spontaneous action potential firing in SCN neurons during the day (Colwell 2011). Decreased SCN neuronal firing rates observed during the daytime in BACHD SCN may potentially be due to a reduction in magnitude of FDR and/or BK currents.

16.3.6 The Molecular Basis for Decreased Firing Rate of SCN Neurons in HD Is Unclear

It is unclear if decreased SCN electrical activity is due to changes in synaptic signaling and/or alterations of intrinsic membrane currents; however, there is a third very appealing hypothesis—disrupted mitochondrial function may also reduce neuronal electrical activity (Correia et al. 2012; Schapira 2012). Membrane repolarization following action potential generation is extremely energetically demanding. The ATP-dependent sodium–potassium pump (Na^+/K^+-ATPase) is critical for maintaining neuronal resting membrane potential, and this pump is more active during the day than at the night (Wang and Huang 2006). The use of this pump is very energy expensive, and insufficient ATP availability would depolarize membrane potential, leading to an inability of neurons to generate action potentials. Na^+/K^+-ATPase activity appear to be critical to the daily rhythm in membrane potential of SCN neurons, and therefore deficits in its activity due to insufficient ATP availability may underlie aberrant physiology in BACHD SCN.

Another mechanistic candidate underlying decreased electrical activity is mutant HTT-driven increases in oxidative damage in SCN neurons. Mutant huntingtin aggregation has been found to decrease mitochondrial efficiency by inhibition of mitochondrial trafficking (Trushina et al. 2004) and increasing reactive oxygen species (ROS) production by decreasing complex I-dependent mitochondrial respiration (Shimohata et al. 2000; Sugars et al. 2004). Notably, circadian transcriptional machinery coordinates antioxidant stress by controlling the transcription of antioxidant enzymes in many cells (Hardeland et al. 2003). Disruption of the intracellular circadian clock may thereby increase cellular susceptibility to oxidative stress and subsequently compromise function and survival (Antoch and Kondratov 2010). This hypothesis is supported by observations of disrupted ROS homeostasis in *Clock* mutant mice (Kondratov et al. 2006). Based on observations of disrupted clock gene expression in the R6/2 mouse model, a similar mechanism may reduce ROS homeostasis and thereby alter neuronal function (Morton et al. 2005). Unlike the R6/2 model, we did not observe deficits in circadian gene expression in BACHD mice, but we still expect to see an increase in ROS levels due to mutant HTT-related mitochondrial inefficiency. Oxidative stress damages a range of cellular processes, but due to K^+ channel sensitivity to oxidative damage, it is particularly relevant to neuronal pacemaking properties (Sesti et al. 2010; Cotella et al. 2012). In the SCN, increased daytime firing rates are dependent on K^+ channel currents, and therefore oxidative damage to the channels mediating these currents may underlie reduced daytime firing rates observed in the BACHD model.

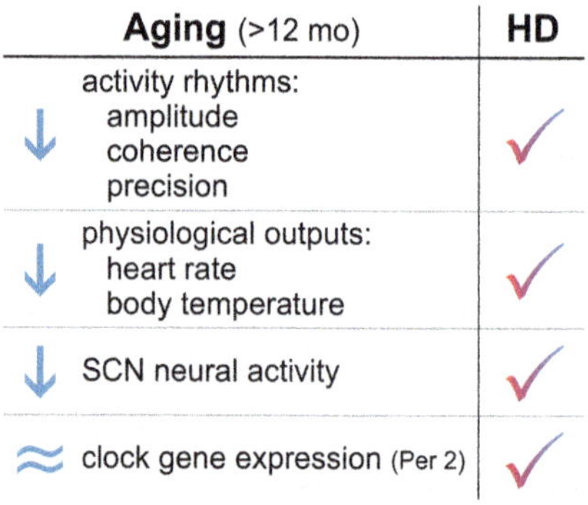

Aging (>12 mo)	HD
↓ activity rhythms: amplitude coherence precision	✓
↓ physiological outputs: heart rate body temperature	✓
↓ SCN neural activity	✓
≈ clock gene expression (Per 2)	✓

Fig. 16.5 Circadian phenotype in BACHD mice compared to middle-aged WT mice. In our laboratory, we measured the circadian phenotype of WT mice at 12 months of age (Nakamura et al. 2011). We found significant reduction in activity output including reduced amplitude, coherence, and precision of wheel-running activity rhythms. Physiological outputs of heart rate and body temperature are also reduced at this age. Finally, we found that the amplitude of SCN neural activity rhythms but not PER2 expression was reduced at middle age. The phenotype of the BACHD mice at 3 months of age is similar to that seen in a middle-aged WT mouse

Based on the literature, the most promising explanation for the decreased SCN firing frequency we observed in the BACHD model is reduced mitochondrial function and increased oxidative stress (Fig. 16.5). Future examination of these targets may provide new avenues for pharmacological intervention and in turn may provide new insights into disease progression and disease prevention.

16.4 Future Directions

Reduced output from the SCN circuit would likely have profound consequences on patient health (Hastings et al. 2003; Takahashi et al. 2008; Bechtold et al. 2010; Reddy and O'Neill 2010). Many rhythms researchers believe that a robust SCN circuit is essential to daily and circadian rhythms of physiology and therefore good health. In recent years, a wide range of studies have demonstrated that disruption of the circadian system leads to a common cluster of symptoms, including metabolic and cardiovascular disease, as well as cognitive deficits. Many of these symptoms are also seen in HD patients. Based on the effects of circadian disruption in healthy humans and animal models, we argue that circadian dysfunction is not just a symptom of HD, but also likely to be a key part of the disease mechanism (Fig. 16.6). We have good reason to think so, because sleep/circadian disruptions occur early in the HD patients and HD mouse models. If this hypothesis is correct, circadian disruption may make the symptoms of HD worse, while stabilizing the rhythm could delay the symptoms. Recent work suggests that interventions that stabilize deteriorating daily rhythms can delay cognitive and motor symptom progression in the R6/2 line (Pallier et al. 2007). Whether this intervention improves rhythms in SCN output or is effective in other mouse models has yet to be determined.

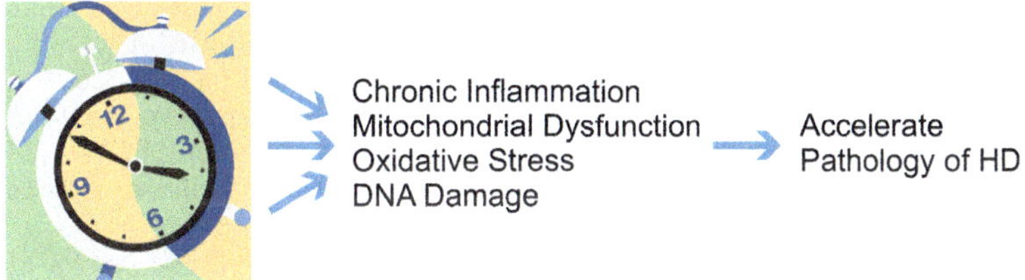

Fig. 16.6 Speculation that circadian dysfunction could accelerate the pathology of HD. Work from several groups indicates that the mouse models of HD show a circadian phenotype. The circadian clockwork regulates mitochondrial function, reactive oxygen species homeostasis, DNA repair, and immune response. Dysfunction of this timing system is likely to contribute to chronic inflammation, mitochondrial dysfunction, and DNA damage. These processes are all thought to contribute to the pathology of HD and contribute to age-related changes in the brain. Therefore, we raise the possibility that circadian dysfunction due to genetic or environmental perturbations can accelerate the pathology of HD

Based on present understanding of the pathophysiology of the master oscillator in the BACHD mouse, future work needs to focus on understanding how mutant huntingtin alters the electrical activity within the SCN and development of interventions that can treat these disruptions. Identifying new therapeutic targets to treat sleep and circadian deficits in HD patients may improve quality of life and potentially slow disease progression for this currently incurable disease.

References

Abrahamson EE, Moore RY (2001) Suprachiasmatic nucleus in the mouse: retinal innervations, intrinsic organization and efferent projections. Brain Res 6(102):172–191

Akita M, Kuwahara M, Tsubone H et al (1998) ECG changes during furosemide-induced hypokalemia in the rat. J Electrocardiol 31(1):45–49

Albrecht U, Sun ZS, Eichele G et al (1997) A differential response of two putative mammalian circadian regulators, mper1 and mper2, to light. Cell 91(7):1055–1064

Andrich J, Schmitz T, Saft C et al (2002) Autonomic nervous system function in Huntington's disease. J Neurol Neurosurg Psychiatry 72(6):726–731

Antle MC, Silver R (2005) Orchestrating time: arrangements of the brain circadian clock. Trends Neurosci 18(3):145–151

Antle MC, Smith VM, Sterniczuk R et al (2009) Physiological responses of the circadian clock to acute light exposure at night. Rev Endocr Metab Disord 10(4):279–291

Antoch MP, Kondratov RV (2010) Circadian proteins and genotoxic stress response. Circ Res 106(1):68–78

Aziz NA, Anguelova GV, Marinus J et al (2010a) Sleep and circadian rhythm alterations correlate with depression and cognitive impairment in Huntington's disease. Parkinsonism Relat Disord 16(5):345–350

Aziz NA, Pijl H, Frolich M et al (2010b) Growth hormone and ghrelin secretion are associated with clinical severity in Huntington's disease. Eur J Neurol 17(2):280–288

Baggs JE, Hogenesch JB (2010) Genomics and systems approaches in the mammalian circadian clock. Curr Opin Genet Dev 20(6):581–587

Bechtold DA, Gibbs JE, Loudon AS (2010) Circadian dysfunction in disease. Trends Pharmacol Sci 31(5):191–198

Bigger JT Jr, Fleiss JL, Steinman RC et al (1992) Frequency domain measures of heart period variability and mortality after myocardial infarction. Circulation 85(1):164–171

Bonelli RM, Beal MF (2012) Huntington's disease. Handb Clin Neurol 106:507–526

Buccelletti E, Gilardi E, Scaini E et al (2009) Heart rate variability and myocardial infarction: systemic literature review and metanalysis. Eur Rev Med Pharmacol Sci 13(4):299–307

Busino L, Bassermann F, Maiolica A et al (2007) SCFFbxl3 controls the oscillation of the circadian clock by directing the degradation of cryptochrome proteins. Science 316:900–904

Carruthers SG, McCall B, Cordell BA et al (1987) Relationships between heart rate and PR interval during physiological and pharmacological interventions. Br J Clin Pharmacol 23(3):259–265

Cepeda C, Hurst RS, Calvert CR et al (2003) Transient and progressive electrophysiological alterations in the corticostriatal pathway in a mouse model of Huntington's disease. J Neurosci 23(3):961–969

Chiu E, Alexander L (1982) Causes of death in Huntington's disease. Med J Aust 1(4):153

Colwell CS (2011) Linking neural activity and molecular oscillations in the SCN. Nat Rev Neurosci 12(10):553–569

Correia SC, Santos RX, Perry G et al (2012) Mitochondrial importance in Alzheimer's, Huntington's and Parkinson's diseases. Adv Exp Med Biol 724:205–221

Cotella D, Hernandez-Enriquez B, Wu X et al (2012) Toxic role of K^+ channel oxidation in mammalian brain. J Neurosci 32(12):4133–4144

Crumbley C, Wang Y, Kojetin DJ et al (2010) Characterization of the core mammalian clock component, NPAS2, as a REV-ERBalpha/RORalpha target gene. J Biol Chem 285(46):35386–35392

Cummings DM, Andre VM, Uzgil BO et al (2009) Alterations in cortical excitation and inhibition in genetic mouse models of Huntington's disease. J Neurosci 29(33):10371–10386

Cuturic M, Abramson RK, Vallini D et al (2009) Sleep patterns in patients with Huntington's disease and their unaffected first-degree relatives: a brief report. Behav Sleep Med 7(4):245–254

Dani VS, Chang Q, Maffei A et al (2005) Reduced cortical activity due to a shift in the balance between excitation and inhibition in a mouse model of Rett syndrome. Proc Natl Acad Sci USA 102(35):12560–12565

Deboer T, Vansteensel MJ, Detari L et al (2003) Sleep states alter activity of suprachiasmatic nucleus neurons. Nat Neurosci 6:1086–1090

Deery MJ, Maywood ES, Chesham JE et al (2009) Proteomic analysis reveals the role of synaptic vesicle cycling in sustaining the suprachiasmatic circadian clock. Curr Biol 19:2031–2036

Dibner C, Schibler U, Albrecht U (2010) The mammalian circadian timing system: organization and coordination of central and peripheral clocks. Annu Rev Physiol 72:517–549

Duff K, Paulsen JS, Beglinger LJ, Predict-HD Investigators of the Huntington Study Group et al (2007) Psychiatric symptoms in Huntington's disease before diagnosis: the Predict-HD study. Biol Psychiatry 62:1341–1346

Duong HA, Robles MS, Knutti D et al (2011) A molecular mechanism for circadian clock negative feedback. Science 332(6036):1436–1439

Foley D, Ancoli-Israel S, Britz P et al (2004) Sleep disturbances and chronic disease in older adults: results of the 2003 National Sleep Foundation *Sleep in America* survey. J Psychosom Res 56(5):497–502

Gallego M, Virshup DM (2007) Post-translational modifications regulate the ticking of the circadian clock. Nat Rev Mol Cell Biol 8:139–148

Gekakis N, Staknis D, Nguyen H et al (1998) Role of the CLOCK protein in the mammalian circadian mechanism. Science 280(5369):1564–1569

Godinho SI, Maywood ES, Shaw L et al (2007) The afterhours mutant reveals a role for Fbxl3 in determining mammalian circadian period. Science 316:897–900

Gogolla N, Leblanc JJ, Quast KB, Sudhof TC, Fagiolini M, Hensch TK (2009) Common circuit defect of excitatory-inhibitory balance in mouse models of autism. J Neurodev Disord 1(2):172–181

Golombek DA, Rosenstei RE (2010) Physiology of circadian entrainment. Physiol Rev 90(3):1063–1102

Goodman AO, Barker RA (2010) How vital is sleep in Huntington's disease? J Neurol 257(6):882–897

Goodman AO, Rogers L, Pilsworth S et al (2011) Asymptomatic sleep abnormalities are a common early feature in patients with Huntington's disease. Curr Neurol Neurosci Rep 11(2):211–217

Gray M, Shirasaki DI, Cepeda C et al (2008) Full-length human mutant huntingtin with a stable polyglutamine repeat can elicit progressive and selective neuropathogenesis in BACHD mice. J Neurosci 28(24):6182–6195

Hamada T, Antle MC, Silver R (2004) Temporal and spatial expression patterns of canonical clock genes and clock-controlled genes in the suprachiasmatic nucleus. Eur J Neurosci 19(7):1741–1748

Hardeland R, Coto-Montes A, Poeggeler B (2003) Circadian rhythms, oxidative stress, and anti-oxidative defense mechanisms. Chronobiol Int 20(6):921–962

Harms E, Kivimae S, Young MW et al (2004) Posttranscriptional and posttranslational regulation of clock genes. J Biol Rhythms 19(5):361–373

Hastings M, Reddy A, Maywood E (2003) A clockwork web: circadian timing in brain and periphery, in health and disease. Nat Rev Neurosci 4:649–661

Hinton SC, Paulsen JS, Hoffmann RG et al (2007) Motor timing variability increases in preclinical Huntington's disease patients as estimated onset of motor symptoms approaches. J Int Neuropsychol Soc 13:539–543

Hogenesch JB, Gu YZ, Jain S et al (1998) The basic-helix-loop-helix-PAS orphan MOP3 forms transcriptionally active complexes with circadian and hypoxia factors. Proc Natl Acad Sci USA 95(10):5474–5479

Hogenesch JB, Panda S, Kay S et al (2003) Circadian transcriptional output in the SCN and liver of the mouse. Novartis Found Symp 253:171–180

Hossain JL, Shapiro CM (2002) The prevalence, cost implications, and management of sleep disorders: an overview. Sleep Breath 6(2):85–102

Huntington's Disease Collaborative Research Group (1993) A novel gene containing a trinucleotide repeat that is expanded and unstable on Huntington's disease chromosomes. Cell 72:971–983

Iranzo A, Santamaria J, Serradell M et al (2006) Rapid-eye-movement sleep behaviour disorder as an early marker for a neurodegenerative disorder: a descriptive study. Lancet Neurol 5(7):572–577

Jackson AC, Yao GL, Bean BP (2004) Mechanism of spontaneous firing in dorsomedial suprachiasmatic nucleus neurons. J Neurosci 24(37):7985–7998

Julein C, Thompson J, Wild S et al (2007) Psychiatric disorders in preclinical Huntington's disease. J Neurol Neurosurg Psychiatry 78:939–943

Kalsbeek A, Palm IF, La Fleur SE et al (2006) SCN outputs and the hypothalamic balance of life. J Biol Rhythms 21:458–469

Kobal J, Meglic B, Mesec A et al (2004) Early sympathetic hyperactivity in Huntington's disease. Eur J Neurol 11(12):842–848

Kobal J, Melik Z, Cankar K et al (2010) Autonomic dysfunction in presymptomatic and early symptomatic Huntington's disease. Acta Neurol Scand 121(6):392–399

Kondratov RV, Kondratova AA, Gorbacheva VY et al (2006) Early aging and age-related pathologies in mice deficient in BMAL1, the core component of the circadian clock. Genes Dev 20:1868–1873

Kononenko NI, Medina I, Dudek FE (2004) Persistent subthreshold voltage-dependent cation single channels in suprachiasmatic nucleus neurons. Neuroscience 129(1):85–92

Kornmann B, Schaad O, Bujard H et al (2007) System-driven and oscillator-dependent circadian transcription in mice with a conditionally active liver clock. PLoS Biol 5(2):e34

Kremer H, Roos R, Dingjan G et al (1990) Atrophy of the hypothalamic lateral tuberal nucleus in Huntington's disease. J Neuropathol Exp Neurol 49:371–382

Kriegsfeld LJ, Leak RK, Yackulic CB et al (2004) Organization of suprachiasmatic nucleus projections in Syrian hamsters (Mesocricetus auratus): an anterograde and retrograde analysis. J Comp Neurol 468(3):361–379

Kudo T, Schroeder A, Loh DH et al (2011) Dysfunctions in circadian behavior and physiology in mouse models of Huntington's disease. Exp Neurol 228(1):80–90

Lanska DJ, Lavine L, Lanska MJ et al (1988) Huntington's disease mortality in the United States. Neurology 38(5):769–772

Lavin P, Bone I, Sheridan P (1981) Studies of hypothalamic function in Huntington's chorea. J Neurol Neurosurg Psychiatry 44:414–418

Lee C, Etchegaray JP, Cagampang FRA et al (2001) Posttranslational mechanisms regulate the mammalian circadian clock. Cell 107(7):855–867

Margolis RL, Ross CA (2003) Diagnosis of Huntington disease. Clin Chem 49(10): 1726–1732

Maywood E, Fraenkel E, McAllister C et al (2010) Disruption of peripheral circadian timekeeping in a mouse model of Huntington's disease and its restoration by temporally scheduled feeding. J Neurosci 30:10199–10204

McCurry SM, Logsdon RG, Teri L et al (2007) Sleep disturbances in caregivers of persons with dementia: contributing factors and treatment implications. Sleep Med Rev 11(2):143–153

Meijer JH, Schaap J, Watanabe K et al (1997) Multiunit activity recordings in the suprachiasmatic nuclei: in vivo versus in vitro models. Brain Res 753:322–327

Miller BR, Walker AG, Shah AS et al (2008) Dysregulated information processing by medium spiny neurons in striatum of freely behaving mouse models of Huntington's disease. J Neurophysiol 100(4):2205–2216

Milnerwood AG, Raymond LA (2010) Early synaptic pathophysiology in neurodegeneration: insights from Huntington's disease. Trends Neurosci 33(11):513–523

Mohawk JA, Takahashi JS (2011) Cell autonomy and synchrony of suprachiasmatic nucleus circadian oscillators. Trends Neurosci 34(7):349–358

Morin LP, Allen CN (2006) The circadian visual system, 2005. Brain Res Rev 51(1):1–60

Morton AJ (2013) Circadian and sleep disorder in Huntington's disease. Exp Neurol 243:34–44

Morton A, Wood N, Hastings M et al (2005) Disintegration of the sleep-wake cycle and circadian timing in Huntington's disease. J Neurosci 25:157–163

Nakamura W, Yamazaki S, Takasu NN et al (2005) Differential response of Period 1 expression within the suprachiasmatic nucleus. J Neurosci 25(23):5481–5487

Nakamura TJ, Nakamura W, Yamazaki S, Kudo T, Cutler T, Colwell CS, Block GD (2011) Age-related decline in circadian output. J Neurosci 31(28):10201–10205

Nelson SB, Turrigiano GG (2008) Strength through diversity. Neuron 60(3):477–482

Pallier P, Maywood E, Zheng Z et al (2007) Pharmacological imposition of sleep slows cognitive decline and reverses dysregulation of circadian gene expression in a transgenic mouse model of Huntington's disease. J Neurosci 27:7869–7878

Panda S, Antoch MP, Miller BH et al (2002) Coordinated transcription of key pathways in the mouse by the circadian clock. Cell 109(3):307–320

Paulsen JS, Langbehn DR, Stout JC, Predict HD Investigators and Coordinators of the Huntington Study Group et al (2008) Detection of Huntington's disease decades before diagnosis: the Predict HD study. J Neurol Neurosurg Psychiatry 79:874–880

Peavy GM, Jacobson MW, Goldstein JL et al (2010) Cognitive and functional decline in Huntington's disease: dementia criteria revisited. Mov Disord 25(9):1163–1169

Pennartz CM, Bierlaagh MA, Geurtsen AM (1997) Cellular mechanisms underlying spontaneous firing in rat suprachiasmatic nucleus: involvement of a slowly inactivation component of sodium current. J Neurophysiol 78(4):1811–1825

Petersén A, Gil J, Maat-Schieman M et al (2005) Orexin loss in Huntington's disease. Hum Mol Genet 14:39–47

Petersén A, Hult S, Kirik D (2009) Huntington's disease – new perspectives based on neuroendocrine changes in rodents models. Neurodegener Dis 6(4):154–164

Preitner N, Damiola F, Lopez-Molina L et al (2002) The orphan nuclear receptor REVERB alpha controls circadian transcription within the positive limb of the mammalian circadian oscillator. Cell 110:251–260

Rebec GV, Conroy SK, Barton SJ (2006) Hyperactive striatal neurons in symptomatic Huntington R6/2 mice: variations with behavioral state and repeated ascorbate treatment. Neuroscience 137(1):327–336

Reddy AB, O'Neill JS (2010) Healthy clocks, healthy body, healthy mind. Trends Cell Biol 20(1):36–44

Reddy AB, Karp NA, Maywood ES et al (2006) Circadian orchestration of the hepatic proteome. Curr Biol 16:1107–1115

Sato F, Kawamoto T, Fujimoto K et al (2004) Functional analysis of the basic helix-loop-helix transcription factor DEC1 in circadian regulation. Interaction with BMAL1. Eur J Biochem 271(22):4409–4419

Schaap J, Meijer JH (2001) Opposing effects of behavioral activity and light on neurons of the suprachiasmatic nucleus. Eur J Neurosci 13:1955–1962

Schapira AH (2012) Mitochondrial diseases. Lancet 379(9828):1825–1834

Schroeder AM, Loh DH, Jordan MC et al (2011) Baroreceptor reflex dysfunction in the BACHD mouse model of Huntington's disease. PLoS Curr 3, RRN1266

Schulte J, Littleton JT (2011) The biological function of the Huntingtin protein and its relevance to Huntington's Disease pathology. Curr Trends Neurol 5:65–78

Sesti F, Liu S, Cai SQ (2010) Oxidation of potassium channels by ROS: a general mechanism of aging and neurodegeneration. Trends Cell Biol 20(1):45–51

Shearman LP, Zylka MJ, Weaver DR et al (1997) Two period homologs: circadian expression and photic regulation in the suprachiasmatic nuclei. Neuron 19(6):1261–1269

Shepherd GM, Katz DM (2011) Synaptic microcircuit dysfunction in genetic models of neurodevelopmental disorders: focus on Mecp2 and Met. Curr Opin Neurobiol 21(6):827–833

Shimohata T, Nakajima T, Yamada M et al (2000) Expanded polyglutamine stretches interact with TAFII130, interfering with CREB-dependent transcription. Nat Genet 26(1):29–36

Siepka SM, Yoo SH, Park J et al (2007) Circadian mutant Overtime reveals F-box protein FBXL3 regulation of cryptochrome and period gene expression. Cell 129:1011–1023

Soneson C, Fontes M, Zhou Y, Huntington Study Group PREDICT-HD Investigators et al (2010) Early changes in the hypothalamic region in prodromal Huntington disease revealed by MRI analysis. Neurobiol Dis 40(3):531–543

Storch KF, Lipna OIL, Viswanathan N et al (2002) Extensive and divergent circadian gene expression in liver and heart. Nature 417:78–83

Sugars KL, Brown R, Cook LJ et al (2004) Decreased cAMP response element-mediated transcription: an early event in exon 1 and full-length cell models of Huntington's disease that contributes to polyglutamine pathogenesis. J Biol Chem 279(6):4988–4999

Sun ZS, Albrecht U, Zuchenko O et al (1997) RIGUI, a putative mammalian ortholog of the Drosophila period gene. Cell 90(6):1003–1011

Takahashi J, Hong H, Ko C et al (2008) The genetics of mammalian circadian order and disorder: implications for physiology and disease. Nat Rev Genet 9:764–775

Takahata S, Sogawa K, Kobayashi A et al (1998) Transcriptionally active heterodimer formation of an Arnt-like PAS protein, Arnt3, with HIF-1a, HLF, and clock. Biochem Biophys Res Commun 794(248):789–794

Thayer JF, Yamamoto SS, Brosschot JF (2010) The relationship of autonomic imbalance, heart rate variability and cardiovascular disease risk factors. Int J Cardiol 141(2):122–131

Thresher RJ, Vitaterna MH, Miyamoto Y et al (1998) Role of mouse cryptochrome blue-light photoreceptor in circadian photoresponses. Science 282(5393):1490–1494

Trushina E, Dyer RB, Badger JD 2nd et al (2004) Mutant huntingtin impairs axonal trafficking in mammalian neurons in vivo and in vitro. Mol Cell Biol 24(18):8195–8209

Vaccarino AL, Sills T, Anderson KE et al (2011) Assessment of depression, anxiety and apathy in prodromal and early Huntington disease. PLoS Curr 3, RRN1242

Van den Pol AN (1980) The hypothalamic suprachiasmatic nucleus of rat: intrinsic anatomy. J Comp Neurol 191(4):661–702

van der Horst GT, Muijtjens M, Kobayashi K et al (1999) Mammalian Cry1 and Cry2 are essential for maintenance of circadian rhythms. Nature 398(6728):627–630

Videnovic A, Leurgans S, Fan W et al (2009) Daytime somnolence and nocturnal sleep disturbances in Huntington disease. Parkinsonism Relat Disord 15(6):471–474

Walker AG, Miller BR, Fritsch JN et al (2008) Altered information processing in the prefrontal cortex of Huntington's disease mouse models. J Neurosci 28(36):8973–8982

Wallick DW, Martin PJ, Masuda Y et al (1982) Effects of autonomic activity and changes in heart rate on atrioventricular conduction. Am J Physiol 243(4):H523–H527

Wang YC, Huang RC (2006) Effects of sodium pump activity on spontaneous firing in neurons of the rat suprachiasmatic nucleus. J Neurophysiol 96(1):109–118

Welsh DK, Takahashi JS, Kay SA (2010) Suprachiasmatic nucleus: cell autonomy and network properties. Annu Rev Physiol 72:551–577

Willison LD, Kudo T, Loh DH et al (2013) Circadian dysfunction may be a key component of the non-motor symptoms of Parkinson's disease: insights from a transgenic mouse model. Exp Neurol 243:57–66

Wood NI, Goodman AO, van der Burg JM et al (2008) Increased thirst and drinking in Huntington's disease and the R6/2 mouse. Brain Res Bull 76(102):70–79

Yamazaki S, Kerbeshian MC, Hocker CG et al (1998) Rhythmic properties of the hamster suprachiasmatic nucleus in vivo. J Neurosci 18:10709–10723

Yan L, Okamura H (2002) Gradients in the circadian expression of Per1 and Per2 genes in the rat suprachiasmatic nucleus. Eur J Neurosci 15(7):1153–1162

Chapter 17
Is It Possible to Modify Clock Genes to Improve Health?

Gabriela Domínguez-Monzón and José Segovia

Abstract Disturbances of circadian rhythms due to clock gene alterations—such as single nucleotide polymorphisms (SNP), genetic ablation, gene overexpression, and epigenetic inactivation—have been associated with several diseases, like cancer, metabolic syndrome, sleep disorders, and neurological and psychiatric disorders. A considerable body of evidence suggests that correcting the molecular alterations or mutations in clock gene expression could provide alternative therapeutic strategies to treat disorders involving the biological clock. Gene therapy allows us to modify the levels of expression of different genes. This can be accomplished, among other methods, by overexpressing a gene that is downregulated or absent in a disease or by silencing a mutated gene, or one that is expressed in an abnormally high level, and is an underlying cause of a disease. There are very few instances of the use of gene therapy strategies in chronobiology to regulate gene expression. In this chapter, we will analyze the role of clock genes alterations in different diseases and discuss the potential use of gene therapy to treat them.

17.1 The Clock of the Brain

Physiological function and behavior of mammals are not exclusively regulated by spatially distinct brain areas and organs, but also follow a strict temporal order. The control of daily modulations is governed by a central circadian clock located in the suprachiasmatic nucleus (SCN) of the hypothalamus. The SCN comprises about 20,000 neurons, which generate circadian rhythms both by electrical activity and neurohormone release. These endogenous rhythms are synchronized with the environmental cycle by processing light information conveyed from specialized cells in

G. Domínguez-Monzón
División de Neurociencias, Instituto de Fisiología Celular, Universidad Nacional Autónoma de México, México, DF 04510, México

J. Segovia (✉)
Departamento de Fisiología, Biofísica y Neurociencias, Centro de Investigación y de Estudios Avanzados del Instituto Politécnico Nacional (CINVESTAV), México, DF 07360, México
e-mail: jsegovia@fisio.cinvestav.mx

© Springer International Publishing Switzerland 2015
R. Aguilar-Roblero et al. (eds.), *Mechanisms of Circadian Systems in Animals and Their Clinical Relevance*, DOI 10.1007/978-3-319-08945-4_17

the retina. The cyclic expression of a family of "clock genes," oscillating within an intracellular transcription–translation negative feedback loop, is viewed as the molecular clockwork generating the approximate 24 h rhythmicity (Ko and Takahashi 2006). However, recent evidence of the dependency of the molecular clock on a functional excitable membrane (Maywood et al. 2006a) and on intracellular rhythmic factors like Ca^{2+} (Lundkvist et al. 2005; Aguilar-Roblero et al. 2007; Mercado et al. 2009) and cAMP (O'Neill et al. 2008) suggests that mutual interactions among different levels of organization are needed to generate a circadian rhythm. Challenges from the environment, such as jet lag (Davidson et al. 2006) and shift work (Schernhammer et al. 2011; Schulz et al. 2008), or from within the organism, such as aging (Duncan 2007) and disease (Wulff et al. 2010), can alter the function of the circadian clock and lead to chronic and serious diseases (Maywood et al. 2006b; Okamura et al. 2010).

17.2 Genetic Variations in Clock Genes and Their Consequences

Clock genes have been linked to sleep/circadian disorders and metabolic, cardiovascular, gastrointestinal, neurological, immune, and endocrine alterations. In this chapter we will describe various mutations in the clock genes and their impact on health.

17.2.1 Sleep/Circadian Disorders

It has been reported that the long and short allele variants of *PER3* are linked to morning and evening chronotypes, respectively (Ebisawa et al. 2001). Accordingly, polymorphisms in *PER2* and *PER3* are related with diurnal activity (Ojeda et al. 2013), whereas mutations in a transcriptional repressor protein (DEC2) are associated with human short time sleep phenotype (He et al. 2009), and variations in the intronic sequences of the *CLOCK* gene are associated with sleep duration (Allebrandt et al. 2010).

17.2.2 Psychiatric and Neurological Diseases

Sleep and rhythm disturbances are frequently reported as major characteristics of psychiatric disorders such as major depression (MD), seasonal affective disorder (SAD), bipolar disorder (BD), and schizophrenia (Goodwin and Jamison 2007). Up to 90 % of MD patients report alterations in sleep patterns (Reynolds and Kupfer 1987). These facts do not allow us to determine whether rhythm and sleep disturbances are *causes* or *consequences* of these disorders, which is information necessary to design future chronotherapeutic strategies. Importantly, it has been shown that repeated insomnia increases the risk of relapse into a new depressive episode (Pigeon et al. 2008). In contrast, controlled sleep deprivation (SD) improves the depressive state, but the effect is not long lasting. Circadian misalignment has been reported in MD and SAD, in both of which the degree of misalignment is correlated with the severity of depression (Gorwood 2010; Lewy et al. 2006, 2009).

It has been suggested that the onset of bipolar affective disorder (BPAD) is associated with a variable number of tandem-repeat (VNTR) polymorphisms in the coding region of the *PER3* gene (Benedetti et al. 2008). Moreover, several different polymorphisms in ten circadian clock genes (*ARNTL, CLOCK, CRY2, CSNK1ε, DBP, GSK3β, NPAS2, PER1, PER2,* and *PER3*) are associated to BPAD (Nievergelt et al. 2006). Also, polymorphisms in *BMAL1, CLOCK, PER1, PER2, PER3, CRY1, CRY2,* and *TIMELESS* have been associated with schizophrenia (Mansour et al. 2006). Furthermore, a single nucleotide polymorphism of the *CLOCK* gene has been linked to elevated transmission of dopamine neurons of the ventral tegmental area (McClung et al. 2005) and schizophrenia (Takao et al. 2007). In major depressive disorder (MDD), the subjects present a disruption of synchronicity between circadian rhythms and external cycles and, also, disrupted phase relationship between individual circadian genes (Edgar and McClung 2013). In aging-related neurodegenerative disease such as Alzheimer's disease (AD), there is a report of the association of the polymorphism 4580704 G/C of the *CLOCK* gene with AD and its relationship with increased risk of AD (Chen et al. 2013b). In the case of Parkinson's disease, the expression of the clock genes is anomalous, and it has been suggested that this is caused by an abnormal methylation in the *CRY1* and *NPAS2* promoters (Lin et al. 2012).

17.2.3   Cancer

The relationship of cancer and clock genes is at the level of the cell cycle; in this case, proteins involved in the progress of the cycle are under the regulation of clock genes, showing circadian rhythms in every cell cycle phase. When there is a loss in the organization of the core clock genes, the chances of developing cancer increase (Borgs et al. 2009). For example, the loss or deregulations of *PER1* and *PER2* gene expression have been observed in several types of human cancers (Gery et al. 2006; Fu et al. 2002; Yang et al. 2009). Other studies have suggested that hypermethylation on the promoters of *PER1, PER2, PER3,* or *CRY1* affects tumor development (Chen et al. 2005; Shih et al. 2006; Kuo et al. 2009). Polymorphisms of the *clock* and *cry1* genes are associated with higher risk of developing breast cancer (Rana and Mahmood 2010).

It is worthwhile highlighting that although there are an increasing number of gene polymorphisms that are associated with different circadian disturbances, long-term environmental factors may also have effects. Epigenetic regulation is an additional key player in circadian systems.

17.3   Epigenetics and the Clock

Epigenetics is the study of the changes in gene expression that do not implicate a modification of the DNA sequence. The changes in gene expression are due to chemical reactions that activate or deactivate part of the genome in a specific site and in a precise time. These reactions are DNA methylation and posttranslational modifications of the histones. DNA methylation and N-acetylation of histones influence the accessibility of DNA for transcription. When histones are acetylated, the DNA is open and transcription can be performed; in contrast, when histones are methylated or phosphorylated, they can either activate or repress transcription (Jenuwein and Allis 2001). On the other hand, DNA methylation leads to condensation of chromatin inhibiting gene expression. Next, we will discuss some reports of clock genes involved in epigenetic activity or others which are regulated by epigenetic mechanisms.

The expression of *c-fos* and *per1* is modulated by epigenetic mechanisms. Environmental stimuli such as light exposure induce chromatin modification in the SCN neurons, caused by the phosphorylation of Ser10 in histone 3 tail (H3-S10), leading to *c-fos* and *per1* transcription (Crosio et al. 2000). The transcription factors Clock-Bmal1 directly participate in protein acetylation and chromatin modification. The circadian expression of the *per* and *cry* genes is associated with chromatin remodeling, so that Clock-Bmal1 can acetylate H3 and attract RNA polymerase II to the promoter regions leading to a circadian-modulated transcription of these genes (Etchegaray et al. 2003). Also, two histone methyltransferases Mll1 and Mll3 modify the chromatin and methylate H3 in a circadian manner, thus allowing the transcription of clock genes (Katada and Sassone-Corsi 2010).

As methyltransferases have an important role in gene expression, the histone lysine demethylase Jarid1a enhances *per2* gene transcription by inhibiting histone deacetylase and consequently increasing histone 3 acetylation, among other functions (DiTacchio et al. 2011). Another histone deacetylase, HDAC3 (histone deacetylase 3) is involved in modulating gene expression with circadian rhythmicity (Alenghat et al. 2008). On the other hand, Sirt1, a histone deacetylase, acts as a molecular regulator of the acetylation activity of Clock and also modulates the acetylation levels of Bmal1 in the liver. Accordingly, the Clock-Bmal1/Sirt1 complex is an important circadian regulator of CCG (clock-controlled genes) expression (Nakahata et al. 2008; Asher et al. 2008).

Taken together, this information sheds light in the relevance of epigenetic mechanisms regulating the expression of clock genes and, since many mRNAs, which oscillate in peripheral tissues, encode for proteins that have critical roles in metabolic processes, indicates that these processes might be related to the etiology of different diseases. Thus, the regulation of gene expression by different means, including expression of transgenic cDNA, miRNA, or pharmacological agents, may lead to the development of new therapeutic strategies to treat these diseases.

17.4   Using Gene Therapy to Impact the Function of Clock Genes

17.4.1   Gene Therapy

Core clock genes regulate the expression of many genes in the cell. It has been previously shown that either the overexpression, downregulation, or polymorphisms in these genes are involved in the development of different diseases, and thus the question on hand is whether gene therapy is a realistic option to treat these diseases. Gene therapy is a set of procedures by which genetic material is introduced into specific cell populations without affecting healthy or neighboring cells to induce beneficial effects in patients (Benitez and Segovia 2003). This therapy can be performed either in vivo or ex vivo. When using the in vivo approach, the genetic material is directly transferred into the organism receiving the treatment, and when ex vivo, cells from the patients (or from other sources) are taken, and the gene transfer procedure is performed in cultures of the cells, and then the genetically modified cells are administered to patients, with the objective of restoring the molecular changes that caused the disease.

There are several methods to transfer transgenes to cells, which can be divided into viral and nonviral methods. It is generally considered that viral methods are more efficient to transfer transgenes, when compared with nonviral methods, particularly liposomes, but they have certain disadvantages. Among them, we find mutational insertions, expression of viral proteins, and immune response to the vector. Several viral vectors have been used in gene therapy, including retrovirus, adenovirus, adeno-associated virus, and lentivirus, and less frequently naked/plasmid DNA has been employed.

Adenovirus, as of today, is the most used viral vector (23 %). Adenovirus is genetically modified to reduce toxicity in patients. These vectors can infect both dividing and nondividing cells efficiently; an important characteristic of this vector is that it can accommodate relative large transgenes when compared with other virus (Capasso et al. 2014). Adeno-associated virus (AAV), a DNA virus, is capable of sustaining expression of transgenes for long periods in infected cells and induces a milder immune response from the host than adenoviruses (Hareendran et al. 2013).

Retrovirus is an RNA virus that integrates the therapeutic gene into the genome of proliferating cells by a process involving reverse transcription of the RNA viral genome and integration of the newly synthesized DNA into the cell's genome, thus allowing long-term expression of the transgenes; this characteristic made it one of the most used vectors to treat human diseases with gene therapy (Kobayashi et al. 2014). However, insertional mutations (Nienhuis et al. 2006) that induced leukemia in pediatric patients (Hacein-Bey-Abina et al. 2003) have almost completely halted the use of these vectors in clinical protocols. In principle lentivirus has two advantages over retrovirus: first, they are capable of infecting nondividing cells in vivo—such as neurons and those of the liver and eye—and, second, are safer than retrovirus because of gene modifications of the vector packaging cassette that prevent the generation of replication competent lentiviruses (RCLs) and insertional mutagenesis (Cockrell and Kafri 2007). There are two strategies to modify lentiviral vectors to decrease insertional mutagenesis: one is by directing the insertion of the lentiviral vector to specific sites within the host genome, and the second is by exploiting non-integrating HIV-1 vector episomes to deliver therapeutic transgenes (Tan et al. 2006). More information regarding adenovirus, retrovirus, and lentivirus vectors for research is available on the National Gene Vector Biorepository (NGVB 2014).

On the other hand, the use of naked DNA in gene therapy is also an alternative. It has been reported that naked DNA can be delivered to cells and express therapeutic genes in vivo (Herweijer and Wolff 2003). However, the efficacy of gene transfer is relatively low; the use of naked DNA appears to be safer than that of recombinant virus and of DNA complexed to lipids (Herweijer and Wolff 2003). Recently almost 18 % of all the clinical trials report the use of naked plasmid DNA (pDNA) (Gene therapy clinical trial worldwide 2014). Plasmids are administered by injection into specific tissues, such as the muscle, skin, and cardiac tissue, and into tumors and the liver and systemically by intravenous injections.

Depending on the particular disease, and the knowledge of its molecular basis, different gene therapy approaches can be used. For instance, when there is a lack of expression of a gene, transfer of a cDNA coding for the missing protein can be used (Fig. 17.1); in contrast the expression of a small interference RNA (siRNA) can be used to silence or downregulate a mutant protein (Fig. 17.2), or expression of a microRNA (miRNA) can also be employed to regulate protein levels (Fig. 17.3).

As of today, most gene therapy clinical trials have been directed to treat cancer, but there are also trials for other neurological diseases (Gene therapy clinical trial worldwide 2014).

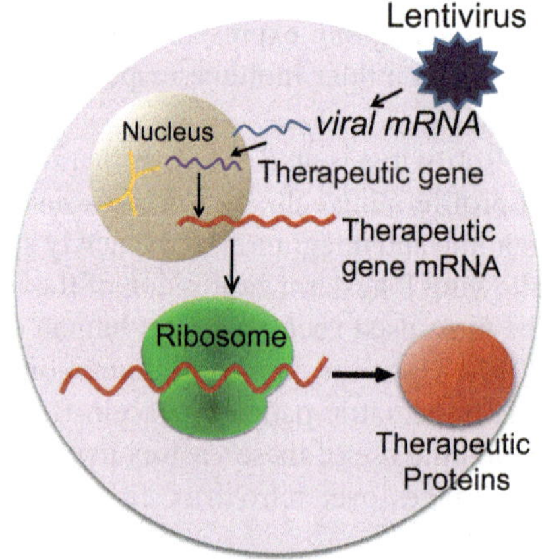

Fig. 17.1 Retroviral mechanism of infection. The retroviral vector infects and integrates the therapeutic gene (clock gene) into the genome of the host cell, and then the cell transcribes the transgene and synthesizes the therapeutic protein

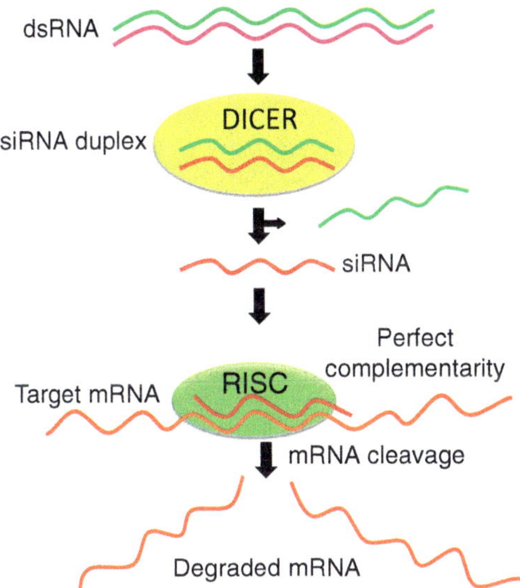

Fig. 17.2 Mechanism of mRNA silencing using small interfering RNA (siRNA). Information coding for a siRNA is transferred using a vector to human cells to degrade specific mRNA sequences to which they are homologous, in this manner silencing the altered gene. Double-stranded RNA (dsRNA) is cleaved by type III RNAse DICER in two smaller fragments of siRNA, sense and antisense strands. Then endonuclease RNA-induced silencing complex (RISC) uses the antisense strand of the siRNA to bind and degrade the corresponding mRNA

17.4.2   Gene Therapy for Clock Genes

Lentiviral and adeno-associated vectors are used as gene transfer systems to produce small interfering RNAs (siRNAs) to diminish or silence the expression of specific genes (Ryther et al. 2005), and they can also be used to express a cDNA coding for proteins or transcription factors like clock genes, lacking in the affected cells (Gene therapy clinical trial worldwide 2014). An important question is whether these methods can be employed in the CNS, without causing secondary damages, due to the vector, due to the method of administration, or by other effects on brain circuitry and functioning. To the best of our knowledge, there are few instances in the field of chronobiology that have used siRNA to silence genes; and in most of these experiments, liposomes were employed as the delivery system. For example, the expression of Bmal1 was downregulated using a siRNA, which reduced the levels of the transcription factor NF-YA (nuclear factor Y A), one of the activators of Bmal1 transcription (Xiao et al. 2013); there is another report using a siRNA to downregulate *Rev-erb alpha* in pancreatic cells (Vieira et al. 2013), and in luteinizing granulosa cells, Bmal1 was knocked down with an interference RNA (Chen et al. 2013c). Gene delivery is a crucial aspect that needs to be improved for gene therapy to be employed to regulate clock gene expression and activity.

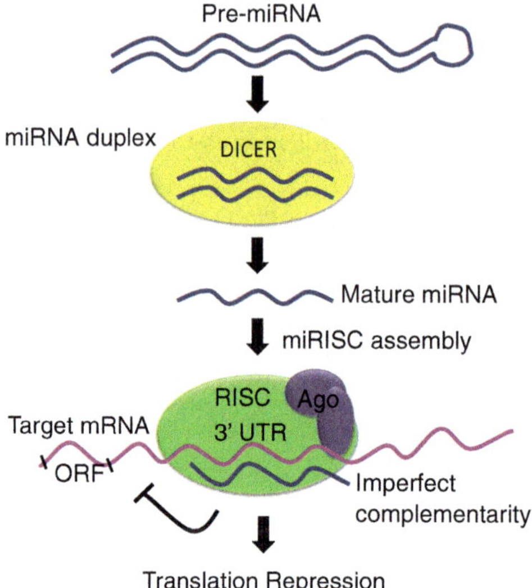

Fig. 17.3 Mechanism of posttranscriptional repression of mRNA by microRNA (miRNA). Information coding for particular miRNAs (small noncoding RNA molecules) is transferred into cells to specifically recognize and precisely regulate mRNA levels, by cleavage of the mRNA strand, and thus promotes repression of mRNA translation. Double-stranded RNA (dsRNA) is processed in the cytoplasm by DICER generating short RNA (22 nucleotides in length). This short fragment is integrated into the RISC complex which contains members of the Argonaute protein family AGO1-4 and becomes an active miRNA. Thereafter miRNAs exert their regulatory effects by binding to imperfect complementary sites within the 3′ untranslated regions (UTRs) of the target mRNA. The formation of the double-stranded RNA, resulting from the binding of the miRNA, leads to translational repression

Recent studies have demonstrated that microRNAs (miRNAs) which have key roles in regulating both synaptic plasticity and brain development can repress the expression of genes implicated in cellular mechanisms that regulate mood, cognition, and circadian functions (Hunsberger et al. 2009). For example, in bipolar disorders, the presence of miRNA-134 in circulating blood can be used as a peripheral marker that reflects acute manic episodes and that can be modulated by pharmacological mood stabilizers (Rong et al. 2011). Specifically, Clock and Bmal1 activate the transcription of miR-219, which attenuates the intracellular calcium response to NMDA activation in the SCN (Cheng et al. 2007), and *Bmal1* is regulated by miR-142-3p (Shende et al. 2013). Also, miR-24, miR-29a, and miR-30a affect the circadian clock through the regulation of Per1 and Per2 in two ways, first in mRNA stability and second in protein translation (Chen et al. 2013a). In contrast, miR-132 increases calcium influx after depolarization (Cheng et al. 2007). These findings implicate that targeting clock gene-associated miRNAs may be a novel way to regulate clock gene expression.

17.5 Conclusion

There are many questions that need to be resolved in order to develop new treatments, like gene therapy, for neurological and degenerative diseases. As mentioned above, polymorphisms in the *CLOCK, BMAL1, PER3*, and *TIMELESS* genes have been associated with mood disorders; is it then possible to develop a gene therapy strategy to regulate the activity of clock genes to normalize their expression and the levels and activity of the proteins associated with circadian rhythm disturbances? Can we associate the influences of some nutrients with the regulation of the clock genes expression? Are the polymorphisms in clock genes a key that can explain the tolerance developed to drugs used to treat these illnesses?

As of now, we have no answers for many of these questions, but this highlights the importance of this field of research since our internal clock needs to function correctly to maintain chronostasis (see Chap. 12).

As previously discussed, there is little experience of gene therapy directed to modify circadian rhythms. Although the literature of experimental strategies based on gene therapy to treat neurological diseases is very abundant, most of the clinical trials of CNS diseases have been directed to treat tumors, particularly gliomas. Furthermore the concept of modifying behavior by gene therapy is highly controversial. An interesting aspect, however, is whether by inducing the recovery of circadian rhythmicity, it would be possible to impact on other diseases, MD, for instance, or whether the use of gene therapy for those diseases would improve circadian rhythmicity.

References

Aguilar-Roblero R, Mercado C, Alamilla J et al (2007) Ryanodine receptor Ca2+-release channels are an output pathway for the circadian clock in the rat suprachiasmatic nuclei. Eur J Neurosci 26(3):575–582

Alenghat T, Meyers K, Mullican SE et al (2008) Nuclear receptor corepressor-histone deacetylase 3 governs circadian metabolic physiology. Nature 456(7224):997–1000

Allebrandt KV, Teder-Laving M, Akyol M et al (2010) *CLOCK* gene variants associate with sleep duration in two independent populations. Biol Psychiatry 67:1040–1047

Asher G, Gatfield D, Stratmann M et al (2008) SIRT1 regulates circadian clock gene expression through PER2 deacetylation. Cell 134:317–328

Benedetti F, Dallaspezia S, Colombo C et al (2008) A length polymorphism in the circadian clock gene Per3 influences age at onset of bipolar disorder. Neurosci Lett 445:184–187

Benitez JA, Segovia J (2003) Gene therapy targeting in the central nervous system. Curr Gene Ther 3:127–145

Borgs L, Beukelaers P, Vandenbosch R et al (2009) Cell "circadian" cycle, new role for mammalian core clock genes. Cell Cycle 8(6):832–837

Capasso C, Garofalo M, Hirvinen M et al (2014) The evolution of adenoviral vectors through genetic and chemical surface modifications. Viruses 6:832–855

Chen ST, Choo KB, Hou MF et al (2005) Deregulated expression of the PER1, PER2 and PER3 genes in breast cancers. Carcinogenesis 26:1241–1246

Chen R, D'Alessandro M, Lee C (2013a) miRNAs are required for generating a time delay critical for the circadian oscillator. Curr Biol 23:1959–1968

Chen HF, Huang CQ, You C et al (2013b) Polymorphism of CLOCK gene rs 4580704 COG is associated with susceptibility of Alzheimer's disease in a Chinese population. Arch Med Res 44:203–207

Chen H, Zhao L, Kumazawa M et al (2013c) Down-regulation of core clock gene *Bmal1* attenuates expression of progesterone and prostaglandin biosynthesis-related genes in rat luteinizing granulosa cells. Am J Physiol Cell Physiol. doi:10.1152/ajpcell.00008.2013

Cheng H-YM, Papp JW, Varlamova O et al (2007) MicroRNA modulation of circadian clock period and entrainment. Neuron 54(5):813–829

Cockrell AS, Kafri T (2007) Gene delivery by lentivirus vectors. Mol Biotechnol 36(3):184–204

Crosio C, Cermakian N, Allis CD et al (2000) Light induces chromatin modification in cells of the mammalian circadian clock. Nat Neurosci 3(12):1241–1247

Davidson AJ, Sellix MT, Daniel J et al (2006) Chronic jet-lag increases mortality in aged mice. Curr Biol 16:R914–R916

DiTacchio L, Le HD, Vollmers C et al (2011) Histone lysine demethylase JARID1a activates CLOCK-BMAL1 and influences the circadian clock. Science 333(6051):1881–1885

Duncan MJ (2007) Aging of the mammalian circadian timing system: changes in the central pacemaker and its regulation by photic and nonphotic signals. Neuroembryol Aging 4:85–101

Ebisawa T, Uchiyama M, Kajimura N et al (2001) Association of structural polymorphisms in the human period3 gene with delayed sleep phase syndrome. MBO Rep 2(4):342–346

Edgar N, McClung CA (2013) Major depressive disorder: a loss of circadian synchrony? Bioessays 35(11):940–944

Etchegaray JP, Lee C, Wade PA et al (2003) Rhythmic histone acetylation underlies transcription in the mammalian circadian clock. Nature 421:177–182

Fu L, Pelicano H, Liu J et al (2002) The circadian gene Period2 plays an important role in tumor suppression and DNA damage response in vivo. Cell 111:41–50

Gene therapy clinical trial worldwide. Wiley. 2014. http://www.abedia.com/wiley/genes.php. Accessed 2 May 2014

Gery S, Komatsu N, Baldjyan L et al (2006) The circadian gene per1 plays an important role in cell growth and DNA damage control in human cancer cells. Mol Cell 22:375–382

Goodwin FK, Jamison KR (2007) Manic-depressive illness: bipolar disorders and recurrent depression. Oxford University Press, New York

Gorwood P (2010) Restoring circadian rhythms: a new way to successfully manage depression. J Psychopharmacol 24:15–19

Hacein-Bey-Abina S, Von Kalle C, Schmidt M et al (2003) LMO2-associated clonal T cell proliferation in two patients after gene therapy for SCID-X1. Science 302(5644):415–419

Hareendran S, Balakrishnan B, Sen D et al (2013) Adeno-associated virus (AAV) vectors in gene therapy: immune challenges and strategies to circumvent them. Rev Med Virol 23:399–413

He Y, Jones CR, Fujiki N et al (2009) The transcriptional repressor DEC2 regulates sleep length in mammals. Science 325:866–870

Herweijer H, Wolff JA (2003) Progress and prospects: naked DNA gene transfer and therapy. Gene Ther 10:453–458

Hunsberger JG, Austin DR, Chen G et al (2009) MicroRNAs in mental health: from biological underpinnings to potential therapies. Neuromolecular Med 11:173–182

Jenuwein T, Allis CD (2001) Translating the histone code. Science 293:1074–1080

Katada S, Sassone-Corsi P (2010) The histone methyltransferase MLL1 permits the oscillation of circadian gene expression. Nat Struct Mol Biol 17:1414–1421

Ko CH, Takahashi JS (2006) Molecular components of the mammalian circadian clock. Hum Mol Genet 15 Spec No 2:R271–R277

Kobayashi E, Kishi H, Ozawa T et al (2014) Retroviral vectors for homologous recombination provide efficient cloning and expression in mammalian cells. Biochem Biophys Res Commun 444(3):319–324

Kuo SJ, Chen ST, Yeh KT et al (2009) Disturbance of circadian gene expression in breast cancer. Virchows Arch 454:467–474

Lewy AJ, Lefler BJ, Emens JS et al (2006) The circadian basis of winter depression. Proc Natl Acad Sci U S A 103:7414–7419

Lewy AJ, Emens JS, Songer JB et al (2009) Winter depression: integrating mood, circadian rhythms, and the sleep/wake and light/dark cycles into a bio-psycho-social-environmental model. Sleep Med Clin 4:285–299

Lin Q, Ding H, Zheng Z et al (2012) Promoter methylation analysis of seven clock genes in Parkinson's disease. Neurosci Lett 507:147–150

Lundkvist GB, Kwak Y, Davis EK et al (2005) A calcium flux is required for circadian rhythm generation in mammalian pacemaker neurons. J Neurosci 25:7682–7686

Mansour HA, Wood J, Logue T et al (2006) Association study of eight circadian genes with bipolar I disorder, schizoaffective disorder and schizophrenia. Genes Brain Behav 5(2):150–157

Maywood ES, Reddy AB, Wong GK et al (2006a) Synchronization and maintenance of timekeeping in Suprachiasmatic circadian clock cells by neuropeptidergic signaling. Curr Biol 16:599–605

Maywood ES, O'Neill J, Wong GK et al (2006b) Circadian timing in health and disease. Prog Brain Res 153:253–269

McClung CA, Sidiropoulou K, Vitaterna M et al (2005) Regulation of dopaminergic transmission and cocaine reward by the Clock gene. Proc Natl Acad Sci U S A 102(26):9377–9381

Mercado C, Díaz-Muñoz M, Alamilla J et al (2009) Ryanodine-sensitive intracellular Ca2+ channels in rat suprachiasmatic nuclei are required for circadian clock control of behavior. J Biol Rhythms 24(3):203–210

Nakahata Y, Kaluzova M, Grimaldi B et al (2008) The NAD+-dependent deacetylase SIRT1 modulates CLOCK-mediated chromatin remodeling and circadian control. Cell 134(2):329–340

National Gene Vector Biorepository. National Heart, Lung, and Blood Institute rough a grant to the Indiana University School of Medicine Department of Medical and Molecular Genetics. https://www.ngvbcc.org/ReagentRepository.action. Accessed 5 May 2014

Nienhuis AW, Dunbar CE, Sorrentino BP (2006) Genotoxicity of retroviral integration in hematopoietic cells. Mol Ther 13:1031–1049

Nievergelt CM, Kripke DF, Barrett TB et al (2006) Suggestive evidence for association of the circadian genes PERIOD3 and ARNTL with bipolar disorder. Am J Med Genet B Neuropsychiatr Genet 141B:234–241

O'Neill JS, Maywood ES, Chesham JE et al (2008) cAMP-dependent signaling as a core component of the mammalian circadian pacemaker. Science 320:949–953

Ojeda DA, Claudia S, Perea CS et al (2013) A novel association of two non-synonymous polymorphisms in PER2 and PER3 genes with specific diurnal preference subscales. Neurosci Lett 553:52–56

Okamura H, Doi M, Fustin JM et al (2010) Mammalian circadian clock system: molecular mechanisms for pharmaceutical and medical sciences. Adv Drug Deliv Rev 62:876–884

Pigeon WR, Hegel M, Unutzer J et al (2008) Is insomnia a perpetuating factor for late-life depression in the IMPACT cohort? Sleep 31:481–488

Rana S, Mahmood S (2010) Circadian rhythm and its role in malignancy. J Circadian Rhythms 8:3

Reynolds CF III, Kupfer DJ (1987) Sleep research in affective illness: state of the art circa 1987. Sleep 10:199–215

Rong H, Liu TB, Yang KJ et al (2011) MicroRNA-134 plasma levels before and after treatment for bipolar mania. J Psychiatr Res 45:92–95

Ryther RC, Flynt AS, Phillips JA 3rd et al (2005) siRNA therapeutics: big potential from small RNAs. Gene Ther 12:5–11

Schernhammer ES, Razavi P, Li TY et al (2011) Rotating night shifts and risk of skin cancer in the nurses' health study. J Natl Cancer Inst 103:602–606

Schulz DJ, Temporal S, Barry DM et al (2008) Mechanisms of voltage-gated ion channel regulation: from gene expression to localization. Cell Mol Life Sci 65:2215–2231

Shende VR, Neuendorff N, Earnest DJ (2013) Role of miR-142-3p in the post-transcriptional regulation of the clock gene Bmal1 in the mouse SCN. PLoS One 8(6):e65300. doi:10.1371/journal.pone.0065300

Shih MC, Yeh KT, Tang KP et al (2006) Promoter methylation in circadian genes of endometrial cancers. Mol Carcinog 45:732–740

Takao T, Tachikawa H, Kawanishi Y et al (2007) CLOCK gene T3111C polymorphism is associated with Japanese schizophrenics: a preliminary study. Eur Neuropsychopharmacol 17:273–276

Tan W, Dong Z, Wilkinson TA et al (2006) Human immunodeficiency virus type 1 incorporated with fusion proteins consisting of integrase and the designed polydactyl zinc finger protein E2C can bias integration of viral DNA into a predetermined chromosomal region in human cells. J Virol 80(4):1939–1948

Vieira E, Marroquí L, Figueroa AFC et al (2013) Involvement of the clock gene Rev-erb alpha in the regulation of glucagon secretion in pancreatic alpha-cells. PLoS One 8(7):e69939

Wulff K, Gatti JG, Wettstein JG et al (2010) Sleep and circadian rhythm disruption in psychiatric and neurodegenerative disease. Nat Rev Neurosci 11:589–599

Xiao J, Zhou Y, Lai H et al (2013) Transcription factor NF-Y is a functional regulator of the transcription of core-clock gene Bmal1. J Biol Chem 288(44):31930–31936

Yang X, Wood PA, Oh EY et al (2009) Downregulation of circadian clock gene Period 2 accelerates breast cancer growth by altering its daily growth rhythm. Breast Cancer Res Treat 117(2):423–431

Glossary

Acrophase Time (phase) of maximum amplitude of a cosine curve fitted to the raw data of a rhythm.

Active element Component in a biological system which is capable of generating self-sustained oscillations.

Actogram A plot of behavioral activity with successive days arranged vertically (the first on top of next one and so on) that allows to visualize large amounts of time series data. The period and phase of the rhythm can be appreciated at a glimpse.

After-effects Long-term transients, often lasting for 100 days or more, after a rhythm is released from a light–dark cycle into constant conditions where it can adopt its spontaneous free-running period.

Amplitude Difference between the mean value (mesor) and the acrophase of a cosine curve fitted to the raw data of a rhythm. Also the difference between the minimal and maximal values of a rhythm (or oscillation). See also range of oscillation.

Batiphase Time (phase) of minimum amplitude of a cosine curve fitted to the raw data of a rhythm. See also Nadir.

Biologic rhythm A periodic fluctuation in the value of a biologic variable.

Biological clocks Self-sustained oscillators which generate biologic rhythms at genetic level in absence of an external periodic input.

Chronobiology Part of biology that studies the timing systems of organisms. The term derived from *chrono* (time), *bios* (life), and *logos* (science).

Chronotherapy Medical treatment that considers biological rhythms in absorption, distribution, and elimination of drugs and the availability of its receptors in target cells, to find the best time to enhance the desired pharmacologic effect and/or reduce undesirable side effects of drugs.

Circadian rhythm Endogenous biological rhythm, generated by a self-sustained biological clock, which in the natural environment of the organisms is entrained to the 24-h period of the day length.

© Springer International Publishing Switzerland 2015

R. Aguilar-Roblero et al. (eds.), *Mechanisms of Circadian Systems in Animals and Their Clinical Relevance*, DOI 10.1007/978-3-319-08945-4

Circadian The term that describes a rhythm with a period close to 24 h (from 20 to 28 h). From *circa* (close to) and *diem* (day).

Circadian time Time unit corresponding to 1/24 of the endogenous period of a free-running circadian rhythm.

Circalunar rhythm Endogenous biological rhythm which in the organisms' natural environment is entrained to the period of the moon (28 days).

Circannual rhythm Endogenous biological rhythm which in the organisms' natural environment is entrained to the period of the annual seasonal variations in the environment.

Circa-rhythms Biological endogenous rhythms with a period close to the period of the environmental cycle to which they are normally synchronized (*see* Circadian, Circatidal, Circalunar, Circannual rhythms).

Circatidal rhythm Self-sustained biological rhythm which in the organism's natural environment is normally entrained to the period of the ocean tides (typically 12.4 h).

"Clock gene" Set of genes that form a transcription–translational self-regulating loop which forms the core molecular circadian oscillator.

Clock-controlled gene (*ccg*) A gene whose expression is controlled by the core molecular circadian clock. It refers to those genes containing E-Boxes in their promoter regions.

Cosinor A mathematical method to detect and describe a rhythm by fitting the raw time series of data to a sine function. The procedure tests the null hypothesis that amplitude is not statistically different to zero.

Cry See cryptochrome. In vertebrates and some invertebrates a "clock gene" and protein.

Cryptochrome A blue light receptor that uses pterine and flavine as chromophores.

Desynchronization Loss of synchronization between two or more rhythms, so that they show independent periods and therefore lose the phase relation among them. It could produce spontaneously because of changes in the environment or in some pathological states.

Endogenous rhythm Self-sustained rhythm generated within an organism.

Entraining agent Periodic environmental stimulus that resets the endogenous clocks, so period of the environmental cycle and that of the biological rhythm are the same. Synonymous with zeitgeber.

Entrainment Resetting of an endogenous rhythm by a periodic environmental stimulus. During entrainment the frequencies of the two oscillations are the same or integral multiples of each other (entrainment by frequency demultiplication; see below).

Exogenous rhythm Biological rhythm generated by the influence of an environmental cycle on the organism.

External desynchronization Loss of synchronization between an endogenous rhythm and its zeitgeber.

Forced internal desynchronization Internal desynchronization (see below) produced by subjecting an organism to a zeitgeber with a period outside or just in the limits of entrainment, which entrains only those biological rhythms with an endogenous period close to that of the zeitgeber.

Free run(ning) Expression of the endogenous period of a biological rhythmic under constant environment conditions, free of entrainment signals.

Frequency demultiplication Hypothesis that postulates entrainment of a biological rhythm by a zeitgeber with a period, outside its range of entrainment, which is an integral fraction of the endogenous period.

Frequency The number of cycles of a periodic signal or biological rhythm occurring in a given time; it is the reciprocal of the period.

Infradian Biological rhythm with a frequency of less than a cycle per day, therefore with a period longer than that of circadian rhythms. From *infra* (below) and diem (day).

Internal desynchronization Loss of synchronization between two or more endogenous rhythms so that they free-run with different periods within the same organism.

Jet lag Internal desynchronization of the organism timing system ocurring after a trans-meridian flight.

Masking Effects of an external rhythmic agent on the expression of an overt rhythm by direct control of the effector of the parameter, without affecting the period or phase of the underlying pacemaker.

Mean value Arithmetic mean of all instantaneous values of an oscillating variable within one cycle. See also mesor.

Mediator Neural or endocrine signal which through its oscillations can transmit period and phase information so as to synchronize the rhythms in a target tissue.

Mesor Central tendency measure of the values of a rhythm adjusted to a cosine function. See also mean value.

Metabolism Set of biochemical reactions that transform molecules and energy within the cellular milieu.

Nadir Time at which the lowest value of a biological rhythm occurs. See also batiphase.

Pacemaker Functional entity capable of imposing phase and period on the rhythms of the organisms. It refers to self-sustaining oscillators or biological clocks that drive overt rhythms.

Passive element Component in a biological system which is not capable of generating self-sustained oscillations and is rhythmic only if driven by a pacemaker or oscillator.

Period Duration of one complete cycle in a rhythmic fluctuation. Time interval between recurrences of a defined phase of a rhythm.

Phase Instantaneous state of an oscillation. It refers to the time of the circadian rhythm at which a particular event (or value) of the cyclic variable occurs. If the period of the rhythm is normalized to a circumference ($360°$) it may be expressed in arc angles rather than hours.

Phase marker Particular event used as a reference to estimate the phase of a rhythm or cyclic phenomenon. Some examples include activity onset, time of feeding, sleep onset, acrophase, sunrise, and sunset.

Phase angle Difference in the phase between two circadian (biological) rhythms expressed as degrees.

Phase control Effect on the phase of an endogenous rhythm by a zeitgeber due to resetting of the period of the endogenous oscillator.

Phase relationship Time relation between two rhythmic variables. Synonymous with phase angle but expressed in units of time.

Phase response curve Plot of the intensity of the response (shift) of a particular phase marker to discrete (minutes to hours) entraining stimulus applied at different times of a free-running circadian rhythm.

Phase shift Displacement of the time of occurrence of phase marker from a rhythm induced by an entrainment signal.

Photoperiod Duration of the interval of light in a light–dark cycle.

Range of entrainment Range of periods of environmental cyclic stimulus within which an endogenous clock can be entrained.

Range of oscillation Difference between the maximum and minimum value (independent of shape of oscillation). See also amplitude.

Relative coordination Modulation in the period of a rhythm which results when the organism is exposed to a zeitgeber too weak to entrain the rhythm.

SCN Suprachiasmatic nuclei of the ventral hypothalamus. Group of hypothalamic neurons situated above the optic chiasm that functions as the circadian pacemaker or hierarchical circadian clock in mammals.

Scotoperiod Duration of the interval of darkness in a light–dark cycle.

Secondary or slave oscillator Oscillator within an organism capable of generating rhythmicity but which usually has less stability and persistence than a pacemaker, is not entrained directly by zeitgebers, and may not necessarily synchronize other oscillators.

Shift work Transient or permanent change in work schedule in relation to the social environment.

Subjective day The interval of a free-running circadian cycle which exhibits the overt rhythmicity that normally occurs during the day in a light–dark cycle.

Subjective night The interval of a free-running circadian cycle which exhibits the overt rhythmicity that normally occurs during the night in a light–dark cycle.

Suprachiasmatic nucleus See SCN.

Synchronizer Environmental cycle that adjust the period of a biological rhythm by resetting its phase to a particular value. See also zeitgeber and entraining agent.

τ Period of a free-running circadian rhythm.

T Period of a zeitgeber.

Temperature compensation Essential property of circadian rhythms that prevents the speed of the clock to be influenced by changes in the temperature in the environment.

Tim See timeless.

Timeless A gene that is an essential element in the circadian molecular mechanism in animals.

Transducer Component in a biological system which detects environmental energy and changes it into a signal that can be used by the organism.

Transient Transition cycles between two steady states of an overt biological rhythm.

Ultradian Biological rhythm with a frequency greater than a cycle per day, therefore with a period shorter than that of circadian rhythms. From *ultra* (beyond) and *diem* (day).

Zeitgeber time The time unit corresponding to 1/24 of the period of a zeitgeber. In a 12:12 h light–dark cycle lights on corresponds to ZT 0 and lights off corresponds to ZT 12.

Zeitgeber Environmental cycle that adjusts the period of a biological rhythm by resetting its phase to a particular value. See also synchronizer and entraining agent.

ARNTL Aryl hydrocarbon receptor nuclear translocator-like, transcription factor of basic helix-loop-helix PAS domain.

bHLH-PAS Basic helix-loop-helix, protein structural motif of "clock genes" *per, arnt, sim.*

Bmal-1 Gene of BMAL-1 transcription factor, binds on the basic helix-loop-helix promoter domain.

CLOCK-BMAL1 Transcription factor dimmer of mammalian "clock genes" that binds to an E-box on the basic helix-loop-helix promoter domain.

DEC Family of negative regulators of transcription factor CLOCK/BMAL1.

Per Family of gene *period*, transcriptional repressor of the basic helix-loop-helix promoter domain. Essential genes of the molecular circadian clock in animals.

REV-ERB α Orphan member of the nuclear hormone receptor protein, transcriptional repressor of the Bmal 1 transcription through the RAR-related orphan receptor (ROR).

Index

© Springer International Publishing Switzerland 2015

R. Aguilar-Roblero et al. (eds.), *Mechanisms of Circadian Systems in Animals and Their Clinical Relevance*, DOI 10.1007/978-3-319-08945-4

9 783319 089447